Birkhäuser

Advances in Mathematical Fluid Mechanics

The *Advances in Mathematical Fluid Mechanics* series is a forum for the publication of high-quality, peer-reviewed research monographs and edited collections on the mathematical theory of fluid mechanics, with special regards to the Navier-Stokes equations and other significant viscous and inviscid fluid models. Titles in this series consider theoretical, numerical, and computational methods, as well as applications to science and engineering. Works in related areas of mathematics that have a direct bearing on fluid mechanics are also welcome. All manuscripts are peer-reviewed to meet the highest standards of scientific literature.

More information about this series at http://www.springer.com/series/5032

Tujin Kim · Daomin Cao

Equations of Motion for Incompressible Viscous Fluids

With Mixed Boundary Conditions

Birkhäuser

Tujin Kim
Institute of Mathematics
State Academy of Sciences
Pyongyang
Democratic People's Republic of Korea

Daomin Cao
Guangzhou University and Institute
of Applied Mathematics
Chinese Academy of Sciences
Guanzhou, Beijing, China

ISSN 2297-0320 ISSN 2297-0339 (electronic)
Advances in Mathematical Fluid Mechanics
ISBN 978-3-030-78658-8 ISBN 978-3-030-78659-5 (eBook)
https://doi.org/10.1007/978-3-030-78659-5

Mathematics Subject Classification: 35Q30, 76D03, 76D05, 49J40, 80A20, 47J20, 35A02, 35A15, 76D07, 35R35, 35J87

This book is published under the imprint Birkhäuser, www.birkhauser-science.com by the registered company Springer Nature Switzerland AG
The registered company address is: Gewerbestrasse 11, 6330 Cham, Switzerland

Preface

The study of the equations appearing in fluid mechanics is a core topic in the fields of partial differential equations, fluid mechanics and applied sciences. Lots of real phenomena of fluid are described by those equations with different kinds of boundary conditions. We are concerned with various boundary conditions, and nowadays for mathematical analysis of phenomena of fluid, the fluid equations with new boundary conditions are studied. In practice, we are concerned with mixture of various boundary conditions. Now there are many papers dealing with fluid equations with mixed boundary conditions; however, almost all monographs deal with the equations with Dirichlet boundary condition and a few monographs deal with mixture of Dirichlet and stress or Dirichlet and a kind of friction boundary conditions.

The purpose of this book is to introduce the recent results and research methods in the field of fluid equations with mixed boundary conditions via one book. For readers broadly ranging from students to engineers and mathematicians involved in the fluid equations, first the fluid equations and the boundary conditions for the Navier-Stokes equations are outlined in the level of senior university students. Then, the recent results for the Navier-Stokes equations and the equations of motion for fluid conducting heat with the mixed boundary conditions originated mainly from authors' results and the first author's lectures for postgraduate students in the Institute of Mathematics, State Academy of Sciences, DPR Korea are described in the level of postgraduate students.

The style of this book is rather different from other mathematical monographs for the Navier-Stokes equations in the following sense:

1. Many monographs for the fluid equations concentrate on mathematical properties of the equations, whereas this book first intends to show how the fluid equations appearing in the journals are obtained and what kinds of boundary conditions are suitable in various practical situations.
2. In the existing monographs, according to the bilinear forms used for variational formulations of the problems, category of mixture of boundary conditions for fluid is determined, whereas this book almost freely deals with mixture of the

boundary conditions for fluid relying on some properties on the boundaries of vector fields given on domains, which is the main point to make this book different from others.
3. Many monographs concentrate on mathematical investigation of the formulated variational problems, whereas this book gives attention to equivalence between the original partial differential equations with mixture of various boundary conditions and the corresponding variational problems, especially variational inequalities with one unknown, which is not obvious due to mixture of complicated boundary conditions.

The book consists of 10 chapters.

Trying to make our book self-contained, in Chap. 1, we give some basic knowledge and known results in analysis used in the book. We do not give the best result for every subject, but sometimes we give results more than just required for subsequent parts of the book to help readers to understand other literature. Readers need not read the whole chapter, and can use the chapter as data to consult when something is quoted in the book. The theory of pseudo-monotone operator in Sect. 1.5 and convex functional in Sect. 1.6 is needed only in Chaps. 7, 9 and 10.

In Chap. 2, we first show how the Navier-Stokes equations and the equations of motion for fluid under consideration of heat are derived from the fundamental physical laws. Though the aim of this book is to study the equations of motion for the incompressible fluids, the equations for compressible fluids are discussed together to widen reader's understanding. Next, we outline some boundary conditions for the Navier-Stokes equations, which would be helpful for readers to understand the meaning of the boundary conditions in this book and to improve the ability to apply these in practice. Then, we consider the widely used existing three kinds of variational formulations for the Navier-Stokes problems with mixed boundary conditions. Also, equivalence between the variational formulations and the original PDE problems is discussed, which is a preparation to understand equivalence between PDE problems with more complicated boundary conditions and their variational formulations in the subsequent chapters. This chapter can serve as an introduction for the students and postgraduate students in the fields of equations describing fluids, mathematical modeling and numerical simulation.

In Chap. 3, we first study some properties on the boundary surfaces of vector fields given on domains. To this end, a little knowledge of differential geometry is used, which is contained in the chapter. This is preparation to embed wide various mixed boundary conditions together into variational formulations and the foundation for the whole book, which is the peculiar features of this book different from others. Based on the results, we can consider mixtures of boundary conditions which were not studied in other literature, for example, mixture of the boundary conditions for stress and pressure which is important in practice. Thus, we study the existence and uniqueness of solutions to the steady Navier-Stokes system with mixed boundary conditions which may include seven kinds of boundary conditions together. According to whether the static pressure or total pressure is included in the boundary conditions, the research methods and results are different, and so these cases are distinguished in this and next chapters.

In Chap. 4, we study the existence and uniqueness of a solution to the non-steadyNavier-Stokes system with mixed boundary conditions corresponding to the steady problem in Chap. 3.

In Chap. 5, we study the steady Navier-Stokes system with mixed boundary conditions including friction-type conditions, which may include 11 kinds of boundary conditions together. Unlike Chaps. 3 and 4, due to the friction boundary conditions, the problems are described by the variational inequalities. Sometimes variational inequalities are used as approximate mathematical models for real phenomena; however, in this part equivalence between the variational inequalities with one unknown and the original boundary value problem of PDE with 11 kinds of boundary conditions is studied concretely. Relying on the results for the variational inequalities, we study the existence and uniqueness of solutions to the problems under consideration.

In Chap. 6, we study the non-steadyNavier-Stokes system with mixed boundary conditions corresponding to the steady problem of Chap. 5. The problems are reduced to the non-steady variational inequalities.

Chapter 7 is devoted to the study of the steady Boussinesq system for heat convection of fluid under mixed boundary conditions including friction conditions. The boundary conditions for fluid are the same as the one in Chap. 5 and the boundary conditions for temperature may include Dirichlet, Neumann and Robin conditions together. The problem is formulated by a simultaneous system of a steady variational inequality for velocity of fluid and a variational equation for temperature.

Chapter 8 is devoted to the study of the non-steady problem corresponding to the steady problem in Chap. 7.

In Chap. 9, we are concerned with the equation for steady flow of heat conducting incompressible Newtonian fluids with dissipative heating by Joule effect under mixed boundary conditions. The boundary conditions are the same as in Chap. 7.

Chapter 10 is devoted to the non-steady problem corresponding to Chap. 9, but the boundary conditions for fluid are restricted to the case of total pressure.

At the end of every chapter from Chap. 2, the bibliographical remark for the problem in the chapter is given, which is to provide more information on the research of the problem studied in the chapter.

We hope the book is helpful for readers to study the research methods and results for the equations for fluid with mixed boundary conditions and to improve the ability to apply the knowledge.

Pyongyang, DPR Korea Tujin Kim
Guangzhou, P.R. China Daomin Cao
May 2020

Contents

Chapter 1
Miscellanea of Analysis

In this chapter, we outline some knowledge of analysis: Banach space, fixed point, Lebesgue and Sobolev spaces, operator and operator-differential equations and convex functional, which will be used in the main part of this book. We do not describe the best results, but to help readers' understanding sometimes we say more than necessary. The readers who are already acquainted with the elements of functional analysis can skip this chapter and may consult the necessary parts when reading main part of this book.

1.1 Banach Space, Fixed Point and Basics of Mapping

1.1.1 Banach Space

Let X be a normed space. A sequence $\{x_n\}$ in X is called a *Cauchy sequence* iff for every $\varepsilon > 0$ there exists an integer N such that $\|x_n - x_m\|_X < \varepsilon$ holds whenever $m, n > N$. A sequence $\{x_n\}$ in X is convergent to the element x_0 if and only if $\lim_{n\to\infty} \|x_n - x_0\|_X = 0$. A space X is said to be complete if every Cauchy sequence in X converges to an element in X. A complete normed space is called a *Banach space*. A subset S of a normed space X is said to be *dense* in $M \subseteq X$ if each $x \in M$ is the limit of a sequence of elements of S. The normed space X is called *separable* if it has a countable dense subset.

Denote by X^* the space of linear continuous functionals on a normed space X, which is called the dual of X. A norm on the dual X^* of a normed space X can be defined by setting

$$\|x^*\|_{X^*} = \sup_{\|x\|_X=1} \langle x^*, x\rangle,$$

T. Kim and D. Cao, *Equations of Motion for Incompressible Viscous Fluids*,
Advances in Mathematical Fluid Mechanics,
https://doi.org/10.1007/978-3-030-78659-5_1

where and in what follows $\langle x^*, x\rangle$ means the value of a functional $x^* \in X^*$ at an element $x \in X$. Since the fields of real and complex numbers $\mathbb{R}$ and $\mathbb{C}$ are complete, with the topology induced by this norm X^* is a Banach space (no matter whether X is complete).

A natural linear injection of a normed space X into its second dual space $X^{**} = (X^*)^*$ is provided by the mapping J whose value Jx at $x \in X$ is given by

$$\langle Jx, y\rangle = \langle y, x\rangle \quad \forall y \in X^*,$$

J is an isometric isomorphism of X into X^{**}, and so $X \subset X^{**}$.

If the range of the isomorphism J is the entire space X^{**}, the space X is called *reflexive*. A reflexive space must be complete and hence a Banach space.

For a sequence $\{x_n\}$ of a normed space X if

$$\langle f, x_n - x_0\rangle \to 0 \quad \forall f \in X^* \quad \text{as } n \to \infty,$$

then $\{x_n\}$ is said to be *weakly convergent* to x_0. We denote norm convergence of a sequence $\{x_n\}$ to x_0 by $x_n \to x_0$, and weak convergence by $x_n \rightharpoonup x_0$.

A sequence $\{x_n\}$ in a normed space X is called a *weak Cauchy sequence* iff for every $\varepsilon > 0$ and $f \in X^*$ there exists an integer N such that $|\langle f, x_n - x_m\rangle| < \varepsilon$ holds whenever $m, n > N$. The space X is called weakly complete if every weak Cauchy sequence in X weakly converges to an element in X.

For a sequence $\{f_n\}$ of a normed space X^*

$$\langle g, f_n - f_0\rangle \to 0 \quad \forall g \in X^{**} \quad \text{as } n \to \infty$$

means that $\{f_n\}$ weakly converges to f_0 in X^*. For a sequence $\{f_n\}$ of a normed space X^* if

$$\langle x, f_n - f_0\rangle \to 0 \quad \forall x \in X \quad \text{as } n \to \infty,$$

then $\{f_n\}$ is said to be $*$-*weakly convergent* to f_0 and it is denoted by $f_n \overset{*}{\rightharpoonup} f$.

Theorem 1.1 *Let M be a convex closed subset of a Banach space X and a sequence $\{x_n\} \subset M$ weakly converges to a limit x_0. Then x_0 belongs to M, i.e. M is weakly closed.*

(See Theorem 3.7 of [1].)

Theorem 1.2 *If a sequence $\{x_n\}$ in a Banach space X weakly converges to an element x, then*

$$\|x\|_X \le \liminf_{n\to\infty} \|x_n\|_X. \tag{1.1}$$

(See Theorem 1, Sect. 1, Chap. V of [2].)

A weak Cauchy sequence of a Banach space may not have a weak limit in the space, and the Banach space may not be weakly complete. But the following theorem holds.

Theorem 1.3 *Reflexive spaces are weakly complete. In every bounded subset of a reflexive Banach space there exists a weakly convergent sequence.*

(See Theorem 7, Sect. 1 and Theorem 1, Sect. 2, Chap. V of [2].)

Theorem 1.4 *If all weakly convergent subsequences of a sequence $\{x_n\}$ of a reflexive space converge to the same element x, then the sequence $\{x_n\}$ weakly converges to x.*

(See Lemma 5.4, Chap. 1 of [3].)

Imbedding of a normed space X into a linear normed space Y means that set-theoretic inclusion $X \subset Y$, for which the following inequality is valid

$$\|x\|_X \leq C\|x\|_Y \quad \exists C, \forall x \in X.$$

We denote by $X \hookrightarrow Y$ the imbedding. If X is dense in Y, then $f \in Y^*$ may be identified with an element in X^*, and so we have $Y^* \hookrightarrow X^*$. If X is reflexive, then this imbedding is dense (see Remark 5.14, Chap. 1 of [3]).

1.1.2 Fixed-Point Theorems

Theorem 1.5 (Banach fixed-point theorem) *Let (X, d) be a complete metric space, M a closed nonempty set of X, and $T : M \subset X \to M$ be a map such that*

$$d(Tx, Ty) \leq kd(x, y) \quad \exists k (0 \leq k < 1), \forall x, y \in M.$$

Then, there exists a unique point $x \in M$ such that $Tx = x$.

Let X be a topological space and $M \subseteq X$. The set M is called *compact* iff every open covering of M contains a finite subcover, i.e. finitely many of these open sets already cover M. The set M is called *relatively compact* iff its closure $\overline{M}$ is compact. Let X be a complete metric space and $M \subseteq X$. Then M is relatively compact if and only if for every $\varepsilon > 0$ there is a finite ε-net for M. Let X be a metric space. Then $M \subseteq X$ is relatively compact if and only if for any sequence $\{x_n\}_{n=1}^{\infty} \subset M$ there is a convergent subsequence (see Proposition 1.2.1 of [4]).

Definition 1.1 Let X, Y be normed spaces, and $T : M \subseteq X \to Y$ an operator. T is called compact if it maps every bounded subset of M into a relatively compact set of Y. If, in addition, T is continuous on M, then it is called *completely continuous* on this set.

When X, Y are normed spaces, for linear operators T the concepts of a compact and a completely continuous operators are the same since every compact linear operator is continuous.

Remark 1.1 Note that in some literatures the completely continuous operator is called compact.

Theorem 1.6 (The Schauder fixed-point theorem) *If a completely continuous operator T maps a bounded closed convex set M of a normed linear space X into itself, then there exists at least one point $x \in M$ such that $Tx = x$.*

(See Theorem 5.1.11 of [4].)

1.1.3 Basics of Mappings

Let T be a single-valued mapping from a set M to N. Mapping $T : M \to N$ is said to be *surjective* if each element of N is the image of a certain element of M; *injective* if different elements of M are mapped into different elements of N; *bijective* if the mapping is both surjective and injective simultaneously. If $T : M \to N$ is bijective, then the mapping is an one-to-one map between M and N.

If $T : M \to N$ is continuous and bijective, and if its inverse map $T^{-1} : N \to M$ is also continuous, then T is called a *homeomorphism* and M and N are said to be homeomorphic.

Definition 1.2 Let X, Y be Banach spaces and $f : U(x) \subseteq X \to Y$ be a map, where $U(x)$ is a neighborhood of x.

(1) The map f is said to be Gâteaux differentiable at x iff there exists a linear continuous operator $T(x) : X \to Y$ such that

$$f(x+th) - f(x) = tT(x)h + o(t) \quad (t \to 0) \tag{1.2}$$

for all h with $\|h\|_X = 1$ and all real numbers t in some neighborhood of zero. If it exists, this $T(x)$ is called *the Gâteaux derivative* at x. Put $T(x) = f'(x)$. The Gâteaux differential at x is defined by $d_G f(x; h) = f'(x)h$.

(2) The map f is said to be Fréchet differentiable at x iff there exists a linear continuous operator $T(x) : X \to Y$ such that

$$f(x+h) - f(x) = T(x)h + o(\|h\|_X) \quad (h \to 0_X). \tag{1.3}$$

$T(x)$ is called *the Fréchet derivative* at x and $T(x)h$ is called the Fréchet differential at x.

The Gâteaux derivative $f'(x)$ of f at x can be defined equivalently by

$$f'(x)h = \lim_{t \to 0} \frac{f(x+th) - f(x)}{t}.$$

Definition 1.3 Let X, Y be Banach spaces, $f : U \subset X \to Y$ and $r \in \mathbb{N}$ (the set of natural numbers).

(1) If U is open, then f is called a C^r-map iff f has continuous Fréchet derivatives up to order r on U.

(2) If U is arbitrary, then f is called a C^r-map iff for each $x \in U$ there exists an open neighborhood $V(x)$ such that f can be extended to a C^r-map on $V(x)$.

Definition 1.4 Let M and N be arbitrary sets in the Banach spaces X and Y, respectively. Let $0 \leq r \leq \infty$.

(1) The mapping $f : M \to N$ is called a C^r-*diffeomorphism* iff f is bijective and both f and f^{-1} are C^r-maps.

(2) A local C^r-diffeomorphism at x_0 is a C^r-diffeomorphism from some neighborhood $U(x_0)$ in X onto some neighborhood $U(f(x_0))$ in Y.

Theorem 1.7 (Local inverse mapping theorem) (Theorem 4.F of [5]) *Let $f : U(x_0) \subseteq X \to Y$ is a C^1-mapping, where both X and Y are Banach spaces. Then f is a local C^1-diffeomorphism at x_0 iff $f'(x_0) : X \to Y$ is bijective.*

Theorem 1.8 (Theorem 3.4 of [6]) *Let X, Y be Banach spaces, T_0 be a one-to-one linear continuous operator from X onto Y, T_c be a compact linear operator from X into Y. Then $T_0 + T_c$ is a one-to-one operator from X into Y if and only if it is an operator from X onto Y.*

1.2 Lebesgue Space, Convergence

In this section we consider real-valued functions defined on a (Lebesgue) measurable subset $\Omega \subset \mathbb{R}^l$, where $\mathbb{R}^l$ is l-dimensional Euclidean space.

1.2.1 Lebesgue Space

A function $u : \Omega \to \mathbb{R} \cup \{\pm\infty\}$ is itself called (Lebesgue) *measurable* if the set $\{x : u(x) > a\}$ is measurable for every real a.

A function $u : \Omega \to \mathbb{R} \cup \{\pm\infty\}$ is said to be countably valued if it assumes at most a countable set of values, assuming each value is on a measurable subset. A countably valued function is written as $u(x) = \sum_i a_i \chi_{A_i}$, where χ_{A_i} is the characteristic function of the measurable set A_i.

A function $u(x)$ is measurable if and only if it is uniform limit almost everywhere of a sequence of countably valued functions. (See Theorem 2, Sect. 5, Chap. V of [7] or p. 30 of Sect. 2.0 of [8].)

A countably valued function $u(x) = \sum_i a_i \chi_{A_i}$ is integrable if $\sum_i a_i mes\,(A_i) < \infty$, where $mes\,(A_i)$ is (Lebesgue) measure of A_i, and the integral of $u(x)$ is defined by

$$I(u) = \int_\Omega u(x)\,dx = \sum_i a_i\, mes\,(A_i).$$

A measurable function $u(x)$ is said to be (Lebesgue) *integrable* if there exists a sequence $\{u_n(x)\}$ of integrable countably valued functions such that $u_n(x) \to u(x)$ uniformly outside of a null-measure set and

$$\int_\Omega |u_n(x) - u_m(x)|\,dx \to 0 \quad \text{as } m, n \to \infty.$$

We define (Lebesgue) *integral* by

$$\int_\Omega u(x)\,dx = \lim_{n\to\infty} \int_\Omega u_n(x)\,dx.$$

The definition of integral (or of integrability) does not depend on the particular sequence $\{u_n\}$ chosen.

The class of integrable functions on Ω is denoted by $L^1(\Omega)$.

Theorem 1.9 (Lebesgue) *Let $\Omega \subset \mathbb{R}^l$ be measurable and let $\{u_k\}$ be a sequence of measurable functions converging to u a.e in $x \in \Omega$. If there exists a function $v(x) \in L^1(\Omega)$ such that*

$$|u_k(x)| \le v(x), \quad k = 1, 2, \cdots,$$

then u is integrable and

$$\int_\Omega u(x)\,dx = \lim_{k\to\infty} \int_\Omega u_k(x)\,dx.$$

Definition 1.5 Let Ω be an open subset of $\mathbb{R}^l$ and let p be a positive real number. The space $L^p(\Omega)$ consists of all (Lebesgue) measurable real functions $u(x)$ defined on Ω such that

$$\int_\Omega |u(x)|^p\,dx < \infty.$$

The space $L^\infty(\Omega)$ consists of all measurable functions $u(x)$ defined on Ω such that

$$|u(x)| \le M \quad \exists M > 0, \quad \text{for a.e. } x \in \Omega$$

$L^p(\Omega)$, $(1 \le p < \infty)$ and $L^\infty(\Omega)$ are Banach spaces, respectively, with norms

$$\|u\|_p = \left(\int_\Omega |u(x)|^p\,dx\right)^{1/p} \quad \text{and} \quad \|u\|_\infty = \operatorname*{ess\,sup}_{x\in\Omega} |u(x)|.$$

(See Theorem 4.8 of [1].)

Let Ω be an open subset of $\mathbb{R}^l$. The support of a continuous function u is defined by

$$\operatorname{supp} u = \overline{\{x; u(x) \neq 0\}} \cap \Omega.$$

The space $C_0^\infty(\Omega)$ consists of all infinitely differentiable functions that have compact support in Ω. If $1 \leq p < \infty$, then $C_0^\infty(\Omega)$ is dense in $L^p(\Omega)$ and $L^p(\Omega)$ is separable.

Theorem 1.10 (The Riesz representation theorem for $L^p(\Omega)$) *Let* $1 \leq p < \infty$. *Then for any* $f \in \left(L^p(\Omega)\right)^*$ *there exists* $v \in L^q(\Omega)$, $1/p + 1/q = 1$, *such that*

$$\langle f, u \rangle = \int_\Omega v(x) u(x)\, dx \quad \forall u \in L^p(\Omega)$$

and $\|f\|_{(L^p(\Omega))^*} = \|v\|_{L^q(\Omega)}$. *The correspondence between* $f \in \left(L^p(\Omega)\right)^*$ *and* $v \in L^q(\Omega)$ *is linear.*

$L^p(\Omega)$ is reflexive if and only if $1 < p < \infty$. (See Sect. 4.3 of [1].)

Theorem 1.11 (Hölder's inequality) *Let* $1 < p < \infty$ *and let* q *denote the conjugate exponent defined by*

$$\frac{1}{p} + \frac{1}{q} = 1.$$

If $u \in L^p(\Omega)$ *and* $v \in L^q(\Omega)$, *then* $uv \in L^1(\Omega)$, *and*

$$\int_\Omega |u(x)v(x)|\, dx \leq \|u\|_p \|v\|_q.$$

Equality holds if and only if $|u(x)|^p$ *and* $|u(x)|^q$ *are proportional a.e. in* Ω.

Theorem 1.12 (cf. Theorem 2.11 of [9]) (complex interpolation of Lebesgue spaces) *Let* $1 \leq p < q < r$, *so that*

$$\frac{1}{q} = \frac{1-\theta}{r} + \frac{\theta}{p}$$

for some θ *satisfying* $0 < \theta < 1$. *If* $u \in L^p(\Omega) \cap L^r(\Omega)$, *then* $u \in L^q(\Omega)$ *and*

$$\|u\|_{(L^r, L^q)_{[\theta]}} \equiv \|u\|_q \leq \|u\|_r^{1-\theta} \, \|u\|_p^\theta.$$

Here the functor $(\ ,\ )_{[\theta]}$ means the complex interpolation (see Definition 2.1.3 of [10] or Sect. 4.1 of [11]).

1.2.2 Convergence of Sequences of Functions

An open and connected subset of $\mathbb{R}^l$ is called a *domain*. Let Ω be a domain.

Lemma 1.1 (Lemma 1.19, Chap. 2 of [3]) *Let $1 \le p < \infty$ and a sequence $\{u_n\} \subset L^p(\Omega)$ be such that*

$$u_n \rightharpoonup u \quad in\ L^p(\Omega),$$
$$u_n(x) \to v(x) \quad a.e.\ in\ \Omega.$$

Then $u = v$.

Lemma 1.2 *Let v_n and v be functions of $L^q(\Omega)$, $1 < q < \infty$, such that*

$$\|v_n\|_{L^q(\Omega)} \le C, \quad v_n(x) \to v(x) \quad a.e.\ in\ \Omega.$$

Then, $v_n \rightharpoonup v$ in $L^q(\Omega)$.

Proof Since $L^q(\Omega)$, $1 < q < \infty$, is reflexive, by Theorem 1.3 there exists a subsequence $\{v_k\} \subset \{v_n\}$ such that $v_k \rightharpoonup u$ in $L^q(\Omega)$. By Lemma 1.1 the limits of all subsequences of $\{v_n\}$ coincide with v. Thus, by Theorem 1.4 we come to the conclusion. □

A set of functions $F \subset L^1(\Omega)$ is called uniformly integrable if

$$\lim_{C\to+\infty} \sup_{f\in F} \int_{|f|>C} |f|\, d\mu = 0.$$

A set consisting of a single integrable function is uniformly integrable since the Lebesgue integral is absolutely continuous. Hence, for any integrable function f_0, the set of all measurable functions f with $|f| \le |f_0|$ is uniformly integrable.

The sequence of functions $\{f_n\}$ is said to converge in measure to a measurable function f if, for every $c > 0$, one has

$$\lim_{n\to\infty} mes\{x : |f(x) - f_n(x)| \ge c\} = 0.$$

In Lebesgue space Vitali's convergence theorem (see Theorem 4.5.4 of [12]) is written as follows.

Theorem 1.13 *Suppose that f is a measurable function and $\{f_n\}$ is a sequence of integrable functions. Then, the following assertions are equivalent:*

(1) the sequence $\{f_n\}$ converges to f in measure and is uniformly integrable;

(2) the function f is integrable and the sequence $\{f_n\}$ converges to f in the space $L^1(\Omega)$.

Lemma 1.3 *Let Ω be a bounded domain of $\mathbb{R}^l$. If $\{u_n\}$ is such that*

$$\{u_n\} \text{ is bounded in } L^\infty(\Omega), \quad u_n(x) \to u(x) \text{ a.e. in } \Omega,$$

then

$$u_n \to u \text{ in } L^p(\Omega) \ \forall p, 1 \leq p < \infty.$$

Proof Sine Ω is bounded, u_n is integrable. By the conditions, $\{u_k\}$ converges to u in measure and is uniformly integrable. By Theorem 1.13,

$$\int_\Omega |u_n - u|^p \, dx \leq M \int_\Omega |u_n - u| \, dx \to 0.$$ □

Corollary 1.1 *Let Ω be a bounded domain of $\mathbb{R}^l$ and $\mu(\xi), \xi \in \mathbb{R}$, be a bounded continuous function. If $\{u_n\}$ and $\{v_n\}$ are such that*

$$u_n(x) \to u(x) \text{ a.e. in } \Omega, \quad v_n \rightharpoonup v \text{ in } L^p(\Omega), 1 < p < \infty,$$

then

$$\mu(u_n)v_n \rightharpoonup \mu(u)v \text{ in } L^p(\Omega).$$

Proof By Lemma 1.3,

$$\mu(u_n) \to \mu(u) \text{ in } L^q(\Omega), \ \frac{1}{p} + \frac{1}{q} = 1.$$

Thus, $\mu(u_n)v_n \rightharpoonup \mu(u)v$ in $L^1(\Omega)$, that is,

$$\int_\Omega \mu(u_n(x))v_n(x)\phi(x) \, dx \to \int_\Omega \mu(u(x))v(x)\phi(x) \, dx \quad \forall \phi \in C_0^\infty(\Omega).$$

Since $\{\|\mu(u_n)v_n\|_{L^p}\}$ is bounded and $C_0^\infty(\Omega)$ is dense in $L^p(\Omega)$, from above we get the asserted conclusion (see Theorem 3, Chap. 5 of [2]). □

Corollary 1.2 *Let Ω be a bounded domain of $\mathbb{R}^l$ and $\mu(\xi), \xi \in \mathbb{R}$, be a positive bounded continuous function. If $\{u_n\}$ and $\{v_n\}$ are such that*

$$u_n(x) \to u(x) \text{ a.e. in } \Omega, \quad v_n \rightharpoonup v \text{ in } L^2(\Omega),$$

then

$$\big(\mu(u(x))v(x), v(x)\big) \leq \liminf_{n\to\infty} \big(\mu(u_n(x))v_n(x), v_n(x)\big).$$

Proof By Corollary 1.1

$$\sqrt{\mu(u_n)}v_n \rightharpoonup \sqrt{\mu(u)}v \text{ in } L^2(\Omega).$$

Then,

$$\|\sqrt{\mu(u)}v\|_{L^2(\Omega)} \le \liminf_{n\to\infty} \|\sqrt{\mu(u_n)}v_n\|_{L^2(\Omega)},$$

which implies the asserted conclusion. □

Lemma 1.4 *Let Ω be a bounded domain of $\mathbb{R}^l$. If*

$$v_\varepsilon \rightharpoonup v \quad \textit{in } L^2(\Omega) \quad \textit{as } \varepsilon \to 0,$$

then

$$\frac{v_\varepsilon}{\sqrt{1+\varepsilon v_\varepsilon^2}} \rightharpoonup v \quad \textit{in } L^2(\Omega).$$

Proof Since

$$\int_\Omega \left|\frac{1}{\sqrt{1+\varepsilon v_\varepsilon^2}} - 1\right|^2 dx \le \int_\Omega \left(2+\varepsilon v_\varepsilon^2 - 2\sqrt{1+\varepsilon v_\varepsilon^2}\right) dx \le \varepsilon \int_\Omega v_\varepsilon^2\, dx,$$

we have

$$\frac{1}{\sqrt{1+\varepsilon v_\varepsilon^2}} \to 1 \quad \text{in } L^2(\Omega). \tag{1.4}$$

The sequence $\left\{\frac{v_\varepsilon}{\sqrt{1+\varepsilon v_\varepsilon^2}}\right\}$ is bounded and $C_0^\infty(\Omega)$ is dense in $L^2(\Omega)$, and to get the conclusion, it is enough to prove that

$$\int_\Omega \Big(\frac{v_\varepsilon}{\sqrt{1+\varepsilon v_\varepsilon^2}} - v\Big)\varphi\, dx \to 0 \quad \forall \varphi \in C_0^\infty(\Omega) \quad \text{as } \varepsilon \to 0.$$

Using (1.4), we have

$$\int_\Omega \Big(\frac{v_\varepsilon}{\sqrt{1+\varepsilon v_\varepsilon^2}} - v\Big)\varphi\, dx = \int_\Omega v_\varepsilon\varphi\Big(\frac{1}{\sqrt{1+\varepsilon v_\varepsilon^2}} - 1\Big)\, dx + \int_\Omega (v_\varepsilon - v)\varphi\, dx \to 0$$
$$\forall \varphi \in C_0^\infty(\Omega) \quad \text{as } \varepsilon \to 0,$$

which implies the asserted conclusion. □

Lemma 1.5 *Let Ω be a domain of $\mathbb{R}^l$. If for a $p > 1$*

$$v_\varepsilon(x) \to v(x) \in L^p(\Omega) \quad \textit{a.e. in } \Omega \quad \textit{as } \varepsilon \to 0,$$

then

$$\frac{1}{1+\varepsilon v_\varepsilon^2} \rightharpoonup 1 \quad \textit{in } L^r(\Omega)\ \forall r\ (1 < r < \infty).$$

Proof Since $\{\frac{1}{1+\varepsilon v_\varepsilon^2}\}$ is bounded, and there exists a subsequence weakly convergent to a limit. On the other hand,

$$\frac{1}{1+\varepsilon v_\varepsilon^2} \to 1 \quad \text{a.e. in } \Omega.$$

Thus, by Lemma 1.1 and Theorem 1.4 we come to the conclusion. □

1.3 Sobolev Space

To simplify the discussion, in this section, we also work with real-valued functions. Let Ω be an open subset of $\mathbb{R}^l$ with boundary $\partial\Omega$ and $\{\rho_\alpha\}$ be a family continuous functions defined on Ω. The family of supports $\{\text{supp}\,\rho_\alpha\}$ is said to be locally finite on Ω if any compact subset of Ω intersects only finitely many elements of $\{\text{supp}\,\rho_\alpha\}$.

Let $\alpha = (\alpha_1, \cdots, \alpha_l)$, α_i—nonnegative integer, $|\alpha| = \alpha_1 + \cdots + \alpha_l$, and

$$D^\alpha = \frac{\partial^{|\alpha|}}{\partial x_1^{\alpha_1} \cdots \partial x_l^{\alpha_l}}.$$

Define the family of semi-norm on $C_0^\infty(\Omega)$

$$p_\rho(\varphi) = \sum_\alpha \sup_{x\in\Omega} |\rho_\alpha(x) D^\alpha \varphi(x)|, \quad \varphi \in C_0^\infty(\Omega), \tag{1.5}$$

where $\rho = \{\rho_\alpha\}$ are taken all families of continuous function supports of which are locally finite. By the family of semi-norms, $C_0^\infty(\Omega)$ becomes a locally convex space, which is denoted by $\mathscr{D}(\Omega)$.

Let $\mathscr{D}^*(\Omega)$ denote the space of linear continuous functional defined on $\mathscr{D}(\Omega)$, often called the space of distributions on Ω. We denote by $\langle\cdot,\cdot\rangle$ the duality pairing between $\mathscr{D}^*(\Omega)$ and $\mathscr{D}(\Omega)$ and we remark that when f is a locally integrable function, then f can be identified with a distribution by

$$\langle f, \phi\rangle = \int_\Omega f(x)\phi(x)\,dx \quad \forall \phi \in \mathscr{D}(\Omega).$$

$\mathscr{D}^*(\Omega)$ is given the $*$-weak topology as the dual of $\mathscr{D}(\Omega)$, i.e. the fact that $f_n \to f$ in $\mathscr{D}^*(\Omega)$ means

$$\langle f_n - f, \varphi\rangle \to 0 \quad \forall \varphi \in \mathscr{D}(\Omega).$$

For $f \in \mathscr{D}^*(\Omega)$, we define $D^\alpha f \in \mathscr{D}^*(\Omega)$ by

$$\langle D^\alpha f, \varphi\rangle = (-1)^{|\alpha|}\langle f, D^\alpha \varphi\rangle \quad \forall \varphi \in \mathscr{D}(\Omega).$$

If $f_n \to f$ in $\mathscr{D}^*(\Omega)$, then for $\varphi \in \mathscr{D}(\Omega)$

$$\lim_{n\to\infty} \langle D^\alpha f_n - D^\alpha f, \varphi\rangle = \lim_{n\to\infty} (-1)^{|\alpha|}\langle f_n - f, D^\alpha \varphi\rangle = 0 \quad \forall \varphi \in \mathscr{D}(\Omega).$$

Thus, mapping $f \to D^\alpha f$ is continuous in $\mathscr{D}^*(\Omega)$.

1.3.1 Definition of Sobolev Space

Definition 1.6 Let Ω be an open subset of $\mathbb{R}^l$, s—nonnegative real number and $1 \le p \le \infty$. We define the Sobolev space $W^{s,p}(\Omega)$ as follows.

(i) $s = k$ (a nonnegative integer), $1 \le p \le \infty$.

$$W^{k,p}(\Omega) = \{u \in L^p(\Omega) : D^\alpha u \in L^p(\Omega) \ \forall |\alpha| \le k\},$$
$$\|u\|_{W^{k,p}(\Omega)} = \Big(\sum_{|\alpha|\le k} \|D^\alpha u\|_{L^p}^p\Big)^{1/p} (1 \le p < \infty),$$
$$\|u\|_{W^{k,\infty}(\Omega)} = \Big(\sum_{|\alpha|\le k} \operatorname*{ess\,sup}_{x\in\Omega} |D^\alpha u|\Big).$$

(ii) $s = k + \sigma,\ 0 < \sigma < 1$, k—nonnegative integer, $1 \le p < \infty$.

$$W^{s,p}(\Omega) = \left\{u \in W^{k,p}(\Omega) : \iint_{\Omega\times\Omega} \frac{|D^\alpha u(x) - D^\alpha u(y)|^p}{|x-y|^{n+\sigma p}}\,dxdy < \infty, \quad |\alpha| = k\right\},$$
$$\|u\|_{W^{s,p}(\Omega)} = \left(\|u\|^p_{W^{k,p}(\Omega)} + \sum_{|\alpha|=k} \iint_{\Omega\times\Omega} \frac{|D^\alpha u(x) - D^\alpha u(y)|^p}{|x-y|^{n+\sigma p}}\,dxdy\right)^{1/p}.$$

(iii) $s = k + \sigma,\ 0 < \sigma < 1$, k—nonnegative integer, $p = \infty$.

$$W^{s,\infty}(\Omega) = \left\{u \in L^\infty : u \in W^{k,\infty}(\Omega),\ \max_{|\alpha|=k} \operatorname*{ess\,sup}_{x,y\in\Omega,\, x\ne y} \frac{|D^\alpha u(x) - D^\alpha u(y)|}{|x-y|^\sigma} < \infty\right\},$$
$$\|u\|_{W^{s,\infty}(\Omega)} = \left(\|u\|^p_{W^{k,p}(\Omega)} + \max_{|\alpha|=k} \operatorname*{ess\,sup}_{x,y\in\Omega,\, x\ne y} \frac{|D^\alpha u(x) - D^\alpha u(y)|}{|x-y|^\sigma}\right\}.$$

The Sobolev spaces are Banach spaces for the norms.

For $s > 0$, we denote by $W_0^{s,p}(\Omega)$ the closure of $C_0^\infty(\Omega)$ in $W^{s,p}(\Omega)$ and by $W^{-s,q}(\Omega)$ the dual space of $W_0^{s,p}(\Omega)$, where $p^{-1} + q^{-1} = 1$.

If a function f is k times continuously differentiable and its derivatives of order k are Lipschitz continuous, then we say it belongs to the class $C^{k,1}$.

Definition 1.7 (*cf. Definition 1.2.1.1 of* [13]) Let Ω be a bounded open subset of $\mathbb{R}^l$. We say that its boundary $\partial\Omega$ is continuous (resp. Lipschitz continuous, of class C^k, of class $C^{k,1}$ for some integer $k \geq 0$) if for every $x \in \partial\Omega$ there exists a neighborhood $\mathscr{O}$ of x in $\mathbb{R}^l$ and new orthogonal coordinates $y = (y', y_n)$ where $y' = (y_1, \cdots, y_{n-1})$, such that

(i) $\mathscr{O}$ is a hypercube in the new coordinates:

$$\mathscr{O} = \{y : |y_i| < a_i,\ 1 \leq i \leq n\};$$

(ii) there exists a continuous (resp. Lipschitz continuous, of class C^k, of class $C^{k,1}$) function ϕ defined in

$$\mathscr{O}' = \{y' : |y_i| < a_i,\ 1 \leq i \leq n-1\}$$

that satisfies

$$|\phi(y')| \leq \frac{a_n}{2} \quad \forall y' \in \mathscr{O}',$$
$$\Omega \cap \mathscr{O} = \{y : y_n < \phi(y')\},$$
$$\partial\Omega \cap \mathscr{O} = \{y : y_n = \phi(y')\}.$$

Remark 1.2 Differently from Definition 1.18, Chap. 2 of [3], in Definition 1.2.1.1 of [13] and Definition 1.1 of [14] they did not demand boundedness of Ω. It is not enough for Lipschitz continuity and class $C^{k,1}$ of $\partial\Omega$. If Ω is bounded, there exist finite family of neighborhood covering $\partial\Omega$, and from local properties we can get Lipschitz continuity and $C^{k,1}$ of $\partial\Omega$.

To examine the boundary values of functions in $W^{s,p}(\Omega)$, we consider the spaces of functions defined on open subset Γ of boundary $\partial\Omega$. We assume that Ω is a bounded open subset of $\mathbb{R}^l$ with a boundary $\partial\Omega$ that is at least Lipschitz continuous. We can view Γ locally as an $n-1$-dimensional submanifold of $\mathbb{R}^l$ by means of the mapping

$$\Phi(y') = (y', \phi(y')) \tag{1.6}$$

from $\mathscr{O}'$ onto $\Gamma \cap \mathscr{O}$, where ϕ, $\mathscr{O}'$ and $\mathscr{O}$ are the ones in Definition 1.7.

We assume that ψ is at least a bi-Lipschitz mapping from Ω_1 onto Ω_2 where Ω_1 and Ω_2 are bounded open subsets of $\mathbb{R}^l$. This hypothesis ensures that Lebesgue measure is mapped by ψ or ψ^{-1} to an equivalent image measure.

Let $u \in W^{s,p}(\Omega_2)$; assume that ψ and ψ^{-1} are of class $C^{k,1}$, where k is an integer $\geq s-1$. Then $u \circ \Phi \in W^{s,p}(\Omega_1)$ (Lemma 1.3.3.1 of [13]). It is a justification for the following definition.

Definition 1.8 (*Definition 1.3.3.2 of* [13]) Let Ω be a bounded open subset of $\mathbb{R}^l$ with a boundary $\partial\Omega$ of class $C^{k,1}$, where k is a nonnegative integer. Let Γ be an open subset of $\partial\Omega$. A distribution u on Γ belongs to $W^{s,p}(\Gamma)$ with $|s| \leq k+1$ if $u \circ \Phi \in W^{s,p}\big(\mathscr{O}' \cap \Phi^{-1}(\Gamma \cap \mathscr{O})\big)$ for all possible $\mathscr{O}$ and ϕ fulfilling the assumptions in Definition 1.7.

Let Φ_j is defined by (1.6) with ϕ_j instead of ϕ. Let $(\mathcal{O}_j, \Phi_j)_{j=1}^J$ be any atlas of Γ such that each couple $(\mathcal{O}_j, \Phi_j)$ satisfies the assumptions of the above definition. Then one possible Banach norm for $W^{s,p}(\Gamma)$ is the functional:

$$u \to \|u\|_{W^{s,p}(\Gamma)} = \Big(\sum_{j=1}^{J} \|u \circ \Phi_j\|_{W^{s,p}\left(V_j' \cap \Phi_j^{-1}(\Gamma \cap V_j)\right)}\Big)^{1/p}. \tag{1.7}$$

In the particular case when $s \in (0, 1)$, one can easily check that any of the norms defined above is equivalent to

$$\left\{\int_\Gamma |u|^p \, d\sigma + \iint_{\Gamma\times\Gamma} \frac{|u(x) - u(y)|^p}{|x - y|^{n-1+sp}} \, d\sigma(x)d\sigma(y)\right\}^{1/p}.$$

1.3.2 Density and Continuation

Unless otherwise stated we assume that $1 < p < \infty$ in the following subsections of this section.

Let Ω be open subset of $\mathbb{R}^l$ and

$$\mathcal{D}(\overline{\Omega}) = \{\phi|_\Omega : \phi \in \mathcal{D}(\mathbb{R}^l)\}.$$

Theorem 1.14 (Theorem 1.4.2.1 of [13]) *If the boundary of Ω is continuous, then $\mathcal{D}(\overline{\Omega})$ is dense in $W^{s,p}(\Omega)$ for all $s > 0$.*

When s is a nonnegative integer and the boundary is Lipschitz continuous, this theorem is valid for $p = 1$ (see Theorem 1.2, Chap. 1 of [14]).

Theorem 1.15 (Theorem 1.4.2.4 of [13]) *Let Ω be a bounded open subset of $\mathbb{R}^l$ with a Lipschitz boundary. Then $\mathcal{D}(\Omega)$ is dense in $W^{s,p}(\Omega)$ for $0 < s \le 1/p$.*

For every $s > 0$, we denote by $W^{s,p}(\overline{\Omega})$ the space of all distributions in Ω which are restrictions of elements of $W^{s,p}(\mathbb{R}^l)$ (Definition 1.3.2.4 of [13]).

In other words, $W^{s,p}(\overline{\Omega})$ is the space of all $u|_\Omega$ where $u \in W^{s,p}(\mathbb{R}^l)$ and $u|_\Omega$ is defined by $\langle u|_\Omega, \varphi\rangle = \langle u, \tilde{\varphi}\rangle$ for all $\varphi \in \mathcal{D}(\Omega)$, where $\tilde{\varphi}$ is the continuation of φ by zero, outside Ω. A Banach norm on $W^{s,p}(\overline{\Omega})$ is defined by

$$\|u\|_{W^{s,p}(\overline{\Omega})} = \inf_{\substack{U \in W^{s,p}(\mathbb{R}^l)\\ U|_\Omega = u}} \|U\|_{W^{s,p}(\mathbb{R}^l)}. \tag{1.8}$$

As obvious consequences of the definition, we have the following inclusions:

$$W^{s,p}(\overline{\Omega}) \subset W^{s,p}(\Omega)$$

((1.3.2.5) of [13]). The following theorem shows a case that $W^{s,p}(\overline{\Omega}) = W^{s,p}(\Omega)$.

Theorem 1.16 (Theorem 1.4.3.1 of [13]) *Let Ω be a bounded open subset of $\mathbb{R}^l$ with a Lipschitz boundary. Then for every $s > 0$ there exists a continuous linear operator P_s from $W^{s,p}(\Omega)$ into $W^{s,p}(\mathbb{R}^l)$ such that*

$$P_s u|_\Omega = u.$$

For every positive s, we denote by $\widetilde{W}^{s,p}(\Omega)$ the space of all $u \in W^{s,p}(\Omega)$ such that $\tilde{u} \in W^{s,p}(\mathbb{R}^l)$, where $\tilde{u}$ is the extension of u by zero outside Ω.

Theorem 1.17 (Corollary 1.4.4.5 of [13]) *Let Ω be a bounded open subset of $\mathbb{R}^l$ with a Lipschitz boundary. Then when $s - 1/p$ is not integer, we have*

$$\widetilde{W}^{s,p}(\Omega) = W_0^{s,p}(\Omega), \tag{1.9}$$

and furthermore, when $0 < s < 1/p$ we have

$$\widetilde{W}^{s,p}(\Omega) = W_0^{s,p}(\Omega) = W^{s,p}(\Omega). \tag{1.10}$$

Theorem 1.18 (Corollary 1.4.4.10 of [13]) *Let Ω be a bounded open subset of $\mathbb{R}^l$ with a Lipschitz boundary. Then for all $s > 0$, we have*

$$\widetilde{W}^{s,p}(\Omega) = \left\{u \in W_0^{s,p}(\Omega) : \rho^{-\sigma} D^\alpha u \in L^p(\Omega), |\alpha| = m\right\}$$

, where $\rho(x)$ is the distance from x to the boundary $\partial\Omega$ of Ω and $s = m + \sigma$, m integer, $\sigma \in [0, 1)$.

Following Sect. 11.5, Chap. 1 of [15], we denote by $H_{00}^{1/2}(\Omega)$, $H_0^{1/2}(\Omega)$ and $H^{1/2}(\Omega)$, respectively, the spaces $\widetilde{W}^{1/2,2}(\Omega)$, $W_0^{1/2,2}(\Omega)$ and $W^{1/2,2}(\Omega)$. By Theorem 1.15

$$H_0^{1/2}(\Omega) \equiv H^{1/2}(\Omega). \tag{1.11}$$

The space $H_{00}^{1/2}(\Omega)$ is strictly contained in $H_0^{1/2}(\Omega)$ with a strictly finer topology

$$\|u\|_{H_{00}^{1/2}(\Omega)} = \left(\|u\|^2_{H^{1/2}(\Omega)} + \|\rho^{-1/2} u\|^2_{L^2(\Omega)}\right)^{1/2}.$$

Since $C_0^\infty(\Omega) \subset H_{00}^{1/2}(\Omega)$, $H_{00}^{1/2}(\Omega)$ is dense in $H^{1/2}(\Omega)$.

Remark 1.3 Let $s \in \mathbb{R}$ and $1 < p \leq \infty$. We denote by $H_p^s(\mathbb{R}^l)$ the space of all distributions in $\mathbb{R}^l$ such that

$$\mathscr{F}^{-1}\left((1 + |\xi|^2)^{s/2} \mathscr{F} u\right) \in L^p(\mathbb{R}^l),$$

where $\mathscr{F}$ is the Fourier transform operator defined by

$$(\mathscr{F}u)(\xi) = \frac{1}{(2\pi)^{n/2}} \int_{\mathbb{R}^l} e^{-ix\xi} u(x)\, dx.$$

The space H_p^s is called with different names, i.e. the Bessel potential space [9], the fractional order Sobolev space [9], the generalized Sobolev space [11], the inhomogeneous Sobolev space [16], and in some literature the notation for Sobolev is used for that.

It is known that

$$W^{s,p}(\mathbb{R}^l) = \begin{cases} H_p^s(\mathbb{R}^l) & s : \text{all real number},\ p = 2, \\ H_p^s(\mathbb{R}^l) & s : \text{integer},\ 1 < p < \infty. \end{cases} \tag{1.12}$$

Furthermore, it is proved that for any real s

$$\begin{aligned} W^{s,p}(\mathbb{R}^l) &\subseteq H_p^s(\mathbb{R}^l) \quad 1 < p \le 2, \\ H_p^s(\mathbb{R}^l) &\subseteq W^{s,p}(\mathbb{R}^l) \quad 2 \le p < \infty. \end{aligned}$$

(See Sect. 1.3.1 of [13].)

Define the space $H_p^s(\Omega)$ by

$$H_p^s(\Omega) = \{u|_\Omega : u \in H_p^s(\mathbb{R}^l)\}. \tag{1.13}$$

Then by Theorem 1.16 and (1.12), for a bounded open subset Ω with a Lipschitz boundary, we have

$$W^{s,p}(\Omega) = \begin{cases} H_p^s(\Omega) & s : \text{all real number},\ p = 2, \\ H_p^s(\Omega) & s : \text{integer},\ 1 < p < \infty. \end{cases} \tag{1.14}$$

Instead of (1.13) in some literatures $H_p^s(\Omega)$ is defined by complex interpolation of $L^p(\Omega)$ and $W^{m,p}(\Omega)$. Similarly to Theorem 1.16, under some condition for boundary of Ω, there exists a continuous linear operator P_s from $H_p^s(\Omega)$ into $H_p^s(\mathbb{R}^l)$. (Noting $H_p^s(\Omega) = F^{s,p,2}(\Omega)$, refer to Sect. 7.69 of [9].) Thus under such definition (1.14) is valid.

1.3.3 Imbedding

If imbedding operator I of a linear normed space into another linear normed space is compact, then the imbedding is called *compact*.

Theorem 1.19 (Theorem 1.4.4.1 of [13]) *The following imbeddings hold:*

$$W^{s,p}(\mathbb{R}^l) \hookrightarrow W^{t,q}(\mathbb{R}^l) \tag{1.15}$$

for $t \le s$ *and* $q \ge p$ *such that* $s - n/p = t - n/q$*; and*

$$W^{s,p}(\mathbb{R}^l) \hookrightarrow C^{k,\alpha}(\mathbb{R}^l) \tag{1.16}$$

for $k < s - n/p < k + 1$*, where* $\alpha = s - k - n/p$*,* k *a nonnegative integer.*

Taking into account (1.8), as a consequence we have the conclusions $W^{s,p}(\overline{\Omega}) \hookrightarrow W^{t,q}(\overline{\Omega})$ and $W^{s,p}(\overline{\Omega}) \hookrightarrow C^{k,\alpha}(\overline{\Omega})$ instead of (1.15) and (1.16), where $C^{k,\alpha}(\overline{\Omega})$ is the space of functions which are k times continuously differentiable and whose derivatives of order k fulfill a uniform Hölder condition of order α throughout Ω.

In other words, (1.15) is valid when $\frac{1}{p} - \frac{s-t}{n} = \frac{1}{q}$. On the other hand, we see that for a bounded open subset Ω

$$W^{t,q}(\Omega) \hookrightarrow W^{t,q_1}(\Omega) \quad \text{for } 1 < q_1 < q.$$

Then taking into account the fact above, by Theorem 1.16, we have

Theorem 1.20 (Sobolev's imbedding theorem) *Let* Ω *be a bounded open subset of* $\mathbb{R}^l$ *with a Lipschitz boundary and* $0 \le t < s$*,* $1 < q < \infty$*.*

If $\frac{1}{p} - \frac{s-t}{n} \le \frac{1}{q}$*, then*

$$W^{s,p}(\Omega) \hookrightarrow W^{t,q}(\Omega). \tag{1.17}$$

If $\frac{1}{p} - \frac{s-k}{n} < 0$*, where* k *is a nonnegative integer, then*

$$W^{s,p}(\Omega) \hookrightarrow C^k(\overline{\Omega}), \tag{1.18}$$

and if, in addition, $s - \frac{n}{p} < k + 1$*, then*

$$W^{s,p}(\Omega) \hookrightarrow C^{k,\alpha}(\overline{\Omega}), \quad \alpha = s - k - n/p. \tag{1.19}$$

Remark 1.4 When $q = \infty$, (1.17) is not valid (cf. Remark 9.2, Chap. 1 of [15]). Theorem 4.47, Corollary 4.53 and Theorem 4.57 of [17] assert that when Ω is $\mathbb{R}^l$ or a unbounded open set Ω with a Lipschitz boundary and $\frac{1}{p} - \frac{s}{n} = 0$, $W^{s,p}(\Omega) \hookrightarrow L^q(\Omega)$ for every $q < \infty$. Under the condition $p \le q < \infty$ it is right (see Sect. 4.12 of [9]).

Theorem 1.21 (Theorem 1.4.3.2 of [13]) *Let* $s_1 > s_2 \ge 0$ *and assume that* Ω *is a bounded open subset of* $\mathbb{R}^l$ *with a Lipschitz boundary. Then the imbedding of* $W^{s_1,p}(\Omega)$ *in* $W^{s_2,p}(\Omega)$ *is compact.*

Theorem 1.22 *Let* Ω *be a bounded open subset of* $\mathbb{R}^l$ *with a Lipschitz boundary. Let* $0 < s$ *and* $1 < p < \infty$*.*

If $\frac{1}{p} - \frac{s}{n} < \frac{1}{q}$*, then the imbedding* $W^{s,p}(\Omega) \hookrightarrow L^q(\Omega)$ *is compact.*

(See Theorem 4.58 of [17].)

Theorem 1.23 (Theorem 1.4.4.2 of [13]) *Let $s_1 \geq s$ and $s_2 \geq s$ be such that either*

$$s_1 + s_2 - s \geq n\left(\frac{1}{p_1} + \frac{1}{p_2} - \frac{1}{p}\right) \geq 0, \quad s_j - s > n\left(\frac{1}{p_j} - \frac{1}{p}\right), \quad j = 1, 2$$

or

$$s_1 + s_2 - s > n\left(\frac{1}{p_1} + \frac{1}{p_2} - \frac{1}{p}\right) \geq 0, \quad s_j - s \geq n\left(\frac{1}{p_j} - \frac{1}{p}\right), \quad j = 1, 2,$$

then $(u, v) \to u \cdot v$ is a continuous bilinear map from $W^{s_1,p_1}(\mathbb{R}^l) \times W^{s_2,p_2}(\mathbb{R}^l)$ into $W^{s,p}(\mathbb{R}^l)$.

A similar statement holds for Sobolev spaces defined on a bounded open subset of $\mathbb{R}^l$ with a Lipschitz boundary (see Sect. 1.4.4 of [13]).

Theorem 1.24 (Theorem 1.4.4.6 of [13]) *Let Ω be a bounded open subset of $\mathbb{R}^l$ with a Lipschitz boundary. Then D^α, $|\alpha| = 1$, is a linear continuous operator from $W^{s,p}(\Omega)$ into $W^{s-1,p}(\Omega)$ unless $s = 1/p$. (s-real number)*

1.3.4 Trace

When u is a continuous function on $\overline{\Omega}$, let γ be the operator such that $\gamma u = u|_{\partial\Omega}$. The following *trace theorem* extends the operator γ to functions in $W^{s,p}(\Omega)$ (cf. Definition, Sect. 5.5 of [18]).

Theorem 1.25 (Theorem 1.5.1.2 of [13]) *Let Ω be a bounded open subset of $\mathbb{R}^l$ with a $C^{k,1}$ boundary. Assume that $s - 1/p$ is not an integer, $s \leq k + 1$, $s - \frac{1}{p} = l + \sigma$, $0 < \sigma < 1$, l an integer ≥ 0. Then mapping*

$$u \to \left\{\gamma u, \gamma\frac{\partial v}{\partial u}, \cdots, \gamma\frac{\partial^l u}{\partial v^l}\right\}$$

which is defined for $u \in C^{k,1}(\overline{\Omega})$ has a unique continuous extension

$$Tr : \textit{from } W^{s,p}(\Omega) \quad \textit{onto} \prod_{j=0}^{k} W^{s-j-\frac{1}{p},p}(\partial\Omega).$$

Theorem 1.26 (Corollary 1.5.1.6 of [13]) *Let Ω be a bounded open subset of $\mathbb{R}^l$ with a $C^{k,1}$ boundary. Assume that $s \leq k + 1$ and $s - 1/p$ is not an integer. Let $s - 1/p = l + \sigma$, $0 < \sigma < 1$, l an integer ≥ 0. Then $u \in W_0^{s,p}(\Omega)$ if and only if $u \in W^{s,p}(\Omega)$ and*

$$\gamma u = \gamma\frac{\partial u}{\partial v} = \cdots = \gamma\frac{\partial^l u}{\partial v^l} = 0.$$

By Theorem 1.26, we see that

$$Ker(Tr) = W_0^{s,p}(\Omega).$$

Thus, by the open-mapping theorem (see Sect. 5, Chap. 2 of [2]) there exists a continuous linear operator

$$Tr^{-1} : \text{from } \prod_{j=0}^{k} W^{s-j-\frac{1}{p},p}(\partial\Omega) \quad \text{onto } W^{s,p}(\Omega) \big/ W_0^{s,p}(\Omega) \text{ (quotient space)}. \tag{1.20}$$

Theorem 1.27 (Theorem 1.5.1.10 of [13]) *Let Ω be a bounded open subset of $\mathbb{R}^l$ with a Lipschitz boundary. Then there exists a constant K such that*

$$\int_{\partial\Omega} |u|^p \, d\sigma \leq K \left[\varepsilon^{1-1/p} \int_\Omega |\nabla u|^p \, dx + \varepsilon^{-1/p} \int_\Omega |u|^p \, dx \right] \quad \forall u \in W^{1,p}(\Omega), \forall \varepsilon \in (0, 1).$$

1.3.5 Some Inequalities

Theorem 1.28 (Friedrichs' inequality) *Let Ω be a bounded domain of $\mathbb{R}^l$ with a Lipschitz boundary and $\Gamma \subset \partial\Omega$, $mes(\Gamma) \neq 0$. Then for $u \in W^{1,p}(\Omega)$ $(p \geq 1)$*

$$\int_\Omega |u|^p \, dx \leq C \left\{ \int_\Omega \left(\sum_{i=1}^{n} \left(\frac{\partial u}{\partial x_i} \right)^2 \right)^{p/2} dx + \left| \int_\Gamma u \, d\sigma \right|^p \right\}.$$

(See Theorem 1.32 of [19] or Lemma 1.36, Chap. 2 of [3].)

Let $u = (u_1, \cdots, u_n) \in \mathbf{W}^{1,p}(\Omega) \equiv W^{1,p}(\Omega)^n$ and $\varepsilon_{ij}(u) = \frac{1}{2}\left(\frac{\partial u_i}{\partial x_j} + \frac{\partial u_j}{\partial x_i} \right)$.

Theorem 1.29 (The first Korn's inequality) *Let Ω be a bounded domain of $\mathbb{R}^l$ with a Lipschitz boundary and Γ be a nonempty open subset of $\partial\Omega$. Then for $u \in \mathbf{W}_\Gamma^{1,p}(\Omega) := \{v \in \mathbf{W}^{1,p}(\Omega) : v|_\Gamma = 0\}$,*

$$\sum_{i,j} \|\varepsilon_{ij}(u)\|_{L^p(\Omega)}^p \geq C \|u\|_{\mathbf{W}^{1,p}(\Omega)}^p.$$

(See Theorem 6.3.4 of [20] for the case that $n = 3$, $p = 2$, and Corollary 4.1 of [21] for the general case.)

1.4 Space of Abstract Functions

1.4.1 Abstract Functions and Its Derivatives

Let X be a Banach space, S be an interval on $\mathbb{R}$ and $u : S \to X$.

A function $u : S \to X$ is said to be finitely valued if it is constant on each of a finite number of disjoint (Lebesgue) measurable sets B_j and equal to 0_X on $S \setminus \cup_j B_j$. A function $u : S \to X$ is said to be *simple* if it is finitely valued and if the set for which $\|u(s)\| > 0$ is of finite measure. $u : S \to X$ is said to be countably valued if it assumes at most a countable set of values in X, assuming each value is different from 0_X on a measurable subset.

Definition 1.9 A function $u(s)$ is said to be (Bochner) *strongly measurable* (simply measurable) if there exists a sequence of simple functions strongly convergent to $u(s)$ at a.e. $s \in S$

$u : S \to X$ is called weakly measurable if, for any $f \in X^*$, the numerical function $\langle f, u(s)\rangle$ is (Lebesgue) measurable.

$v : S \to X^*$ is called $*$-weakly measurable if, for any $g \in X$, the numerical function $\langle g, v(s)\rangle$ is (Lebesgue) measurable.

Remark 1.5 There are some different definitions for strongly measurable function, and some of them are listed below:

(i) A function $u(s)$ is said to be strongly measurable if it is the strong limit of a sequence of finitely valued functions almost everywhere. (p. 1681 of [22])
(ii) A function $u(s)$ is said to be strongly measurable if there exists a sequence of countably valued functions converging almost everywhere. (Definition 3.5.4 of [23])
(iii) A function $u(s)$ is called measurable function if there exists a sequence $\{u_n(s)\}$ of countably valued functions such that $u_n(s) \to u(s)$ uniformly outside a null set (i.e. the set with null measure). (p. 174 of [8])

Definition (i) above is a special case of (ii).

$u(s)$ is called almost separably valued if there exists a measurable set B_0 of measure zero such that $\{u(s); s \in S - B_0\}$ is separable.

By Theorem (Pettis) of Sect. 4, Chap. 5 in [2], $u(s)$ is strongly measurable in sense of Definition 1.9 iff it is weakly measurable and almost separably valued. By Theorem 3.5.3 of [23], for $u(s)$ to be strongly measurable in the sense of Definition (ii), so is it. Thus, Definition 1.9 and Definition (ii) are equivalent.

By Corollary 1 of Theorem 3.5.3 of [23], Definitions (ii) and (iii) above are equivalent.

Therefore, all definitions in Remark 1.5 are equivalent to Definition 1.9.

Theorem 1.30 (Theorem 5.0.11 of [8]) *Let X be separable. Then $u : S \to X$ is weakly measurable if and only if it is strongly measurable.*

If X^* is separable, then strong measurability of X^*-valued function $v : S \to X^*$ is equivalent to $*$-weak measurability,

Remark 1.6 An X^*- valued function $v(s)$ is said to be weakly measurable if $\langle f, v(s)\rangle$ is measurable for all $f \in X^{**}$. The weak measurability of X^*-valued function is in general stronger than $*$-weak measurability.

For an example of $*$-weakly measurable but not measurable X^*-valued function we refer to Example 5.0.10 of [8].

Theorem 1.30 is valid when X is a normed space.

Let $u(s) : S \to X$ be a finitely valued function. Let $u(s) = x_i$ on B_i, where B_i are disjoint and $mes(B_i) < \infty$ for $i = 1, \cdots, n$ and $u(s) = 0$ on $S - \cup_i B_i$. Then we can define the (Bochner) integral $\int_S u(s)\,ds$ of $u(s)$ over S by $\sum_{i=1}^n x_i mes(B_i)$.

Define the integral of more general functions.

Definition 1.10 A function $u(s) : S \to X$ is said to be Bochner integrable, if there exists a sequence of finitely valued functions $\{u_n(s)\}$ which converges to $u(s)$ a.e. in S and

$$\lim_{n\to\infty} \int_S \|u(s) - u_n(s)\|\,ds = 0.$$

For any measurable set $B \subset S$, the *Bochner integral* of $u(s)$ over B is defined by

$$\int_B u(s)\,ds = \lim_{n\to\infty} \int_S \chi_B(s)u_n(s)\,ds,$$

where χ_B is the characteristic function of the set B.

The definition is justified since the limit on the right above exists and the value of this limit is independent of the approximating sequence of functions $\{u_n\}$ (see Sect. 5, Chap. 5 of [2]).

If $u(\cdot)$ is strongly measurable, then $\|u(\cdot)\|$ is measurable, and $u(\cdot)$ is Bochner integrable if and only if $\|u(\cdot)\|$ is Lebesgue integrable, both integrals standing in the relation

$$\left\|\int_S u(s)\,ds\right\| \le \int_S \|u(s)\|\,ds.$$

Theorem 1.31 (Theorem 1.8, Chap. 4 of [3]) *Let $u : S \to X$ be Bochner integrable and $B \subset S$ be a Lebesgue measurable set. Then for any $f \in X^*$,*

$$\int_B \langle f, u(s)\rangle\,ds = \langle f, \int_B u(s)\,ds\rangle.$$

(See Theorem 8, Appendix E of [18] or Corollary 2, Sect. 5 of Chap. V in [2].)

Differently from real functions, not every Banach space valued absolutely continuous functions is the indefinite integral of an integrable function (see Example 5.0.13 of [8]). But the next holds.

Theorem 1.32 *If $u(\cdot) : S \to X$ is Bochner integrable in S, then the function $v : S \to X$ defined by*

$$v(t) = \int_{t_0}^{t} u(s)\,ds, \quad t_0 \in S$$

is differentiable almost everywhere on S and

$$v'(t) = u(t) \quad \textit{for a.e. } t \in S.$$

(See p. 176 of [8] or Theorem 1.7, Chap. 4 of [3].)

Remark 1.7 A Banach space X is called a Gelfand space if every X-valued absolutely continuous function is differentiable almost everywhere. Let X be either (a) reflexive or (b) separable and the dual of another Banach space. Then X is a Gelfand space (Theorem 5.0.14 of [8]).

Definition 1.11 The space $L^p(S; X)$, $(1 \le p < \infty)$ consists of all X-valued strongly measurable functions $u(\cdot)$ defined in S and such that

$$\int_S \|u(s)\|^p\,ds < \infty.$$

The space $L^\infty(S; X)$ consists of all X-valued strongly measurable functions $u(\cdot)$ defined in S and such that

$$\|u(s)\| \le M \quad \exists M > 0, \quad \text{a.e. } s \in S.$$

$L^p(S; X)$, $(1 \le p < \infty)$ and $L^\infty(S; X)$ are Banach spaces, respectively, with norms

$$\|u\|_{L^p(S;X)} = \left(\int_S \|u(s)\|^p\,ds\right)^{1/p} \quad \text{and} \quad \|u\|_{L^\infty(0,T;X)} = \operatorname*{esssup}_{s\in S} \|u(s)\|.$$

Remark 1.8 Let $1 \le p < \infty$. If Y is a dense subspace of X, then the set $\{\varphi = \phi(t)u : \phi \in \mathscr{D}(0, T),\ u \in Y\}$ is dense in $L^p(0, T; X)$ (see Theorem 5.0.27 of [8]). Denote by $C^k([0, T]; X)$ the space of k-times continuously differentiable functions on $[0, T]$, $0 < T < \infty$. Then, the set of polynomials $\{p : p(s) = \sum_{k=0}^{m} a_k s^k,\ s \in [0, T],\ a_k \in X,\ m = 0, 1, \cdots\}$ is dense in the set $C([0, T]; X)$ (Weierstrass' approximation theorem, see Theorem 1.3, Sect. 1, Chap. 4 of [3]). Therefore, the set $\{\varphi = \phi(t)u : \phi \in C^1[0, T],\ u \in X\}$ is dense in $L^p(0, T; X)$. Also, we can see that the set $\{\varphi = \phi(t)u : \phi \in C^1[0, T], \phi(T) = 0,\ u \in X\}$ is dense in $\{\varphi \in C^1([0, T]; X) : \varphi(T) = 0\}$.

Theorem 1.33 *Let X_0, X_1 be Banach spaces such that $X_0 \hookrightarrow X_1$, $1 \le p_0 < \infty$ and $0 < \theta < 1$. Then,*

$$L^{p_0}(S; X_0) \cap L^\infty(S; X_1) \hookrightarrow L^p(S; (X_0, X_1)_{[\theta]}) \tag{1.21}$$

with $p = p_0/(1-\theta)$, *where the functor* $(\ ,\ )_{[\theta]}$ *means the complex interpolation (see Definition 2.1.3 of [10]).*

Proof Since $X_0 \hookrightarrow X_1$, $u \in L^p(S; X_0) \cap L^\infty(S; X_1)$ is $(X_0, X_1)_{[\theta]}$-value strongly measurable and $u(t) \in (X_0, X_1)_{[\theta]}$ at a.e. $t \in S$. Then,

$$\|u(t)\|_{(X_0,X_1)_{[\theta]}} \le \|u(t)\|_{X_0}^{1-\theta} \|u(t)\|_{X_1}^{\theta}$$

(see Corollary 2.1.8 of [10]), which yields (1.21). □

Complex interpolation between simple Sobolev spaces is as follows.

Let Ω be a bounded domain with Lipschitz boundary, $s_0 \neq s_1$ and $0 < \theta < 1$. Then,

$$\left(W^{s_0,2}(\Omega), W^{s_1,2}(\Omega)\right)_{[\theta]} = W^{s^*,2}(\Omega), \tag{1.22}$$

where

$$s^* = (1-\theta)s_0 + \theta s_1.$$

(cf. Theorem 6.4.5 of [11]. We used that when $p = 2$, Sobolev space $W^{s,p}$ equals to Bessel potential space H_p^s (see (1.12)). All spaces in [11] are defined on $\mathbb{R}^3$; however, the results are valid for the spaces continuously expandable to the space defined on $\mathbb{R}^3$, for example, the space defined on a bounded domain with Lipschitz boundary, since from the interpolation results for the spaces defined on $\mathbb{R}^3$ we can reduce the results for the spaces defined on the restricted domain. We refer to Sect. 1.3.2.)

Remark 1.9 In [11, 24] a space $L^\infty(S; X)$ is defined as follows:

If χ_B *is the characteristic function of the measurable set* $B \subset S$, *then* $L^\infty(S; X)$ *is defined as a completion of*

$$v(x) = \sum_{j=1}^{N} a_j\, \chi_{B_j}(s),$$

where B_j *are disjoint sets, in the norm* $\|v\|_\infty = \operatorname{ess\,sup}_{x\in S}\|v(x)\|_X$. *If above only sets with* $mes(B_j) < \infty$ *are admitted, then the completion of those elements is denoted by* $L^0_\infty(S; X)$.

Then, Remark 3 of Theorem 1.18.4 of [24] and Theorem 5.1.2 of [11] assert without detailed proof the following.

Theorem 1.34 *Assume that* X_0 *and* X_1 *are Banach spaces and that* $1 \le p_0 < \infty$, $0 < \theta < 1$. *Then*

$$(L^{p_0}(S; X_0), L^0_\infty(S; X_1))_{[\theta]} = L^p(S; (X_0, X_1)_{[\theta]})$$

with $p = p_0/(1-\theta)$.

Let us consider whether the space $L^\infty(S; X)$ in [11, 24] is the same as the one in Definition 1.11 and $L^0_\infty(S; X)$ coincides with $L^\infty(S; X)$ when $S = [0, T]$, $T < \infty$. The following counterexample shows that the answer is negative.

Counterexample (the case that $S = [0, T]$, $T < \infty$)

Let l^2 be the space of sequences $(a_1, a_2, \cdots, a_i, \cdots)$ such that $\sum_i a_i^2 < \infty$ and $e_i = (0, \cdots, a_i, \cdots)$ with $a_i = 1$. Let $S_i = [\frac{(2^{i-1}-1)T}{2^{i-1}}, \frac{(2^i-1)T}{2^i})$, $i = 1, 2, \cdots$. Define a function $u : S \to l^2$ by

$$u(s) = e_n \quad \forall s \in S_n, \quad u(T) = e_1.$$

It is easy to verify that when $X = l^2$, this function belongs to $L^\infty([0, T]; X)$ in the sense of Definition 1.11.

Let us consider a sequence $\{f_n\}$ of finite-value functions

$$f_n(s) = \sum_{i=1}^{k_n} x_i^{(n)} \chi_{B_i^{(n)}}(s) \quad x_i^{(n)} \in X, \;\; B_i^{(n)} - \text{disjoint measurable subsets of } [0, T].$$

For every n there exists an $i_n \le k_n$ such that $B_{i_n}^{(n)}$ intersects with infinite set of $S_{k_{i_n}}$ on measurable set with positive measure. Thus,

$$\forall N, \; \exists m_N \ge N \quad mes(S_{m_N} \cap B_{i_n}^{(n)}) > 0. \tag{1.23}$$

On the other hand, since $x_{i_n}^{(n)} \in l^2$,

$$x_{i_n}^{(n)} = \sum_{l=1}^{\infty} \alpha_l^{(i_n)} e_l, \qquad \lim_{l\to\infty} \alpha_l^{(i_n)} = 0.$$

Thus,

$$\exists N, \; \forall l \ge N \quad |\alpha_l^{(i_n)}| < \frac{1}{2}. \tag{1.24}$$

Now let m_N of (1.23) be the number corresponding to N of (1.24) and $E_{m_N}^{(n)} \equiv S_{m_N} \cap B_{i_n}^{(n)}$. Then, at $s \in E_{m_N}^{(n)}$ we have

$$\|u(s) - f_n(s)\|_{l^2} = \| \sum_{l \ne m_N} \alpha_l^{(i_n)} e_l + (1 - \alpha_{m_N}^{(i_n)}) e_{m_N} \|_{l^2} \ge \frac{1}{2}.$$

By (1.23) $mes(E_{m_N}^{(n)}) > 0$, and so any sequence $\{f_n\}$ cannot converge to u in the sense of essential supremum norm. Thus, $L^0_\infty([0, T]; X) \subsetneqq L^\infty([0, T]; X)$.

Let us show that $C([0, T]; X_1) \subset L^0_\infty([0, T]; X_1)$.

If $T < \infty$, then the set of polynomials $\{p : p(s) = \sum_{k=0}^{m} a_k s^k,\ s \in [0, T],\ a_k \in X_1\}$ is dense in the set $C([0, T]; X_1)$, where X_1 is a Banach space. Thus, for any $\varepsilon > 0$ there exists a polynomial $p(s) = \sum_{k=0}^{m_\varepsilon} a_k s^k$ such that $\sup_{s\in[0,T]} \|u(s) - p(s)\|_{X_1} \le \varepsilon$. We can make a partition $B_i = [s_i, s_{i+1}]$ of $[0, T]$ such that

$$\max_{k\in\{0,1,\cdots,m_\varepsilon\}} \max_i |s_i^k - s_{i+1}^k| \le \frac{\varepsilon}{\sum_{k=0}^{m_\varepsilon} \|a_k\|_{X_1}}.$$

Let us consider a simple function $\sum_i b_i \chi_{B_i}(s)$, where $b_i = \sum_{k=0}^{m_\varepsilon} a_k s_i^k$. Then, we have

$$\begin{aligned}\|u(s) - \sum_i b_i \chi_{B_i}(s)\| &= \Big\|u(s) - \sum_{k=0}^{m_\varepsilon} a_k\Big(\sum_i s_i^k \chi_{B_i}(s)\Big)\Big\| \\ &\le \|u(s) - p(s)\| + \Big\|\sum_{k=0}^{m_\varepsilon} a_k s^k - \sum_{k=0}^{m_\varepsilon} a_k \sum_i s_i^k \chi_{B_i}(s)\Big\| \\ &\le \varepsilon + \Big\|\sum_{k=0}^{m_\varepsilon} a_k(s^k - \sum_i s_i^k \chi_{B_i}(s))\Big\| \le 2\varepsilon,\end{aligned}$$

which shows the asserted conclusion.

Let X be a real Banach space. By Theorem 12.9.2 and (12.9.29) of [8] we get

Theorem 1.35 *Let $1 < p < \infty$, X^* a separable Banach space. Then to every $f \in \big(L^p(0, T; X)\big)^*$ corresponds an unique $v_f \in L^q(0, T; X^*)$, $1/p + 1/q = 1$, such that*

$$\langle u, f\rangle = \int_S \langle u(s), v_f(s)\rangle\, ds, \quad \forall u \in L^p(0, T; X)$$

and this correspondence is algebraically and metrically isomorphic.

By Theorem 12.2.11 and (12.2.13) of [8] we get

Theorem 1.36 *Let X^* be separable Banach space. Then to every $f \in \big(L^1(0, T; X)\big)^*$ corresponds an unique $v_f \in L^\infty(0, T; X^*)$ such that*

$$\langle u, f\rangle = \int_S \langle u(s), v_f(s)\rangle\, ds, \quad \forall u \in L^1(0, T; X)$$

and this correspondence is algebraically and metrically isomorphic.

Remark 1.10 The space $X = L^1(0, 1)$ is separable, but its dual space $X^* = L^\infty(0, 1)$ is not separable (Example 5.0.31 of [8]). But if the dual X^* of a Banach space X is separable, then X is separable (Lemma of Sect. 2, Chap. 5 of [2]). Thus, if X be a separable reflective Banach space, then from the fact that $X^{**} = X$ is separable it follows that X^* is separable. Therefore, Theorems 1.35, 1.36 are valid when X is a separable reflective Banach space. When X^* is a Gelfand space or a Banach space, for generalizations of Theorems 1.35, 1.36 we refer to Chap. 12 of [8].

Let S be an open interval, $\mathscr{D}(S)$ be the local convex space made from $C_0^\infty(S)$ by introducing topology with the semi-norms (1.5) and X_w denote a Banach space X with weak topology.

Definition 1.12 The set $\mathscr{L}(\mathscr{D}(S), X_w)$ of linear continuous operators from $\mathscr{D}(S)$ to X_w is denoted by $\mathscr{D}^*(S; X)$ and its element is called a distribution on S with value in X.

Definition 1.13 The distribution $f' \in \mathscr{D}^*(S; X)$ defined by

$$f'(\varphi) = -f(\varphi') \quad \forall \varphi \in \mathscr{D}(S)$$

is called derivative of $f \in \mathscr{D}^*(S; X)$.

If $f_n \to f$ in $\mathscr{D}^*(S; X)$, then for $y \in X^*$ and $\varphi \in \mathscr{D}(S)$

$$\lim_{n\to\infty} \langle y, f_n'(\varphi) - f'(\varphi)\rangle = \lim_{n\to\infty} \langle y, f(\varphi') - f_n(\varphi')\rangle = 0.$$

Thus, mapping $f \to f'$ is continuous in $\mathscr{D}^*(S; X)$.

Let V be a real separable reflective Banach space and H be a Hilbert space such that $V \hookrightarrow H$ and V is dense in it. Identifying H and H^*, we have

$$V \hookrightarrow H \hookrightarrow V^*.$$

Let $1 < p \le p_0 < \infty$ and

$$X = L^p(S; V) \cap L^{p_0}(S; H), \quad \|u\|_X = \|u\|_{L^p(S;V)} + \|u\|_{L^{p_0}(S;H)}. \tag{1.25}$$

Then,

$$X^* = L^q(S; V) + L^{q_0}(S; H), \quad \frac{1}{p} + \frac{1}{q} = 1, \quad \frac{1}{p_0} + \frac{1}{q_0} = 1,$$

$$\|f\|_{X^*} = \inf_{\substack{f_1 \in L^q(S;V^*) \\ f_2 \in L^{q_0}(S;H) \\ f_1+f_2=f}} \max\{\|f_1\|_{L^q(S;V^*)}, \|f_2\|_{L^{q_0}(S;H)}\}.$$

Define a space W by

$$W = \{u : u \in X, \, u' \in X^*\}, \quad \|u\|_W = \|u\|_X + \|u'\|_{X^*}. \tag{1.26}$$

Theorem 1.37 *$W \subset C(S; H)$. If S is a compact interval, then the imbedding of W to $C(S; H)$ is continuous. The following integration by parts holds. For any $u, v \in W$,*

$$\big(u(t), v(t)\big)_H - \big(u(s), v(s)\big)_H = \int_s^t \langle u'(\tau), v(\tau)\rangle \, d\tau + \int_s^t \langle u(\tau), v'(\tau)\rangle \, d\tau \quad s, t \in S.$$

From the theorem above we have that when $v = u$,

$$\frac{1}{2}\left(\|u(t)\|_H^2 - \|u(s)\|_H^2\right) = \int_s^t \langle u'(\tau), u(\tau)\rangle \, d\tau. \tag{1.27}$$

Remark 1.11 If $u' = f_1 + f_2$, $f_1 \in L^q(S; V)$, $f_2 \in L^{q_0}(S; H)$, then

$$\int_s^t \langle u'(\tau), v(\tau)\rangle \, d\tau = \int_s^t \langle f_1(\tau), v(\tau)\rangle_V \, d\tau + \int_s^t (f_2(\tau), v(\tau))_H \, d\tau.$$

(See Lemma 7.3 of [19] for the case of $p = p_0$, and Theorem 1.17, Chap. 4 of [3] for the general case.)

1.4.2 Compactness

Let B be a Banach space. For a positive integer m and $1 \le p \le \infty$ denote

$$W^{m,p}(0, T; B) = \left\{ f : f, \cdots, \frac{\partial^m f}{\partial t^m} \in L^p(0, T; B) \right\}.$$

For $0 < \sigma < 1$ and $1 \le p \le \infty$ denote

$$W^{\sigma,p}(0, T; B) = \left\{ f : f \in L^p(0, T; B), \int_0^T \int_0^T \frac{\|f(t) - f(s)\|_B^p}{|t - s|^{\sigma p+1}} dt ds < \infty \right\}$$

with the usual change if $p = \infty$ (see (iii) of Definition 1.6). For a positive integer m and $0 < \sigma < 1$ denote

$$W^{m+\sigma,p}(0, T; B) = \left\{ f : f, \cdots, \frac{\partial^{m-1} f}{\partial t^{m-1}} \in L^p(0, T; B), \frac{\partial^m f}{\partial t^m} \in W^{\sigma,p}(0, T; B) \right\}.$$

If f is defined on $[0, T]$, then the translated function $(\tau_h f)(t) = f(t + h)$ for $h > 0$ is defined on $[-h, T - h]$, where $0 < T < \infty$.

Let X, B, Y be Banach spaces such that $X \hookrightarrow B \hookrightarrow Y$ and the imbedding $X \to B$ is compact.

Theorem 1.38 (Theorem 5 of [25]) *Let $1 \le p \le \infty$ and F be a bounded set of $L^p(0, T; X)$. If*

$$\|\tau_h f - f\|_{L^p(0,T-h;Y)} \to 0 \quad as\ h \to 0, \quad uniformly\ for\ f \in F,$$

then F is relatively compact in $L^p(0, T; B)$ (and in $C(0, T; B)$ if $p = \infty$).

Theorem 1.39 (Corollary 5 of [25]) *Let $1 \le p \le \infty$, $1 \le r \le \infty$ and F be a bounded set of $L^p(0,T;X) \cap W^{s,r}(0,T;Y)$, where $s > 0$ if $r \ge p$ and where $s > 1/r - 1/p$ if $r \le p$.*

Then F is relatively compact in $L^p(0,T;B)$ (and in $C(0,T;B)$ if $p = \infty$).

1.5 Operator Equations and Operator-Differential Equations

1.5.1 Monotone Operator Equation

Let X be a separable real reflective Banach space such that $X \hookrightarrow H$, where H is a real separable Hilbert space..

Definition 1.14 The operator $A : X \to X^*$ is said to be *radial continuous* if $s \to \langle A(u+sv), v\rangle$ is continuous on $[0,1]$ for any fixed $u, v \in X$, *hemi-continuous* if $s \to \langle A(u+sv), h\rangle$ is continuous on $[0,1]$ for any fixed $u, v, h \in X$, *demi-continuous* if from $u_n \to u$ in X it follows that $Au_n \rightharpoonup Au$ in X^*, and *Lipschitz continuous* if

$$\|Au - Av\|_{X^*} \le M\|u - v\|_X \quad \exists M > 0, \quad \forall u, v \in X.$$

Definition 1.15 Let u, v be any elements of X. The operator $A : X \to X^*$ is said to be *monotone* if

$$\langle Au - Av, u - v\rangle \ge 0,$$

strictly monotone if

$$\langle Au - Av, u - v\rangle > 0 \quad \text{for } u \ne v,$$

strongly monotone if

$$\langle Au - Av, u - v\rangle \ge \alpha\|u - v\|_X^2 \quad \alpha > 0$$

and *coercive* if

$$\frac{\langle Au, u\rangle}{\|u\|_X} \to +\infty \quad \text{as } \|u\|_X \to \infty.$$

Definition 1.16 The operator $A : X \to X^*$ is called bounded iff the image of every bounded set of X is bounded in X^*.

Theorem 1.40 *Let $A : X \to X^*$ be radially continuous, monotone and coercive. Then for any $f \in X^*$ the set of solutions to*

$$Au = f$$

is nonempty, weakly closed and convex.

Theorem 1.41 *Let $A : X \to X^*$ be radially continuous, strictly monotone and coercive. Then there exists $A^{-1} : X^* \to X$ which is strictly monotone, bounded and demicontinuous.*

Theorem 1.42 *Let $A : X \to X^*$ be radially continuous, strongly monotone. Then there exists $A^{-1} : X^* \to X$ which is Lipschitz continuous. If, in addition, A is Lipschitz continuous, then A^{-1} is strongly monotone.*

(For Theorems 1.40–1.42 we refer to Theorems 2.1, 2.2 and Corollary 2.3 of Chap. 3 of [3] or Theorem 26.A of [26]. If A is continuous and strictly monotone, then for Theorem 1.40 we refer to Theorem 5.16 of [1].)

Remark 1.12 If A is linear, strongly monotone and Lipschitz continuous on Hilbert space X, Theorem 1.42 means the Lax-Milgram lemma (Theorem 1, Sect. 6.2 of [18]).

Let X be real Hilbert space and a nonlinear functional be defined by

$$a(\,\cdot\,;\,\cdot\,,\,\cdot\,) : (w, u, v) \in X \times X \times X \to a(w; u, v) \in \mathbb{R},$$

where, for $w \in X$, the mapping $(u, v) \to a(w; u, v)$ is a bilinear continuous form on $X \times X$.

Let us consider the following problem:

Find $u \in X$ such that

$$a(u; u, v) = \langle f, v\rangle \quad \forall v \in X, \quad \text{where } f \in X^*. \tag{1.28}$$

Theorems 1.2 and 1.3 of Chap. 4 of [14] are rewritten as follows.

Theorem 1.43 *Assume that the following hypotheses hold:*

(i) there exists a constant $\alpha > 0$ such that

$$a(v; v, v) \geq \alpha \|v\|_X^2 \quad \forall v \in X.$$

(ii) the space X is separable and, for all $v \in X$, the mapping

$$u \to a(u; u, v)$$

is sequentially weakly continuous on X, i.e.

$$weak - \lim_{m\to\infty} u_m = u \text{ in } X \text{ implies } \lim_{m\to\infty} a(u_m; u_m, v) = a(u; u, v)\,\forall v \in X.$$

Then, problem (1.28) has at least one solution $u \in X$.

Theorem 1.44 *Assume that the following hypotheses hold:*

(i) there exists a constant $\alpha > 0$ such that

$$a(w; v, v) \geq \alpha \|v\|_X^2 \quad \forall v, w \in X.$$

(ii) there exists a continuous and monotonically increasing function $L : \mathbb{R}^+ \to \mathbb{R}^+$ such that for all $\mu > 0$

$$|a(w_1; u, v) - a(w_2; u, v)| \leq L(\mu)\|u\|_X \|v\|_X \|w_1 - w_2\|_X \\ \forall u, v \in X, \forall w_1, w_2 \in \mathscr{O}_\mu(0_X),$$

where $\mathscr{O}_\mu(0_X)$ is μ-neighborhood of 0_X in X. Then, under the condition

$$\frac{\|f\|_{X^*}}{\alpha^2} L\big(\|f\|_{X^*}/\alpha\big) < 1$$

problem (1.28) *has a unique solution $u \in X$.*

1.5.2 Pseudo-Monotone Operator Equation

Theorem 1.45 *Let $A : X \to X^*$ be an operator on the real, separable and reflexive Banach space X. Let A be coercive and bounded. If for every sequence $\{x_n\}$ such that*

$$\begin{aligned} &x_n \rightharpoonup x \text{ in } X, \\ &\limsup_{n\to\infty} \langle Ax_n, x_n - x\rangle \leq 0 \end{aligned} \tag{1.29}$$

there exists a subsequence $\{x_k\}$ such that

$$\liminf_{k\to\infty} \langle Ax_k, x_k - v\rangle \geq \langle Ax, x - v\rangle \quad \forall v \in X, \tag{1.30}$$

then for any $f \in X^*$ there exists a solution to

$$Au = f.$$

Remark 1.13 If (1.29) implies (not for a subsequence)

$$\liminf_{n\to\infty} \langle Ax_n, x_n - v\rangle \geq \langle Ax, x - v\rangle \quad \forall v \in X,$$

then A is called pseudo-monotone. For the coercive, bounded and pseudo-monotone operators A, the asserted conclusion was proved (see Theorem 27.A of [26] or Theorem 2.7, Chap. 2 of [27]). However, proofs of the facts that A has property (M) (Proposition 2.5, Chap. 2 in [27]) and A is demi-continuous (footnote of (2.27), Chap. 2 of [27]), which guarantee the existence of a solution to $Au = f$, hold with (1.30) for subsequence, and so we have the conclusion.

1.5.3 Operator-Differential Equations

Definition 1.17 Let X_1, X_2 be spaces and $S = [0, T], T > 0$. Mapping

$$D(F) \subset (S \to X_1),$$
$$F : D(F) \to (S \to X_2),$$

where $(S \to X_i)$ is a set of functions from S to X_i, is called Volterra operator iff

$$u(s) = v(s) \text{ a.e. } s \in [0, t],\ t \in S \quad \text{implies} \quad (Fu)(t) = (Fv)(t) \text{ a.e. } t \in S.$$

Let V be a real separable reflective Banach space and H be a Hilbert space such that $V \hookrightarrow H$ and V is dense in it. Now, let the spaces X and W be the ones in (1.25) and (1.26).

Theorem 1.46 *Let $\mathscr{A} : X \to X^*$ be a radial continuous, monotone, coercive and Volterra operator. Then, for any $f \in X^*$ and $a \in H$, there exists a unique solution $u \in W \subset C([0, T]; H)$ to*

$$u' + \mathscr{A}u = f, \quad u(0) = a,$$

and the mapping $a \to u$ is continuous from H to $C([0, T]; H)$.

(See Theorem 1.1, Chap. 6 of [3] or Sect. 8.1 of [19].)

Definition 1.18 Let A be multivalued operator from $D(A) \subset X$ to 2^{X^*}(the family of subsets of X^*) and its graph $G(A) = \{(x, y) \in X \times X^*; y \in Ax\}$. The operator is said to be monotone if

$$\langle y_1 - y_2, x_1 - x_2 \rangle \geq 0 \ \ \forall (x_1, y_1), (x_2, y_2) \in G(A).$$

A monotone operator A is said to be maximal monotone if its graph is not properly contained in graph of any other monotone operator from X to 2^{X^*}.

Note that if A is a single-valued operator such that $D(A) \subseteq X \to X^*$ it is monotone if

$$\langle Au - Av, u - v \rangle \geq 0 \ \ \forall u, v \in D(A)$$

and it is maximal monotone if, in addition,

$$\langle f - Av, u - v \rangle \geq 0 \ \forall v \in D(A) \quad \text{implies} \quad u \in D(A) \text{ and } Au = f.$$

Example 1.1

Let V be a real separable reflective Banach space and H be a Hilbert space such that $V \hookrightarrow H$ and V is dense in H. Identifying H and H^*, we have

$$V \hookrightarrow H \hookrightarrow V^*.$$

Let $1 < p < \infty$ and

$$X = L^p(0, T; V), \quad \|u\|_X = \|u\|_{L^p(0,T;V)}.$$

Then,

$$X^* = L^q(0, T; V^*), \quad \frac{1}{p} + \frac{1}{q} = 1.$$

Put

$$W = \{u : u \in X,\ u' \in X^*\}, \quad \|u\|_W = \|u\|_X + \|u'\|_{X^*}.$$

Then, the operator $L : D(L) \subset X \to X^*$ defined by

$$L = \frac{d}{dt}, \quad D(L) = \{u \in W : u(0) = 0\}$$

is a maximal monotone linear operator (see Sect. 2.1 of Chap. 3 in [27]).

Theorem 1.47 *Let X be a real reflexive Banach space and norms of X and X^* be strictly convex. Let a linear operator L, where $D(L)$ is a dense linear subspace of X, be maximal monotone. Let the operator $A : D(L) \to X^*$ satisfy the following:*

(i) For every sequence such that $x_n \rightharpoonup x$ in X, $x_n, x \in D(L)$, $Lx_n \rightharpoonup Lx$ in X^* and $\limsup_{n\to\infty}\langle Ax_n, x_n - x\rangle \le 0$, there exists a subsequence x_k such that

$$\liminf_{k\to\infty}\langle Ax_k, x_k - v\rangle \ge \langle Ax, x - v\rangle \quad \forall v \in X; \tag{1.31}$$

(ii) There exists a function $\psi : \mathbb{R}^+ \equiv [0, \infty) \to \mathbb{R}^+$, which is bounded on any compact interval, and a number θ, $0 \le \theta < 1$, such that

$$\|A(x)\|_{X^*} \le \psi(\|x\|_X) + \theta\|Lx\|_{X^*} \quad \forall x \in D(L); \tag{1.32}$$

(iii)

$$\frac{\langle A(x), x\rangle}{\|x\|_X} \to \infty, \quad \text{as } \|x\|_X \to \infty. \tag{1.33}$$

Then, for any $f \in X^*$ there exists a solution to

$$Lx + Ax = f.$$

Remark 1.14 When

$$\liminf_{n\to\infty}\langle Ax_n, x_n - v\rangle \geq \langle Ax, x - v\rangle \quad \forall v \in X,$$

instead of (1.31) holds, the asserted conclusion was proved in [28]. In Theorem 1.2, Chap. 3 of [27] it is proved by using Theorem 2.7, Chap. 2 of [27]. However, existence of solutions to the equation holds with (1.31) for subsequence, and so we have the conclusion (cf. Remark 1.13).

1.6 Convex Functional

Throughout this section, X is a real Banach space with dual X^*.

Definition 1.19 A functional $f : X \to \overline{\mathbb{R}} \equiv \mathbb{R} \cup +\infty$ is said to be *proper* if it is not identically equal to ∞. If $f(x) \in (-\infty, +\infty)$ $\forall x \in X$, then it is said to be *finite*.

If a function $\varphi : X \to \overline{\mathbb{R}}$ satisfies the inequality

$$\varphi\big((1-\lambda)x + \lambda y\big) \leq (1-\lambda)\varphi(x) + \lambda\varphi(y) \quad \forall x, y \in X, \quad \forall \lambda \in [0, 1]$$

then it is said to be convex.

The functional $\varphi : X \to \overline{\mathbb{R}}$ is said to be lower *semi-continuous* (l.s.c.) on X if

$$\liminf_{u\to v} \varphi(u) \geq \varphi(v) \quad \forall x \in X$$

or, equivalently, every *level subset* $\{x \in X : \varphi(x) \leq \lambda\}$ is closed.

The functional $\varphi : X \to \overline{\mathbb{R}}$ is said to be *weakly lower semi-continuous* if it is lower semi-continuous on the space X endowed with weak topology.

Because every level set of a convex functional is convex and every closed convex set is weakly closed (see Theorem 1.1), we may therefore conclude that a proper convex functional is lower semi-continuous if and only if it is weakly lower semi-continuous (see Sect. 1.2 of [29]).

As a special case of Definition 1.2 we have

Definition 1.20 Given a functional $f : X \to \mathbb{R}$,

$$Df(x, h) = \lim_{\lambda\downarrow 0} \frac{f(x + \lambda h) - f(x)}{\lambda} \quad x, h \in X \tag{1.34}$$

(if it exists) is called the *directional derivative* of f at x in direction h.

If $Df(x, h)$ is linear and continuous with respect to h, that is, there exists $\nabla f(x) \in X^*$ such that

$$Df(x, h) = \langle \nabla f(x), h \rangle \quad \forall h \in X,$$

the functional $f : X \to \mathbb{R}$ is said to be Gâteaux differentiable at $x \in X$ and $\nabla f(x) \in X^*$ is called the Gâteaux derivative at x.

If $Df(x, h)$ in (1.34) is linear and continuous with respect to h and convergence is uniform in h on bounded subsets, then f is said to be Fréchet differentiable and $\nabla f(x)$ is called the Fréchet derivative of f at x.

Remark 1.15 For a finite functional $f : X \to \mathbb{R}$, if there exists the Gâteaux derivative at all $x \in X$, then the operator $\nabla f : X \to X^*$ defined by

$$x \to \nabla f(x)$$

is called *gradient* of f. When a finite functional f has its gradient ∇f, the functional f is convex if and only if $\nabla f : X \to X^*$ is monotone (see Theorem 4.4 of [19] or Lemma 4.10, Chap. 3 of [3]). Every monotone gradient operator of a finite functional is demi-continuous (see the footnote[9] in p. 113 of [19] or Lemma 4.12, Chap. 3 of [3]).

Definition 1.21 Given a l.s.c., convex, proper function $\varphi : X \to \overline{\mathbb{R}}$, the mapping $\partial\varphi : X \to 2^{X^*}$ defined by

$$\partial\varphi(x) = \{x^* \in X^* : \varphi(y) - \varphi(x) \geq \langle x^*, y - x \rangle \quad \forall y \in X\}$$

is called the *subdifferential* of φ.

In general, $\partial\varphi$ is a multivalued operator not everywhere defined. We denote by $D(\partial\varphi)$ the set of all $x \in X$ for which $\partial\varphi(x) \neq \emptyset$. If a functional φ is convex and there exists Gâteaux derivative $\nabla\varphi(x)$ at x, then $\partial\varphi(x) = \nabla\varphi(x)$ (see Sect. 1.2 of [29]). By the definition of $\partial\varphi(x)$ we see that $\varphi(x) = \inf\{\varphi(y); \, y \in X\}$ if and only if $0_{X^*} \in \partial\varphi(x)$.

Example 1.2

Let Ω be a bounded domain, $K(\Omega)$ be a nonempty convex closed subset of a Sobolev space $H^1(\Omega)$, Γ be a open subset of $\partial\Omega$ and $g(x) \in L^2(\Gamma)$. Let a functional $\Phi(u) : H^1(\Omega) \to \overline{\mathbb{R}}$ be defined by

$$\Phi(u) = \begin{cases} \displaystyle\int_\Gamma g(x) u \, dx & \forall u \in K(\Omega), \\ +\infty & \forall u \notin K(\Omega). \end{cases}$$

Then the functional Φ is proper, convex and weakly lower semi-continuous.

Theorem 1.48 (Theorem 2.8 of [29]) *Let X be a real Banach space and let $\Phi : X \to \overline{\mathbb{R}}$ be a l.s.c. proper convex function. Then $\partial\Phi$ is a maximal monotone operator such that $X \to 2^{X^*}$.*

In what follows of this section for simplicity let X be a real Hilbert space and X^* be identified with X. Let A be a maximal monotone operator in X. For every $x \in X$ the equation

$$0_X \in (x_\varepsilon - x) + \varepsilon A x_\varepsilon$$

has a unique solution x_ε. Thus

$$x_\varepsilon = (I + \varepsilon A)^{-1} x,$$

where I is the unit operator on X. The operator $A_\varepsilon : X \to X$ defined by

$$A_\varepsilon x = \frac{1}{\varepsilon}(x - x_\varepsilon) \equiv \frac{1}{\varepsilon}\left(I - (I + \varepsilon A)^{-1}\right) x$$

is called the *Yosida approximation* of A.

Let the functional $\Phi : X \to \overline{\mathbb{R}}$ be proper, convex and weakly lower semi-continuous (l.s.c.). For every $0 < \varepsilon < 1$, define a functional Φ_ε by

$$\Phi_\varepsilon(y) = \inf\left\{\frac{\|y - u\|_X^2}{2\varepsilon} + \Phi(u);\ u \in X\right\}, \quad y \in X, \tag{1.35}$$

which is called the *Moreau-Yosida approximation* (or regularization) of Φ.

When $\partial\Phi : X \to 2^X$ is the subdifferential of Φ in the Hilbert space X, taking into account Theorem 1.48 let

$$J_\varepsilon = (I + \varepsilon\partial\Phi)^{-1}.$$

Then the Yosida approximation of $\partial\Phi$ is as follows:

$$(\partial\Phi)_\varepsilon = \varepsilon^{-1}(I - J_\varepsilon) \quad \forall \varepsilon > 0.$$

From Theorem 2.9 of [29], we have

Theorem 1.49 *Let X be a real Hilbert space and let $\Phi : X \to \overline{\mathbb{R}}$ be a l.s.c. proper convex functional. Then the functional Φ_ε is convex, continuous, Fréchet differentiable, and $\nabla\Phi_\varepsilon = (\partial\Phi)_\varepsilon$. Moreover*

$$\Phi_\varepsilon(y) = \frac{\|y - J_\varepsilon y\|_X^2}{2\varepsilon} + \Phi(J_\varepsilon y) \quad \forall y \in X, \tag{1.36}$$

$$\lim_{\varepsilon \to 0} \Phi_\varepsilon(y) = \Phi(y), \quad \Phi(J_\varepsilon y) \le \Phi_\varepsilon(y) \le \Phi(y) \quad \forall y \in X. \tag{1.37}$$

Remark 1.16 The operator $\nabla\Phi_\varepsilon$ is Lipschitz continuous with the constant $1/\varepsilon$ (see Theorem 2.9 and Proposition 2.3 of [29]).

The Moreau-Yosida approximation Φ_ε of a l.s.c. proper convex functional Φ is finite, convex, Fréchet differentiable, and so $\nabla\Phi_\varepsilon$ is monotone (see Remark 1.15).

If a function $u(t) : [0, T] \to X$ is differentiable at $t \in [0, T]$, then since $\nabla\Phi_\varepsilon$ is Lipschitz continuous and monotone (for notation we refer to Remark 1.17),

$$\begin{aligned}\langle(\nabla\Phi_\varepsilon(u(t))', u(t)'\rangle &= \langle\lim_{h\to 0}\frac{\nabla\Phi_\varepsilon\big(u(t+h)\big) - \nabla\Phi_\varepsilon\big(u(t)\big)}{h}, \lim_{h\to 0}\frac{u(t+h)-u(t)}{h}\rangle \\ &= \lim_{h\to 0}\frac{1}{h^2}\langle\nabla\Phi_\varepsilon\big(u(t+h)\big) - \nabla\Phi_\varepsilon\big(u(t)\big), u(t+h)-u(t)\rangle \\ &\geq 0.\end{aligned} \tag{1.38}$$

(cf. (5.41), Chap. 1 of [30] or p. 116 of [31])

Remark 1.17 Let V, H be real Hilbert spaces such that $V \hookrightarrow H$ and V is dense in H. Then, we have that $H^* \hookrightarrow V^*$ (see Sect. 1.1.1). Denote by $\langle f, v\rangle$ the value of $f \in V^*$ at $v \in V$. Identifying H and H^* by Riesz's theorem, we have

$$V \hookrightarrow H \hookrightarrow V^*. \tag{1.39}$$

In this case, $H \hookrightarrow V^*$ means that an element u of H is regarded as an element of V^* in the sense that

$$\langle u, v\rangle = (u, v)_H \quad \forall v \in V. \tag{1.40}$$

On the other hand, by Riesz's theorem, there exists an element $Ru = g_u \in V$ corresponding to $u \in V^*$ such that

$$\langle u, v\rangle = (Ru, v)_V = (g_u, v)_V \quad \forall v \in V. \tag{1.41}$$

According to inner products of V, the functions g_u are different, however always

$$(g_u, v)_V = (u, v)_H \quad \forall v \in V.$$

Sometimes, identifying V and V^* by Riesz's theorem, one finds a continuous linear functional f on V as an element of V, and so

$$\langle f, v\rangle = (f, v)_V \quad \forall v \in V. \tag{1.42}$$

If $f \in V$ is not a functional and is an element of V, then it is also an element of H, and is an element of V^* by (1.39). In this case, the role of $f \in V^*$ is represented as

$$\langle f, v\rangle = (f, v)_H \quad \forall v \in V, \tag{1.43}$$

which is different from (1.42). To avoid confusion, one can write the left-hand side of (1.42) as $\langle R^{-1} f, v\rangle$. However, if it is obvious that f represents a functional on V and is not the functional via (1.39), then for simplicity we use (1.42).

1.7 Some Elementary Inequalities

Theorem 1.50 (Young's inequality) *Suppose that a function $f(x)$ is continuous, strictly monotone increasing in $x \geq 0$, and $f(0) = 0$. Denote the inverse function of f by f^{-1}. For $a, b > 0$, we have*

$$ab \leq \int_0^a f(x)\,dx + \int_0^b f^{-1}(x)\,dx$$

and the equality holds only if $b = f(a)$. (See Sect. 8.3 of [9].)

In particular, for $f(x) = x^{p-1} (1 < p < \infty)$, we have that for $s, t \geq 0$

$$st \leq \frac{t^p}{p} + \frac{s^q}{q}, \quad \frac{1}{p} + \frac{1}{q} = 1.$$

By this inequality, we have that for $u \in L^p(\Omega)$ and $v \in L^q(\Omega)$

$$\int_\Omega |u(x)v(x)|\,dx \leq \frac{\varepsilon}{p}\|u\|_p^p + \frac{1}{q\varepsilon^{q-1}}\|v\|_q^q \qquad \forall \varepsilon > 0.$$

Theorem 1.51 (Bellman-Gronwall's inequality) *Let $u(t)$ and $n(t)$ be real continuous functions on $J = [0, T]$, $f(t) \geq 0$ be integrable on J, and*

$$u(t) \leq n(t) + \int_\alpha^t f(s)u(s)\,ds.$$

Then for all $t \in J$

$$u(t) \leq n(t) + \int_0^t f(s)n(s)\big[\exp \int_s^t f(r)\,dr\big]\,ds. \tag{1.44}$$

In addition, if $n(t)$ is nondecreasing, then

$$u(t) \leq n(t) \exp \int_0^t f(s)\,ds, \tag{1.45}$$

which is called Gronwall's inequality

Remark 1.18 For the proof of (1.45) in the case that $f \equiv c \geq 0$ and $n(t)$ is a real and nonnegative function, we refer to Lemma 1.3, Chap. 5 of [3]. For the proof of

(1.45) in the case that $u(t)$, $f(t)$ are continuous and nonnegative functions and $n(t)$ is a continuous, positive and nondecreasing function, we refer to Theorem 1.1.4 of [32]. Following the proof of Theorem 1.2.1 of [32], we can prove the general case.

References

1. H. Brézis, *Functional Analysis, Sobolev Spaces and Partial Differential Equations* (Springer, 2011)
2. K. Yosida, *Functional Analysis* (Springer, 1995)
3. H. Gajewski, K. Gröger, K. Zacharias, *Nichtlineare Operatorgleichungen und Operatordifferentialgleichungen* (Academic Press, Berlin, 1974) (Russian 1978)
4. P. Drábek, J. Milota, *Methods of Nonlinear Analysis* (Birkhäser, 2007)
5. E. Zeidler, *Nonlinear Functional Analysis and Its Applications I* (Springer, 1986)
6. P. Kučera, Basic properties of solution of the non-steady Navier-Stokes equations with mixed boundary conditions in a bounded domain. Ann. Univ. Ferrara **55**, 289–308 (2009)
7. A.N. Kolmogorov, S.V. Fomin, *Element of Theory of Functions and Functional Analysis* (Moscow "Science", Russian, 1981)
8. H.O. Fattorini, *Infinite Dimensional Optimization and Control Theory* (Cambridge University Press, Cambridge, 1999)
9. R.A. Adams, J.J.F. Fournier, *Sobolev Spaces* (Academic Press, 2005)
10. A. Lunardi, *Interpolation Theory*, 2nd edn. (Edizioni della Normale, Pisa, 2009)
11. J. Bergh, J. Löfström, *Interpolation Spaces* (Springer, 1976)
12. V.I. Bogachev, *Measure Theory I, II* (Springer, 2007)
13. P. Grisvard, *Elliptic Problems in Nonsmooth Domains* (SIAM, Philadelphia, 1985)
14. V. Girault, P.A. Raviart, *Finite Element Methods for Navier-Stokes Equations* (Springer, 1986)
15. J.L. Lions, E. Magenes, *Non-Homogeneous Boundary Value Problems and Applications I* (Springer, Berlin, 1972)
16. L. Grafakos, *Modern Fourier Analysis* (Springer, 2009)
17. F. Demengel, G. Demengel, *Functional Spaces for the Theory of Elliptic Partial Differential Equations* (Springer, 2012)
18. L.C. Evans, *Partial Differential Equations* (American Mathematical Society, 2010)
19. T. Roubíček, *Nonlinear Partial Differential Equations with Applications* (Birkhäuser Verlag, 2005)
20. P.G. Ciarlet, *Mathematical Elasticity I: Three-Dimensional Elasticity* (North-Holland, 1988)
21. W. Pompe, Korn's first inequality with variable coefficients and its generalization. Comment. Math. Univ. Carol. **44**, 57–70 (2003)
22. The Mathematical Society of Japan, *Encyclopedic Dictionary of Mathematics*, 2edn. (The MIT Press, Cambridge, 1993)
23. E. Hille, R.S. Phillips, *Functional Analysis and Semi-Groups* (American Mathematical Society Providence, Rhode Island, 1957)
24. H. Triebel, *Interpolation Theory, Function Spaces, Differential Operators* (Johann Ambrosius Barth Verlag, 1995)
25. J. Simon, Compact sets in the space $L^p(0, T; B)$. Ann. Math. Pure Appl. **146**(4), 65–96 (1987)
26. E. Zeidler, *Nonlinear Functional Analysis and Its Applications II/B* (Springer, 1990)
27. J.L. Lions, *Quelques Méthodes de Résolution des Problèmes aux limites Non linéaires* (Dunod, Gauthier-Villars, Paris, 1969) (Russian 1972)
28. H. Brézis, Perturbation non lineaire d'opérateurs maximaux montones. C. R. Acad. Sci. Paris Ser. **I**(269), 566–569 (1968)

29. V. Barbu, *Nonlinear Differential Equations of Monotone Types in Banach Spaces* (Springer, 2010)
30. G. Duvaut, J.L. Lions, *Inequalities in Mechanics and Physics* (Springer, Berlin, 1976)
31. J. Naumann, On evolution inequalities of Navier-Stokes type in three dimensions. Ann. Math. Pure Appl. **124**(4), 107–125 (1980)
32. Y. Qin, *Integral and Discrete Inequalities and Their Applications I* (Birkhäuser, 2016)

Chapter 2
Fluid Equations

In this chapter, we first show how the Navier-Stokes equations and the equations of motion for fluid under consideration of heat are derived. Next, we outline some boundary conditions for the Navier-Stokes equations, mainly being concerned with the ones dealt with in this book. Last, we consider three kind of bilinear forms for the Stokes and Navier-Stokes equations, variational formulations for the Navier-Stokes problems with mixed boundary conditions and establish the equivalence between the variational formulations and the original PDE problems.

2.1 Derivation of Equations for Fluid Motion

In this section, on the basis of conservation laws of mass, momentum and energy in classical mechanics and physics, we derive the equations of motion for fluid. The purpose of this section is for readers to be familiar with different kinds of equations of motion for incompressible fluids as well as for compressible fluids.

2.1.1 Navier-Stokes Equations

2.1.1.1 Acceleration of Fluid Particle

Let (x, t) be coordinates in $\mathbb{R}^l \times (-\infty, +\infty), l = 2, 3$. There are two ways of describing a fluid motion: Lagrangian (material) description and Eulerian (spatial) description.

T. Kim and D. Cao, *Equations of Motion for Incompressible Viscous Fluids*,
Advances in Mathematical Fluid Mechanics,
https://doi.org/10.1007/978-3-030-78659-5_2

In *Lagrangian description*, a *kinematic equation of fluid particles* $x = \Phi(x_0, t) = \big(\phi_1(x_0, t), \cdots, \phi_l(x_0, t)\big)$ is given, which represents the location of a particle at t whose position was x_0 at the initial instant $t = 0$. The velocity and acceleration of a fluid particle in the Lagrangian description are given, respectively, by

$$u(x, t) = \frac{\partial \Phi(x_0, t)}{\partial t} \quad \text{and} \quad \frac{du(x, t)}{dt} = \frac{\partial^2 \Phi(x_0, t)}{\partial t^2}.$$

In *Eulerian description*, velocity of fluid particle passing a spatial point x at an instant t, that is, velocity field $v(x, t) = (v_1(x, t), \cdots, v_l(x, t))$ is given.

In the two descriptions, velocities of fluid particles passing x at instant t are the same, that is, $u(x, t) = v(x, t)$, and acceleration is calculated by

$$\begin{aligned}\frac{\partial^2 \Phi(x_0, t)}{\partial t^2} = \frac{du(x, t)}{dt} &= \frac{dv(\Phi(x_0, t), t)}{dt} \\ &= \frac{\partial v(x, t)}{\partial t} + \sum_i \frac{\partial v(x, t)}{\partial x_i} \cdot \frac{\partial \phi_i(x_0, t)}{\partial t} \\ &= \frac{\partial v(x, t)}{\partial t} + (v \cdot \nabla) v(x, t).\end{aligned}$$

The total derivative

$$\frac{dv}{dt} = \frac{\partial v}{\partial t} + (v \cdot \nabla) v \tag{2.1}$$

is called *material derivative* of v to emphasize the fact that the derivative is taken following a fluid element. Thus, acceleration of a fluid particle equals to material derivative of velocity field $v(x, t)$. Similarly, we can consider material derivative of other fields, for example, density field $\rho(x, t)$, which is the rate of change of the quantity following a fluid element.

The instantaneous curves that are everywhere tangent to the velocity vector are called the *streamlines* of flow. For unsteady flows the streamline pattern changes with time. If $\frac{\partial v}{\partial t} = 0$, then the flow is called *steady flow*. The *path line* is the trajectory of a fluid particle of fixed identity over a period of time. Path lines and streamlines are identical in a steady flow, but not in an unsteady flow.

The Eulerian description is used in most problems of fluid flows, but the Lagrangian description is used occasionally, i.e. when one is interested in finding path lines of fixed particles. In this book we use Eulerian description.

If $\Phi(x_0, t)$ of the kinematic equation is continuous with respect to their independent variables, then particles filling a domain Ω_t at instant t fill a new domain Ω_{t_1} at instant t_1. A domain moving with fluid as this is called a *material domain* and volume of material domain is called *material volume*.

For material domain the following *transport theorem* is valid.

Theorem 2.1 (Sect. 2.1 of [1]) *Let Ω_t be a material domain moving with fluid of velocity field $v(x, t)$ and $f(x, t)$ be scalar or vector function. Then*

$$\frac{d}{dt}\int_{\Omega_t} f(x,t)\,dx = \int_{\Omega_t}\left[\frac{\partial}{\partial t}f(x,t) + v\cdot\nabla f(x,t) + f(x,t)\,div\,v(x,t)\right]dx. \tag{2.2}$$

2.1.1.2 Conservation of Mass

Let us fix a subdomain W of a domain Ω occupied by a fluid. Let $m(W,t)$ be the mass of fluid in W at instant t and $\rho(x,t)$—density of fluid at (x,t).

In this section, we assume that $\rho(x,t)$ and $v(x,t)$ are differentiable with respect to x and t. Then, the rate of change of fluid mass in the fixed subdomain W is as follows.

$$\frac{d}{dt}m(W,t) = \frac{d}{dt}\int_W \rho(x,t)\,dx = \int_W \frac{\partial \rho}{\partial t}(x,t)\,dx.$$

According to the conservation law of mass, the rate of change of fluid mass in the fixed subdomain W at t equals to the total sum of mass passing through the boundary of W in duration of unit time, and

$$\int_W \frac{\partial \rho}{\partial t}(x,t)\,dx = -\int_{\partial W}\rho(x,t)\big(v(x,t)\cdot n\big)\,ds,$$

where n is outward unit vector normal to the boundary and ds is surface element of the boundary. Thus, we have

$$\int_W\left[\frac{\partial\rho(x,t)}{\partial t} + \mathrm{div}\,\big(\rho(x,t)v(x,t)\big)\right]dx = 0.$$

Since W is arbitrary, from above we get the differential form of law of conservation of mass

$$\frac{\partial\rho(x,t)}{\partial t} + \mathrm{div}\,\big(\rho(x,t)v(x,t)\big) = 0, \tag{2.3}$$

which is called the *continuity equation.* In view of material derivative, it can be written as

$$\frac{d\rho(x,t)}{dt} + \rho(x,t)\,\mathrm{div}\,v(x,t) = 0. \tag{2.4}$$

If volume of any material domain is constant at all time t, the fluid is called *incompressible*. The rate of change of volume equals to the total sum of volume of fluid passing the boundary of subdomain W in duration of unit time. Since the rate of change of any volume Ω_t of incompressible fluid equals to zero, we have

$$\frac{d}{dt}\int_{\Omega_t} dx = \int_{\partial\Omega_t} v(x,t)\cdot n\,ds = \int_{\Omega_t}\mathrm{div}\,v(x,t)\,dx = 0.$$

Thus, incompressibility is equivalent to

$$\operatorname{div} v(x,t) = 0. \tag{2.5}$$

The continuity equation (2.3) for incompressible fluid is rewritten

$$\frac{d\rho(x,t)}{dt} = 0,$$

which means that the density of particles of incompressible fluid is not changed for duration of motion. Thus, if the density of an incompressible fluid at initial instant equals over whole domain, that is, $\rho(x,0) = \rho_0 = const$ (homogeneous), then for any $t \geq 0$ the density of fluid $\rho(x,t)$ equals to ρ_0. The problem

$$\begin{cases} \dfrac{d\rho(x,t)}{dt} = 0, \\ \rho(x,0) = \text{const} \end{cases}$$

is trivial, and for homogeneous incompressible fluids continuity equation (2.4) is equivalent to (2.5).

Note that incompressibility of a fluid does not mean the fluid is homogeneous and for inhomogeneous fluid, incompressibility condition (2.5) is not equivalent to continuity equation (2.3).

2.1.1.3 Conservation of Momentum

To derive equations of motion for fluid relying on the conservation of momentum, let us clarify the forces acting on fluid elements. The forces acting on a fluid element can be divided conveniently into two classes, namely, *body forces* and *surface forces*.

Body forces are those that arise from "action at a distance", without physical contact. They result from the medium being placed in a certain force field, which can be gravitational, magnetic, electrostatic or electromagnetic in origin. They are distributed throughout the mass of the fluid and are proportional to the mass. Body forces are expressed per unit mass, which is denoted by f.

Surface forces are those that are exerted on an area element by the surroundings through direct contact. The surface force acting on a small element of area in a continuous medium depends not only on the magnitude of the area but also on the orientation of the area. The *stress*, which is force per unit area on a plane, depends not only on the magnitude of the force and orientation of the plane but also on the direction of the force. Thus, specification of stress at a point requires two vectors, one perpendicular to the plane on which the force is acting and the other in the direction of the force. Then state of stress at a point is, in fact, specified by a *stress tensor*, which has nine components. Let $\{s_{ij}(x)\}$ be stress tensor and n be the outward unit normal vector at point x of the boundary surface of a fluid element. Then the stress on the area element of boundary including x is given by

$$\sigma(x) = S \cdot n,$$

which can be decomposed into normal and tangent components.

Stress depends on deformation of a fluid element, which is described in terms of the *strain rate tensor* or the *deformation rate tensor*

$$\mathcal{E}(v(x)) = \left(\varepsilon_{ij}(v)\right) = \begin{pmatrix} \frac{\partial v_1}{\partial x_1} & \frac{1}{2}\left(\frac{\partial v_1}{\partial x_2} + \frac{\partial v_2}{\partial x_1}\right) & \frac{1}{2}\left(\frac{\partial v_1}{\partial x_3} + \frac{\partial v_3}{\partial x_1}\right) \\ \frac{1}{2}\left(\frac{\partial v_2}{\partial x_1} + \frac{\partial v_1}{\partial x_2}\right) & \frac{\partial v_2}{\partial x_2} & \frac{1}{2}\left(\frac{\partial v_2}{\partial x_3} + \frac{\partial v_3}{\partial x_2}\right) \\ \frac{1}{2}\left(\frac{\partial v_3}{\partial x_1} + \frac{\partial v_1}{\partial x_3}\right) & \frac{1}{2}\left(\frac{\partial v_3}{\partial x_2} + \frac{\partial v_2}{\partial x_3}\right) & \frac{\partial v_3}{\partial x_3} \end{pmatrix}. \tag{2.6}$$

The relation between the stress and deformation in a continuum is called a *constitutive equation*. When the constitutive equation is the first-order equation, that is, the stress tensor $S(v, p)$ is given by

$$s_{ij} = \mu 2\varepsilon_{ij}(v) + C\delta_{ij},$$

where δ_{ij} is the Kronecker symbol and μ is the *dynamic viscosity coefficient*, the fluid is called *Newtonian fluid*. For thermodynamical reasons, C is of the form $C = \lambda \operatorname{div} v - p$, where p denotes the pressure. Therefore, the stress tensor of Newtonian fluid $S(v)$ is given by

$$s_{ij} = 2\mu\varepsilon_{ij}(v) + \lambda \operatorname{div} v\, \delta_{ij} - p\delta_{ij}. \tag{2.7}$$

(cf. Sect. 5.3 of [2]). By the second law of thermodynamics it is known that $\mu \geq 0$, $2\mu + 3\lambda \geq 0$ (see Exercise 2.11 of [1]). A fluid is called *perfect* if $\mu = \lambda = 0$. ($\lambda + \frac{2}{3}\mu$ is called the *coefficient of bulk viscosity*. For many applications, the *Stokes assumption*

$$\lambda + \frac{2}{3}\mu = 0$$

is found to be sufficiently accurate, and can also be supported from the kinetic theory of monatomic gases. In [3], the fluid such that $\lambda = -\frac{2}{3}\mu$ holds is called the Newtonian fluid.)

The surface force acting on a fluid element W by the surroundings is

$$\int_{\partial W} S(v, p) \cdot n\, ds$$

and the conservation law of momentum on fluid element W is

$$\int_W \rho \left[\frac{\partial v(x,t)}{\partial t} + (v \cdot \nabla)v(x,t)\right] dx = \int_W \rho f(x,t)\, dx + \int_{\partial W} S(v, p) \cdot n\, ds. \tag{2.8}$$

From (2.8) we get the *differential equation of conservation law of momentum*

$$\rho\left[\frac{\partial v(x,t)}{\partial t}+(v\cdot\nabla)v(x,t)\right]=\rho f(x,t)+\operatorname{div}S(v,p), \tag{2.9}$$

where $\operatorname{div}S(v,p)$ is the vector $\left(\operatorname{div}S_1(v,p),\operatorname{div}S_2(v,p),\operatorname{div}S_3(v,p)\right)^T$,

$$S_i(v,p)=\left(s_{i1}(v,p),s_{i2}(v,p),s_{i3}(v,p)\right),\quad i=1,2,3.$$

By (2.7) for Newtonian fluid we have

$$\rho\left[\frac{\partial v}{\partial t}+(v\cdot\nabla)v\right]=\rho f(x,t)+\operatorname{div}\left(2\mu\mathcal{E}(v)\right)+\lambda\nabla(\operatorname{div}v)-\nabla p. \tag{2.10}$$

Since for vectors w, u

$$w\otimes u=\begin{pmatrix}w_1u_1 & w_1u_2 & w_1u_3\\ w_2u_1 & w_2u_2 & w_2u_3\\ w_3u_1 & w_3u_2 & w_3u_3\end{pmatrix},$$

we have

$$\operatorname{div}(w\otimes u)=w\operatorname{div}u+(u\cdot\nabla)w.$$

Then,

$$\operatorname{div}(\rho v\otimes v)=\operatorname{div}(v\otimes\rho v)=v\operatorname{div}(\rho v)+\rho(v\cdot\nabla)v. \tag{2.11}$$

In view of (2.11) and continuity equation (2.3), Eq. (2.10) is rewritten as follows:

$$\frac{\partial(\rho v)}{\partial t}+\operatorname{div}\left((\rho v)\otimes v\right)=\rho f(x,t)+\operatorname{div}\left(2\mu\mathcal{E}(v)\right)+\lambda\nabla(\operatorname{div}v)-\nabla p. \tag{2.12}$$

Since for $\mathcal{E}_i(v)=(\varepsilon_{i1}(v),\varepsilon_{i2}(v),\varepsilon_{i3}(v))$ $(i=1,2,3)$

$$\operatorname{div}\left(2\mathcal{E}_i(v)\right)=\Delta v_i+\frac{\partial}{\partial x_i}\operatorname{div}v \tag{2.13}$$

(see (2.77)), in the case that $\mu=const$ (2.10) and (2.12) become, respectively,

$$\rho\left[\frac{\partial v}{\partial t}+(v\cdot\nabla)v\right]=\rho f(x,t)+\mu\Delta v+(\mu+\lambda)\nabla(\operatorname{div}v)-\nabla p \tag{2.14}$$

and

$$\frac{\partial(\rho v)}{\partial t}+\operatorname{div}\left((\rho v)\otimes v\right)=\rho f(x,t)+\mu\Delta v+(\mu+\lambda)\nabla(\operatorname{div}v)-\nabla p. \tag{2.15}$$

2.1.1.4 The Navier-Stokes Equations

Now, from (2.14), (2.15) and (2.3) we get the system of *equations of motion for the Newtonian fluid*

$$\begin{cases} \rho\left[\dfrac{\partial v}{\partial t}+(v\cdot\nabla)v\right]=\rho f(x,t)+\operatorname{div}\big(2\mu\mathcal{E}(v)\big)+\lambda\nabla(\operatorname{div}v)-\nabla p, \\ \dfrac{\partial\rho}{\partial t}+\operatorname{div}(\rho v)=0 \end{cases} \tag{2.16}$$

or

$$\begin{cases} \dfrac{\partial(\rho v)}{\partial t}+\operatorname{div}\big((\rho v)\otimes v\big)=\rho f(x,t)+\mu\Delta v+(\mu+\lambda)\nabla(\operatorname{div}v)-\nabla p, \\ \dfrac{\partial\rho}{\partial t}+\operatorname{div}(\rho v)=0. \end{cases} \tag{2.17}$$

From (2.16) we have the system of *equations of motion for the incompressible Newtonian fluids*

$$\begin{cases} \rho\left[\dfrac{\partial v}{\partial t}+(v\cdot\nabla)v\right]=\rho f(x,t)+\operatorname{div}\big(2\mu\mathcal{E}(v)\big)-\nabla p, \\ \operatorname{div}v=0, \\ \dfrac{\partial\rho(x,t)}{\partial t}+\nabla\rho(x,t)\cdot v(x,t)=0. \end{cases} \tag{2.18}$$

For the incompressible fluid with a constant density, continuity equation (2.3) is equivalent to (2.5), and from (2.18) we have

$$\begin{cases} \dfrac{\partial v}{\partial t}-\operatorname{div}\big(2\nu\mathcal{E}(v)\big)+(v\cdot\nabla)v+\dfrac{1}{\rho}\nabla p=f(x,t), \\ \operatorname{div}v=0, \end{cases} \tag{2.19}$$

where $\nu=\mu/\rho$ is called *kinematic viscosity coefficient*. In view of (2.13), we have the equations of motion for incompressible Newtonian fluids with constant kinematic viscosity and constant density, the *Navier-Stokes equations*

$$\begin{cases} \dfrac{\partial v}{\partial t}-\nu\Delta v+(v\cdot\nabla)v+\dfrac{1}{\rho}\nabla p=f(x,t), \\ \operatorname{div}v=0. \end{cases} \tag{2.20}$$

On the other hand, in the case that $l=2$, for convenience, vector $v=(v_1(x_1,x_2), v_2(x_1,x_2))$ is identified with $\bar{v}=(v_1(x_1,x_2),v_2(x_1,x_2),0)$. Then, we have

$$\operatorname{rot} v = \operatorname{rot} \bar{v} \equiv \left(0, 0, \frac{\partial v_2}{\partial x_1} - \frac{\partial v_1}{\partial x_2}\right),$$

and in the case that $l = 2, 3$

$$\begin{aligned} -\Delta v &= \operatorname{rot}\operatorname{rot} v - \nabla(\operatorname{div} v), \\ (v \cdot \nabla)v &= \operatorname{rot} v \times v + \frac{1}{2}\nabla |v|^2. \end{aligned} \tag{2.21}$$

By using the relation above, (2.20) is rewritten as

$$\begin{cases} \dfrac{\partial v}{\partial t} + \nu \operatorname{rot}\operatorname{rot} v + (v \cdot \nabla)v + \dfrac{1}{\rho}\nabla p = f(x,t), \\ \operatorname{div} v = 0, \end{cases}$$

$$\begin{cases} \dfrac{\partial v}{\partial t} - \nu\Delta v + \operatorname{rot} v \times v + \dfrac{1}{\rho}\nabla q = f(x,t), \\ \operatorname{div} v = 0, \end{cases} \tag{2.22}$$

where $q = \frac{1}{2}\rho|v|^2 + p$ is called *total pressure* or *Bernoulli's pressure*. To avoid confusion, sometimes p is called the *static pressure*.

Remark 2.1 Due to the conservation law of momentum we get a dynamic equation (2.9). There is another fundamental law of dynamics in classical mechanics, the law of *conservation of angular momentum*. For an arbitrary continuous medium obeying the continuity equation (2.3) and the dynamical equation (2.9), the symmetry of stress tensor is equivalent to that the equation by conservation of angular momentum holds (see Theorem 2.1 of [1]). Thus, for Newtonian fluid with symmetry stress tensor, a dynamical equation by conservation of angular momentum is equivalent to (2.9). Symmetry of stress tensor is true, however, under a crucial assumption that all internal momenta of the continuum are neglected which might not be sufficient for some materials. Therefore, for modeling of liquid crystals, micropolar fluid and ferrofluid an extra equation for conservation of angular momentum is used (see [4–6] and references therein).

Remark 2.2 Any displacement of a rigid body can be expressed by composition of a translation and a rotation. A velocity field v can be regarded as a velocity field of a rigid body motion if and only if $\mathcal{E}(v(x,t)) = 0$ (see Theorem 5.1 of [2]).

Remark 2.3 Laws of conservation of momentum and angular momentum are universal, but constitutive laws (equations) depend on physical properties of materials under consideration. Whenever different nonlinear constitutive equations are taken for materials, then, correspondingly, different equations for motion of non-Newtonian fluids are obtained.

2.1.2 *Equations of Motion for Fluid Under Consideration of Heat*

2.1.2.1 Conservation of Energy

Let $f(x,t)$ be a scalar function and $\rho(x,t)$ a density field. Then by the transport theorem we have

$$\begin{aligned}\frac{d}{dt}\int_{\Omega_t}\rho f\,dx &= \int_{\Omega_t}\Big[\frac{\partial}{\partial t}(\rho f)(x,t)+\operatorname{div}(\rho f v)(x,t)\Big]\,dx\\ &= \int_{\Omega_t}\Big[\frac{\partial\rho}{\partial t}f+\rho\frac{\partial f}{\partial t}+\operatorname{div}(\rho v)f+\rho v\cdot\nabla f\Big]\,dx\\ &= \int_{\Omega_t}\Big[\Big(\frac{\partial\rho}{\partial t}+\operatorname{div}(\rho v)\Big)f+\rho\Big(\frac{\partial f}{\partial t}+v\cdot\nabla f\Big)\Big]\,dx.\end{aligned}$$

Hence, in view of the continuity equation, we get

$$\frac{d}{dt}\int_{\Omega_t}\rho f\,dx=\int_{\Omega_t}\Big[\rho\Big(\frac{\partial f}{\partial t}+v\cdot\nabla f\Big)\Big]\,dx=\int_{\Omega_t}\rho\frac{d}{dt}f\,dx. \tag{2.23}$$

Let $e(x,t)$ be internal energy of unit mass of fluid at (x,t). The total energy of a material domain Ω_t at time t is the sum of its kinetic energy and of its internal energy: $\int_{\Omega_t}\rho\big(e+\frac{1}{2}|v|^2\big)\,dx$. By (2.23) the rate of change of the total energy of Ω_t is as follows:

$$\frac{d}{dt}\int_{\Omega_t}\rho\big(e+\frac{1}{2}|v|^2\big)\,dx=\int_{\Omega_t}\rho\frac{d}{dt}\big(e+\frac{1}{2}|v|^2\big)\,dx.$$

This equals to sum of the work done by body force and by surface force, heat flux through the boundary and heat radiated by the inside heat source for the duration of unit time. Let $g(x,t)$ be mass density of inside heat source and χ be a surface density of heat. It is known that χ is expressed by

$$\chi=-q\cdot n,$$

where n is the unit outward normal on the boundary and q is the *heat flux vector* at (x,t).

Therefore, taking into account symmetry of S, we have

$$\begin{aligned}\int_{\Omega_t} \rho \frac{d}{dt}\Big(e + \frac{1}{2}|v|^2\Big)\, dx \\ &= \int_{\Omega_t} \rho f \cdot v\, dx + \int_{\partial\Omega_t} (S(v)n) \cdot v\, ds - \int_{\partial\Omega_t} q \cdot n\, ds + \int_{\Omega_t} \rho g\, dx \\ &= \int_{\Omega_t} \rho f \cdot v\, dx + \int_{\Omega_t} \operatorname{div}(S_1 \cdot v, \cdots, S_l \cdot v)\, dx - \int_{\Omega_t} \operatorname{div} q\, dx + \int_{\Omega_t} \rho g\, dx.\end{aligned} \tag{2.24}$$

Taking into account (2.1) and making the $\mathbb{R}^l$-inner product of (2.9) with v, we have

$$\rho \frac{1}{2}\frac{d}{dt}|v|^2 = \rho f(x,t) \cdot v + \operatorname{div} S(v) \cdot v. \tag{2.25}$$

Substituting (2.25) into (2.24), we have

$$\begin{aligned}\int_{\Omega_t} \rho \frac{d}{dt} e\, dx = &- \int_{\Omega_t} \operatorname{div} S(v) \cdot v\, dx \\ &+ \int_{\Omega_t} \operatorname{div}(S_1 \cdot v, \cdots, S_l \cdot v)\, dx - \int_{\Omega_t} \operatorname{div} q\, dx + \int_{\Omega_t} \rho g\, dx.\end{aligned} \tag{2.26}$$

By symmetry of (2.6),

$$\operatorname{div}(S_1 \cdot v, \cdots, S_l \cdot v) = \operatorname{div} S(v) \cdot v + \sum_{ij} s_{ij} \frac{\partial v_i}{\partial x_j} = \operatorname{div} S(v) \cdot v + \sum_{ij} s_{ij} \varepsilon_{ij}(v), \tag{2.27}$$

and from (2.26) and (2.27) we get a *differential equation for internal energy*

$$\rho \left[\frac{\partial e}{\partial t} + (v \cdot \nabla) e\right] = \sum_{ij} s_{ij} \varepsilon_{ij}(v) - \operatorname{div} q + \rho g. \tag{2.28}$$

In view of (2.7), we have from (2.28) a *differential equation for internal energy of Newtonian fluid*

$$\rho \left[\frac{\partial e}{\partial t} + (v \cdot \nabla) e\right] = 2\mu \mathcal{E}(v) : \mathcal{E}(v) + \lambda (\operatorname{div} v)^2 - p \operatorname{div} v - \operatorname{div} q + \rho g, \tag{2.29}$$

where $\mathcal{E}(v) : \mathcal{E}(v) = \sum_{ij} \varepsilon_{ij}\varepsilon_{ij}$. The first three terms on the right-hand side express change of internal energy per unit time by deformation of fluid.

Relying on the rate of change of total energy on a material domain Ω_t, we got (2.24). Now using a fixed element ω, not a material domain, we can get another equation. For this we additionally need to take care of the flow of total energy through the boundary of fluid element $\partial\omega$. Then, we have

$$
\begin{aligned}
\frac{d}{dt}\int_\omega \rho(e+\frac{1}{2}|v|^2)\,dx &= \int_{\partial\omega} \rho[(e+\frac{1}{2}|v|^2)](v\cdot n)\,ds + \int_\omega \rho f\cdot v\,dx \\
&\quad + \int_{\partial\omega} (S(v)n)\cdot v\,ds - \int_{\partial\Omega_t} q\cdot n\,ds + \int_\omega \rho g\,dx \\
&= \int_\omega \operatorname{div}[\rho v(e+\frac{1}{2}|v|^2)]\,dx + \int_\omega \rho f\cdot v\,dx \\
&\quad + \int_\omega \operatorname{div}(S_1\cdot v,\cdots,S_n\cdot v)\,dx - \int_\omega \operatorname{div} q\,dx + \int_\omega \rho g\,dx,
\end{aligned}
$$

which yields the differential equation

$$
\begin{aligned}
&\frac{\partial}{\partial t}\big[\rho\big(e+\frac{1}{2}|v|^2\big)\big] + \operatorname{div}\big[\rho\big(e+\frac{1}{2}|v|^2\big)v\big] \\
&\qquad = \rho f\cdot v + \operatorname{div}(S_1\cdot v,\cdots,S_n\cdot v) - \operatorname{div} q + \rho g.
\end{aligned}
\tag{2.30}
$$

In view of (2.27), we have from (2.30) the *differential equation for total energy*

$$
\begin{aligned}
&\frac{\partial}{\partial t}\big[\rho\big(e+\frac{1}{2}|v|^2\big)\big] + \operatorname{div}\big[\rho\big(e+\frac{1}{2}|v|^2\big)v\big] \\
&\qquad = \rho f\cdot v + \operatorname{div} S(v)\cdot v + \sum_{ij} s_{ij}\varepsilon_{ij}(v) - \operatorname{div} q + \rho g.
\end{aligned}
\tag{2.31}
$$

For Newtonian fluid, taking into account (2.7), we have from (2.31)

$$
\begin{aligned}
&\frac{\partial}{\partial t}\big[\rho\big(e+\frac{1}{2}|v|^2\big)\big] + \operatorname{div}\big[\rho v\big(e+\frac{1}{2}|v|^2\big)v\big] \\
&\qquad = \rho f\cdot v + \operatorname{div} S(v)\cdot v + 2\mu\mathcal{E}(v):\mathcal{E}(v) + \lambda(\operatorname{div} v)^2 - p\operatorname{div} v - \operatorname{div} q + \rho g.
\end{aligned}
\tag{2.32}
$$

Using (2.7), we can rewrite (2.32) as

$$
\begin{aligned}
&\frac{\partial}{\partial t}\big[\rho\big(e+\frac{1}{2}|v|^2\big)\big] + \operatorname{div}\big[\rho v\big(e+\frac{1}{2}|v|^2\big)v\big] \\
&\qquad = \rho f\cdot v + \operatorname{div}(2\mu\mathcal{E}(v))\cdot v - \nabla p\cdot v + \lambda\nabla(\operatorname{div} v)\cdot v \\
&\qquad\quad + 2\mu\mathcal{E}(v):\mathcal{E}(v) + \lambda(\operatorname{div} v)^2 - p\operatorname{div} v - \operatorname{div} q + \rho g,
\end{aligned}
$$

which yields the differential equation for *total energy of Newtonian fluid*

$$
\begin{aligned}
&\frac{\partial}{\partial t}\big[\rho\big(e+\frac{1}{2}|v|^2\big)\big] + \operatorname{div}\big[\rho v\big(e+\frac{1}{2}|v|^2\big) + pv\big] \\
&\qquad = \rho f\cdot v + \operatorname{div}(2\mu\mathcal{E}(v))\cdot v + \lambda\nabla(\operatorname{div} v)\cdot v \\
&\qquad\quad + 2\mu\mathcal{E}(v):\mathcal{E}(v) + \lambda(\operatorname{div} v)^2 - \operatorname{div} q + \rho g.
\end{aligned}
\tag{2.33}
$$

Let us show that (2.33) is equivalent to (2.29). To this end, we rewrite the terms on the left-hand side of (2.33).

$$
\begin{aligned}
\frac{\partial}{\partial t}\big[\rho\big(e+\frac{1}{2}|v|^2\big)\big] &= \frac{\partial\rho}{\partial t}e+\rho\frac{\partial e}{\partial t}+\frac{\partial\rho}{\partial t}\frac{|v|^2}{2}+\rho v\frac{\partial v}{\partial t},\\
\operatorname{div}\big[\rho v\big(e+\frac{1}{2}|v|^2\big)+pv\big] &= \operatorname{div}(\rho v)e+\rho(v\cdot\nabla)e+\operatorname{div}(\rho v)\frac{1}{2}|v|^2\\
&\quad+\rho v\cdot(v\cdot\nabla)v+\nabla p\cdot v+p\operatorname{div} v.
\end{aligned}
\tag{2.34}
$$

Taking into account (2.3) and

$$
\begin{aligned}
\frac{dv}{dt} &= \frac{\partial v}{\partial t}+(v\cdot\nabla)v,\\
\frac{de}{dt} &= \frac{\partial e}{\partial t}+(v\cdot\nabla)e
\end{aligned}
$$

(see (2.2)) and adding two equations of (2.34), we have

$$
\begin{aligned}
&\frac{\partial}{\partial t}\big[\rho\big(e+\frac{1}{2}|v|^2\big)\big]+\operatorname{div}\big[\rho v\big(e+\frac{1}{2}|v|^2\big)+pv\big]\\
&\quad=\rho\frac{de}{dt}+\rho v\frac{dv}{dt}=\rho\Big(\frac{\partial e}{\partial t}+(v\cdot\nabla)e\Big)+\rho\frac{1}{2}\frac{d|v|^2}{dt}+\nabla p\cdot v+p\operatorname{div} v.
\end{aligned}
\tag{2.35}
$$

By (2.25) and (2.7), we have

$$
\begin{aligned}
\rho\frac{1}{2}\frac{d}{dt}|v|^2 &= \rho f(x,t)\cdot v+\operatorname{div} S(v)\cdot v\\
&= \rho f\cdot v+\operatorname{div}(2\mu\mathcal{E}(v))\cdot v-\nabla p\cdot v+\lambda\nabla(\operatorname{div} v)\cdot v.
\end{aligned}
$$

Thus, we get (2.29) from (2.33) taking into account the above and (2.35).

The relation between the heat flux vector and temperature in thermodynamics is also called a *constitutive equation*. We take *Fourier's law* as the constitutive equation

$$
q=-k\nabla\theta, \tag{2.36}
$$

where k is thermal conductivity and $\theta(x,t)$ is temperature of fluid. Denoting the specific heat of fluid by γ and assuming the specific internal energy of the fluid is proportional to its temperature, we have

$$
e=\gamma\theta. \tag{2.37}
$$

Specific heat, thermal conductivity and viscosity of fluid, in general, depend on the temperature of fluid.

Taking into account (2.37), for the incompressible Newtonian fluid with constant density we have from (2.29)

$$\frac{\partial}{\partial t}\big(\gamma(\theta)\theta(x,t)\big) - \mathrm{div}\,\big(\kappa(\theta)\nabla\theta\big) + \big(v\cdot\nabla(\gamma(\theta)\theta)\big) - 2\nu(\theta)\mathcal{E}(v) : \mathcal{E}(v) = g, \tag{2.38}$$

where $\kappa = \frac{k}{\rho}$. In the left-hand side of (2.38) the second term expresses heat flow by conducting, the third one expresses heat transport by fluid movement (convection) and the fourth one represents dissipation of kinetic energy by viscous friction of fluid (the Joule effect).

As in [3], in this book we shall reserve the term *convection* to describe heat transport by fluid movements, and so $(v\cdot\nabla)v$ describing movement of fluid from place to place is called *"advection"* (see pp. 55–56 of [3]). Such terminology is suitable with the terms: the convective heat transfer, convection diffusion, heat convection and so on.

2.1.2.2 Equations of Motion for Fluid Under Consideration of Heat

From (2.16), (2.17) and (2.29) we get the system of *equations of Newtonian fluid under consideration of internal energy*

$$\begin{cases} \rho\left[\dfrac{\partial v}{\partial t} + (v\cdot\nabla)v\right] = \rho f(x,t) + \mathrm{div}\,\big(2\mu\mathcal{E}(v)\big) + \lambda\nabla(\mathrm{div}\,v) - \nabla p, \\ \dfrac{\partial\rho}{\partial t} + \mathrm{div}\,(\rho v) = 0, \\ \rho\left[\dfrac{\partial e}{\partial t} + (v\cdot\nabla)e\right] = 2\mu\mathcal{E}(v) : \mathcal{E}(v) + \lambda(\mathrm{div}\,v)^2 - p\,\mathrm{div}\,v - \mathrm{div}\,q + \rho g \end{cases} \tag{2.39}$$

or, equivalently,

$$\begin{cases} \dfrac{\partial(\rho v)}{\partial t} + \mathrm{div}\,\big((\rho v)\otimes v\big) = \rho f(x,t) + \mu\Delta v + (\mu+\lambda)\nabla(\mathrm{div}\,v) - \nabla p, \\ \dfrac{\partial\rho}{\partial t} + \mathrm{div}\,(\rho v) = 0, \\ \rho\left[\dfrac{\partial e}{\partial t} + (v\cdot\nabla)e\right] = 2\mu\mathcal{E}(v) : \mathcal{E}(v) + \lambda(\mathrm{div}\,v)^2 - p\,\mathrm{div}\,v - \mathrm{div}\,q + \rho g. \end{cases} \tag{2.40}$$

Remark 2.4 If we use the equation of total energy instead of the equation of internal energy, we get the system of *equations of motion of Newtonian fluids under consideration of total energy*

$$
\begin{cases}
\dfrac{\partial(\rho v)}{\partial t}+\mathrm{div}\,\big((\rho v)\otimes v\big)=\rho f(x,t)+\mu\Delta v+(\mu+\lambda)\nabla(\mathrm{div} v)-\nabla p,\\
\dfrac{\partial\rho}{\partial t}+\mathrm{div}\,(\rho v)=0,\\
\dfrac{\partial}{\partial t}\big[\rho\big(e+\frac{1}{2}|v|^2\big)\big]+\mathrm{div}\,\big[\rho v\big(e+\frac{1}{2}|v|^2\big)+pv\big]=\rho f\cdot v+\mathrm{div}\,(2\mu\mathcal{E}(v))\cdot v\\
\qquad\qquad +\lambda\nabla(\mathrm{div}\,v)\cdot v+2\mu\mathcal{E}(v):\mathcal{E}(v)+\lambda(\mathrm{div}\,v)^2-\mathrm{div}\,q+\rho g.
\end{cases}
\tag{2.41}
$$

The total number of equations constituting (2.39) (or (2.40)), (2.36) and (2.37) is less than the total number of unknown functions, and so one usually needs, in addition, a relation in terms of pressure, density and temperature. Then, the left-hand side of the new system obtained from (2.41) is written as

$$\frac{\partial\mathbf{u}}{\partial t}+\mathrm{div}\,F(\mathbf{u}),$$

which is convenient for study of equations for compressible fluid.

When the density of fluid is constant and it is independent of temperature, by (2.19) and (2.38), we have the system of *equations of motion of incompressible Newtonian fluids under consideration of heat*

$$
\begin{cases}
\dfrac{\partial v}{\partial t}-\mathrm{div}\,\big(2\nu(\theta)\mathcal{E}(v)\big)+(v\cdot\nabla)v+\dfrac{1}{\rho}\nabla p=f(x,t),\\
\mathrm{div}\,v=0,\\
\dfrac{\partial}{\partial t}\big(\gamma(\theta)\theta(x,t)\big)-\mathrm{div}\,(\kappa(\theta)\nabla\theta)+(v\cdot\nabla(\gamma(\theta)\theta))-2\nu(\theta)\mathcal{E}(v):\mathcal{E}(v)=g(x,t).
\end{cases}
\tag{2.42}
$$

If a fluid is in the gravitational field and density is not constant, then there is buoyancy. For slightly compressible homogeneous fluids in gravitational field, we simplify the equations for the mass and momentum conservation laws as follows:

(i) ρ is assumed to be constant ($=\rho_0$) everywhere in the equations, except in the gravity term;
(ii) In the term of gravity, ρ is replaced by a linear function of θ

$$\rho=\rho_0-\alpha(\theta-\theta_0),$$

where ρ_0 and θ_0 are the average value of the density and temperature.

If there is not any body force other than gravity, then f is the gravitational acceleration g_g. Then density of buoyancy is $-\alpha(\theta-\theta_0)g_g$, and we have an approximation of the conservation of momentum

$$\frac{\partial v}{\partial t}+(v\cdot\nabla)v-\nu\Delta v+\frac{1}{\rho_0}\nabla p=\frac{1}{\rho_0}[\rho_0-\alpha(\theta-\theta_0)]g_g.$$

Replacing the first equation of (2.42) with the above and neglecting energy dissipation by viscous friction of fluid, we have

$$\begin{cases} \dfrac{\partial v}{\partial t} + (v \cdot \nabla)v - \nu \Delta v + \dfrac{1}{\rho_0} \nabla p = [1 - \alpha_1(\theta - \theta_0)] g_g, \\ \operatorname{div} v = 0, \\ \dfrac{\partial}{\partial t}\big(\gamma\theta(x,t)\big) - \operatorname{div}(\kappa \nabla \theta) + (v \cdot \nabla(\gamma\theta)) = g, \end{cases} \tag{2.43}$$

which is used for heat convection in gravitational field and is called *Boussinesq equations*.

Putting

$$[1 - \alpha_1(\theta - \theta_0)] g_g = f(1 - \alpha_0\theta)$$

and adding the work $\alpha_1\theta f \cdot v$ done by the force $\alpha_1\theta f$ due to heat expansion to the heat equation, we have

$$\begin{cases} \dfrac{\partial v}{\partial t} - \operatorname{div}\big(2\nu(\theta)\mathcal{E}(v)\big) + (v \cdot \nabla)v + \dfrac{1}{\rho_0} \nabla p = f(1 - \alpha_0\theta), \\ \operatorname{div} v = 0, \\ \dfrac{\partial}{\partial t}\big(\gamma\theta(x,t)\big) - \operatorname{div}(\kappa \nabla \theta) + (v \cdot \nabla(\gamma\theta)) - 2\nu\mathcal{E}(v) : \mathcal{E}(v) = \alpha_1\theta f \cdot v + g, \end{cases} \tag{2.44}$$

which is a system of equations of motion for incompressible Newtonian fluid under consideration of dissipation of energy in gravitational field [7].

2.2 Boundary Conditions for the Navier-Stokes Equations

When solving the Navier-Stokes equations, appropriate initial and boundary conditions need to be applied. The derivation of suitable boundary conditions for flow problems is not obvious and depends on the physics which is to be modeled.

In this section, without deriving boundary conditions from point of view of physics, we outline those concerned in this book for the Stokes equations

$$\frac{\partial v}{\partial t} - \mu \Delta v + \nabla p = f, \quad \operatorname{div} v = 0 \quad \text{in } \Omega \subset \mathbb{R}^l,\ l = 2, 3, \tag{2.45}$$

and the Navier-Stokes equations

$$\frac{\partial v}{\partial t} - \mu \Delta v + (v \cdot \nabla)v + \nabla p = f, \quad \operatorname{div} v = 0 \quad \text{in } \Omega. \tag{2.46}$$

Note that Eq. (2.46) is the one with $\rho = 1$ in (2.20), and $p + \frac{1}{2}|v|^2$ is the total pressure instead of $p + \rho\frac{1}{2}|v|^2$.

Let $v = (v_1, \cdots, v_l)$ be a solution to (2.45) or (2.46). The strain tensor $\mathcal{E}(v)$ is the matrix with components

$$\varepsilon_{ij}(v) = \frac{1}{2}(\partial_{x_j} v_i + \partial_{x_i} v_j)$$

and the stress tensor $S(v, p)$ is the one with components

$$s_{ij} = -p\delta_{ij} + 2\mu\varepsilon_{ij}(v).$$

In the present book the boundary surfaces concerned by us are pieces of boundary of 3-D or 2-D bounded connected domains, and so we can assume the surfaces are oriented. Thus, let n be an outward unit normal vector on a boundary and τ be unit vector tangent to the boundary. Then the *stress vector* on the boundary is

$$\sigma(v, p) = S \cdot n \equiv -pn + 2\mu\mathcal{E}(v) \cdot n, \tag{2.47}$$

the value of normal component of the stress vector is

$$\sigma_n(v, p) = \sigma \cdot n$$

and its tangent component is

$$\sigma_\tau = \sigma - \sigma_n \cdot n.$$

Replacing the static pressure p in the stress tensor by the total pressure (Bernoulli's pressure) $p + \frac{1}{2}|v|^2$, we have the *total stress tensor* $S^t(v, p)$ with components

$$s^t_{ij} = -(p + \frac{1}{2}|v|^2)\delta_{ij} + 2\mu\varepsilon_{ij}(v). \tag{2.48}$$

Then the *total stress vector* on the boundary is

$$\sigma^t(v, p) = S^t \cdot n,$$

the value of normal component of the total stress is

$$\sigma^t_n(v, p) = \sigma^t \cdot n$$

and its tangent component is

$$\sigma^t_\tau = \sigma^t - \sigma^t_n \cdot n.$$

Note that the tangent components of stress and total stress are equal, i.e. $\sigma_\tau = \sigma_\tau^t$. Denote the value of normal component $v \cdot n$ of velocity v by v_n, and the tangent component of v is $v_\tau = v - v_n n$.

In some literatures, the terms "stress" and "total stress", respectively, are used for (2.6) and (2.47), but in this book we will use the term total stress for (2.48).

From mathematical point of view, there are four kinds of boundary conditions:

(i) Dirichlet boundary condition assigning variable value (velocity, pressure).
(ii) Neumann boundary condition assigning (normal) gradient of variable.
(iii) Combination of Dirichlet and Neumann boundary condition.
(iv) Periodic boundary conditions assuming equality between values of a variable on distant parallel two planes.

From physical point of view, there are two kinds of boundary conditions:

(i) Natural boundary conditions.
These are boundary conditions on the real boundary of a domain occupied by fluid, for example, no-slip condition on the wall, threshold slip condition on the wall and so on.
(ii) Artificial boundary conditions.
These are boundary conditions on imaginary boundaries of fluid domain, for example, a boundary condition on symmetric planes usually taken to reduce computational burden, Navier slip-with-friction boundary condition on a surface close to a real rough wall and so on.

According to relationship between fluid domain under consideration and surroundings, the boundaries of fluid domain may be classified as follows:

(i) Wall,
(ii) Symmetric plane,
(iii) Inlet and outlet,
(iv) Free surface.

Boundary conditions first of all must be able to reflect to a mathematical formulation of problem and must be consistent with real phenomena. Some boundary condition involved to a mathematical formulation has not a correct physical meaning, which is used as an artificial boundary condition for numerical practice. In such a case for mathematical result to approximate the real phenomenon it is important where the artificial boundary is drawn.

We outline what kinds of boundary conditions are used on every class of boundaries above when the domain of fluid is not variable along with time t.

2.2.1 Boundary Conditions on the Walls

In practice, some domain of fluid has a solid wall boundary and changes of momentum, velocity and scalar quantities near the wall are serious. Thus, boundary condition on the wall must be given in accordance with physical situations.

2.2.1.1 Stick Boundary Condition on the Walls

Most solid surfaces are impermeable to fluid and the fluid sticks to the surfaces. Thus, there is no slip and no penetration, and the fluid particles on the wall move with the velocity of wall w:

$$v = w.$$

If the walls do not move, then *stick condition* on the wall $\Gamma_D \subset \partial\Omega$ is the homogeneous Dirichlet condition

$$v = 0 \quad \text{on } \Gamma_D. \tag{2.49}$$

Pressure on the wall is seriously variable according to the flow situation, and usually the pressure boundary condition on the wall is not given.

2.2.1.2 Slip Boundary Conditions on the Walls

When a wall is impermeable to fluid and fluid is not stick, a slip boundary condition may be used. Accurate experiments clearly demonstrate that the no-slip boundary condition postulated in most hydrodynamics investigations may be far to be fulfilled (see [8]).

If a fluid slips without friction on the wall Γ, then *Navier slip boundary condition*

$$v_n = 0, \quad \sigma_\tau(v) = 0 \quad \text{on } \Gamma \tag{2.50}$$

is used. Sometimes free slip on a surface Γ is expressed by a *vorticity boundary condition*

$$v_n = 0, \quad \operatorname{rot} v \times n = 0 \quad \text{on } \Gamma. \tag{2.51}$$

In some papers the boundary condition (2.51) is called Navier slip-type condition [9–13], slip condition [14] or Hodge boundary condition [15, 16]. On flat portions of the boundary, the vorticity boundary condition (2.51) coincides with the classical Navier slip boundary condition (2.50), but usually these are different as a quantity depending on the shape of boundary surface (cf. Sect. 6 of [10], Remark 1.1 of [14], Remark 3.1 of [17]). The vorticity boundary condition (2.51) is also used on a portion of boundary of a tank containing fluid, which is closed by a membrane (see [18]).

C.-L. Navier claimed that the tangential component of the stress at the boundary should be proportional to the tangential velocity. This condition is expressed by

$$v_n = 0, \quad \sigma_\tau(v) + \alpha v = 0, \quad \alpha : \text{constant or function on } \Gamma, \tag{2.52}$$

which is called *Navier slip-with-friction boundary condition* (see [19–21]). The Navier slip-with-friction boundary condition was rigorously justified as a homogenization of the no-slip condition on a rough boundary and might be more realistic than the classical no-slip boundary condition when the wall is rough. The condition is used for simulations of flows near rough boundaries, such as in aerodynamics, in weather forecast, in haemodynamics, etc. When an artificial boundary condition near the real boundary with periodic micro-roughness is used, usually α is a matrix, which maps each vector tangent to boundary into such a vector. From now on, when we are concern with (2.52) with matrix α, we will always regard the matrix as being such a one. If the shape operator of the boundary Γ (see Sect. 3.1) is used instead $\alpha/2$ in (2.52), then the condition (2.52) becomes (2.51) (see [13] and Sect. 3.1).

2.2.1.3 Threshold Slip Boundary Conditions on the Walls

According to Tresca friction law between two solids, there is not any slip on the surface when the magnitude of tangent stress is less than a threshold, but when the magnitude of tangent stress reaches the threshold, slipping can occur, the magnitude of tangent stress is not greater than the threshold and the direction of velocity is opposite to the tangent stress. Such phenomena between walls and fluids are expressed by

$$v_n = 0, \quad |\sigma_\tau(v)| \le g_\tau, \quad \sigma_\tau(v) \cdot v_\tau + g_\tau |v_\tau| = 0 \quad \text{on } \Gamma, \tag{2.53}$$

where g_τ is a threshold of tangent stress for slip. This condition is called *Tresca friction boundary condition* or *threshold slip boundary condition*. Using subdifferential operator, we can rewrite (2.53) as

$$v_n = 0, \quad -\sigma_\tau(v) \in g_\tau \partial |v_\tau|, \tag{2.54}$$

where

$$\partial |z| = \begin{cases} \frac{z}{|z|} & z \ne 0, \ z \in \mathbb{R}^l, \\ \{w \in \mathbb{R}^l, |w| \le 1\} & z = 0, \ z \in \mathbb{R}^l. \end{cases}$$

Physical and experimental backgrounds of such boundary conditions are mentioned in several papers [22–24].

2.2.1.4 Leak Boundary Conditions on the Walls

Similarly to the threshold slip boundary condition, leak boundary conditions of friction type are used when one wants to model a flow problem involving a leak of flow through the boundary.

Assume that the tangent component of velocity on a potion of boundary vanishes, there is not any leak through the surface when the magnitude of normal stress on a portion of boundary surface is less than a given threshold, leak through the surface can occur if the magnitude of normal stress reaches the threshold and the magnitude of normal stress is not greater than the threshold. Such phenomena on the boundary surface are expressed by

$$v_\tau = 0, \quad |\sigma_n(v)| \le g_n, \quad \sigma_n(v)v_n + g_n|v_n| = 0 \quad \text{on } \Gamma, \tag{2.55}$$

where $g_n(> 0)$ is a threshold of normal stress to show leak [24–27]. Using subdifferential operator, we can rewrite (2.55) as

$$v_\tau = 0, \quad -\sigma_n(v) \in g_n \partial |v_n|. \tag{2.56}$$

For physical and experimental backgrounds of such boundary conditions, we refer to [24, 27]. The boundary condition (2.55) or (2.56) is called *threshold leak boundary condition*.

Another threshold leak boundary condition, which is based on the normal total stress,

$$v_\tau = 0, \quad |\sigma_n^t(v)| \le g_n, \quad \sigma_n^t(v)v_n + g_n|v_n| = 0 \quad \text{on } \Gamma \tag{2.57}$$

may be also used. (Note that since $\sigma_\tau(v, p) = \sigma_\tau^t(v, p) = 2\mu\varepsilon_{n\tau}(v)$, there is not another slip boundary condition based on $\sigma_\tau^t(v, p)$.)

The boundary conditions (2.55) and (2.57) mean that according to direction of normal stress, fluid may penetrate out or into through boundary. The stress on a boundary surface is density of force applied to the boundary from surroundings. Note n is outward unit vector on a boundary. Therefore, if $\sigma_n(v) > 0$, which means that direction of force acting on the boundary surface of domain of fluid by surrounding is outward, then the value of density of force reacting on the boundary by fluid equals to $-\sigma_n(v) < 0$, and (2.55) implies $v_n < 0$, which means that fluid penetrates into through the boundary. If $\sigma_n(v) < 0$, which means that direction of force acting on the boundary surface of domain of fluid by surrounding is inward, then the value of density of force reacting on the boundary by fluid equals to $-\sigma_n(v) > 0$, which means that the reacting force is outward. On the other hand (2.55) implies $v_n > 0$, which means that fluid penetrates out through the boundary.

In practice, we are encountered with one-sided leak of fluid, for example, sand layers, a semipermeable membranes and so on.

Assume that the tangent component of velocity on a portion of boundary vanishes, the fluid can only leak out through boundary, there is not any flow through the boundary when $-\sigma_n(v)$ is less than a threshold $g_{+n}(> 0)$, a leak out can occur if $-\sigma_n(v)$ reaches $g_{+n}(> 0)$ and $-\sigma_n(v)$ cannot be greater than the threshold. We can describe such a phenomenon by

$$v_\tau = 0, \quad v_n \geq 0, \quad \sigma_n(v) + g_{+n} \geq 0, \quad (\sigma_n(v) + g_{+n})v_n = 0, \tag{2.58}$$

which is called *threshold leak out boundary condition* [28, 29].

In contrast, assume that a fluid can only leak into through boundary, there is not any flow through the boundary when $-\sigma_n(v)$ is greater than $-g_{-n}$ (for a threshold $g_{-n} > 0$), but there can be a leak into if $-\sigma_n(v)$ is same as $-g_{-n}$, and $-\sigma_n(v)$ cannot be less than $-g_{-n}$. Such a phenomenon on the boundary surface is expressed by

$$v_\tau = 0, \quad v_n \leq 0, \quad \sigma_n(v) - g_{-n} \leq 0, \quad (\sigma_n(v) - g_{-n})v_n = 0, \tag{2.59}$$

which is called *threshold leak into boundary condition.*

Other threshold one-sided leak boundary conditions

$$\begin{aligned} &v_\tau = 0, \quad v_n \geq 0, \quad \sigma_n^t(v) + g_{+n} \geq 0, \quad (\sigma_n^t(v) + g_{+n})v_n = 0,\\ &v_\tau = 0, \quad v_n \leq 0, \quad \sigma_n^t(v) - g_{+n} \leq 0, \quad (\sigma_n^t(v) - g_{+n})v_n = 0 \end{aligned} \tag{2.60}$$

based on the normal total stress may be also used.

The threshold leak into or out boundary condition is called the *one-sided leak boundary conditions*. The threshold slip, leak and one-sided leak boundary conditions are called the *boundary conditions of friction type.*

2.2.2 Boundary Conditions on Symmetric Planes

If flow field and geometry of domain of fluid is symmetric with respect to a plane, then to reduce the amount of computation for the problem a half of domain with the symmetry plane as an imaginary boundary may be under consideration. The normal velocity and normal gradients of all variables on a symmetry plane are zero, and tangent stress on the symmetric plane vanishes. Therefore, on the symmetric plane, the boundary conditions

$$v_n = 0, \quad \sigma_\tau(v) = 0 \tag{2.61}$$

or

$$v_n = 0, \quad \operatorname{rot} v \times n = 0 \tag{2.62}$$

can be used.

2.2.3 Boundary Conditions on Inlets and Outlets

If there are inlets and outlets on the boundary of fluid domain, then Dirichlet boundary conditions for velocity may be used. But *pressure boundary condition* is more common because it is difficult to know velocity profile except special cases. In some real-life situations, it is natural to prescribe the value of the pressure on some part of the boundary, as, for instance, in case of pipelines, blood vessels, different hydraulic systems involving pumps, etc. Thus, it seems perfectly reasonable to impose the pressure boundary condition on inlets and outlets for the Navier-Stokes system [30].

In practice according to measurement instruments, we can obtain the static pressure p or total pressure $\frac{1}{2}|v|^2 + p$ (Bernoulli's pressure), and both the static pressure [30, 31] and total pressure [32, 33] may be used for boundary conditions on the real inlets and outlets.

When pressure boundary conditions are used, direction of fluid flow (usually flow is orthogonal to the boundary) is given together. But it is known that the static pressure boundary condition and direction of flow are not enough to determine a velocity field of the Navier-Stokes equations. Moreover, when one of static pressure (instead of total pressure), stress (instead of total stress) or the outflow boundary conditions (see the next subsection) is given on a portion of boundary, for the initial boundary value problems of the Navier-Stokes equations only existence of a unique local-in-time solution and a unique solution on a given interval for small given data (in what follows we call it a solution for small data) are proved. From the mathematical point of view, the main difficulty of such problems results from the fact that in the process of a priori estimation of velocity, the term

$$\left((v \cdot \nabla)v, v\right)_\Omega$$

arising from the nonlinear (advection) term $(v \cdot \nabla)v$ is not canceled since the normal component of velocity on the portion of boundary does not vanish. From the mechanical point of view, it is explained by the fact that the kinetic energy of fluid (with density 1) $\frac{1}{2}\int_\Omega |v(t,x)|^2\,dx$ is not controlled by the data of problem and uncontrolled "backward flow" can take place at the portion of boundary (see Preface in [34]).

If the total pressure condition is given, then mathematical treatment is more easy because by

$$(v \cdot \nabla)v = \operatorname{rot} v \times v + \frac{1}{2}\nabla\,|v|^2$$

the Navier-Stokes equations (2.46) are rewritten as

$$\frac{\partial v}{\partial t} - \mu\Delta v + \operatorname{rot} v \times v + \nabla(p + \frac{1}{2}|v|^2) = f, \;\; \nabla \cdot v = 0 \;\; \text{in } \Omega, \tag{2.63}$$

and in the process of a priori estimation of solutions the term

$$\left(\operatorname{rot} v \times v, v\right)_\Omega$$

arising from the nonlinear term rot $v \times v$ vanishes by a property of the mixed product of vectors.

As a boundary condition on in/out-stream surfaces, a combination of the tangent component of the velocity and the normal component of stress

$$\begin{aligned} v_\tau &= 0, \\ \sigma_n(v, p) &= f \end{aligned} \tag{2.64}$$

is also used [35].

If the velocity field v of the Navier-Stokes equations on a domain is obtained, then by de Rham Theorem (Proposition 2.3), pressure p is determined with different constants. If, moreover, pressure condition on a portion of boundary is given, then the constant is determined and pressure is uniquely determined when the solution is smooth enough for existence of traces of velocity and pressure on the portion of boundary. Also, if a boundary condition including pressure p as addend (stress, total stress, normal stress) on a portion of boundary is given, then the pressure on the domain is determined uniquely when the solution is smooth enough (see Theorem 4.1 of [36] and Theorems 5.1, 5.3).

2.2.4 Outflow Boundary Conditions on Imaginary Boundary

To reduce amount of computation, imaginary boundaries may be drawn inside flow. On the portions of imaginary boundary where flow is out, velocity and pressure boundary conditions are not used since the velocity and pressure on the potion of imaginary boundary are not known prior to prescription of solution.

In such portions of imaginary boundary, the *free outflow boundary conditions*

$$\mu\frac{\partial v}{\partial n} - pn = 0 \tag{2.65}$$

[34, 37–47], which is called "*do nothing*" boundary condition, and

$$\mu\mathcal{E}(v)n - pn = 0 \tag{2.66}$$

[47, 48] are used. "Do nothing" boundary condition (2.65) results from variational formulation based on Dirichlet bilinear form $(\nabla v, \nabla u)$ obtained by integrating by parts $(\Delta v, u)$ (see (2.86)) and does not have a real physical meaning, but is rather used in truncating large physical domains to get smaller computational domains. The condition is appropriate where the exit flow is close to a fully developed one and the normal gradient for velocity is close to zero. Thus, the artificial boundary for the outflow boundary condition must be placed downstream fully and stream line similar to parallel (see [37]).

2.2.5 Boundary Conditions on Free Surfaces

Sometimes, we are encountered with a boundary between two fluids, which is called *free surface*. A common example occurs when a liquid film flows down an inclined plane. The surface of the liquid film in contact with the surrounding gas is a fluid-fluid interface. Other examples include the interface between a liquid drop and the surrounding continuous phase or that between two liquid layers. Free surfaces are unknowns to be determined and require two boundary conditions to be applied.

(i) A *kinematic condition*: This condition relates the motion of the free interface to the fluid velocities at the free surface.

Let $v = (v_1, v_2, v_3)$ be the velocity field of one fluid and the position of a free surface be written in an implicit form $f(x, t) = 0$. Fluid particles on the free surface do not move across the free surface, therefore we can write

$$\frac{\partial f}{\partial t} + \sum_i v_i \frac{\partial f}{\partial x_i} = \frac{\partial f}{\partial t} + v \cdot \nabla f = 0, \tag{2.67}$$

which is a kinematic condition. When a free surface is written in the form $x_3 = h(x_1, x_2, t)$, the kinematic boundary condition is

$$v_3 = \frac{\partial f}{\partial t} + v_1 \frac{\partial f}{\partial x_1} + v_2 \frac{\partial f}{\partial x_2}.$$

For steady problems, we have $\frac{\partial f}{\partial t} = 0$, and the kinematic condition is written as

$$v \cdot n = 0,$$

where n is normal unit vector on the surface, since ∇f is a normal vector on the surface given by $f(x) = 0$. This condition means that there is no flow through the free surface. Note that there can be a flow tangent to the free surface.

(ii) A *dynamic condition*: This condition is concerned with force balance on the free surface.

When one is concerned with a free surface, if the capillary (surface) force is neglected, then the fluid is called a *heavy fluid* and if the capillary force is essential, then the fluid is called a *capillary fluid*.

For a heavy fluid, the traction exerted by fluid (1) onto fluid (2) is equal and opposite to the traction exerted by fluid (2) onto fluid (1). Since the outward normal vectors on the boundary between two domains of fluids are opposite, we have the dynamic boundary condition

$$\sigma^{(1)} = -\sigma^{(2)}.$$

If n is the normal unit vector oriented from fluid (1) to fluid (2) on the surface, then by (2.47) we have

$$\sigma^{(1)} \equiv S^{(1)} \cdot n = -\sigma^{(2)} \equiv S^{(2)} \cdot n, \tag{2.68}$$

where $S^{(i)}$ is the stress tensor of fluid (i). If fluid (2) is air, then the dynamic condition on the free surface of fluid (1) is rewritten as (see (1.16), Sect. 3.1 of [49])

$$\sigma \equiv -pn + 2\mu\mathcal{E}(v) \cdot n = -p_a n, \tag{2.69}$$

where p_a is the air pressure.

For free surface of capillary fluids, surface tensions must be taken into account. The pressure jump induced by surface tension is given by

$$sk,$$

where s is the surface tension and k is twice the mean curvature of the free surface (see Sect. 3.1.1). Surface tension acts like a tensioned membrane at the free surface and tries to minimize the surface area. Hence the pressure inside of fluid including the center of curvature tends to be higher than other. Thus, the dynamic boundary condition for the capillary fluid on free surfaces is

$$S^{(1)} \cdot n + skn = S^{(2)} \cdot n, \tag{2.70}$$

where $k > 0$ if the centers of curvature lie inside fluid (1) when the free surface is a sphere or a circle. Moreover, if a free surface is convex (concave) (see Definition 3.2.) with respect to the normal unit vector from fluid (1) to fluid (2) on the surface, then $k \geq 0$ ($k \leq 0$) (see Chap. 4 of [3] and Sect. 3.1.1).

In what follows let us further consider heavy fluids. If a free surface to be determined is steady and a fluid may move freely along the free surface, then we have a free boundary condition

$$\begin{aligned} v \cdot n &= 0, \\ \sigma(v, p) &= 0, \end{aligned} \tag{2.71}$$

which is useful in many physical problems, especially in mathematical modeling of coating flows, flows of melted semiconductors (see [50] and references therein).

When one fluid is air, sometimes free surfaces are regarded as fixed surfaces. Then, the kinematic boundary condition vanishes and the boundary condition on the fixed surface is (2.69). Neglecting the press of air, we get

$$\sigma(v, p) = 0,$$

which in some papers is called Neumann boundary condition (see [51] and references therein, [15]).

If there is no flow across a fixed surface and the fluid flows along the surface, then a boundary condition

$$\begin{aligned} v \cdot n &= 0, \\ \sigma_\tau(v, p) &= f_\tau \end{aligned} \tag{2.72}$$

may be used.

In practice we deal with mixture of some kinds of boundary conditions. For a channel flow a mixture of Dirichlet condition $v = 0$ on the wall and "do nothing" condition on the outlet is used. But for a channel flow with a rough boundary surface a mixture of Dirichlet condition, the Navier slip-with-friction boundary condition and "do nothing" condition may be used. For a flow in a vessel with in/outlet a mixture of Dirichlet condition $v = 0$ on the wall and pressure conditions on the inlet/outlet is used. But for the flow in a vessel with in/outlet and a free surface a mixture of Dirichlet condition, a Neumann condition $2\nu\varepsilon(v)n - pn = 0$ and pressure conditions may be used.

There is a vast body of literature for the Stokes and Navier-Stokes problems with mixed boundary conditions, and several variational formulations have been used for them.

2.3 Bilinear Forms for Hydrodynamics

In this section, we consider three types of bilinear form reduced from $(\Delta v, u)$ by integration by parts and outline the variational formulations of the Navier-Stokes problems with mixed boundary conditions.

From now on, in this book we use the following notations.

Let Ω be an open subset of $\mathbb{R}^l$, $l = 2, 3$.

When X is a Banach space, $\mathbf{X} = X^l$ and $\mathbf{X}^*$ is the dual of $\mathbf{X}$. Let $W^{k,\alpha}(\Omega)$ be Sobolev spaces, $H^k(\Omega) = W^{k,2}(\Omega)$, and so $\mathbf{H}^k(\Omega) = \{H^k(\Omega)\}^l$, $\mathbf{H}_0^k(\Omega) = \{W_0^{k,2}(\Omega)\}^l$.

An inner product and norm in the space $\mathbf{L}^2(\Omega)$ or $L^2(\Omega)$ are, respectively, denoted by $(\cdot\,,\cdot)$ and $\|\cdot\|$. $\langle\cdot\,,\cdot\rangle$ means the duality pairing between a Sobolev space X and its dual one.

Also, $(\cdot\,,\cdot)_{\Gamma_i}$ is an inner product in the $\mathbf{L}^2(\Gamma_i)$ or $L^2(\Gamma_i)$, and $\langle\cdot\,,\cdot\rangle_{\Gamma_i}$ means the duality pairing between $\mathbf{H}^{\frac{1}{2}}(\Gamma_i)$ and $\mathbf{H}^{-\frac{1}{2}}(\Gamma_i)$ or between $H^{\frac{1}{2}}(\Gamma_i)$ and $H^{-\frac{1}{2}}(\Gamma_i)$ (with an exception in Remark 5.1 concerned with the duality pairing between $H_{00}^{\frac{1}{2}}(\Gamma_i)$ and $(H_{00}^{\frac{1}{2}}(\Gamma_i))^*$).
The inner product and norms in $\mathbb{R}^l$, respectively, are denoted by $(\cdot\,,\cdot)_{\mathbb{R}^l}$ and $|\cdot|$. Sometimes $a \cdot b$ or ab is used for inner product in $\mathbb{R}^l$ between a and b. For simplicity, when A is a matrix and n is vector, sometimes we write $A \cdot n$ by An.

2.3.1 Bilinear Forms

2.3.1.1 Dirichlet Bilinear Form

Unless otherwise specified, let Ω be a bounded domain with Lipschitz boundary of $\mathbb{R}^l, l = 2, 3$, and $\phi \in H^{1/2}(\partial\Omega)$. Then there exists a continuous linear operator $H^{1/2}(\partial\Omega) \to H^1(\Omega)$ (lift operator), and for every $\phi \in H^{1/2}(\partial\Omega)$ there exists a $w \in H^1(\Omega)$ with

$$w|_{\partial\Omega} = \phi, \quad \|w\|_{H^1(\Omega)} \le c\|\phi\|_{H^{1/2}(\partial\Omega)}, \tag{2.73}$$

where c is independent of ϕ (Theorem 1.5.1.2 of [52] or (1.3.15)).

Let us introduce the space

$$H(\mathrm{div}; \Omega) = \left\{v \in \mathbf{L}^2(\Omega);\ \mathrm{div}\, v \in L^2(\Omega)\right\},$$

which is a Hilbert space for the inner product

$$(v, u)_{H(\mathrm{div};\Omega)} = (v, u) + (\mathrm{div}\, v, \mathrm{div}\, u).$$

If Ω is an open subset with Lipschitz boundary of $\mathbb{R}^l$ (not necessarily bounded), then the space $\mathscr{D}(\bar{\Omega})^l$ is dense in $H(\mathrm{div}; \Omega)$ (see Theorem 2.4, Chap. 1 of [53]).

For $u \in H(\mathrm{div}; \Omega)$, let us consider the functional on $\phi \in H^{1/2}(\partial\Omega)$ defined by

$$F_u(\phi) = (\mathrm{div}\, u, w) + (u, \nabla w).$$

By (2.73) we have

$$|F_u(\phi)| \le c\big(\|u\|_{\mathbf{L}^2} + \|\mathrm{div}\, u\|_{L^2}\big)\|\phi\|_{H^{1/2}(\partial\Omega)},$$

which shows that the functional F_u is linear and continuous on $H^{1/2}(\partial\Omega)$. Since

$$(\mathrm{div}\, u, w) + (u, \nabla w) = (u \cdot n, \phi)_{\partial\Omega} \quad \text{for } u \in \mathbf{C}^\infty(\Omega),$$

where n is the outward unit normal vector on $\partial\Omega$, we can define a trace of $u \cdot n$

$$\gamma_n u \in H^{-1/2}(\partial\Omega)$$

as a generalization of $u \cdot n$ of smooth functions u. Thus, we have

Proposition 2.1 *Let Ω be a bounded domain with Lipschitz boundary of $\mathbb{R}^l$. The mapping $\gamma_n : v \to v \cdot n|_{\partial\Omega}$ defined on $\mathscr{D}(\bar{\Omega})^l$ can be extended by continuity to a linear and continuous mapping, from $H(div; \Omega)$ into $H^{-1/2}(\partial\Omega)$.*

Moreover, the following Green's formula holds:

$$(div\, u, w) + (u, \nabla w) = \langle\gamma_n u, w\rangle_{\partial\Omega} \quad \forall u \in H(div; \Omega),\ \ \forall w \in H^1(\Omega). \tag{2.74}$$

(cf. Theorem 2.5 and (2.17), Chap. 1 of [53]).

By Proposition 2.1, for $v \in \mathbf{H}^1(\Omega)$ with $\Delta v \in \mathbf{L}^2(\Omega)$ and $w \in \mathbf{H}^1(\Omega)$, we have

$$-(\Delta v, w) = \sum_i (\nabla v_i, \nabla w_i) - \sum_i \langle \partial_n v_i, w_i \rangle_{\partial\Omega}, \tag{2.75}$$

where $\partial_n v_i = \gamma_n(\nabla v_i)$. Having in mind the fact that $\gamma_n(\nabla v_i) = \frac{\partial v_i}{\partial n}\big|_{\partial\Omega}$ for smooth functions v_i, in what follows we write (2.75) as

$$-(\Delta v, w) = (\nabla v, \nabla w) - \langle \partial_n v, w \rangle_{\partial\Omega}. \tag{2.76}$$

We call $(\nabla v, \nabla w)$ *Dirichlet bilinear form.*

2.3.1.2 Strain Bilinear Form

When $v \in \mathbf{H}^1(\Omega)$, let us prove

$$\operatorname{div}\big(2\mathcal{E}_i(v)\big) = \Delta v_i + \frac{\partial}{\partial x_i}\operatorname{div} v \quad \text{in } H^{-1}(\Omega), \tag{2.77}$$

where $\mathcal{E}_i(v) = (\varepsilon_{i1}(v), \varepsilon_{i2}(v), \varepsilon_{i3}(v))$, $i = 1, 2, 3$. If $q \in C_0^\infty(\Omega)$, then we have

$$2\int_\Omega (div\, \mathcal{E}_i(v))q\, d\omega = -2\int_\Omega \mathcal{E}_i(v)\cdot \nabla q\, d\omega = -\int_\Omega \sum_j \Big(\frac{\partial v_i}{\partial x_j} + \frac{\partial v_j}{\partial x_i}\Big)\frac{\partial q}{\partial x_j}\, d\omega$$
$$= -\int_\Omega \nabla v_i \cdot \nabla q\, d\omega - \int_\Omega \frac{\partial v}{\partial x_i}\cdot \nabla q\, d\omega = \int_\Omega \Delta v_i \cdot q\, d\omega + \int_\Omega div\, \frac{\partial v}{\partial x_i}\cdot q\, d\omega.$$

Since $C_0^\infty(\Omega)$ is dense in $H_0^1(\Omega)$, from above we get (2.77).

Let $v \in \mathbf{H}^2(\Omega)$ and $u \in \mathbf{H}^1(\Omega)$. By (2.74) and (2.77), we have

$$-(\Delta v_i, u_i) = 2(\mathcal{E}_i(v), \nabla u_i) - 2(\mathcal{E}_i(v)\cdot n, u_i)_{\partial\Omega} + (\operatorname{div}\frac{\partial v}{\partial x_i}, u_i). \tag{2.78}$$

Since the tensor $\mathcal{E}(v)$ is symmetric, we have

$$\begin{aligned}\big(\mathcal{E}(v), \nabla u\big) &= \sum_i \big(\mathcal{E}_i(v), \nabla u_i\big) = \sum_i\sum_j \big(\varepsilon_{ij}(v), \frac{\partial u_i}{\partial x_j}\big)\\ &= \sum_i\sum_j \big(\varepsilon_{ij}(v), \frac{1}{2}\big(\frac{\partial u_i}{\partial x_j} + \frac{\partial u_j}{\partial x_i}\big)\big) = \big(\mathcal{E}(v), \mathcal{E}(u)\big).\end{aligned} \tag{2.79}$$

Taking into account (2.79), we have from (2.78)

$$
\begin{aligned}
-(\Delta v, u)_\Omega &= 2(\mathcal{E}(v), \mathcal{E}(u)) + (\nabla(\operatorname{div} v), u) - 2(\mathcal{E}(v)n, u)_{\partial\Omega} \\
&= 2(\mathcal{E}(v), \mathcal{E}(u)) - (\operatorname{div} v, \operatorname{div} u) - 2(\mathcal{E}(v)n, u)_{\partial\Omega} + (\operatorname{div} v, u \cdot n)_{\partial\Omega}.
\end{aligned}
\tag{2.80}
$$

If $v \in \mathbf{H}^1(\Omega)$, $\Delta v \in \mathbf{L}^2(\Omega)$ and $\operatorname{div} v = 0$, then by (2.77) $\operatorname{div}\big(\mathcal{E}_i(v)\big) \in L^2(\Omega)$, and there exists $\gamma_n(\mathcal{E}_i(v)) \in H^{-1/2}(\partial\Omega)$.

Therefore for $v \in \mathbf{H}^1(\Omega)$ with $\Delta v \in \mathbf{L}^2(\Omega)$, $\operatorname{div} v = 0$, and $u \in \mathbf{H}^1(\Omega)$, we have

$$
-(\Delta v, u)_\Omega = 2(\mathcal{E}(v), \mathcal{E}(u)) - 2(\mathcal{E}(v)n, u)_{\partial\Omega}, \tag{2.81}
$$

where $\mathcal{E}(v)n|_{\partial\Omega} = (\gamma_n(\mathcal{E}_1(v)), \gamma_n(\mathcal{E}_2(v)), \gamma_n(\mathcal{E}_3(v)) \in \mathbf{H}^{-1/2}(\partial\Omega)$.

We call $(\mathcal{E}(v), \mathcal{E}(u))$ *the strain bilinear form.*

2.3.1.3 Vorticity Bilinear Form

Let us introduce the space

$$
H(\operatorname{rot}; \Omega) = \big\{v \in \mathbf{L}^2(\Omega); \operatorname{rot} v \in \mathbf{L}^2(\Omega)\big\}.
$$

When Ω is a domain with Lipschitz boundary of $\mathbb{R}^l$ (not necessarily bounded), $\mathscr{D}(\bar{\Omega})^l$ is dense in $H(\operatorname{rot}; \Omega)$ (Theorem 2.10, Chap. 1 of [53]).

Proposition 2.2 *Let Ω be a domain with Lipschitz boundary of $\mathbb{R}^3$ (not necessarily bounded). The mapping $\gamma_\tau : u \to u \times n|_{\partial\Omega}$ defined on $\mathscr{D}(\bar{\Omega})^3$ can be extended by continuity to a linear and continuous mapping, from $H(rot; \Omega)$ into $\mathbf{H}^{-1/2}(\partial\Omega)$.*

Moreover, the following Green's formula holds:

$$
(\mathit{rot}\, u, w)_\Omega - (u, \mathit{rot}\, w)_\Omega = -\langle\gamma_\tau u, w\rangle_{\partial\Omega} \quad \forall w \in \mathbf{H}^1(\Omega). \tag{2.82}
$$

(cf. Theorem 2.11, Chap. 1 of [53]). In what follows for convenience we write $\gamma_\tau u$ by $u \times n$.

Whenever we are concerned with vorticity of vector fields with $l = 2$, for convenience as before, vector $u = (u_1(x_1, x_2), u_2(x_1, x_2))$ is identified with $\bar{u} = (u_1(x_1, x_2), u_2(x_1, x_2), 0)$.

Thus, for $u = (u_1(x_1, x_2), u_2(x_1, x_2))$, $v = (v_1(x_1, x_2), v_2(x_1, x_2))$ and $n = (n_1, n_2)$, we think that

$$\begin{aligned}
&\operatorname{rot} u = \operatorname{rot} \bar{u} \equiv \left(0, 0, \frac{\partial u_2}{\partial x_1} - \frac{\partial u_1}{\partial x_2}\right), \\
&u \times n = \bar{u} \times \bar{n} = (0, 0, u_1 n_2 - u_2 n_1), \\
&\operatorname{rot} v \times n = \operatorname{rot} \bar{v} \times \bar{n} = \left(-n_2\left(\frac{\partial v_2}{\partial x_1} - \frac{\partial v_1}{\partial x_2}\right), n_1\left(\frac{\partial v_2}{\partial x_1} - \frac{\partial v_1}{\partial x_2}\right), 0\right), \\
&(\operatorname{rot} v, \operatorname{rot} u)_\Omega = \int_\Omega \operatorname{rot} \bar{v} \cdot \operatorname{rot} \bar{u}\, dx, \\
&(\operatorname{rot} v \times n, u)_{\partial\Omega} = \int_{\partial\Omega} (\operatorname{rot} \bar{v} \times \bar{n}) \cdot \bar{u}\, d\sigma.
\end{aligned}$$

Corollary 2.1 *Let Ω be a bounded domain with Lipschitz boundary of $\mathbb{R}^l$, $l = 2, 3$. If $v \in \mathbf{H}^1(\Omega)$, $\Delta v \in \mathbf{L}^2(\Omega)$ and $\operatorname{div} v = 0$, then there exists $\gamma_\tau(\operatorname{rot} v) \equiv \operatorname{rot} v \times n \in \mathbf{H}^{-1/2}(\partial\Omega)$ and the following holds:*

$$-(\Delta v, u)_\Omega = (\operatorname{rot} v, \operatorname{rot} u)_\Omega - \langle \operatorname{rot} v \times n, u\rangle_{\partial\Omega}. \tag{2.83}$$

Proof Since

$$-\Delta v = \operatorname{rot} \operatorname{rot} v - \operatorname{grad}(\operatorname{div} v)$$

(see (2.21)) and $\operatorname{div} v = 0$, we have $\operatorname{rot} \operatorname{rot} v \in \mathbf{L}^2(\Omega)$.

If $l = 3$, then by Proposition 2.2 we come to the conclusion.

If $l = 2$. Let n be the outward unit normal on $\partial\Omega$ and $\bar{n} = (n_1, n_2, 0)$. On the domain $\tilde{\Omega} = \Omega \times (0, 1) \subset \mathbb{R}^3$ let us consider vector fields $\bar{v}, \bar{u}$. Then by Proposition 2.2 we have

$$-(\Delta\bar{v}, \bar{u})_{\tilde{\Omega}} = (\operatorname{rot} \bar{v}, \operatorname{rot} \bar{u})_{\tilde{\Omega}} - (\operatorname{rot} \bar{v} \times \tilde{n}, \bar{u})_{\partial\tilde{\Omega}}, \tag{2.84}$$

where

$$\partial\tilde{\Omega} = (\partial\Omega \times (0, 1)) \bigcup (\Omega \times \{0\}) \bigcup (\Omega \times \{1\})$$

and $\tilde{n}$ is the outward normal on $\partial\tilde{\Omega}$. Since both vectors $\operatorname{rot} \bar{v}$ and $\tilde{n}$ are orthogonal to $\Omega \times \{0\} \bigcup \Omega \times \{1\}$, $\operatorname{rot} \bar{v} \times \tilde{n} = 0$ on $\Omega \times \{0\} \bigcup \Omega \times \{1\}$. Then, noting that $\tilde{n} = \bar{n}$ on $\partial\Omega \times (0, 1)$, we have

$$(\operatorname{rot} \bar{v} \times \tilde{n}, \bar{u})_{\partial\tilde{\Omega}} = \int_{\partial\Omega\times(0,1)} (\operatorname{rot} \bar{v} \times \bar{n}) \cdot \bar{u}\, d\sigma = \int_{\partial\Omega} (\operatorname{rot} \bar{v} \times \bar{n}) \cdot \bar{u}\, d\sigma = (\operatorname{rot} v \times n, u)_{\partial\Omega}.$$

Also,

$$\begin{aligned}
&(\Delta\bar{v}, \bar{u})_{\tilde{\Omega}} = \int_{\tilde{\Omega}} \Delta\bar{v} \cdot \bar{u}\, dx = \int_\Omega \Delta\bar{v} \cdot \bar{u}\, dx = (\Delta v, u)_\Omega, \\
&(\operatorname{rot} \bar{v}, \operatorname{rot} \bar{u})_{\tilde{\Omega}} = \int_{\tilde{\Omega}} \operatorname{rot} \bar{v} \cdot \operatorname{rot} \bar{u}\, dx = \int_\Omega \operatorname{rot} \bar{v} \cdot \operatorname{rot} \bar{u}\, dx = (\operatorname{rot} v, \operatorname{rot} u)_\Omega.
\end{aligned}$$

Therefore, from (2.84) we get (2.83) for $l = 2$. □

We call $(\operatorname{rot} v, \operatorname{rot} u)$ *the vorticity bilinear form.*

2.3.2 Variational Formulations for Mixed Boundary Value Problems of the Navier-Stokes Equations

Now, we will briefly outline the main points of variational formulations for the Stokes and Navier-Stokes problems with mixed boundary conditions.

In this subsection from now on, let Ω be a bounded domain with Lipschitz boundary of $\mathbb{R}^l$, $l = 2, 3$. Let us introduce the following spaces of divergence-free functions:

$$\mathscr{V} = \{\phi \in \mathscr{D}(\Omega)^l;\ \operatorname{div}\phi = 0\}, \quad \mathbf{V} = \{v \in \mathbf{H}_0^1(\Omega);\ \operatorname{div} v = 0\}.$$

The space $\mathscr{V}$ is dense in $\mathbf{V}$ for the norm of $\mathbf{H}^1(\Omega)$ (Corollary 2.5, Chap. 1 of [53]).

The following proposition which is a simplified version of *de Rham's theorem* (see Proposition 1.1, Chap. 1 of [54]) holds.

Proposition 2.3 (Theorem 2.3, Chap. 1 of [53]) *If $f \in \mathbf{H}^{-1}(\Omega)$ satisfies*

$$\langle f, \phi\rangle = 0 \quad \forall\phi \in \mathscr{V},$$

then there exists $p \in L^2(\Omega)$ such that

$$f = \nabla p.$$

If Ω is connected, then p is unique up to an additive constant.

2.3.2.1 Variational Formulation Based on Dirichlet Bilinear Form

Let $\partial\Omega = \overline{\Gamma}_D \cup \overline{\Gamma}_N$, where Γ_D, Γ_N are open subsets of $\partial\Omega$ such that $\Gamma_D \cap \Gamma_N = \emptyset$, and

$$\mathbf{V}_D = \{v \in \mathbf{H}^1(\Omega);\ \operatorname{div} v = 0,\ v|_{\Gamma_D} = 0\}.$$

Let us consider the problem

$$\begin{cases} -\mu\Delta v + (v\cdot\nabla)v + \nabla p = f & \text{in } \Omega,\\ \operatorname{div} v = 0 & \text{in } \Omega,\\ v|_{\Gamma_D} = 0, & \\ (\mu\dfrac{\partial v}{\partial n} - pn)|_{\Gamma_N} = g. & \end{cases} \tag{2.85}$$

Assuming that (v, p) is smooth and making duality product with $u \in \mathbf{V}_D$ in the first equation, by (2.76) we get

$$\mu(\nabla v, \nabla u) + \langle (v \cdot \nabla)v, u\rangle = \langle f, u\rangle + \langle g, u\rangle_{\Gamma_N} \quad \forall u \in \mathbf{V}_D. \tag{2.86}$$

Thus, when $f \in \mathbf{V}_D^*$, $g \in \mathbf{H}^{-1/2}(\Gamma_N)$ are given, a function $v \in \mathbf{V}_D$ satisfying (2.86) is called a solution to problem (2.85).

To show that the concept of solution is well defined, we will prove that if a solution v is smooth ($v \in \mathbf{H}^2(\Omega)$, $f \in \mathbf{L}^2(\Omega)$), then there exists a p such that (v, p) satisfies (2.85). By (2.76) we get from (2.86)

$$-\mu(\Delta v, u) + \langle (v \cdot \nabla)v, u\rangle + \mu\langle \partial_n v, u\rangle_{\Gamma_N} = \langle f, u\rangle + \langle g, u\rangle_{\Gamma_N} \quad \forall u \in \mathbf{V}_D. \tag{2.87}$$

Taking any $u \in \mathscr{V}$, we have from (2.87)

$$(-\mu\Delta v + (v \cdot \nabla)v - f, u) = 0.$$

By Proposition 2.3, there exists a unique function $P \in L^2(\Omega)$ such that $\int_\Omega P\,dx = 0$ and

$$-\mu\Delta v + (v \cdot \nabla)v - f = -\nabla P. \tag{2.88}$$

Since $v \in \mathbf{H}^2(\Omega)$, $f \in \mathbf{L}^2(\Omega)$ and $(v \cdot \nabla)v \in \mathbf{L}^2(\Omega)$, we have that $\nabla P \in \mathbf{L}^2(\Omega)$, which shows that $P \in H^1(\Omega)$. Substituting (2.88) into (2.87) and using integration by parts, we have

$$\langle \mu\partial_n v - Pn - g, u\rangle_{\Gamma_N} = 0. \tag{2.89}$$

For any tangent vector field $\phi \in \mathbf{H}_{00}^{1/2}(\Gamma_N)$ on Γ_N there exists its continuation $\tilde{\phi} \in \mathbf{H}^{1/2}(\partial\Omega)$ such that $\tilde{\phi} = 0$ on $\partial\Omega \backslash \Gamma_N$ (see Theorem 1.18). There exists a solution $u \in \mathbf{H}^1(\Omega)$ to the Stokes problem

$$\begin{cases} -\Delta u + \nabla p = 0, \\ \operatorname{div} u = 0, \\ u\,|_{\partial\Omega} = \tilde{\phi} \end{cases}$$

(see Theorem IV.1.1 of [55]), and $u \in \mathbf{V}_D$. Taking such u in (2.89), we see that the tangent components of $\mu\partial_n v - Pn - g$ vanish. Similarly, for any normal vector field $\phi \in \mathbf{H}_{00}^{1/2}(\Gamma_N)$ such that $\int_{\Gamma_N} \phi \cdot n\,ds = 0$ there exists a $u \in \mathbf{V}_D$ such that $u|_{\Gamma_N} = \phi$. Using such u in (2.89), we can verify that the values of the normal components of $\mu\partial_n v - Pn - g$ must be the same. Therefore, we have

$$\mu\frac{\partial v}{\partial n} - Pn - g = cn \quad \text{on } \Gamma_N, \tag{2.90}$$

where c is some constant. Putting $p = P + c$, we get the boundary condition on Γ_N. Thus, taking into account $v \in \mathbf{V}_D$, we know that (v, p) satisfies (2.85).

If $\Gamma_D \neq \emptyset$, then by Friedrichs' inequality the norm $\|v\|_{\mathbf{V}_D}$ is equivalent to $\|\nabla v\|_{\mathbf{L}^2}$ and

$$(\nabla v, \nabla v) \geq c\|v\|^2_{\mathrm{H}^1(\Omega)}, \tag{2.91}$$

which is fundamental for study of the problem (2.86).

2.3.2.2 Variational Formulation Based on Strain Bilinear Form

Let $\partial\Omega = \overline{\Gamma}_D \cup \left(\sum_{i=1}^3 \overline{\Gamma}_i\right)$, where Γ_D, Γ_i are open subsets of $\partial\Omega$ such that $\Gamma_D \cap \Gamma_i = \emptyset$, $\Gamma_i \cap \Gamma_j = \emptyset$ for $i \neq j$ and

$$\mathbf{V}_D = \left\{v \in \mathbf{H}^1(\Omega);\ \mathrm{div}\, v = 0,\ v|_{\Gamma_D} = 0,\ v_\tau|_{\Gamma_1} = 0,\ v_n|_{\Gamma_2} = 0\right\}.$$

Let us consider the problem

$$\begin{cases} -\mu\Delta v + (v\cdot\nabla)v + \nabla p = f & \text{in } \Omega, \\ \mathrm{div}\, v = 0 & \text{in } \Omega, \\ v|_{\Gamma_D} = 0, \\ v_\tau|_{\Gamma_1} = 0,\ (-p + 2\mu\varepsilon_{nn}(v))|_{\Gamma_1} = \phi_1, \\ v_n|_{\Gamma_2} = 0,\ 2(\mu\varepsilon_{n\tau}(v) + \alpha v_\tau)|_{\Gamma_2} = \phi_2, \\ (-pn + 2\mu\varepsilon_n(v))|_{\Gamma_3} = \phi_3, \end{cases} \tag{2.92}$$

where $\varepsilon_n(v) = \mathcal{E}(v)\cdot n$, $\varepsilon_{nn}(v) = (\mathcal{E}(v)\cdot n, n)_{\mathbb{R}^l}$, $\varepsilon_{n\tau}(v) = \mathcal{E}(v)\cdot n - \varepsilon_{nn}(v)n$.

Assuming that (v, p) is smooth and making duality product with $u \in \mathbf{V}_D$ in the first equation, by (2.81) we get

$$\begin{aligned} &2\mu(\mathcal{E}(v), \mathcal{E}(u)) + \langle(v\cdot\nabla), u\rangle + 2(\alpha v_\tau, u)_{\Gamma_2} \\ &\quad = \langle f, u\rangle + \langle\phi_1, u_n\rangle_{\Gamma_1} + \langle\phi_2, u\rangle_{\Gamma_2} + \langle\phi_3, u\rangle_{\Gamma_3} \quad \forall u \in \mathbf{V}_D. \end{aligned} \tag{2.93}$$

Thus, when $f \in \mathbf{V}_D^*$, $\phi_1 \in H^{-1/2}(\Gamma_1)$, $\phi_2 \in \mathbf{H}^{-1/2}(\Gamma_2)$, $\phi_3 \in \mathbf{H}^{-1/2}(\Gamma_3)$ are given, a function $v \in \mathbf{V}_D$ satisfying (2.93) is called a solution to problem (2.92).

If we are concerned with

$$\begin{cases} -\mu\Delta v + (v\cdot\nabla)v + \nabla p = f \quad \text{in } \Omega, \\ \operatorname{div} v = 0 \quad \text{in } \Omega, \\ v|_{\Gamma_D} = 0, \\ v_\tau|_{\Gamma_1} = 0, \ \Big(-(p + \frac{1}{2}|v|^2) + 2\mu\varepsilon_{nn}(v)\Big)\Big|_{\Gamma_1} = \phi_1, \\ v_n|_{\Gamma_2} = 0, \ 2(\mu\varepsilon_{n\tau}(v) + \alpha v_\tau)|_{\Gamma_2} = \phi_2, \\ \Big(-(p + \frac{1}{2}|v|^2)n + 2\mu\varepsilon_n(v)\Big)\Big|_{\Gamma_3} = \phi_3, \end{cases} \tag{2.94}$$

which includes total stress boundary conditions, then by

$$(v\cdot\nabla)v = \operatorname{rot} v \times v + \frac{1}{2}\nabla\,|v|^2$$

we can rewrite the Navier-Stokes equations as

$$\frac{\partial v}{\partial t} - \mu\Delta v + \operatorname{rot} v \times v + \nabla(p + \frac{1}{2}|v|^2) = f, \ \ \operatorname{div} v = 0 \ \text{ in } \Omega, \tag{2.95}$$

and using (2.81) for a smooth solution we get

$$\begin{aligned} &2\mu(\mathcal{E}(v), \mathcal{E}(u)) + \langle \operatorname{rot} v \times v, u\rangle + 2(\alpha v_\tau, u)_{\Gamma_2} \\ &\quad = \langle f, u\rangle + \langle \phi_1, u_n\rangle_{\Gamma_1} + \langle \phi_2, u\rangle_{\Gamma_2} + \langle \phi_3, u\rangle_{\Gamma_3} \quad \forall u \in \mathbf{V}_D. \end{aligned} \tag{2.96}$$

Thus, when $f \in \mathbf{V}_D^*$, $\phi_1 \in H^{-1/2}(\Gamma_1)$, $\phi_2 \in \mathbf{H}^{-1/2}(\Gamma_2)$, $\phi_3 \in \mathbf{H}^{-1/2}(\Gamma_3)$ are given, a function $v \in \mathbf{V}_D$ satisfying (2.96) is called a solution to problem (2.94).

To show that the concepts of solution are well defined, as above we can prove that if a solution v is smooth ($v \in \mathbf{H}^2(\Omega)$, $f \in \mathbf{L}^2(\Omega)$), then there exists a p such that (v, p) satisfies (2.92) or (2.94). For more general cases, we will prove such a fact in Chap. 5.

If $\Gamma_D \neq \emptyset$, then by Korn's and Friedrichs' inequalities the norm $\|v\|_{\mathbf{V}_D}$ is equivalent to $\|\mathcal{E}(v)\|_{\mathbf{L}^2}$ and

$$(\mathcal{E}(v), \mathcal{E}(v)) \geq c\|v\|^2_{\mathbf{H}^1(\Omega)}, \tag{2.97}$$

which is fundamental for study of the problems (2.93) and (2.96).

2.3.2.3 Variational Formulation Based on Vorticity Bilinear Form

Let $\partial\Omega = \overline{\Gamma}_D \cup \Big(\sum_{i=1}^2 \overline{\Gamma}_i\Big)$, where Γ_D, Γ_i, $i = 1, 2$, are open subsets of $\partial\Omega$ such that $\Gamma_D \cap \Gamma_i = \emptyset$, $\Gamma_i \cap \Gamma_j = \emptyset$ for $i \neq j$ and

$$\mathbf{V}_D = \big\{v \in \mathbf{H}^1(\Omega);\, \operatorname{div} v = 0,\, v|_{\Gamma_D} = 0,\, v_\tau|_{\Gamma_1} = 0,\, v_n|_{\Gamma_2} = 0\big\}.$$

Let us consider the problem

$$\begin{cases} -\mu\Delta v + (v\cdot\nabla)v + \nabla p = f & \text{in } \Omega, \\ \operatorname{div} v = 0 & \text{in } \Omega, \\ v|_{\Gamma_D} = 0, \\ v_\tau|_{\Gamma_1} = 0, \ \ -p|_{\Gamma_1} = \phi_1, \\ v_n|_{\Gamma_2} = 0, \ \ (\operatorname{rot} v \times n)|_{\Gamma_2} = \phi_2/\mu. \end{cases} \tag{2.98}$$

Assuming that (v, p) is smooth and making duality product with $u \in \mathbf{V}_D$ in the first equation, by (2.83) we get

$$\mu(\operatorname{rot} v, \operatorname{rot} u) + \langle (v\cdot\nabla)v, u\rangle = \langle f, u\rangle + \langle \phi_1, u_n\rangle_{\Gamma_1} + \langle \phi_2, u\rangle_{\Gamma_1} \quad \forall u \in \mathbf{V}_D, \tag{2.99}$$

where $(\operatorname{rot} v \times n, u)|_{\Gamma_1} = 0$ was used.

Thus, when $f \in \mathbf{V}_D^*$, $\phi_1 \in H^{-1/2}(\Gamma_1)$ and $\phi_2 \in \mathbf{H}^{-1/2}(\Gamma_2)$ are given, a function $v \in \mathbf{V}_D$ satisfying (2.99) is called a solution to problem (2.98).

Like (2.96), we can get a variational formulation for the problem with the total pressure boundary condition instead of the static pressure boundary condition.

For the discussion of that the concepts of solutions are well defined, we refer to Chap. 5.

Let $\Omega \subset \mathbb{R}^3$, $\partial\Omega \in C^{1,1}$ or Ω be a convex polyhedron, $\Gamma_D \neq \emptyset$ and $\bar{\Gamma}_1 \cap \bar{\Gamma}_2 = \emptyset$. Then, by Lemma 4.1 of [56]

$$(\operatorname{rot} v, \operatorname{rot} v) \geq c\|v\|^2_{\mathbf{H}^1(\Omega)}, \tag{2.100}$$

which is fundamental for study of the problem (2.99). When $\Omega \subset \mathbb{R}^3$, $\partial\Omega \in C^2$ and $\Gamma_D \neq \emptyset$, (2.100) is also valid without the condition $\bar{\Gamma}_1 \cap \bar{\Gamma}_2 = \emptyset$ (see Lemma 2 of [57]).

As shown above, relying on the Dirichlet bilinear form

$$a(v, u) = (\nabla v, \nabla u), \tag{2.101}$$

we can get a variational formulation for the Navier-Stokes problem with Dirichlet and the outlet boundary conditions. Relying on strain bilinear form

$$a(v, u) = 2\Sigma_{i,j}(\varepsilon_{ij}(v), \varepsilon_{ij}(u)), \tag{2.102}$$

we can get a variational formulation for the problem with Dirichlet and the stress (total stress) boundary conditions, whereas relying on vorticity bilinear form

$$a(v, u) = (\operatorname{rot} v, \operatorname{rot} u) \tag{2.103}$$

we get one for the problem with Dirichlet, static pressure (total pressure) and vorticity boundary conditions.

2.4 Bibliographical Remarks

2.4.1 Fluid Equations

For Eulerian and Lagrangian descriptions of movement of particles, one can see Chap. 3 of [3] or Chap. 1 of [2]. For proof of the transport theorem, we refer to Chap. 2 of [1] or Chap. 5 of [58], and for derivation of equations using conservation laws we refer to Chap. 4 of [3], Chap. 1 of [59] or [60].

For papers studying (2.16) and (2.17), respectively, we refer to [61–64]. For papers dealing with (2.39) and (2.40), respectively, see [65–68]. For papers discussing (2.41) we refer to [69–71] and refer to [72] for the incompressible fluid. In [73] Eq. (2.19) is considered and most papers for incompressible Newtonian fluid are concerned with Eq. (2.20).

For papers investigating (2.42), we refer to [72, 74–81]. There are many papers dealing with (2.43), for example, [82–84]. For papers dealing with (2.44), we refer to [7, 85].

2.4.2 Boundary Conditions of the Navier-Stokes Equations

We refer to Chap. 4 of [3] and Chap. 2 of [59] for explanation of some boundary conditions.

The stick boundary condition was formulated by G. Stokes in 1845 and the Navier slip-with-friction boundary condition was suggested by C.-L. Navier in 1823. (cf. Introduction of [86]). In [87] the Navier slip-with-friction boundary condition has been derived rigorously from the boundary condition at the kinetic level (Boltzmann equation) for compressible fluids. For papers dealing with vorticity boundary condition see [9–11, 13, 14, 18, 88, 89].

By introducing the concept of energy-preserving boundary conditions, in [90] a rational derivation of a large class of nonstandard boundary conditions containing several different artificial boundary conditions is given.

In 1990s, the leak and slip boundary conditions with threshold have been introduced by H. Fujita (see Introduction of [25]). Physical and experimental backgrounds of the threshold slip boundary conditions are mentioned in several papers (see [8, 22–24]). For physical backgrounds of the threshold leak boundary conditions, we refer to [24, 27, 91]. For non-Newtonian fluid equations with friction slip boundary conditions, we refer to [92–95].

In [96], the Navier-Stokes problem with one-sided threshold leak boundary condition based on the total stress was considered as an example of application of a variational inequality. The one-sided threshold leak boundary conditions based on the stress for the Navier-Stokes equations were considered first in [28]. For similar one-sided boundary conditions of elasticity, we refer to [97], Sect. 5.4.1 of Chap. 3 in [98]. "Do nothing" boundary condition was introduced in [99].

2.4.3 Bilinear Forms for Hydrodynamics

For papers applying the Dirichlet bilinear form (2.101), we refer to [43, 48, 91, 100, 101].

For papers utilizing the strain bilinear form (2.102), we refer to [24–27, 35, 73, 91, 102–109]. In [73], another equivalent variational formulation, where strain, pressure, velocity and vorticity are unknown functions, also is given.

For the papers relying on the vorticity bilinear form (2.103), see [18, 30, 32, 33, 56, 57, 110–116]. In Sect. 1 of [114], the Dirichlet bilinear form (2.101) instead of the vorticity bilinear form (2.103) is used since two bilinear forms (2.101) and (2.103) for polygon or polyhedral domain under some boundary conditions are equal (see [117]). When one deals with the boundary condition for pressure or vorticity on a portion of boundary, there are other variational formulations using three unknown functions v, p and ω, where $\omega = \mathrm{rot}\, v$, (see [31, 118–120]) for the two-dimensional case and v, p and a vector potential for the three-dimensional case (see [121]).

References

1. G. Łukaszewicz, P. Kalita, *Navier-Stokes Equations An Introduction with Applications* (Springer, 2016)
2. R. Temam, A. Miranvillie, *Mathematical Modeling in Continuum Mechanics* (Cambridge University Press, 2005)
3. P.K. Kundu, I.M. Cohen, *Fluid Mechanics*, 4th edn. (Elsevier, 2008)
4. Y. Amirat, K. Hamdache, Unique solvability of equations of motion for ferrofluids. Nonlinear Anal. **73**, 471–494 (2010)
5. D. Bian, Y. Xiao, Global solution to the nematic liquid crystal flows with heat effect. J. Diff. Equ. **263**, 5298–5329 (2017)
6. L. Ma, On two-dimensional incompressible magneto-micropolar system with mixed partial viscosity. Nonlinear Anal.?: RWA **40**, 95–139 (2018)
7. Y. Kagei, M. Růžicka, G. Thater, Natural convection with dissipative heating. Commun. Math. Phys. **214**, 287–313 (2000)
8. H. Hervet, L. Léger, Flow with slip at the wall: from simple to complex fluids. C. R. Phys. **4**, 241–249 (2003)
9. H. Beirão da Veiga, F. Crispo, Sharp inviscid limit results under Navier type boundary conditions, an Lp theory. J. Math. Fluid Mech. **12**(3), 397–411 (2010)
10. H. Beirão da Veiga, F. Crispo, A missed persistence property for the Euler equations, and its effect on inviscid limits. Nonlinearity **25**(6), 1661–1669 (2012)

11. H. Bellout, J. Neustupa, P. Penel, On viscosity-continuous solutions of the Euler and Navier-Stokes equations with a Navier-type boundary condition. C. R. Math. Acad. Sci. Paris Ser. I **347**(19–20), 1141–1146 (2009)
12. L.C. Berselli, S. Spirito, On the vanishing viscosity limit of 3D Navier-Stokes equations under slip boundary conditions in general domains. Commun. Math. Phys. **316**, 171–198 (2012)
13. G.-M. Gie, J.P. Kelliher, Boundary layer analysis of the Navier-Stokes equations with generalized Navier boundary conditions. J. Diff. Equ. **253**(6), 1862–1892 (2012)
14. H. Beirão da Veiga, F. Crispo, The 3-D inviscid limit result under slip boundary conditions. A negative answer. J. Math. Fluid Mech. **14**(1), 55–59 (2012)
15. S. Monniaux, Various boundary conditions for Navier-Stokes equations in bounded Lipschitz domains. Discret. Contin. Dyn. Syst. Ser. S **6**, 1355–1369 (2013)
16. S. Monniaux, Z. Shen, Stokes problems in irregular domains with various boundary conditions, in *Handbook of Mathematical Analysis in Mechanics of Viscous Fluids*, ed. by Y. Giga, A. Novotny (Springer, 2016), https://doi.org/10.1007/978-3-319-10151-4-4
17. T. Kim, D. Cao, Some properties on the surfaces of vector fields and its application to the Stokes and Navier-Stokes problems with mixed boundary conditions. Nonlinear Anal. **113**, 94–114 (2015). Erratum, ibid **135**, 249–250 (2016)
18. C. Bernardi, F. Hecht, R. Verfürth, A finite element discretization of the three-dimensional Navier-Stokes equations with mixed boundary conditions. ESAIM: Math. Model. Numer. Anal. **43**, 1185–1201 (2009)
19. D. Iftimie, G. Planas, Inviscid limits for the Navier-Stokes equations with Navier friction boundary conditions. Nonlinearity **19**, 899–918 (2006)
20. D. Iftimie, F. Sueur, Viscous boundary layers for the Navier-Stokes equations with the Navier slip conditions. Arch. Rat. Mech. Anal. **199**, 145–175 (2011)
21. N. Masmoudi, F. Rousset, Uniform regularity for the Navier-Stokes equation with Navier boundary condition. Arch. Rat. Mech. Anal. **203**, 529–575 (2012)
22. M. Ayadi, M.K. Gdoura, T. Sassi, Mixed formulation for Stokes problem with Tresca friction. C. R. Acad. Sci. Paris Ser. I **348**, 1069–1072 (2010)
23. G. Bayada, M. Boukrouche, On a free boundary problem for the Reynolds equation derived from the Stokes system with Tresca boundary conditions. J. Math. Anal. Appl. **282**, 212–231 (2003)
24. H. Fujita, A mathematical analysis of motions of viscous incompressible fluid under leak or slip boundary conditions. RIMS Kokyuroku **888**, 199–216 (1994)
25. H. Fujita, Non-stationary Stokes flows under leak boundary conditions of friction type. J. Comput. Math. **19**, 1–8 (2001)
26. H. Fujita, Variational inequalities and nonlinear semi-groups applied to certain nonlinear problems for the Stokes equation. RIMS Kokyuroku **1234**, 70–85 (2001)
27. H. Fujita, A coherent analysis of Stokes flows under boundary conditions of friction type. J. Comput. Appl. Math. **149**, 57–69 (2002)
28. T. Kim, D. Cao, The steady Navier-Stokes and Stokes systems with mixed boundary conditions including one-sided leaks and pressure. Methods Appl. Anal. **23**, 329–364 (2016)
29. T. Kim, D. Cao, Non-stationary Navier-Stokes equations with mixed boundary conditions. J. Math. Sci. Univ. Tokyo **24**, 159–194 (2017)
30. S. Marušić, On the Navier-Stokes system with pressure boundary condition. Ann. Univ. Ferrara **53**, 319–331 (2007)
31. M. Amara, D. Capatina-Papaghiuc, D. Trujillo, Stabilized finite element method for Navier-Stokes equations with physical boundary conditions. Math. Comput. **76**(259), 1195–1217 (2007)
32. J.M. Bernard, Non-standard Stokes and Navier-Stokes problems: existence and regularity in stationary case. Math. Methods Appl. Sci. **25**, 627–661 (2002)
33. J.M. Bernard, Time-dependent Stokes and Navier-Stokes problems with boundary conditions involving pressure: existence and regularity. Nonlinear Anal.?: RWA **4**, 805–839 (2003)
34. M. Beneš, Mixed initial-boundary value problem for the three-dimensional Navier-Stokes equations in polyhedral domains. Discret. Contin. Dyn. Syst. Suppl. **02**, 135–144 (2011)

35. V. Maz'ya, J. Rossmann, Lp estimates of solutions to mixed boundary value problems for the Stokes system in polyhedral domains. Math. Nachr. **280**(7), 751–793 (2007)
36. T. Kashiwabara, On a strong solution of the non-stationary Navier-Stokes equations under slip or leak boundary conditions of friction. J. Differ. Equ. **254**, 756–778 (2013)
37. M. Beneš, Solutions to the mixed problem of viscous incompressible flows in a channel. Arch. Math. **93**, 287–297 (2009)
38. M. Beneš, Strong solutions to non-stationary channel flows of heat-conducting viscous incompressible fluids with dissipative heating. Acta Appl. Math. **116**, 237–254 (2011)
39. M. Beneš, Solutions of the Navier-Stokes equations with various types of boundary conditions. Arch. Math. **98**, 487–497 (2012)
40. M. Beneš, A note on the regularity of thermally coupled viscous flows with critical growth in nonsmooth domains. Math. Methods Appl. Sci. **36**, 1290–1300 (2013)
41. M. Beneš, A note on regularity and uniqueness of natural convection with effects of viscous dissipation in 3D open channels. Z. Angew. Math. Phys. **65**, 961–975 (2014)
42. M. Beneš, The "do nothing" problem for Boussinesq fluids. Appl. Math. Lett. **31**, 25–28 (2014)
43. M. Beneš, P. Kučera, Non-steady Navier-Stokes equations with homogeneous mixed boundary conditions and arbitrarily large initial condition. Carpathian J. Math. **23**(1–2), 32–40 (2007)
44. M. Beneš, P. Kučera, On the Navier-Stokes flows for heat-conducting fluids with mixed boundary conditions. J. Math. Anal. Appl. **389**, 769–780 (2012)
45. M. Beneš, P. Kučera, Solutions to the Navier-Stokes equations with mixed boundary conditions in two-dimensional bounded domains. Math. Nachr. **289**(2–3), 194–212 (2016)
46. S. Kračmar, J. Neustupa, A weak solvability of a steady variational inequality of the Navier-Stokes type with mixed boundary conditions. Nonlinear Anal. **47**, 4169–4180 (2001)
47. P. Kučera, The time periodic solutions of the Navier-Stokes equations with mixed boundary conditions. Discret. Contin. Dyn. Syst. Ser. S **3**(2), 325–337 (2010)
48. P. Kučera, Basic properties of solution of the non-steady Navier-Stokes equations with mixed boundary conditions in a bounded domain. Ann. Univ. Ferrara **55**, 289–308 (2009)
49. N.D. Kopachevsky, S.G. Krein, *Operator Approach to Linear Problems of Hydrodynamics 1, 2* (Springer Basel AG 2001, 2003)
50. E. Bänsch, K. Deckelnick, Optimal error estimates for the Stokes and Navier-Stokes equations with slip-boundary condition. Math. Model. Numer. Anal. **33**(5), 923–938 (1999)
51. R. Brown, I. Mitrea, M. Mitrea, M. Wright, Mixed boundary value problems for the Stokes system. Trans. Am. Math. Soc. **362**(3), 1211–1230 (2010)
52. P. Grisvard, *Elliptic Problems in Nonsmooth Domains* (SIAM, Philadelphia, 1985)
53. V. Girault, P.A. Raviart, *Finite Element Methods for Navier-Stokes Equations* (Springer, 1986)
54. R. Temam, *Navier-Stokes Equations* (North-Holland, 1985)
55. G.P. Galdi, *An Introduction to the Mathematical Theory of the Navier-Stokes Equations* (Springer, 2011)
56. C. Conca, F. Murat, O. Pironneau, The Stokes and Navier-Stokes equations with boundary conditions involving the pressure. Jpn. J. Math. **20**(2), 279–318 (1994)
57. A.A. Illarionov, A.Yu. Chebotarev, Solvability of a mixed boundary value problem for the stationary Navier-Stokes equations. Differ. Equ. **37**(5), 724–731 (2001)
58. J.N. Reddy, *An Introduction to Continuum Mechanics with Applications* (Cambridge University Press, 2008)
59. M. Köhne, *L_p-Theory for Incompressible Newtonian Flows*, (Springer Spektrum, 2013)
60. J. Málek, V. Průša, Derivation of equations for continuum mechanics and thermodynamics of fluids, in *Handbook of Mathematical Analysis in Mechanics of Viscous Fluids*, ed. by Y. Giga, A. Novotny (Springer, 2016), https://doi.org/10.1007/978-3-319-10151-4-1
61. R. Duan, H. Liu, S. Ukai, T. Yang, Optimal $L^p - L^q$ convergence rates for the compressible Navier-Stokes equations with potential force. J. Diff. Equ. **238**, 220–233 (2007)
62. Y. Kagei, K. Tsuda, Existence and stability of time periodic solution to the compressible Navier-Stokes equation for time periodic external force with symmetry. J. Diff. Equ. **258**, 399–444 (2015)

63. Z. Guo, Z. Li, Global existence of weak solution to the free boundary problem for compressible Navier-Stokes. Kinet. Relat. Model **9**, 75–103 (2016)
64. J. Qian, J. Zhao, Global existence of classical solutions for isentropic compressible Navier-Stokes equations with small initial density. J. Diff. Equ. **259**, 6830–6850 (2015)
65. H. Ma, S. Ukai, T. Yang, Time periodic solutions of compressible Navier-Stokes equations. J. Diff. Equ. **248**, 2275–2293 (2010)
66. J. Qian, H. Yin, On the stationary solutions of the full compressible Navier-Stokes equations and its stability with respect to initial disturbance. J. Diff. Equ. **237**, 225–256 (2007)
67. C. Doua, S. Jiang, Y. Ou, Low Mach number limit of full Navier-Stokes equations in a 3D bounded domain. J. Diff. Equ. **258**, 379–398 (2015)
68. S. Jiang, Y. Ou, Incompressible limit of the non-isentropic Navier-Stokes equations with well-prepared initial data in three-dimensional bounded domains. J. Math. Pures Appl. **96**, 1–28 (2011)
69. R. Danchin, Global existence in critical spaces for flows of compressible viscous and heat-conductive gases. Arch. Rat. Mech. Anal. **160**, 1–39 (2001)
70. R. Danchin, Local theory in critical spaces for compressible viscous and heat-conductive gases. Commun. Part. Differ. Equ. **26**(7 & 8), 1183–1233 (2001)
71. X. Zhao, L. Yao, Global classical solutions of the full compressible Navier-Stokes equations with cylindrical or spherical symmetry. Nonlinear Anal.?: RWA **33**, 139–167 (2017)
72. M. Bulíček, E. Feireisl, J. Málek, A Navier-Stokes-Fourier system for incompressible fluids with temperature dependent material coefficients. Nonlinear Anal.?: RWA **10**, 992–1015 (2009)
73. M. Farhloul, S. Nicaise, L. Paquet, A refined mixed finite-element method for the stationary Navier-Stokes equations with mixed boundary conditions. IMA J. Numer. Anal. **28**, 25–45 (2008)
74. L. Consiglieri, Regularity for the Navier-Stokes-Fourier system. Differ. Equ. Appl. **1**, 583–604 (2009)
75. L. Consiglieri, Existence of proper weak solutions to the Navier-Stokes-Fourier system. Funkc. Ekvacioj **55**, 347–365 (2012)
76. H. Inoue, On heat convection equations with dissipative terms in time dependent domains. Nonlinear Anal. **30**, 4441–4448 (1997)
77. H. Inoue, Strong solutions of two dimensional heat convection equations with dissipating terms. Tokyo J. Math. **22**, 445–472 (1999)
78. H. Inoue, Global and periodic solutions of heat convection equations with internal friction terms. Nonlinear Anal. **47**, 2881–2886 (2001)
79. J. Naumann, Existence of weak solutions to the equations of stationary motion of heat-conducting incompressible viscous fluids, *Progress in Nonlinear Differential Equations and Their Applications*, vol. 66 (Birkhäuser, 2005), pp. 373–390
80. J. Naumann, On the existence of weak solutions to the equations of non-stationary motion of heat-conducting incompressible viscous fluids. Math. Methods Appl. Sci. **29**, 1883–1906 (2006)
81. J. Naumann, An existence theorem for weak solutions to the equations of non-stationary motion of heat-conducting incompressible viscous fluids. J. Nonlinear Convex Anal. **7**(3), 483–497 (2006)
82. H. Morimoto, On non-stationary Boussinesq equations. Proc. Jpn. Acad. Ser. A **67** 159–161 (1991)
83. H. Morimoto, Non-stationary Boussinesq equations. J. Fac. Sci. Univ. Tokyo Sect. IA Math. **39**, 61–75 (1992)
84. H. Morimoto, Heat convection equation with nonhomogeneous boundary conditions. Funkc. Ekvacioj **53**, 213–229 (2010)
85. J. Naumann, J. Wolf, Existence of weak solutions to the equations of natural convection with dissipative heating, *Advances in Mathematical Fluid Mechanics* (Springer, 2010), pp. 367–384

86. C. Amrouche, P. Penel, N. Seloula, Some remarks on the boundary conditions in the theory of Navier-Stokes equations. Ann. Math. Blaise Pascal **20**, 37–73 (2013)
87. F. Coron, Derivation of slip boundary conditions for the Navier-Stokes system from the Boltzmann equation. J. Stat. Phys. **54**, 829–857 (1989)
88. H. Bellout, J. Neustupa, A Navier-Stokes approximation of the 3D Euler equation with the zero flux on the boundary. J. Math. Fluid Mech. **10**(4), 531–553 (2008)
89. H. Bellout, J. Neustupa, P. Penel, On the Navier-Stokes equation with boundary conditions based on vorticity. Math. Nachr. **269**(270), 59–72 (2004)
90. D. Bothe, M. Köhne, J. Prüss, On a class of energy preserving boundary conditions for incompressible Newtonian flows. SIAM J. Math. Anal. **45**, 3768–3822 (2013)
91. R. An, Y. Li, K. Li, Solvability of Navier-Stokes equations with leak boundary conditions. Acta Math. Appl. Sin. Engl. Ser. **25**, 225–234 (2009)
92. M. Boukrouche, I. Boussetouan, L. Paoli, Non-isothermal Navier-Stokes system with mixed boundary conditions and friction law: uniqueness and regularity properties. Nonlinear Anal. **102**, 168–185 (2014)
93. M. Boukrouche, I. Boussentoua, L. Paoli, Existence for non-isothermal fluid flows with Tresca's friction and Cattaneo's heat law. J. Math. Anal. Appl. **427**, 499–514 (2015)
94. M. Boukrouche, G. Łukaszewicz, On global in time dynamics of a planar Bingham flow subject to a subdifferential boundary conditions. Discret. Contin. Dyn. Syst. **34**, 3969–3983 (2014)
95. J.K. Djoko, P.A. Razafimandimby, Analysis of the Brinkman-Forchheimer equations with slip boundary conditions. Appl. Anal. **93**, 1477–1494 (2014)
96. J. Naumann, On evolution inequalities of Navier-Stokes type in three dimensions. Ann. Math. Pure Appl. **124**(4), 107–125 (1980)
97. A. Khludnev, G. Leugering, Unilateral contact problems for two perpendicular elastic structures. J. Anal. Appl. **27**, 157–177 (2008)
98. G. Duvaut, J.L. Lions, *Inequalities in Mechanics and Physics* (Springer, Berlin, 1976)
99. P.M. Gresho, Some current CFD issues relevant to the incompressible Navier-Stokes equations. Comput. Methods Appl. Mech. Eng. **87**, 201–252 (1991)
100. K. Li, R. An, On the rotating Navier-Stokes equations with mixed boundary conditions. Acta Math. Sin. **24**(4), 577–598 (2008)
101. Z. Skalák, P. Kučera, An existence theorem for the Boussinesq equations with non-Dirichlet boundary conditions. Appl. Math. **45**(2), 81–98 (2000)
102. R. An, K. Li, Variational inequality for the rotating Navier-Stokes equations with subdifferential boundary conditions. Comput. Math. Appl. **55**, 581–587 (2008)
103. R. An, K. Li, Existence of weak solution to nonhomogeneous steady Navier-Stokes equations with mixed boundary conditions (Chinese). Acta Math. Appl. Sin. **32**(4), 664–672 (2009)
104. L. Formaggia, A. Moura, F. Nobile, Coupling 3D and 1D fluid-structure interaction models for blood flow simulations. Proc. Appl. Math. Mech. **6**, 27–30 (2006)
105. L. Formaggia, A. Moura, F. Nobile, On the stability of the coupling of 3D and 1D fluid-structure interaction models for blood flow simulations. ESAIM: Math. Model. Numer. Anal. **41**(4), 743–769 (2007)
106. Z. Gao, Y. Ma, Shape gradient of the dissipated energy functional in shape optimization for the viscous incompressible flow. Appl. Numer. Math. **58**, 1720–1741 (2008)
107. P. Kučera, Z. Skalák, Local solutions to the Navier-Stokes equations with mixed boundary conditions. Acta Appl. Math. **54**, 275–288 (1998)
108. V. Maz'ya, J. Rossmann, Mixed boundary value problems for the Navier-Stokes system in polyhedral domains. Arch. Rat. Mech. Anal. **194**, 669–712 (2009)
109. J. Rossmann, Mixed boundary value problems for Stokes and Navier-Stokes systems in polyhedral domains. Oper. Theory: Adv. Appl. **193**, 269–280 (2009)
110. G.V. Alekseev, A.B. Smishliaev, Solvability of the boundary value problems for Boussinesq equations with inhomogeneous boundary conditions. J. Math. Fluid Mech. **3**(1), 18–39 (2001)
111. K. Amedodji, G. Bayada, M. Chambat, On the unsteady Navier-Stokes equations in a time-moving domain with velocity-pressure boundary conditions. Nonlinear Anal. **49**(4), 565–587 (2002)

112. C. Begue, C. Conca, F. Murat, O. Pironneau, A nouveau sur les equations de Stokes et de Navier-Stokes avec des conditions aux limites sur la pression. C. R. Acad. Sci. Paris Ser. I **304**, 23–28 (1987)
113. C. Begue, C. Conca, F. Murat, O. Pironneau, Les equations de Stokes et de Navier-Stokes avec des condition sur la pression. Nonlinear Part. Diff. Equ. Appl. College de France, Seminar IX, 179–384 (1988)
114. P. Ciarlet Jr., System de Stokes avec flux de vitesse et pression imposes. C. R. Acad. Sci. Paris Ser. I **337**, 119–124 (2003)
115. C. Conca, C. Pares, O. Pironneau, M. Thiriet, Navier-Stokes equations with imposed pressure and velocity fluxes. Int. J. Numer. Methods Fluids **20**, 267–287 (1995)
116. G. Łukaszewicz, On the Navier-Stokes equations in time dependent domains and with boundary conditions involving the pressure. J. Math. Sci. Univ. Tokyo **4**(3), 529–550 (1997)
117. M. Costabel, A coercive bilinear form for Maxwell's equations. J. Math. Anal. Appl. **157**, 527–541 (1991)
118. M. Amara, D. Capatina-Papaghiuc, E.C. Vera, D. Trujillo, A three field stabilized finite element method for the Stokes equations. C. R. Acad. Sci. Paris Ser. I **334**, 603–608 (2002)
119. M. Amara, D. Capatina-Papaghiuc, E. Chacón-Vera, D. Trujillo, Vorticity-velocity-pressure formulation for the Stokes problem. Math. Comput. **73**, 1673–1697 (2003)
120. M. Amara, D. Capatina-Papaghiuc, E. Chacón-Vera, D. Trujillo, Vorticity-velocity-pressure formulation for the Stokes equations. Comput. Vis. Sci. **6**, 47–52 (2004)
121. V. Girault, Incompressible finite element methods for Navier-Stokes equations with nonstandard boundary conditions in R^3. Math. Comput. **51**(183), 55–74 (1988)

Chapter 3
The Steady Navier-Stokes System

In this chapter, we are concerned with the steady Navier-Stokes systems with mixed boundary conditions involving Dirichlet, pressure, vorticity, stress and normal derivative of velocity together. As we have seen in Sect. 2.3.2, according to what kinds of bilinear forms for variational formulation are used, types of boundary conditions under consideration together are different. The variational formulations in Sect. 2.3.2 do not reflect, for example, the boundary conditions for stress and pressure together, but this case is important in practice. To include wider boundary conditions together, we first study the relations among strain, vorticity, normal derivative of velocity, and shape of boundary surfaces. Using the relations on the boundary surfaces, we reflect all boundary conditions into variational formulations of the problems. Then we prove the existence and uniqueness of solutions to the problems.

3.1 Properties on the Boundary Surfaces of Vector Fields

In this section, we first recall ways describing the shape of surface and study the relations among strain, vorticity, normal derivative of vector fields and shape of a surface when the vector fields near a surface are tangent or normal on the surface.

T. Kim and D. Cao, *Equations of Motion for Incompressible Viscous Fluids*,
Advances in Mathematical Fluid Mechanics,
https://doi.org/10.1007/978-3-030-78659-5_3

3.1.1 The Second Fundamental Form and Shape Operator of Surface

3.1.1.1 The Second Fundamental Form of a Surface

When $\gamma : (\alpha, \beta) \to \mathbb{R}^l$ is a parametrized curve, if its derivative $\gamma'(t)$ at every $t \in (\alpha, \beta)$ is a unit vector, then γ is said to be a *unit-speed curve*. If the *natural parameter* (length) of curve is taken as the parameter t, then the curve is a unit-speed curve.

A subset M of $\mathbb{R}^3$ is called a *surface* if, for every point $p \in M$, there is an open set $U \subset \mathbb{R}^2$ and an open set $W \subset \mathbb{R}^3$ containing p such that $M \cap W$ is homeomorphic to U. A subset of a surface M of form $M \cap W$, where W is an open subset of $\mathbb{R}^3$, is called an open subset of M. A homeomorphism $\sigma : U \to M \cap W$ as in this definition is called a *surface patch* for *parametrization* of the open subset $M \cap W$ of M.

If function $\sigma : U \to \mathbb{R}^3$ for a surface M is in class C^k (or $C^{k,1}$), then the surface is called class C^k–surface (or $C^{k,1}$-surface), and a class C^1-surface is called smooth.

If a vector is tangent to a curve in M passing through p, then it is called a *tangent vector* to the surface M at the point $p \in M$.

For a surface patch $\sigma(\xi, \zeta)$ for a surface M,

$$\sigma_\xi(\xi_0, \zeta_0) = \frac{\partial \sigma}{\partial \xi}\Big|_{(\xi_0,\zeta_0)} \quad \text{and} \quad \sigma_\zeta(\xi_0, \zeta_0) = \frac{\partial \sigma}{\partial \zeta}\Big|_{(\xi_0,\zeta_0)}$$

are tangent at the point $p(\xi_0, \zeta_0) \in M$ to the surface M. If, moreover, ξ and ζ are, respectively, the lengths of curves $\sigma(\xi, \zeta_0)$ and $\sigma(\xi_0, \zeta)$ (natural parameters), then those are tangent units at the point $p(\xi_0, \zeta_0) \in M$.

The set of all vectors tangent to M at p is called the *tangent space* T_pM of M at p. The tangent space T_pM of a surface M at a point $p \in M$ is completely determined by giving a unit vector orthogonal to it at p, called a unit normal to M at p. There are, of course, two unit normal vectors, but choosing a surface patch $\sigma : U \to \mathbb{R}^3$ containing point p leads to a definite choice, namely,

$$n_\sigma = \frac{\sigma_\xi \times \sigma_\zeta}{\|\sigma_\xi \times \sigma_\zeta\|}.$$

This is called the standard unit normal of the surface patch σ at point p. To be specific, unless otherwise stated we always take the standard unit normal.

If in a neighborhood of $(\xi_0, \zeta_0) \in U$ a surface patch $\sigma : U \to \mathbb{R}^3$ is in class C^1, then a unit normal field n of class C always exists in a neighborhood of $\sigma(\xi_0, \zeta_0) \in M$. But it may not be possible to extend n to all of M, e.g. Möbius band. If a smooth unit normal field n defined on all of M is chosen, then the surface M is said to be *oriented*.

Now we are interested in studying the shape of surfaces in $\mathbb{R}^3$. The objects describing the shape of a surface $M \subset \mathbb{R}^3$ are the second fundamental form and the shape operator (Weingarten map) of the surface.

Suppose that σ is a surface patch for $M \subset \mathbb{R}^3$ with the standard unit normal n. As the parameters (ξ, ζ) of σ change to $(\xi + \Delta\xi, \zeta + \Delta\zeta)$, the surface moves away from the tangent plane through $\sigma(\xi, \zeta)$. Making inner product in $\mathbb{R}^3$ between the vector $\sigma(\xi + \Delta\xi, \zeta + \Delta\zeta) - \sigma(\xi, \zeta)$ and the normal vector $n(\xi, \zeta)$ at point $\sigma(\xi, \zeta) \in M$, we get the deviation of σ along n from its tangent plane through $\sigma(\xi, \zeta) \in M$:

$$\big(\sigma(\xi + \Delta\xi, \zeta + \Delta\zeta) - \sigma(\xi, \zeta)\big) \cdot n(\xi, \zeta).$$

By the two variable form of Taylor's theorem,

$$\begin{aligned}&\sigma(\xi + \Delta\xi, \zeta + \Delta\zeta) - \sigma(\xi, \zeta)\\ &= \sigma_\xi \Delta\xi + \sigma_\zeta \Delta\zeta + \frac{1}{2}\big[\sigma_{\xi\xi}(\Delta\xi)^2 + 2\sigma_{\xi\zeta}\Delta\xi\Delta\zeta + \sigma_{\zeta\zeta}(\Delta\zeta)^2\big] + o\big((\Delta\xi)^2 + (\Delta\zeta)^2\big),\end{aligned}$$

where $o\big((\Delta\xi)^2 + (\Delta\zeta)^2)/((\Delta\xi)^2 + (\Delta\zeta)^2\big)$ tends to zero as $(\Delta\xi)^2 + (\Delta\zeta)^2 \to 0$.

Note that σ_ξ and σ_ζ are tangent to the surface, hence perpendicular to n, and so the deviation of σ in the direction of the unit vector n from its tangent plane is

$$\frac{1}{2}\big[L(\Delta\xi)^2 + 2K\Delta\xi\Delta\zeta + N(\Delta\zeta)^2\big] + o\big((\Delta\xi)^2 + (\Delta\zeta)^2\big), \tag{3.1}$$

where

$$L = \sigma_{\xi\xi}(\xi, \zeta) \cdot n(\xi, \zeta), \quad K = \sigma_{\xi\zeta}(\xi, \zeta) \cdot n(\xi, \zeta), \quad N = \sigma_{\zeta\zeta}(\xi, \zeta) \cdot n(\xi, \zeta). \tag{3.2}$$

At point $\sigma(\xi, \zeta) \in M$

$$\sigma_\xi(\xi, \zeta) \cdot n(\xi, \zeta) = 0, \quad \sigma_\zeta(\xi, \zeta) \cdot n(\xi, \zeta) = 0,$$

which yields by differentiation that

$$\begin{aligned}L &= \sigma_{\xi\xi}(\xi, \zeta) \cdot n(\xi, \zeta) = -\sigma_\xi(\xi, \zeta) \cdot n_\xi(\xi, \zeta),\\ K &= \sigma_{\xi\zeta}(\xi, \zeta) \cdot n(\xi, \zeta) = -\sigma_\xi(\xi, \zeta) \cdot n_\zeta(\xi, \zeta) = -\sigma_\zeta(\xi, \zeta) \cdot n_\xi(\xi, \zeta),\\ N &= \sigma_{\zeta\zeta}(\xi, \zeta) \cdot n(\xi, \zeta) = -\sigma_\zeta(\xi, \zeta) \cdot n_\zeta(\xi, \zeta).\end{aligned} \tag{3.3}$$

The expression

$$L(\Delta\xi)^2 + 2K\Delta\xi\Delta\zeta + N(\Delta\zeta)^2 \tag{3.4}$$

is called *the second fundamental form* of the surface patch σ.

Remark 3.1 If in a neighborhood of a point $p \in M$ the surface is deviated from the tangent plane to the side opposite to the unit normal, then the quadratic form (3.4) at $p \in M$ is negative.

Since all vectors $w \in T_pM$ are expressed as

$$w = \lambda\sigma_\xi + \mu\sigma_\zeta,$$

denote by

$$d\xi(w) = \lambda, \quad d\zeta(w) = \mu \quad \text{for } w \in T_pM.$$

Let us define a symmetric bilinear form on T_pM by

$$\langle v, w\rangle_{p,M} := Ld\xi(v)d\xi(w) + K\big(d\xi(v)d\zeta(w) + d\xi(w)d\zeta(v)\big) + Nd\zeta(v)d\zeta(w) \quad \forall v, w \in T_pM. \tag{3.5}$$

Let a surface patch $\tilde{\sigma}(\tilde{\xi}, \tilde{\zeta})$ be a reparametrization of a surface patch $\sigma(\xi, \zeta)$ with reparametrization map $(\xi, \zeta) = \Phi(\tilde{\xi}, \tilde{\zeta})$. Denote the Jacobian matrix of the reparametrization map $(\tilde{\xi}, \tilde{\zeta}) \to (\xi, \zeta)$

$$\begin{pmatrix} \frac{\partial \xi}{\partial \tilde{\xi}} & \frac{\partial \xi}{\partial \tilde{\zeta}} \\ \frac{\partial \zeta}{\partial \tilde{\xi}} & \frac{\partial \zeta}{\partial \tilde{\zeta}} \end{pmatrix}$$

by $J(\Phi)$. Then we can prove that

$$\begin{pmatrix} \tilde{L} & \tilde{K} \\ \tilde{K} & \tilde{N} \end{pmatrix} = \pm J(\Phi)^t \begin{pmatrix} L & K \\ K & N \end{pmatrix} J(\Phi) \tag{3.6}$$

with the plus sign if $\det(J(\Phi)) > 0$ and the minus sign if $\det(J(\Phi)) < 0$.

From (3.6), we can see that the second fundamental form of a surface patch is unchanged by a reparametrization of the patch preserving its orientation (which means $\det(J(\Phi)) > 0$) (see Exercise 6.1.4 of [1]).

3.1.1.2 The Shape Operator of Surface

Let M be a smooth surface and $f : M \to \mathbb{R}$ be a smooth function on M. This means that $f = f \circ \sigma$ is smooth for all patches $\sigma : U \to M$.

For $p \in M$ and $X \in T_pM$, the directional derivative of f at p in the direction X, denoted by $\nabla_X f$, is defined as follows. Let $\delta : (-\varepsilon, \varepsilon) \to M \subset \mathbb{R}^3$ be any smooth curve in M such that $\delta(0) = p$ and $\delta'(0) = X$. Then,

$$\nabla_X f = \frac{d}{dt} f \circ \delta(t)|_{t=0}.$$

The directional derivative is well defined, i.e. independent of the particular choice of δ, and linear.

A vector field along a surface M is a rule which assigns a vector (need not be tangent to M) to each point of M,

$$p \in M \to Y(p) \in T_p\mathbb{R}^3,$$

where $T_p\mathbb{R}^3$ means the space of 3-D vectors with the origin at p. Then when $X \in T_pM$ and $Y(p) = \big(Y_1(p), Y_2(p), Y_3(p)\big) \in T_pM$, we define an operator

$$X \in T_pM \to \nabla_X Y(p) \in T_p\mathbb{R}^3$$

by

$$X \in T_pM \to \big(\nabla_X Y_1(p), \nabla_X Y_1(p), \nabla_X Y_1(p)\big) \in T_p\mathbb{R}^3.$$

Let M be a surface, $p \in M$ and n be a smooth unit normal vector field defined along a neighborhood $W \subset M$ of p. Let us consider the operator

$$X \in T_pM \to \nabla_X n(p).$$

Since $\big(n(p), n(p)\big) = 1$, we know that

$$\nabla_X (n(p), n(p)) = 0,$$

and

$$(\nabla_X n(p), n(p)) + (n(p), \nabla_X n(p)) = 0,$$

which implies

$$2(\nabla_X n(p), n(p)) = 0.$$

Therefore,

$$\nabla_X n(p) \perp n(p) \quad \text{and} \quad \nabla_X n \in T_pM.$$

Definition 3.1 Let M be a surface, $p \in M$ and n be a smooth unit normal defined along a neighborhood $W \subset M$ of p. The map $S : X \in T_pM \to T_pM$ defined by

$$S(X) = \nabla_X n(p)$$

is called the *shape operator* of M at p (Weingarten map).

Remark 3.2 In some books, the operator $-S$ is called the shape operator of M at p.

Let us introduce a bilinear form $\mathcal{L}$ on T_pM

$$\mathcal{L}(v, w) = -\big(S(v), w\big) = -\big(\nabla_v n(p), w\big) \quad \forall v, w \in T_pM. \tag{3.7}$$

Proposition 3.1 (Proposition 7.2.2 of [1]) *Let p be a point of a surface M, let $\sigma(\xi,\zeta)$ be a surface patch of M with p in its image and let $\langle v, w\rangle_{p,M}$ be the bilinear form defined by* (3.5). *Then, for any $v, w \in T_pM$,*

$$\mathcal{L}(v, w) = \langle v, w\rangle_{p,M}, \tag{3.8}$$

where L, K, N are as in (3.3).

Proof Since both sides of (3.8) define bilinear forms on T_pM and

$$v = d\xi(v)\sigma_\xi + d\zeta(v)\sigma_\zeta,$$
$$w = d\xi(w)\sigma_\xi + d\zeta(w)\sigma_\zeta,$$

it suffices to verify that (3.8) is valid when v and w are σ_ξ or σ_ζ. Recalling that $d\xi(\sigma_\xi) = d\zeta(\sigma_\zeta) = 1$ and $d\xi(\sigma_\zeta) = d\zeta(\sigma_\xi) = 0$, we only need to show that

$$\mathcal{L}(\sigma_\xi, \sigma_\xi) = L, \quad \mathcal{L}(\sigma_\xi, \sigma_\zeta) = \mathcal{L}(\sigma_\zeta, \sigma_\xi) = K, \quad \mathcal{L}(\sigma_\zeta, \sigma_\zeta) = N. \tag{3.9}$$

Taking into account (3.3), we have

$$\mathcal{L}(\sigma_\xi, \sigma_\xi) = -(\nabla_{\sigma_\xi} n(p), \sigma_\xi) = -n_\xi \cdot \sigma_\xi = L.$$

In the same way as above, the other relations in (3.9) can be proved. □

Corollary 3.1 (Corollary 7.2.4 of [1]) *The shape operator is self-adjoint.*

In what follows let ξ, ζ be natural parameters. The curves $\sigma(\xi, \zeta_0)$ and $\sigma(\xi_0, \zeta)$ intersect at point $p(\xi_0, \zeta_0)$, and $\sigma_\xi(\xi, \zeta_0) = e_1$ and $\sigma_\zeta(\xi_0, \zeta) = e_2$ are the tangent unit vectors at point $p(\xi_0, \zeta_0)$. Thus the shape operator is expressed by the following matrix:

$$S = \begin{pmatrix} -L & -K \\ -K & -N \end{pmatrix}, \tag{3.10}$$

where

$$-L = e_1 \cdot \frac{\partial n}{\partial e_1}, \quad -K = e_2 \cdot \frac{\partial n}{\partial e_1} = e_1 \cdot \frac{\partial n}{\partial e_2}, \quad -N = e_2 \cdot \frac{\partial n}{\partial e_2}.$$

Since the shape operator S is self-adjoint, there exists an orthonormal basis $\{e_1, e_2\}$ of T_pM and real numbers λ_1, λ_2 such that

$$Se_1 = \lambda_1 e_1, \quad Se_2 = \lambda_2 e_2,$$

and the matrix for S at $p \in M$ is

$$S = \begin{pmatrix} \lambda_1 & 0 \\ 0 & \lambda_2 \end{pmatrix}.$$

The trace $tr(S)$ of matrix S is invariable with respect to orthogonal transformation. On the other hand, if $\sigma(\xi, \zeta)$ is a surface patch of M with $p = \sigma(\xi_0, \zeta_0) \in M, \sigma_\xi|_p = e_1$ and $\sigma_\zeta|_p = e_2$, then by (3.3)

$$-\sigma_{\xi\xi} \cdot n = \sigma_\xi \cdot n_\xi = \lambda_1, \quad -\sigma_{\zeta\zeta} \cdot n = \sigma_\zeta \cdot n_\zeta = \lambda_2.$$

Since $\sigma_{\xi\xi}$ and $\sigma_{\zeta\zeta}$ are, respectively, curvature vectors at the p of the curves $\sigma(\xi, \zeta_0)$ and $\sigma(\xi_0, \zeta)$, $\sigma_{\xi\xi} \cdot n$ and $\sigma_{\zeta\zeta} \cdot n$ are the normal curvatures. Therefore, for any orthogonal basis $\frac{1}{2}tr(S)(p)$ is called the *mean curvature* at $p \in M$.

Let us consider plane curves and their curvatures. If $\gamma(t)$ is a unit-speed plane curve with parameter t, then its curvature $k(t)$ at the point $\gamma(t)$ is defined to be $|\gamma''(t)|$. Let n be the normal on γ. Since $\gamma'(t) \cdot n = 0$,

$$\gamma''(t) \cdot n = -\gamma'(t) \cdot n'(t) \equiv -e \cdot \frac{dn}{de}, \tag{3.11}$$

where e is the tangent unit vector. Taking into account above, we call $e \cdot \frac{dn}{de}$ the *signed curvature* of γ.

In this book, the surfaces under consideration are pieces of boundary of 3-D or 2-D bounded connected domains, and so the surfaces are oriented and the outsides of surfaces are fixed.

Definition 3.2 If a piece of surface Γ on a neighborhood of $x \in \Gamma$ is on the opposite (same) side of the outward normal vector with respect to tangent plane (line for $l = 2$) at x or coincides with the tangent plane, then the surface is said to be *convex* (*concave*) at x. If the surface is convex (concave) at all $x \in \Gamma$, then Γ is said to be convex (concave).

Lemma 3.1 *If $\Gamma = \bigcup_j \Gamma_j$, Γ_j are convex (concave), S_p is the shape operator of Γ at $p \in \Gamma$ and $k(p)$ is twice the mean curvature at $p \in \Gamma$, then for $v \in T_p\Gamma$ the quadratic forms $(S_p v, v)_{\mathbb{R}^{l-1}}$ and $(k(p)v, v)_{\mathbb{R}^{l-1}}$ are positive (negative).*

Proof Let us consider only the case of convexity, since the case of concavity can be proved in the same way. Let $l = 3$. Since surface Γ_j is convex, then at every point of Γ_j the quadratic form

$$L(\Delta\xi)^2 + 2K\Delta\xi\Delta\zeta + N(\Delta\zeta)^2$$

is negative (cf. Remark 3.1), and so the matrix S is positive at every point of Γ_j. By the argument as above the case that $l = 2$ is proved.

Since $k(p) = tr(S)$, we have the second conclusion. □

3.1.2 Properties on the Boundary Surface of Vector Fields

Let Γ be a surface (curve for $l = 2$) of C^2 and v be a vector field of C^2 on a domain of $\mathbb{R}^l$ near Γ. Let n be the outward unit normal on Γ.

Theorem 3.1 *Suppose that $v \cdot n|_\Gamma = 0$. Then, at every point of the surface Γ the following hold:*

$$(\mathcal{E}(v)n, \tau)_{\mathbb{R}^l} = \frac{1}{2}(rot\, v \times n, \tau)_{\mathbb{R}^l} - (S\tilde{v}, \tilde{\tau})_{\mathbb{R}^{l-1}}, \tag{3.12}$$

$$(rot\, v \times n, \tau)_{\mathbb{R}^l} = \left(\frac{\partial v}{\partial n}, \tau\right)_{\mathbb{R}^l} + (S\tilde{v}, \tilde{\tau})_{\mathbb{R}^{l-1}}, \tag{3.13}$$

$$(\mathcal{E}(v)n, \tau)_{\mathbb{R}^l} = \frac{1}{2}\left(\frac{\partial v}{\partial n}, \tau\right)_{\mathbb{R}^l} - \frac{1}{2}(S\tilde{v}, \tilde{\tau})_{\mathbb{R}^{l-1}}, \tag{3.14}$$

where $\mathcal{E}(v)$ denotes the matrix with the components $\varepsilon_{ij}(v) = \frac{1}{2}(\frac{\partial v_i}{\partial x_j} + \frac{\partial v_j}{\partial x_i})$, τ is tangent vectors on Γ, S is the shape operator of the surface Γ (the matrix (3.10)) for $l = 3$ and the signed curvature of Γ for $l = 2$, and $\tilde{v}$, $\tilde{\tau}$ are expressions of the vectors v, τ in a local orthogonal curvilinear coordinates on Γ.

Proof Let $l = 3$. For vector field $v(x) = (v_1(x), v_2(x), v_3(x))$ denoted in orthogonal coordinates x, let

$$J(v) = \begin{pmatrix} \frac{\partial v_1}{\partial x_1} & \cdots & \frac{\partial v_3}{\partial x_1} \\ \cdots & \cdots & \cdots \\ \frac{\partial v_1}{\partial x_3} & \cdots & \frac{\partial v_3}{\partial x_3} \end{pmatrix}.$$

Since

$$\mathcal{E}(v)n = \frac{1}{2}\left(J(v)^T + J(v)\right)n = \frac{1}{2}\left(J(v)^T - J(v)\right)n + J(v)n, \tag{3.15}$$

$$J(v)^T - J(v) = \text{rot}\, v \times n, \tag{3.16}$$

we have

$$\mathcal{E}(v)n = \frac{1}{2}\text{rot}\, v \times n + J(v)n. \tag{3.17}$$

On the other hand, $n(x) \in C^1$ since $\Gamma \in C^2$. Let us make a vector field on a small enough neighborhood of $p \in \Gamma$ in $\mathbb{R}^3$ as follows. Consider a family of curves orthogonal to Γ none of which intersect with all the others. Assign tangent unit vector to every point of the lines. Then, the vector coincides with the unit normal $n(x)$ at $x \in \Gamma$. Denote the vector field again by $n(x)$. Since $v(x) \cdot n(x)|_\Gamma = 0$, the surface Γ is a contour of the scalar function $v \cdot n$, and so $\nabla(v \cdot n)$ is orthogonal on Γ. Thus,

$$\tau \cdot \nabla(v \cdot n)|_{\Gamma} = 0. \tag{3.18}$$

$\nabla(v \cdot n) = J(v)n + J(n)v$, and from (3.18) we get

$$\tau \cdot J(v)n = -\tau \cdot J(n)v \quad \text{on } \Gamma. \tag{3.19}$$

Since

$$\tau \cdot J(n)v = J(n)^T \tau \cdot v = \nabla_\tau n \cdot v = (S\tilde{\tau}, \tilde{v})_{\mathbb{R}^{l-1}} = (S\tilde{v}, \tilde{\tau})_{\mathbb{R}^{l-1}} \quad \text{on } \Gamma, \tag{3.20}$$

where the fact that the shape operator is self-adjoint (Corollary 3.1) was used, from (3.17), (3.19) and (3.20) we have (3.12).

Since

$$(J(v)^T - J(v))n = \operatorname{rot} v \times n, \quad J(v)^T n = \frac{\partial v}{\partial n},$$

we get

$$(\operatorname{rot} v \times n, \tau) = \left(\frac{\partial v}{\partial n}, \tau\right) - (J(v)n, \tau), \tag{3.21}$$

which, together with (3.19) and (3.20), implies (3.13). Formulas (3.12) and (3.13) imply (3.14).

When $l = 2$, putting $v = (v_1(x_1, x_2), v_2(x_1, x_2), 0)$ and considering a cylindrical surface with the line Γ as a section, we get the conclusion. □

Remark 3.3 The bilinear form $(S\tilde{v}, \tilde{u})_{\mathbb{R}^{l-1}}$ for vectors u, v tangent to the surface is independent of choice of orthogonal curvilinear coordinate system on the surface (see (3.6)).

Theorem 3.2 *On the surface Γ the following holds:*

$$(\mathcal{E}(v)n, n)_{\mathbb{R}^l} = \left(\frac{\partial v}{\partial n}, n\right)_{\mathbb{R}^l}. \tag{3.22}$$

If $v \cdot \tau|_{\Gamma} = 0$, then

$$(\mathcal{E}(v)n, n)_{\mathbb{R}^l} = \left(\frac{\partial v}{\partial n}, n\right)_{\mathbb{R}^l} = -(k(x)v, n)_{\mathbb{R}^l} + \operatorname{div} v, \tag{3.23}$$

where $k(x) = 2 \cdot$ mean curvature.

Proof Let $n(x)$ be the vector field expanded on a domain near Γ as in the proof of Theorem 3.1. Since

$$\mathcal{E}(v)n = \frac{1}{2}J(v)^T n + \frac{1}{2}J(v)n = \frac{1}{2}\frac{\partial v}{\partial n} + \frac{1}{2}J(v)n,$$

$$(J(v)n,n)_{R^l}=\sum_{j=1}^{l}\sum_{i=1}^{l}\frac{\partial v_j}{\partial x_i}n_jn_i=\left(\frac{\partial v}{\partial n},n\right)_{R^l},$$

we have (3.22).

In a neighborhood $W\subset\mathbb{R}^3$ of a point $x_0\in\Gamma$ let us introduce local curvilinear coordinates $(y_1(x_1,x_2,x_3),\ y_2(x_1,x_2,x_3),\ y_3(x_1,x_2,x_3))$ (if $l=2$ then omit y_3) such that the coordinate lines (y_1), (y_2) and (y_3) are orthogonal at all points to each other, the linear (y_3) is the outward unit normal n on Γ and the surface $y_3=0$ coincides with Γ, where $x=(x_1,x_2,x_3)$ is the original orthogonal coordinates. Denote the unit vector of $\nabla_x y_k$ by e_k, then e_1, e_2 are tangent on Γ and $e_3=n$. At $x_0\in\Gamma$ let us calculate div $v(x_0)$. Denote $v_\tau=v-(v\cdot n)n$. Then

$$\begin{aligned}\operatorname{div}v(x_0)&=\frac{\partial v}{\partial e_1}(x_0)\cdot e_1+\frac{\partial v}{\partial e_2}(x_0)\cdot e_2+\frac{\partial v}{\partial n}(x_0)\cdot n\\&=\frac{\partial v_\tau}{\partial e_1}(x_0)\cdot e_1+\frac{\partial v_n n}{\partial e_1}(x_0)\cdot e_1+\frac{\partial v_\tau}{\partial e_2}(x_0)\cdot e_2+\frac{\partial v_n n}{\partial e_2}(x_0)\cdot e_2\\&\quad+\frac{\partial v}{\partial n}(x_0)\cdot n.\end{aligned}\tag{3.24}$$

Since $v\cdot n=0$,

$$\frac{\partial v_\tau}{\partial e_1}(x_0)\cdot e_1+\frac{\partial v_\tau}{\partial e_2}(x_0)\cdot e_2=0.\tag{3.25}$$

Also,

$$\begin{aligned}\frac{\partial v_n n}{\partial e_1}(x_0)\cdot e_1+\frac{\partial v_n n}{\partial e_2}(x_0)\cdot e_2&=\frac{\partial v_n}{\partial e_1}(x_0)n\cdot e_1+v_n\frac{\partial n}{\partial e_1}(x_0)\cdot e_1\\&\quad+\frac{\partial v_n}{\partial e_2}(x_0)n\cdot e_2+v_n\frac{\partial n}{\partial e_2}(x_0)\cdot e_2\\&=\left(\frac{\partial n}{\partial e_1}\cdot e_1+\frac{\partial n}{\partial e_2}\cdot e_2\right)v_n\\&=\operatorname{tr}(S)v\cdot n=2(\text{mean curvature of }\Gamma)\,v\cdot n.\end{aligned}\tag{3.26}$$

(see Lemma 7 of [2]). Taking into account (3.25) and (3.26), we have from (3.24)

$$\left(\frac{\partial v}{\partial n},n\right)_{\mathbb{R}^l}=-(k(x)v,n)_{\mathbb{R}^l}+\operatorname{div}v\quad\text{on }\Gamma,\tag{3.27}$$

which implies (3.23). □

Remark 3.4 If $\Gamma\in C^{2,1}$, then elements of the matrix S belong to $C^1(\overline{\Gamma})$ and so does $k(x)$.

For elements in the divergence-free spaces, Dirichlet, strain and vorticity bilinear forms are not equal in usual. Below we show that for more general bilinear forms in space $\mathbf{H}^1(\Omega)$ when the bilinear forms are equal to each other.

Let Ω be a bounded domain such that $\partial\Omega \in C^{0,1}$, $\partial\Omega = \bigcup_{i=1}^{3} \overline{\Gamma}_i$, $\Gamma_i = \bigcup_j \Gamma_{ij}$, $\Gamma_{ij} \in C^2$ and

$$\mathbf{H}^1_{\Gamma_1}(\Omega) = \left\{u \in \mathbf{H}^1(\Omega) : u\mid_{\Gamma_1}= 0, u \cdot \tau \mid_{\Gamma_2}= 0, u \cdot n \mid_{\Gamma_3}= 0\right\}.$$

Corollary 3.2 *Assume that Γ_{ij} for $i = 2, 3$ are pieces of plane (straight segments for 2-D). Then,*

$$(\nabla v, \nabla u) = (rot\, v, rot\, u) + (div\, v, div\, u) = 2(\mathcal{E}(v), \mathcal{E}(u)) - (div\, v, div\, u) \qquad \forall v, u \in \mathbf{H}^1_{\Gamma_1}(\Omega). \tag{3.28}$$

Proof By density of smooth functions, it suffices to prove for $v, u \in C^2(\bar{\Omega}) \cap \mathbf{H}^1_{\Gamma_1}(\Omega)$.

By integration by parts, we get

$$-(\Delta v, u)_\Omega = (\nabla v, \nabla u) - \left(\frac{\partial v}{\partial n}, u\right)_{\Gamma_2 \cup \Gamma_3}. \tag{3.29}$$

On the other hand, using the facts that

$$-\Delta v = \text{rot rot}\, v - \text{grad}(\text{div}\, v),$$

$$(\text{rot}\, v, u) - (v, \text{rot}\, u) = -(v \times n, u)_{\partial\Omega}$$

(see (2.82)), we get

$$\begin{aligned} -(\Delta v, u)_\Omega &= (\text{rot}\, v, \text{rot}\, u)_\Omega + (\text{div}\, v, \text{div}\, u) - (\text{rot}\, v \times n, u)_{\Gamma_2 \cup \Gamma_3} - (\text{div}\, v, u \cdot n)_{\Gamma_2 \cup \Gamma_3} \\ &= (\text{rot}\, v, \text{rot}\, u)_\Omega + (\text{div}\, v, \text{div}\, u) - (\text{rot}\, v \times n, u)_{\Gamma_3} - (\text{div}\, v, u \cdot n)_{\Gamma_2}, \end{aligned} \tag{3.30}$$

where $u \cdot \tau \mid_{\Gamma_2}= 0$ and $u \cdot n \mid_{\Gamma_3}= 0$ were used. Since $S \equiv 0$ on Γ_3, by (3.13) we have

$$(\text{rot}\, v \times n, u)_{\Gamma_3} = \left(\frac{\partial v}{\partial n}, u\right)_{\Gamma_3}. \tag{3.31}$$

Since $k(x) = tr(S) \equiv 0$ on Γ_2 and $v \cdot \tau|_{\Gamma_2} = 0$, multiplying (3.27) by $(u \cdot n)$, we get

$$(\text{div}\, v, u \cdot n)_{\Gamma_2} = \left(\frac{\partial v}{\partial n}, u\right)_{\Gamma_2}. \tag{3.32}$$

From (3.29)–(3.32), the first equality of (3.28) follows.

Also, using

$$\Delta v_i = 2\text{div}\,\mathcal{E}_i(v) - \text{div}\,\frac{\partial v}{\partial x_i}, \quad \text{where } \mathcal{E}_i(v) = (\varepsilon_{i1}(v), \cdots, \varepsilon_{il}(v)),$$

$$(\Delta v_i, u_i) = -2(\mathcal{E}_i(v), \nabla u_i) + 2(\mathcal{E}_i(v)\cdot n, u_i)_{\partial\Omega} - (\text{div}\,\frac{\partial v}{\partial x_i}, u_i)$$

((2.77) and (2.78)), by the fact that the tensor $\mathcal{E}(v)$ is symmetric we get

$$\begin{aligned}-(\Delta v, u)_\Omega &= 2(\mathcal{E}(v), \mathcal{E}(u)) + (\nabla(\text{div}\,v), u) - 2(\mathcal{E}(v)n, u)_{\Gamma_2\cup\Gamma_3} \\ &= 2(\mathcal{E}(v), \mathcal{E}(u)) - (\text{div}\,v, \text{div}\,u) - 2(\mathcal{E}(v)n, u)_{\Gamma_2\cup\Gamma_3} + (\text{div}\,v, u\cdot n)_{\Gamma_2}.\end{aligned} \tag{3.33}$$

By (3.14) and (3.22), we have

$$-2(\mathcal{E}(v)n, u)_{\Gamma_3} = -\left(\frac{\partial v}{\partial n}, u\right)_{\Gamma_3} \tag{3.34}$$

and

$$-2(\mathcal{E}(v)n, u)_{\Gamma_2} = -2\left(\frac{\partial v}{\partial n}, u\right)_{\Gamma_2}. \tag{3.35}$$

From (3.32)–(3.35), we obtain

$$-(\Delta v, u)_\Omega = 2(\mathcal{E}(v), \mathcal{E}(u)) - (\text{div}\,v, \text{div}\,u) - \left(\frac{\partial v}{\partial n}, u\right)_{\Gamma_2\cup\Gamma_3}. \tag{3.36}$$

Thus, (3.36) and (3.29) imply the second equality of (3.28). □

Remark 3.5 For polygon or polyhedron, the first equality of (3.28) follows from Theorem 4.1 of [3].

3.2 Variational Formulations of the Steady Problems

Let

(i) Ω be a bounded domain of $\mathbb{R}^l$, $l = 2, 3$;
(ii) $\partial\Omega \in C^{0,1}$, $\partial\Omega = \bigcup_{i=1}^{7} \overline{\Gamma}_i$, $\Gamma_i \cap \Gamma_j = \emptyset$ for $i \neq j$;
(iii) $\Gamma_i = \bigcup_j \Gamma_{ij}$, where Γ_{ij} are connected open subsets of $\partial\Omega$ and $\Gamma_{ij} \in C^2$, $i = 2, 3, 4, 5, 7$.

We consider the Stokes equations

$$-\mu\Delta v + \nabla p = f, \quad \text{div}\,v = 0 \quad \text{in } \Omega \tag{3.37}$$

and the Navier-Stokes equations

$$-\mu\Delta v+(v\cdot\nabla)v+\nabla p=f,\quad \operatorname{div}v=0\quad \text{in }\Omega \tag{3.38}$$

with the following boundary conditions:

$$\begin{aligned}
&(1)\quad v|_{\Gamma_1}=h_1,\\
&(2)\quad v_\tau|_{\Gamma_2}=0,\ \ -p|_{\Gamma_2}=\phi_2,\\
&(3)\quad v_n|_{\Gamma_3}=0,\ \ \operatorname{rot}v\times n|_{\Gamma_3}=\phi_3/\mu,\\
&(4)\quad v_\tau|_{\Gamma_4}=h_4,\ \ (-p+2\mu\varepsilon_{nn}(v))|_{\Gamma_4}=\phi_4,\\
&(5)\quad v_n|_{\Gamma_5}=h_5,\ \ 2(\mu\varepsilon_{n\tau}(v)+\alpha v_\tau)|_{\Gamma_5}=\phi_5,\quad \alpha:\text{a matrix},\\
&(6)\quad (-pn+2\mu\varepsilon_n(v))|_{\Gamma_6}=\phi_6,\\
&(7)\quad v_\tau|_{\Gamma_7}=0,\ \Big(-p+\mu\frac{\partial v}{\partial n}\cdot n\Big)\Big|_{\Gamma_7}=\phi_7,
\end{aligned} \tag{3.39}$$

or

$$\begin{aligned}
&(1)\quad v|_{\Gamma_1}=h_1,\\
&(2)\quad v_\tau|_{\Gamma_2}=0,\ \ -(p+\frac{1}{2}|v|^2)|_{\Gamma_2}=\phi_2,\\
&(3)\quad v_n|_{\Gamma_3}=0,\ \ \operatorname{rot}v\times n|_{\Gamma_3}=\phi_3/\mu,\\
&(4)\quad v_\tau|_{\Gamma_4}=h_4,\ \ \Big(-p-\frac{1}{2}|v|^2+2\mu\varepsilon_{nn}(v)\Big)\Big|_{\Gamma_4}=\phi_4,\\
&(5)\quad v_n|_{\Gamma_5}=h_5,\ \ 2(\mu\varepsilon_{n\tau}(v)+\alpha v_\tau)|_{\Gamma_5}=\phi_5,\quad \alpha:\text{a matrix},\\
&(6)\quad \Big(-pn-\frac{1}{2}|v|^2n+2\mu\varepsilon_n(v)\Big)\Big|_{\Gamma_6}=\phi_6,\\
&(7)\quad v_\tau|_{\Gamma_7}=0,\ \Big(-p-\frac{1}{2}|v|^2+\mu\frac{\partial v}{\partial n}\cdot n\Big)\Big|_{\Gamma_7}=\phi_7,
\end{aligned} \tag{3.40}$$

where

(i) $v_n=v\cdot n,\quad v_\tau=v-(v\cdot n)n;$
(ii) $\varepsilon_n(v)=\mathcal{E}(v)\cdot n,\quad \varepsilon_{nn}(v)=(\mathcal{E}(v)\cdot n,n)_{\mathbb{R}^l},\quad \varepsilon_{n\tau}(v)=\mathcal{E}(v)\cdot n-\varepsilon_{nn}(v)n;$
(iii) h_i,ϕ_i,α_{ij} (components of matrix α) are given functions or vectors of functions.

The boundary condition (3.40) is obtained from (3.39) by replacing the static pressure p with the total pressure $p+1/2|v|^2$. Taking into account $(v\cdot\nabla)v=\operatorname{rot}v\times v+\nabla\frac{1}{2}|v|^2$ and rewriting the Navier-Stokes equation as

$$\frac{\partial v}{\partial t}-\mu\Delta v+\operatorname{rot}v\times v+\nabla(p+\frac{1}{2}|v|^2)=f,\quad \operatorname{div}v=0\quad \text{in }\Omega,$$

we are concerned with the total pressure and total stress. Then we study the Stokes equation with boundary conditions (3.39), whereas we study the Navier-Stokes equations with boundary condition (3.39) or (3.40).

For convenience in what follows, the Navier-Stokes problems with boundary conditions (3.39) and (3.40) are, respectively, called *the case of static pressure* and *the case of total pressure*.

Let

$$\mathbf{V} = \{u \in \mathbf{H}^1(\Omega) : \operatorname{div} u = 0,\ u\mid_{\Gamma_1} = 0,\ u_\tau\mid_{(\Gamma_2\cup\Gamma_4\cup\Gamma_7)} = 0,\ u_n\mid_{(\Gamma_3\cup\Gamma_5)} = 0\},$$
$$\mathbf{V}_{\Gamma 237} = \{u \in \mathbf{H}^1(\Omega) : \operatorname{div} u = 0,\ u_\tau\mid_{(\Gamma_2\cup\Gamma_7)} = 0,\ u_n\mid_{\Gamma_3} = 0\}.$$

Let us consider variational formulations based on the bilinear form $(\mathcal{E}(v), \mathcal{E}(u))$.

Taking into account (2.81) and applying Theorems 3.1 and 3.2 on Γ_i $(i = 2, 3, 7)$, for $v \in \mathbf{H}^2(\Omega) \cap \mathbf{V}_{\Gamma 237}$ and $u \in \mathbf{V}$, we have

$$\begin{aligned}
-(\Delta v, u) &= 2(\mathcal{E}(v), \mathcal{E}(u)) - 2(\mathcal{E}(v)\cdot n, u)_{\cup_{i=2}^7 \Gamma_i}\\
&= 2(\mathcal{E}(v), \mathcal{E}(u)) + 2(k(x)v, u)_{\Gamma_2} - (\operatorname{rot} v \times n, u)_{\Gamma_3}\\
&\qquad\qquad + 2(S\tilde{v}, \tilde{u})_{\Gamma_3} - 2(\varepsilon_n(v), u)_{\cup_{i=4}^7 \Gamma_i}\\
&= 2(\mathcal{E}(v), \mathcal{E}(u)) + 2(k(x)v, u)_{\Gamma_2} - (\operatorname{rot} v \times n, u)_{\Gamma_3} + 2(S\tilde{v}, \tilde{u})_{\Gamma_3}\\
&\quad - 2(\varepsilon_{nn}(v), u\cdot n)_{\Gamma_4} - 2(\varepsilon_{n\tau}(v), u)_{\Gamma_5} - 2(\varepsilon_n(v), u)_{\Gamma_6}\\
&\quad - \left(\frac{\partial v}{\partial n}, u\right)_{\Gamma_7} + (k(x)v, u)_{\Gamma_7}.
\end{aligned} \tag{3.41}$$

Also, for $p \in H^1(\Omega)$ and $u \in \mathbf{V}$ we have

$$\begin{aligned}
(\nabla p, u) &= (p, u\cdot n)_{\cup_{i=2}^7 \Gamma_i}\\
&= (p, u_n)_{\Gamma_2} + (p, u_n)_{\Gamma_4} + (pn, u)_{\Gamma_6} + (p, u_n)_{\Gamma_7},
\end{aligned} \tag{3.42}$$

where $u\cdot n\mid_{\Gamma_3\cup\Gamma_5} = 0$ was used. By (3.41) and (3.42), we have

$$\begin{aligned}
&-\mu(\Delta v, u) + (\nabla p, u)\\
&\quad = 2\mu(\mathcal{E}(v), \mathcal{E}(u)) + 2\mu(k(x)v, u)_{\Gamma_2} + 2\mu(S\tilde{v}, \tilde{u})_{\Gamma_3} + \mu(k(x)v, u)_{\Gamma_7}\\
&\qquad + (p, u\cdot n)_{\Gamma_2} - \mu(\operatorname{rot} v \times u, u)_{\Gamma_3} - \big((-p + 2\mu\varepsilon_{nn}(v)), u\cdot n\big)_{\Gamma_4}\\
&\qquad - 2\mu(\varepsilon_{n\tau}(v), u)_{\Gamma_5} - \big((-pn + 2\mu(\varepsilon_n(v)), u\big)_{\Gamma_6} - \Big(\big(-p + \mu\frac{\partial v}{\partial n}\cdot n\big), u_n\Big)_{\Gamma_7}.
\end{aligned} \tag{3.43}$$

Assume that the following holds.

Assumption 3.1 There exists a function $U \in \mathbf{H}^1(\Omega)$ such that

$$\operatorname{div} U = 0,\ U\mid_{\Gamma_1} = h_1,\ U_\tau\mid_{(\Gamma_2\cup\Gamma_7)} = 0,\ U_n\mid_{\Gamma_3} = 0,\ U_\tau\mid_{\Gamma_4} = h_4,\ U_n\mid_{\Gamma_5} = h_5.$$

Also, $f \in \mathbf{V}^*$, $\phi_i \in H^{-\frac{1}{2}}(\Gamma_i), i = 2, 4, 7$, $\phi_i \in \mathbf{H}^{-\frac{1}{2}}(\Gamma_i), i = 3, 5, 6$, $\alpha_{ij} \in L^\infty(\Gamma_5)$, and $\Gamma_1 \neq \emptyset$.

Then, in view of (3.43), we get a variational formulation for the Stokes problem (3.37), (3.39):

Problem 3.1 Find v such that

$$\begin{aligned} & v - U \in \mathbf{V}, \\ & 2\mu(\mathcal{E}(v), \mathcal{E}(u)) + 2\mu(k(x)v, u)_{\Gamma_2} + 2\mu(S\tilde{v}, \tilde{u})_{\Gamma_3} + 2(\alpha(x)v_\tau, u)_{\Gamma_5} + \mu(k(x)v, u)_{\Gamma_7} \\ & \quad = \langle f, u\rangle + \sum_{i=2,4,7} \langle \phi_i, u_n\rangle_{\Gamma_i} + \sum_{i=3,5,6} \langle \phi_i, u\rangle_{\Gamma_i} \quad \forall u \in \mathbf{V}. \end{aligned} \tag{3.44}$$

Also, we get a variational formulation for the Navier-Stokes problem of the case of static pressure:

Problem 3.2 Find v such that

$$\begin{aligned} & v - U \in \mathbf{V}, \\ & 2\mu(\mathcal{E}(v), \mathcal{E}(u)) + \langle (v\cdot\nabla)v, u\rangle + 2\mu(k(x)v, u)_{\Gamma_2} + 2\mu(S\tilde{v}, \tilde{u})_{\Gamma_3} \\ & \qquad + 2(\alpha(x)v_\tau, u)_{\Gamma_5} + \mu(k(x)v, u)_{\Gamma_7} \\ & \quad = \langle f, u\rangle + \sum_{i=2,4,7} \langle \phi_i, u_n\rangle_{\Gamma_i} + \sum_{i=3,5,6} \langle \phi_i, u\rangle_{\Gamma_i} \quad \forall u \in \mathbf{V}. \end{aligned} \tag{3.45}$$

On the other hand, taking $(v\cdot\nabla)v = \operatorname{rot} v \times v + \frac{1}{2}\operatorname{grad}|v|^2$ into account, we get a variational formulation for the Navier-Stokes problem of the case of total pressure:

Problem 3.3 Find v such that

$$\begin{aligned} & v - U \in \mathbf{V}, \\ & 2\mu(\mathcal{E}(v), \mathcal{E}(u)) + \langle \operatorname{rot} v \times v, u\rangle + 2\mu(k(x)v, u)_{\Gamma_2} + 2\mu(S\tilde{v}, \tilde{u})_{\Gamma_3} \\ & \qquad + 2(\alpha(x)v_\tau, u)_{\Gamma_5} + \mu(k(x)v, u)_{\Gamma_7} \\ & \quad = \langle f, u\rangle + \sum_{i=2,4,7} \langle \phi_i, u_n\rangle_{\Gamma_i} + \sum_{i=3,5,6} \langle \phi_i, u\rangle_{\Gamma_i} \quad \forall u \in \mathbf{V}. \end{aligned} \tag{3.46}$$

To show that the formulations above are reasonable, we can prove that if solutions v are smooth ($v \in \mathbf{H}^2(\Omega)$, $f \in \mathbf{L}^2(\Omega)$), then there exists p such that (v, p) satisfies the original problems. We will do it for more general cases including the boundary conditions of friction type in Chap. 5.

Next, let us consider variational formulations using the bilinear form $(\nabla v, \nabla u)$. Assume that $\Gamma_6 = \emptyset$ and the boundary conditions are as follows:

$$\begin{aligned}
&(1)\quad v|_{\Gamma_1} = h_1,\\
&(2)\quad v_\tau|_{\Gamma_2} = 0,\ \ -p|_{\Gamma_2} = \phi_2,\\
&(3)\quad v_n|_{\Gamma_3} = 0,\ \ \operatorname{rot} v \times n|_{\Gamma_3} = \phi_3/\mu,\\
&(4')\quad v_\tau|_{\Gamma_4} = 0,\ \ (-p + \mu\varepsilon_{nn}(v))|_{\Gamma_4} = \phi_4,\\
&(5)\quad v_n|_{\Gamma_5} = 0,\ \ 2(\mu\varepsilon_{n\tau}(v) + \alpha v_\tau)|_{\Gamma_5} = \phi_5,\quad \alpha : \text{a matrix},\\
&(7')\quad (-pn + \mu\frac{\partial v}{\partial n})|_{\Gamma_7} = \phi_7,\quad \phi_7 : \text{vector}.
\end{aligned} \tag{3.47}$$

Let

$$\mathbf{V}_{\Gamma 1-5} = \{u \in \mathbf{H}^1(\Omega) : \operatorname{div} u = 0,\ u\ |_{\Gamma_1}= 0,\ u_\tau\ |_{(\Gamma_2\cup\Gamma_4)}= 0,\ u\cdot n\ |_{(\Gamma_3\cup\Gamma_5)}= 0\},$$
$$\mathbf{V}_{\Gamma 2-5} = \{u \in \mathbf{H}^1(\Omega) : \operatorname{div} u = 0,\ u_\tau\ |_{(\Gamma_2\cup\Gamma_4)}= 0,\ u\cdot n\ |_{(\Gamma_3\cup\Gamma_5)}= 0\}.$$

Applying Theorems 3.1 and 3.2 on Γ_i $(i = 2, 3, 4, 5)$, for $v \in \mathbf{H}^2(\Omega) \cap \mathbf{V}_{\Gamma 2-5}$ and $u \in \mathbf{V}_{\Gamma 1-5}$, we have

$$\begin{aligned}
-(\Delta v, u) &= (\nabla v, \nabla u) - \left(\frac{\partial v}{\partial n}, u\right)_{\partial\Omega}\\
&= (\nabla v, \nabla u) + (k(x)v, u)_{\Gamma_2} - (\operatorname{rot} v \times n, u)_{\Gamma_3} + (S\tilde v, \tilde u)_{\Gamma_3}\\
&\quad - (\varepsilon_n(v), u)_{\Gamma_4} - (\varepsilon_n(v), u)_{\Gamma_5} + (S\tilde v, \tilde u)_{\Gamma_5} - \left(\frac{\partial v}{\partial n}, u\right)_{\Gamma_7}\\
&= (\nabla v, \nabla u) + (k(x)v, u)_{\Gamma_2} - (\operatorname{rot} v \times n, u)_{\Gamma_3} + (S\tilde v, \tilde u)_{\Gamma_3}\\
&\quad - (\varepsilon_{nn}(v), u\cdot n)_{\Gamma_4} - 2(\varepsilon_{n\tau}(v), u)_{\Gamma_5} - (S\tilde v, \tilde u)|_{\Gamma_5} - \left(\frac{\partial v}{\partial n}, u\right)_{\Gamma_7}.
\end{aligned} \tag{3.48}$$

Using (3.48) and (3.42) yields

$$\begin{aligned}
&-\mu(\Delta v, u) + (\nabla p, u)\\
&\quad = \mu(\nabla v, \nabla u) + \mu(k(x)v, u)_{\Gamma_2} + \mu(S\tilde v, \tilde u)_{\Gamma_3} + \mu(k(x)v, u)_{\Gamma_7}\\
&\qquad + (p, u\cdot n)_{\Gamma_2} - (\operatorname{rot} v \times u, u)_{\Gamma_3} - \big((-p + \mu\varepsilon_{nn}(v)), u\cdot n\big)_{\Gamma_4}\\
&\qquad - 2\mu(\varepsilon_{n\tau}(v), u)_{\Gamma_5} - \Big(\big(-pn + \mu\frac{\partial v}{\partial n}\big), u_n\Big)_{\Gamma_7}.
\end{aligned} \tag{3.49}$$

In view of (3.49), we get a variational formulation for the Stokes problem (3.37), (3.47):

Problem 3.4 Find $v \in \mathbf{V}_{\Gamma 2-5}$ such that

$$
\begin{aligned}
&v|_{\Gamma_1} = h_1, \\
&\mu(\nabla v, \nabla u) + \mu(k(x)v_\tau, u)_{\Gamma_2} + \mu(S\tilde{v}, \tilde{u})_{\Gamma_3} + 2(\alpha(x)v_\tau, u)_{\Gamma_5} - \mu(S\tilde{v}, \tilde{u})_{\Gamma_5} \\
&\quad = \langle f, u\rangle + \sum_{i=2,4,7} \langle \phi_i, u_n\rangle_{\Gamma_i} + \sum_{i=3,5} \langle \phi_i, u\rangle_{\Gamma_i} \quad \forall u \in \mathbf{V}_{\Gamma 1-5}.
\end{aligned}
\tag{3.50}
$$

For existence of solutions to the problems of (3.44), (3.45) and (3.46), coercivity of the quadratic form corresponding to the bilinear form

$$2\mu(\mathcal{E}(v), \mathcal{E}(u)) + 2\mu(k(x)v, u)_{\Gamma_2} + 2\mu(S\tilde{v}, \tilde{u})_{\Gamma_3} + \mu(k(x)v, u)_{\Gamma_7}$$

is important (see the next section), and so is coercivity of the quadratic form corresponding to

$$\mu(\nabla v, \nabla u) + \mu(k(x)v, u)_{\Gamma_2} + \mu(S\tilde{v}, \tilde{u})_{\Gamma_3} - \mu(S\tilde{v}, \tilde{u})_{\Gamma_5}$$

for problem (3.50). Thus, from the point of view of coercivity, two bilinear forms $2(\mathcal{E}(v), \mathcal{E}(u))$ and $(\nabla v, \nabla u)$ seem to be similar. However, relying on the bilinear form $(\nabla v, \nabla u)$, we cannot reflect the boundary conditions (6) in (3.39), (3.40) into variational formulations, because we know the relation between strain and normal derivative of vector fields on the boundary only when the vector fields are tangent or orthogonal to the boundary. The conditions (4′), (7′) of (3.47) are slightly different from (4), (7) of (3.39).

Remark 3.6 Since $S \equiv 0$ on the flat surface Γ, the vorticity boundary condition

$$v_n|_\Gamma = 0, \quad \operatorname{rot} v \times n|_\Gamma = \phi/\mu$$

and the Navier slip condition $v_n|_\Gamma = 0, \ \mu\varepsilon_{n\tau}(v)|_\Gamma = \phi$ are equivalent.

Remark 3.7 Condition (7') of (3.47) (with $\phi_7 = 0$) is "do nothing" condition, but (7) of (3.39) (with $\phi_7 = 0$) is rather different from "do nothing" condition and we cannot replace (7) with (7').

Let us consider why (7) of (3.39) cannot be replaced by (7') of (3.47). Above relying on the bilinear form $2(\mathcal{E}(v), \mathcal{E}(u))$ and integrating by parts $(-\mu\Delta v + \nabla p, u)$, we get boundary integral $(-2\mu(\mathcal{E}(v)n, u)_{\partial\Omega} + (p, u \cdot n)_{\partial\Omega}$. Then, in order to reflect the boundary conditions into Problems 3.1 and 3.2, using $v_\tau = 0$ or $v_n = 0$ and applying Theorems 3.1 or 3.2, we transform the boundary integrals on $\Gamma_i, i = 2, 3, 7$. (cf. (3.43)). Concretely, under condition $v_\tau|_{\Gamma_7} = 0$ we have

$$\Big(- pn + \mu\frac{\partial v}{\partial n}, u\Big)_{\Gamma_7} \quad \forall u \text{ with } u_\tau = 0. \tag{3.51}$$

Usually, $v_\tau = 0$ does not imply $\frac{\partial v}{\partial n} \cdot \tau = 0$, but by virtue of the conditions $u_\tau = 0$ and (7) of (3.39) we have

$$\left\langle -pn+\mu\frac{\partial v}{\partial n},u\right\rangle_{\Gamma_7}=\left\langle -p+\mu\frac{\partial v}{\partial n}n,u_n\right\rangle_{\Gamma_7}=\langle \phi_7,u_n\rangle_{\Gamma_7}\quad \forall u \text{ with } u_\tau=0. \tag{3.52}$$

Thus, replacing $\left\langle -p+\mu\frac{\partial v}{\partial n}\cdot n,u_n\right\rangle_{\Gamma_7}$ by $\langle \phi_7,u_n\rangle_{\Gamma_7}$, we reflect the boundary condition (7) of (3.39) into Problems 3.1 and 3.2.

Let us replace (7) of (3.39) by $(-pn+\mu\frac{\partial v}{\partial n})|_{\Gamma_7}=\phi_7$ with a vector ϕ_7. Transforming as above and replacing $\left\langle -pn+\mu\frac{\partial v}{\partial n},u\right\rangle_{\Gamma_7}$ by $\langle \phi_7,u\rangle_{\Gamma_7}$, we can come to a formal variational formulation. But when a solution v is smooth enough, going back from the formal variational formulation to the original problem, we come to

$$(-pn+\mu\frac{\partial v}{\partial n},u)_{\Gamma_7}=\langle \phi_7,u\rangle_{\Gamma_7}\quad \forall u \text{ with } u_\tau=0. \tag{3.53}$$

If we have (3.53) without $u_\tau=0$, then from (3.53) we can get

$$-pn+\mu\frac{\partial v}{\partial n}=\phi_7 \quad \text{on } \Gamma_7.$$

But due to $u_\tau=0$, we get only

$$(-pn+\mu\frac{\partial v}{\partial n},n)_{\Gamma_7}=<\phi_7,n>_{\Gamma_7}.$$

This shows that the formal variational formulation is not equivalent to the original condition on Γ_7 and equivalent to $(-p+\mu\frac{\partial v}{\partial n}n)|_{\Gamma_7}=\phi_7\cdot n$.

Remark 3.8 "Do nothing" boundary condition (7′) with $\phi_7=0$ is used in truncating large physical domains to smaller computational domains by assuming parallel flow. If the flow is parallel near the boundary, then (7) of (3.39) is the same as (7'). In the outlet the boundary condition $\mu\mathcal{E}(v)n-pn=0$ is also used. But, to our knowledge it seems not known whether this condition and "do nothing" condition are equivalent or not. As "do nothing" boundary condition, the boundary condition

$$(4')\quad v_\tau=0,\quad \mu\mathcal{E}(v)n-pn-\phi_4$$

also results from the variational formulation based on $(\nabla v,\nabla u)$. By Theorem 3.2, we know that for divergence-free flows orthogonal to the boundary, two conditions (4′) and (7′) are equivalent in variational formulations above.

3.3 Existence of Solutions to the Steady Problems

When $\mathbf{V}$ is a Banach space, the zero element of $\mathbf{V}$ is denoted by $0_{\mathbf{V}}$ and $\mathscr{O}_M(0_{\mathbf{V}})$ means M-neighborhood of $0_{\mathbf{V}}$.

For the Stokes problem we have

Theorem 3.3 *Assume that the surfaces Γ_{2j}, Γ_{3j}, Γ_{7j} are convex and α is a positive matrix. Then, under Assumption 3.1 there exists a unique solution to Problem 3.1 for the steady Stokes system with mixed boundary condition* (3.39) *for any f and ϕ_i, $i = 2, \cdots, 7$.*

Proof Having in mind Assumption 3.1 and putting $v = w + U$, we get a new problem equivalent to Problem 3.1:

Find $w \in \mathbf{V}$ such that

$$\begin{aligned}
&2\mu(\mathcal{E}(w), \mathcal{E}(u)) + 2\mu(k(x)w, u)_{\Gamma_2} + 2\mu(S\tilde{w}, \tilde{u})_{\Gamma_3} + 2(\alpha(x)w, u)_{\Gamma_5} + \mu(k(x)w, u)_{\Gamma_7} \\
&\quad = -2\mu(\mathcal{E}(U), \mathcal{E}(u)) - 2\mu(k(x)U_\tau, u)_{\Gamma_2} - 2\mu(S\tilde{U}, \tilde{u})_{\Gamma_3} - 2(\alpha(x)U_\tau, u)_{\Gamma_5} \\
&\qquad - \mu(k(x)U, u)_{\Gamma_7} + \langle f, u\rangle + \sum_{i=2,4,7} \langle \phi_i, u \cdot n\rangle_{\Gamma_i} + \sum_{i=3,5,6} \langle \phi_i, u\rangle_{\Gamma_i} \quad \forall u \in \mathbf{V}.
\end{aligned} \tag{3.54}$$

Define a linear operator $A : \mathbf{V} \to \mathbf{V}^*$ by

$$\begin{aligned}
\langle Aw, u\rangle =& 2\mu(\mathcal{E}(w), \mathcal{E}(u)) + 2\mu(k(x)w, u)_{\Gamma_2} + 2\mu(S\tilde{w}, \tilde{u})_{\Gamma_3} \\
&+ 2(\alpha(x)w, u)_{\Gamma_5} + \mu(k(x)w, u)_{\Gamma_7} \quad \forall w, u \in \mathbf{V}.
\end{aligned} \tag{3.55}$$

Then, by Remark 3.4 and Assumption 3.1, we have

$$|\langle Aw, u\rangle| \le K\|w\|_{\mathbf{V}} \cdot \|u\|_{\mathbf{V}} \quad \forall w, u \in \mathbf{V}. \tag{3.56}$$

By Korn's inequality

$$2\mu(\mathcal{E}(w), \mathcal{E}(w)) \ge \delta\|w\|^2_{\mathbf{V}} \quad \exists \delta > 0$$

(see Theorem 1.29), Assumption 3.1 and Lemma 3.1, we have

$$\langle Aw, w\rangle \ge \beta_1\|w\|^2_{\mathbf{V}} \quad \exists \beta_1 > 0. \tag{3.57}$$

Define an element $F \in \mathbf{V}^*$ by

$$\begin{aligned}
\langle F, u\rangle =& -2\mu(\mathcal{E}(U), \mathcal{E}(u)) - 2\mu(k(x)U, u)_{\Gamma_2} - 2\mu(S\tilde{U}, \tilde{u})_{\Gamma_3} - 2(\alpha(x)U_\tau, u)_{\Gamma_5} \\
&- \mu(k(x)U, u)_{\Gamma_7} + \langle f, u\rangle + \sum_{i=2,4,7} \langle \phi_i, u \cdot n\rangle_{\Gamma_i} + \sum_{i=3,5,6} \langle \phi_i, u\rangle_{\Gamma_i} \\
&\forall u \in \mathbf{V}.
\end{aligned} \tag{3.58}$$

Thus, in view of (3.56)–(3.58), by the Lax-Milgram lemma we come to our assertion. □

Remark 3.9 In (3.54), we used $(\alpha(x)w, u)_{\Gamma_5}$ instead of $(\alpha(x)w_\tau, u)_{\Gamma_5}$ since $w = w_\tau$ on Γ_5, and for simplicity from now on we will use such expressions.

Remark 3.10 Since $n(x) \in \mathbf{C}^1(\overline{\Gamma}_i)$, $u \cdot n \in H^{\frac{1}{2}}(\Gamma_i)$ (see Theorem 1.23). Also, $H^{\frac{1}{2}}(\Gamma_i) = H_0^{\frac{1}{2}}(\Gamma_i)$ (see (1.11)), and so for $\phi_i \in \mathbf{H}^{-1/2}(\Gamma_i), i = 3, 5, 6$, and $\phi_i \in H^{-1/2}(\Gamma_i), i = 2, 4, 7$, respectively, $\langle \phi_i, u \rangle_{\Gamma_i}$ and $\langle \phi_i, u_n \rangle_{\Gamma_i}$ are meaningful.

When the surfaces Γ_{2j}, Γ_{3j} and Γ_{7j} are not convex, if the absolute value of twice the mean curvature $k(x)$ and the norm of the shape operator S are small enough, then by Korn's inequality estimate (3.57) is valid, and so is the assertion.

For the Navier-Stokes problem of the case of total pressure, we have

Theorem 3.4 *Suppose that Assumption 3.1 holds and that the surfaces Γ_{2j}, Γ_{3j}, Γ_{7j} are convex, α is a positive matrix and $\|U\|_{\mathbf{L}^3(\Omega)}$ is small enough. Then, there exists a solution to Problem 3.3 for the steady Navier-Stokes system (the case of total pressure) for any f and ϕ_i, $i = 2-7$. If $U = 0$, $\|f\|_{\mathbf{V}^*}$, $\|\phi_i\|_{H^{-\frac{1}{2}}(\Gamma_i)}$, $i = 2, 4, 7$, and $\|\phi_i\|_{\mathbf{H}^{-\frac{1}{2}}(\Gamma_i)}$, $i = 3, 5, 6$, are small enough, then the solution is unique.*

Proof In the same way as in the proof of Theorem 3.3, we get a new problem equivalent to Problem 3.3:

Find $w \in \mathbf{V}$ such that

$$\begin{aligned}
&2\mu(\mathcal{E}(w), \mathcal{E}(u)) + \langle \operatorname{rot} w \times w, u\rangle + \langle \operatorname{rot} U \times w, u\rangle + \langle \operatorname{rot} w \times U, u\rangle \\
&\quad + 2\mu(k(x)w, u)_{\Gamma_2} + 2\mu(S\tilde{w}, \tilde{u})_{\Gamma_3} + 2(\alpha(x)w, u)_{\Gamma_5} + \mu(k(x)w, u)_{\Gamma_7} \\
&= -2\mu(\mathcal{E}(U), \mathcal{E}(u)) - \langle \operatorname{rot} U \times U, u\rangle - 2\mu(k(x)U, u)_{\Gamma_2} - 2\mu(S\tilde{U}, \tilde{u})_{\Gamma_3} \\
&\quad - 2(\alpha(x)U_\tau, u)_{\Gamma_5} - \mu(k(x)U, u)_{\Gamma_7} + \langle f, u\rangle + \sum_{i=2,4,7} \langle \phi_i, u_n\rangle_{\Gamma_i} + \sum_{i=3,5,6} \langle \phi_i, u\rangle_{\Gamma_i} \\
&\qquad \forall u \in \mathbf{V}.
\end{aligned} \tag{3.59}$$

Define a $a(w; v, u) : \mathbf{V} \times \mathbf{V} \times \mathbf{V} \to \mathbb{R}$ by

$$\begin{aligned}
a(w; v, u) =& 2\mu(\mathcal{E}(v), \mathcal{E}(u)) + \langle \operatorname{rot} w \times v, u\rangle + \langle \operatorname{rot} U \times v, u\rangle + \langle \operatorname{rot} w \times U, u\rangle \\
&+ 2\mu(k(x)v, u)_{\Gamma_2} + 2\mu(S\tilde{v}, \tilde{u})_{\Gamma_3} + 2(\alpha(x)v, u)_{\Gamma_5} + \mu(k(x)v, u)_{\Gamma_7} \\
&\qquad \forall w, v, u \in \mathbf{V}.
\end{aligned} \tag{3.60}$$

On the other hand, for any $w \in \mathbf{V}$ we have

$$\begin{aligned}
&\langle \operatorname{rot} w \times w, w\rangle = 0, \quad \langle \operatorname{rot} U \times w, w\rangle = 0, \\
&|\langle \operatorname{rot} w \times U, w\rangle| \le \int_\Omega |(\operatorname{rot} w \times U) \cdot w|\, dx \le \gamma \|w\|_{\mathbf{V}}^2 \cdot \|U\|_{\mathbf{L}^3}.
\end{aligned} \tag{3.61}$$

Also, by Korn's inequality,

$$2\mu(\mathcal{E}(v), \mathcal{E}(v)) \ge \delta \|v\|_{\mathbf{V}}^2. \tag{3.62}$$

Therefore, if $\delta - \gamma\|U\|_{\mathbf{L}^3(\Omega)} = \beta_2 > 0$, then by Assumption 3.1, Lemma 3.1, (3.61) and (3.62), we have

$$a(v; v, v) \geq \beta_2 \|v\|_{\mathbf{V}}^2 \quad \beta_2 > 0. \tag{3.63}$$

Define an element $F \in \mathbf{V}^*$ by

$$\begin{aligned}\langle F, u\rangle = &- 2\mu(\mathcal{E}(U), \mathcal{E}(u)) - \langle \operatorname{rot} U \times U, u\rangle - 2\mu(k(x)U, u)_{\Gamma_2} \\ &- 2\mu(S\tilde{U}, \tilde{u})_{\Gamma_3} - 2(\alpha(x)U_\tau, u)_{\Gamma_5} - \mu(k(x)U, u)_{\Gamma_7} + \langle f, u\rangle \\ &+ \sum_{i=2,4,7} \langle \phi_i, u_n\rangle_{\Gamma_i} + \sum_{i=3,5,6} \langle \phi_i, u\rangle_{\Gamma_i} \quad \forall u \in \mathbf{V}.\end{aligned} \tag{3.64}$$

Now, let us prove that

$$\text{if } w^k \rightharpoonup w \text{ in } \mathbf{V} \text{ as } k \to \infty, \text{ then } a(w^k; w^k, u) \to a(w; w, u) \quad \forall u \in \mathbf{V}. \tag{3.65}$$

First, let us prove that

$$\langle \operatorname{rot} w^k \times w^k, u\rangle \to \langle \operatorname{rot} w \times w, u\rangle \quad \forall u \in \mathbf{V} \quad \text{as } k \to \infty. \tag{3.66}$$

Indeed,

$$\begin{aligned}&\langle \operatorname{rot} w^k \times w^k, u\rangle - \langle \operatorname{rot} w \times w, u\rangle \\ &\qquad = \langle \operatorname{rot} w^k \times (w^k - w), u\rangle + \langle \operatorname{rot} (w^k - w) \times w, u\rangle.\end{aligned} \tag{3.67}$$

Let us estimate the first term on the right-hand side of (3.67).

$$\begin{aligned}\left|\langle \operatorname{rot} w^k \times (w^k - w), u\rangle\right| &\leq \gamma_1 \int_\Omega |\operatorname{rot} w^k| \cdot |(w^k - w)| \cdot |u|\, dx \\ &\leq \gamma_2 \|\operatorname{rot} w^k\|_{\mathbf{L}^2} \cdot \|(w^k - w)\|_{\mathbf{L}^3} \cdot \|u\|_{\mathbf{L}^6}.\end{aligned} \tag{3.68}$$

Since $w^k \to w$ in $\mathbf{L}^3(\Omega)$ and $\operatorname{rot} w^k$ is bounded in $\mathbf{L}^2(\Omega)$, by virtue of (3.68) the first term on the right-hand side of (3.67) converges to zero as $k \to \infty$.

By the imbedding of the space $\mathbf{V}$ into $\mathbf{L}^6(\Omega)$, we have that $w_i u_j \in L^2(\Omega)$, $i, j = 1, \cdots, l$, for any $w, v \in \mathbf{V}$. Also, since $w^k \rightharpoonup w$ in $\mathbf{V}$ as $k \to \infty$, it follows that $\operatorname{rot} w^k \rightharpoonup \operatorname{rot} w$ in $\mathbf{L}^2(\Omega)$. Then, the second term on the right-hand side of (3.67) converges to zero as $k \to \infty$. Thus, we get (3.66).

All terms except $\langle \operatorname{rot} w \times w, u\rangle$ in $a(w; w, u)$ are linear with respect to w, and so it is easy to check their convergence as $k \to \infty$.

Therefore, we have proved (3.65).

By (3.63) and (3.65), there exists a solution to (3.59) (see Theorem 1.43), and so (3.46) has a solution.

Let us now prove uniqueness. Taking into account $U = 0$ and $\langle \operatorname{rot} w \times v, v\rangle = 0$, we get from (3.60)

$$a(w; v, v) \geq \beta \|v\|_{\mathbf{V}}^2 \quad \beta > 0 \quad \forall w, v \in \mathbf{V}.$$

We can get the following estimates:

$$
\begin{aligned}
|\langle \operatorname{rot} w_1 \times u, v\rangle - \langle \operatorname{rot} w_2 \times u, v\rangle| &\le \gamma_3 \|\operatorname{rot} w_1 - \operatorname{rot} w_2\|_{\mathbf{L}^2} \cdot \|u\|_{\mathbf{L}^3} \cdot \|v\|_{\mathbf{L}^6} \\
&\le \gamma_4 \|w_1 - w_2\|_{\mathbf{V}} \cdot \|u\|_{\mathbf{V}} \cdot \|v\|_{\mathbf{V}} \quad \forall u, v, w_1, w_2 \in \mathbf{V},
\end{aligned}
$$

$$
\|F\|_{\mathbf{V}^*} \le M_1 \Big(\|f\|_{\mathbf{V}^*} + \sum_{i=2,4,7} \|\phi_i\|_{H^{-\frac{1}{2}}(\Gamma_i)} + \sum_{i=3,5,6} \|\phi_i\|_{\mathbf{H}^{-\frac{1}{2}}(\Gamma_i)} \Big),
$$

where M_1 depends on mean curvature, shape operator, μ and α.

Thus, if $\frac{\|F\|_{\mathbf{V}^*}}{\beta^2}\gamma_4 < 1$, then by virtue of Theorem 1.44 the solution is unique. □

For the Navier-Stokes problem of the case of static pressure, we have

Theorem 3.5 *Suppose that Assumption 3.1 holds and α is a positive matrix. Let $\Gamma_2 = \Gamma_4 = \Gamma_6 = \Gamma_7 = \emptyset$, the surfaces Γ_{3j} are convex and $\|U\|_{\mathbf{L}^3(\Omega)}$ is small enough. Then, there exists a solution to Problem 3.2 for the steady Navier-Stokes system for any f and ϕ_i, $i = 3, 5$. If, moreover, $\|U\|_{\mathbf{H}^1}$, $\|f\|_{\mathbf{V}^*}$, $\|\phi_i\|_{\mathbf{H}^{-\frac{1}{2}}(\Gamma_i)}$, $i = 3, 5$, are small enough, then the solution is unique.*

Proof In the same way as in the proof of Theorem 3.3, we get a new problem equivalent to Problem 3.2:

Find $w \in \mathbf{V}$ such that

$$
\begin{aligned}
&2\mu(\mathcal{E}(w), \mathcal{E}(u)) + \langle (w\cdot\nabla)w, u\rangle + \langle (U\cdot\nabla)w, u\rangle \\
&\qquad + \langle (w\cdot\nabla)U, u\rangle + 2\mu(S\tilde{w}, \tilde{u})_{\Gamma_3} + 2(\alpha(x)w, u)_{\Gamma_5} \\
&\quad = -2\mu(\mathcal{E}(U), \mathcal{E}(u)) - \langle (U\cdot\nabla)U, u\rangle - 2\mu(S\tilde{U}, \tilde{u})_{\Gamma_3} \\
&\qquad - 2(\alpha(x)U_\tau, u)_{\Gamma_5} + \langle f, u\rangle + \sum_{i=3,5} \langle \phi_i, u\rangle_{\Gamma_i} \quad \forall u \in \mathbf{V}.
\end{aligned}
\tag{3.69}
$$

Define a nonlinear operator $A : \mathbf{V} \to \mathbf{V}^*$ by

$$
\begin{aligned}
\langle Aw, u\rangle =& 2\mu(\mathcal{E}(w), \mathcal{E}(u)) + \langle (w\cdot\nabla)w, u\rangle + \langle (U\cdot\nabla)w, u\rangle \\
&+ \langle (w\cdot\nabla)U, u\rangle + 2\mu(S\tilde{w}, \tilde{u})_{\Gamma_3} + 2(\alpha(x)w, u)_{\Gamma_5} \quad \forall w, u \in \mathbf{V}.
\end{aligned}
\tag{3.70}
$$

Using the fact that $w_n = 0$, $v_n = 0$ on any parts of the boundary, we can prove that

$$
\langle (w\cdot\nabla)u, v\rangle = -\langle (w\cdot\nabla)v, u\rangle \quad \forall w, v \in \mathbf{V}, u \in \mathbf{H}^1(\Omega),
$$

and so we have

$$
\langle (w\cdot\nabla)w, w\rangle = 0, \quad \langle (w\cdot\nabla)U, w\rangle = -\langle (w\cdot\nabla)w, U\rangle.
$$

Then,

$$
|\langle (U\cdot\nabla)w, w\rangle + \langle (w\cdot\nabla)U, w\rangle| \le \gamma_1 \|w\|_{\mathbf{V}}^2 \cdot \|U\|_{\mathbf{L}^3}. \tag{3.71}
$$

If $\delta - \gamma_1 \|U\|_{\mathbf{L}^3} = \beta_3 > 0$, then by Korn's inequality (3.62), Assumption 3.1, Lemma 3.1, (3.70) and (3.71) we have

$$\langle Aw, w \rangle \geq \beta_3 \|w\|_{\mathbf{V}}^2, \ \beta_3 > 0. \tag{3.72}$$

Define $F \in \mathbf{V}^*$ by

$$\begin{aligned}\langle F, u \rangle = &- 2\mu(\mathcal{E}(U), \mathcal{E}(u)) - \langle (U \cdot \nabla)U, u \rangle - 2\mu(S\tilde{U}, \tilde{u})_{\Gamma_3} \\ &- 2(\alpha(x)U_\tau, u)_{\Gamma_5} + \langle f, u \rangle + \sum_{i=3,5} \langle \phi_i, u \rangle_{\Gamma_i} \quad \forall u \in \mathbf{V}.\end{aligned} \tag{3.73}$$

Now, by virtue of (3.72), (3.73) in the same way as in the proof of Theorem 3.4 we have the existence of a solution. Uniqueness can also be proved as in the proof of Theorem 3.4. □

For the Navier-Stokes problem of the case of static pressure, we have

Theorem 3.6 *Suppose that Assumption 3.1 holds, α is a positive matrix, the surfaces $\Gamma_{2j}, \ \Gamma_{3j}, \ \Gamma_{7j}$ are convex and $\|U\|_{\mathbf{H}^1(\Omega)}$ is small enough. Then, when f and ϕ_i, $i = 2, \cdots, 7$, are small enough, there exists a unique solution to Problem 3.2 for the steady Navier-Stokes system in a neighborhood of U in $\mathbf{H}^1(\Omega)$.*

Proof In the same way as in the proof of Theorem 3.5, we get the following new problem equivalent to Problem 3.2:

Find $w \in \mathbf{V}$ such that

$$\begin{aligned}&2\mu(\mathcal{E}(w), \mathcal{E}(u)) + \langle (w \cdot \nabla)w, u \rangle + \langle (U \cdot \nabla)w, u \rangle + \langle (w \cdot \nabla)U, u \rangle \\ &\quad + 2\mu(k(x)w, u)_{\Gamma_2} + 2\mu(S\tilde{w}, \tilde{u})_{\Gamma_3} + 2(\alpha(x)w, u)_{\Gamma_5} + \mu(k(x)w, u)_{\Gamma_7} \\ &= -2\mu(\mathcal{E}(U), \mathcal{E}(u)) - \langle (U \cdot \nabla)U, u \rangle - 2\mu(k(x)U, u)_{\Gamma_2} - 2\mu(S\tilde{U}, \tilde{u})_{\Gamma_3} \\ &\quad - 2(\alpha(x)U_\tau, u)_{\Gamma_5} - \mu(k(x)U, u)_{\Gamma_7} + \langle f, u \rangle + \sum_{i=2,4,7} \langle \phi_i, u_n \rangle_{\Gamma_i} \\ &\quad + \sum_{i=3,5,6} \langle \phi_i, u \rangle_{\Gamma_i} \quad \forall u \in \mathbf{V}.\end{aligned} \tag{3.74}$$

Define operators $A, B : \mathbf{V} \to \mathbf{V}^*$ and an element $F \in \mathbf{V}^*$, respectively, by

$$\begin{aligned}\langle Aw, u \rangle &= 2\mu(\mathcal{E}(w), \mathcal{E}(u)) + \langle (U \cdot \nabla)w, u \rangle + \langle (w \cdot \nabla)U, u \rangle + 2\mu(k(x)w, u)_{\Gamma_2} \\ &\quad + 2\mu(S\tilde{w}, \tilde{u})_{\Gamma_3} + 2(\alpha(x)w, u)_{\Gamma_5} + \mu(k(x)w, u)_{\Gamma_7} \quad \forall w, u \in \mathbf{V}, \\ \langle Bw, u \rangle &= \langle (w \cdot \nabla)w, u \rangle \quad \forall w, u \in \mathbf{V}, \\ \langle F, u \rangle &= -2\mu(\mathcal{E}(U), \mathcal{E}(u)) - \langle (U \cdot \nabla)U, u \rangle - 2\mu(k(x)U, u)_{\Gamma_2} - 2\mu(S\tilde{U}, \tilde{u})_{\Gamma_3} \\ &\quad - 2(\alpha(x)U_\tau, u)_{\Gamma_5} - \mu(k(x)U, u)_{\Gamma_7} + \langle f, u \rangle + \sum_{i=2,4,7} \langle \phi_i, u_n \rangle_{\Gamma_i} \\ &\quad + \sum_{i=3,5,6} \langle \phi_i, u \rangle_{\Gamma_i} \quad \forall u \in \mathbf{V}.\end{aligned}$$

Then, we can rewrite (3.74) as follows:

$$\langle Aw, u\rangle = \langle F, u\rangle - \langle Bw, u\rangle \quad \forall u \in \mathbf{V}. \tag{3.75}$$

We can see that

$$\begin{aligned} &|\langle Aw, u\rangle| \le m\|w\|_{\mathbf{V}} \cdot \|u\|_{\mathbf{V}}, \\ &\langle Aw, w\rangle \ge \beta_4\|w\|^2_{\mathbf{V}}, \quad \exists \beta_4 > 0, \ \forall w, u \in \mathbf{V}. \end{aligned} \tag{3.76}$$

Also

$$\begin{aligned} |\langle Bw_1 - Bw_2, u\rangle| &\le \delta M\|w_1 - w_2\|_{\mathbf{V}} \cdot \|u\|_{\mathbf{V}} \\ &\forall w_i \in \mathscr{O}_M(0_{\mathbf{V}}) \equiv \{v \in \mathbf{V} : \|v\|_{\mathbf{V}} \le M\}, \ \forall u \in \mathbf{V}, \end{aligned} \tag{3.77}$$

$$\|F\|_{\mathbf{V}^*} \le M_1\Big(\|U\|^2_{\mathbf{H}^1} + \|f\|_{\mathbf{V}^*} + \sum_{i=2,4,7} \|\phi_i\|_{H^{-\frac{1}{2}}(\Gamma_i)} + \sum_{i=3,5,6} \|\phi_i\|_{\mathbf{H}^{-\frac{1}{2}}(\Gamma_i)}\Big), \tag{3.78}$$

where M_1 depends on the mean curvature, shape operator, μ and α.

By the Lax-Milgram lemma and (3.76), for any fixed $z \in \mathscr{O}_M(0_{\mathbf{V}})$ there exists a unique solution to the problem

$$Aw = F - Bz, \tag{3.79}$$

and by (3.76) and (3.77) the solution w to (3.79) is estimated as follows:

$$\|w\|_{\mathbf{V}} \le \frac{1}{\beta_4}(\|F\|_{\mathbf{V}^*} + \|Bz\|_{\mathbf{V}^*}) \le \frac{1}{\beta_4}(\|F\|_{\mathbf{V}^*} + \delta M^2).$$

Thus, if $\|F\|_{\mathbf{V}^*}$ and M are small enough, then the map $z \to w$ is strict contract in $\mathscr{O}_M(0_{\mathbf{V}})$, and so there exists a unique solution to (3.75) in $\mathscr{O}_M(0_{\mathbf{V}})$, which shows our assertion. □

In the same way as in Theorem 3.3, for the the Stokes system with boundary condition (3.47), we have

Theorem 3.7 *Assume that $\Gamma_6 = \emptyset$, the surfaces Γ_{2j}, Γ_{3j} are convex, Γ_{5j} is concave and α is a positive matrix. Then, under Assumption 3.1 without the condition $U_\tau\,|_{\Gamma_7}= 0$ there exists a unique solution to Problem 3.4 for the stationary Stokes system with mixed boundary condition* (3.47) *for any f and ϕ_i, $i = 2, \cdots, 5, 7$.*

Remark 3.11 Assuming that the surfaces Γ_{2j}, Γ_{3j} are convex and Γ_{5j} are concave, we can obtain existence and uniqueness of solution to the Navier-Stokes problem with boundary condition (3.47) formulated similarly to Problem 3.4.

3.4 Bibliographical Remark

Let us point out that explanation for the second fundamental form of surface and shape operator comes mainly from [1], but the shape operator in this book has the sign opposite to the one in [1] and so are the signed curvature and mean curvature.

The remaining content of Sect. 3.1 is taken from Sect. 2 of [4]. Formula (3.12) is a generalization of Lemma 2.1 for 2-D in [5] and for three-dimensional case it is known in the form

$$2\,(\mathcal{E}(v)n,\tau)_{\mathbb{R}^3} = (\operatorname{rot} v \times n, \tau)_{\mathbb{R}^3} - 2v \cdot \frac{\partial n}{\partial \tau}$$

in [6, 7]. For a direct elementary proof of (3.12) not using the knowledge of differential geometry in Sect. 3.1, we refer to Theorem 2.1 and Lemma A.1 of [4]. The proof of (3.23) follows from Lemma 7 of [2].

The content of Sect. 3.2 is from Sect. 3 of [4] and the content of Sect. 3.3 is taken from Sect. 4 of [4]. Some mistakes in [4] are corrected here.

Based on the strain bilinear form, the Stokes and Navier-Stokes problems with the mixture of Dirichlet and stress boundary conditions were studied. For such a monograph, we refer to [8]. In a polyhedral domain with $\Gamma_i = \emptyset, i = 2, 3, 7,$, the Stokes problem in [9, 10] and the Navier-Stokes problem in [10, 11] were studied. In [12], the Navier-Stokes problem with $\Gamma_i = \emptyset, i = 2, 3, 4, 5, 7,$ was studied.

In [13], by the potential theory, a mixed boundary value problem for the Navier-Stokes equations in a bounded Lipschitz two-dimensional domain with $\Gamma_i = \emptyset, i = 2, 3, 4, 6, 7,$ was studied. Reducing the original problem to a boundary integral equation, [14] studied the Stokes problem in a class of Lipschitz domains with $\Gamma_i = \emptyset, i = 2, 3, 4, 7$. In [15], smoothness of a solution to the Navier-Stokes problems with $\Gamma_i = \emptyset, i = 2, 4, 5, 6, 7,$ or $\Gamma_i = \emptyset, i = 2, 3, 4, 6, 7,$ was studied.

Applying the vorticity bilinear form, someone studied the Stokes and Navier-Stokes problems with the mixture of Dirichlet, pressure and vorticity boundary conditions. The Stokes and Navier-Stokes problems with $\Gamma_i = \emptyset, i = 4, 5, 6, 7$ in [16–19] and [20], where for the Navier-Stokes problem the total pressure condition is used on Γ_2. In [21, 22], the Navier-Stokes problems with $\Gamma_i = \emptyset, i = 2, 4, 5, 6, 7,$ were studied.

Relying on the Dirichlet bilinear form, someone studied the Navier-Stokes problems with the mixture of Dirichlet, "do nothing" condition (see [23]).

Before [4] no one has considered mixed boundary problems for the Stokes and Navier-Stokes equations with both stress and pressure boundary conditions.

References

1. A. Pressley, *Elementary Differential Geometry* (Springer, 2010)
2. U. Kangro, R. Nicoaides, Divergence boundary conditions for vector Helmholtz equations with divergence constraints. Math. Model. Numer. Anal. **33**(3), 479–492 (1999)
3. M. Costabel, A coecive bilinear form for Maxwell's equations. J. Math. Anal. Appl. **157**, 527–541 (1991)
4. T. Kim, D. Cao, Some properties on the surfaces of vector fields and its application to the Stokes and Navier-Stokes problems with mixed boundary conditions. Nonlinear Anal. **113**, 94–114 (2015). Erratum, ibid **135**, 249–250 (2016)
5. T. Clopeau, A. Mikelić, R. Robert, On the vanishing viscosity limit for the 2D incompressible Navier-Stokes equations with the friction type boundary conditions. Nonlinearity **11**, 1625–1636 (1998)
6. H. Beirão da Veiga, F. Crispo, Sharp inviscid limit results under Navier type boundary conditions, An Lp theory. J. Math. Fluid Mech. **12**(3), 397–411 (2010)
7. H. Beirão da Veiga, F. Crispo, A missed persistence property for the Euler equations, and its effect on inviscid limits. Nonlinearity **25**(6), 1661–1669 (2012)
8. N.D. Kopachevsky, S.G. Krein, *Operator Approach to Linear Problems of Hydrodynamics 1, 2* (Springer Basel AG 2001, 2003)
9. V. Maz'ya, J. Rossmann, Lp estimates of solutions to mixed boundary value problems for the Stokes system in polyhedral domains. Math. Nachr. **280**(7), 751–793 (2007)
10. J. Rossmann, Mixed boundary value problems for Stokes and Navier-Stokes systems in polyhedral domains. Oper. Theory Adv. Appl. **193**, 269–280 (2009)
11. V. Maz'ya, J. Rossmann, Mixed boundary value problems for the Navier-Stokes system in polyhedral domains. Arch. Rational Mech. Anal. **194**, 669–712 (2009)
12. M. Orlt, A.-M. Sändrig, Regularity of viscous Navier-Stokes flows in nonsmooth domains, Boundary value problems and integral equations in nonsmooth domains. Lect. Note Pure Appl. Math. **167**, 185–201 (1995)
13. A. Russo, G. Starita, A mixed problem for the steady Navier-Stokes equations. Math. Comput. Model. **49**, 681–688 (2009)
14. R. Brown, I. Mitrea, M. Mitrea, M. Wright, Mixed boundary value problems for the Stokes system. Trans. Am. Math. Soc. **362**(3), 1211–1230 (2010)
15. C. Ebmeyer, J. Frehse, Steady Navier-Stokes equations with mixed boundary value conditions in three-dimensional Lipschitzian domains. Mathematische Annalen **319**, 349–381 (2001)
16. C. Begue, C. Conca, F. Murat, O. Pironneau, A nouveau sur les equations de Stokes et de Navier-Stokes avec des conditions aux limites sur la pression. C. R. Acad. Sci. Paris Ser. **I**(304), 23–28 (1987)
17. C. Begue, C. Conca, F. Murat, O. Pironneau, Les Equations de Stokes et de Navier-Stokes Avec Des Condition Sur La Pression. Nonlinear Partial Diff. Equat. and Their Appl. College de France, Seminar **IX**, 179–384 (1988)
18. C. Conca, F. Murat, O. Pironneau, The Stokes and Navier-Stokes equations with boundary conditions involving the pressure. Japan J. Math. **20**(2), 279–318 (1994)
19. C. Conca, C. Pares, O. Pironneau, M. Thiriet, Navier-Stokes equations with imposed pressure and velocity fluxes. Inter. J. Numer. Meth. Fluids **20**, 267–287 (1995)
20. A.A. Illarionov, A.Yu. Chebotarev, Solvability of a mixed boundary value problem for the stationary Navier-Stokes equations. Diff. Equat. **37**(5), 724–731 (2001)
21. C. Bernardi, F. Hecht, R. Verfürth, A finite element discretization of the three-dimensional Navier-Stokes equations with mixed boundary conditions. ESAIM: Math. Model. Numer. Anal. **43**, 1185–1201 (2009)
22. V. Girault, Incompressible finite element methods for Navier-Stokes equations with nonstandard boundary conditions in R^3. Math. Comput. **51**(183), 55–74 (1988)
23. M. Beneš, Solutions to the mixed problem of viscous incompressible flows in a channel. Arch. Math. **93**, 287–297 (2009)

Chapter 4
The Non-steady Navier-Stokes System

In this chapter we are concerned with the non-steady Navier-Stokes equations and Stokes equations with mixed boundary conditions including conditions for velocity, pressure, stress, vorticity and Navier slip condition together. As in Sect. 3.2, relying on the result in Sect. 3.1, we embed all these boundary conditions into variational formulations of problems.

For the problem with boundary conditions involving the total pressure and total stress, by a transformation of unknown functions and a penalty method we can turn the problem into an elliptic operator equation for functions defined in the time-spatial domain. In this case we need not assume that the given data are small enough.

For the problem with boundary conditions involving the static pressure and the stress, we prove the existence of a unique solution for small data under a compatibility condition at initial time. We can also prove that if a smooth solution is given, then under the compatibility condition for the small perturbed data there exists a unique solution.

Section 4.1 is devoted to the problems with boundary conditions involving the total pressure and the total stress, and Sect. 4.2 is concerned with the problems with boundary conditions involving the static pressure and the stress.

4.1 Existence of a Solution: The Case of Total Pressure

Let Ω satisfy the following conditions:

(i) Ω be a bounded domain of $\mathbb{R}^l$, $l = 2, 3$,
(ii) $\partial\Omega \in C^{0,1}$, $\partial\Omega = \bigcup_{i=1}^{7} \overline{\Gamma}_i$, $\Gamma_i \cap \Gamma_j = \emptyset$ for $i \neq j$,

T. Kim and D. Cao, *Equations of Motion for Incompressible Viscous Fluids*,
Advances in Mathematical Fluid Mechanics,
https://doi.org/10.1007/978-3-030-78659-5_4

(iii) $\Gamma_i = \bigcup_j \Gamma_{ij}$, where Γ_{ij} are connected open subsets of $\partial\Omega$ and $\Gamma_{ij} \in C^{2,1}$, $i = 2-5, 7$,

and $Q = \Omega \times (0, T)$, $\Sigma_i = \Gamma_i \times (0, T)$, $0 < T < \infty$.

4.1.1 Problem and Variational Formulation

In this section we study the Navier-Stokes problem

$$\begin{cases} \dfrac{\partial v}{\partial t} - \mu \Delta v + (v \cdot \nabla) v + \nabla p = f, \\ \operatorname{div} v = 0, \\ v(0) = v_0 \end{cases} \tag{4.1}$$

with boundary conditions including the total pressure

$$\begin{aligned}
&(1)\quad v|_{\Gamma_1} = h_1, \\
&(2)\quad v_\tau|_{\Gamma_2} = 0, \ -(p + \frac{1}{2}|v|^2)|_{\Gamma_2} = \phi_2, \\
&(3)\quad v_n|_{\Gamma_3} = 0, \ \operatorname{rot} v \times n|_{\Gamma_3} = \phi_3/\mu, \\
&(4)\quad v_\tau|_{\Gamma_4} = h_4, \ \Big(- p - \frac{1}{2}|v|^2 + 2\mu\varepsilon_{nn}(v)\Big)\Big|_{\Gamma_4} = \phi_4, \\
&(5)\quad v_n|_{\Gamma_5} = h_5, \ 2(\mu\varepsilon_{n\tau}(v) + \alpha v_\tau)|_{\Gamma_5} = \phi_5, \ \ \alpha : \text{a matrix}, \\
&(6)\quad \Big(- pn - \frac{1}{2}|v|^2 n + 2\mu\varepsilon_n(v)\Big)\Big|_{\Gamma_6} = \phi_6, \\
&(7)\quad v_\tau|_{\Gamma_7} = 0, \Big(- p - \frac{1}{2}|v|^2 + \mu\frac{\partial v}{\partial n} \cdot n\Big)\Big|_{\Gamma_7} = \phi_7,
\end{aligned} \tag{4.2}$$

where $v_n = v \cdot n$, $v_\tau = v - (v \cdot n)n$, $\varepsilon_n(v) = \mathcal{E}(v)n$, $\varepsilon_{nn}(v) = (\mathcal{E}(v)n, n)_{\mathbb{R}^l}$, $\varepsilon_{n\tau}(v) = \varepsilon(v)n - \varepsilon_{nn}(v)n$ and h_i, ϕ_i, α_{jk} (components of matrix α) are given functions or vectors of functions of x, t on Σ_i.

Define

$\mathbf{V} = \{u \in \mathbf{H}^1(\Omega) : \operatorname{div} u = 0,\ u\,|_{\Gamma_1} = 0,\ u_\tau\,|_{(\Gamma_2 \cup \Gamma_4 \cup \Gamma_7)} = 0,\ u_n\,|_{(\Gamma_3 \cup \Gamma_5)} = 0\}$,
$\mathbf{\Lambda}(Q) = \{u \in \mathbf{C}^2(\bar{Q}) : \operatorname{div} u = 0, u\,|_{\Sigma_1} = 0, u_\tau\,|_{(\Sigma_2 \cup \Sigma_4 \cup \Sigma_7)} = 0, u_n\,|_{(\Sigma_3 \cup \Sigma_5)} = 0\}$,
$\mathbf{V}(Q) = L^2(0, T; \mathbf{V})$,
$\mathbf{W}(Q) = \{$the completion of $\mathbf{\Lambda}(Q)$ in the space $\mathbf{H}^1(Q)\}$.

Assume that the following holds:

Assumption 4.1 There exists a function $U \in \mathbf{H}^1(Q) \cap \mathbf{L}^\infty(Q)$ such that

$$\operatorname{div} U = 0,\ U\mid_{\Sigma_1} = h_1,\ U_\tau\mid_{(\Sigma_2 \cup \Sigma_7)} = 0,\ U_n\mid_{\Sigma_3} = 0,\ U_\tau\mid_{\Sigma_4} = h_4,\ U_n\mid_{\Sigma_5} = h_5.$$

Also, $f \in L^2(0, T; \mathbf{V}^*)$, $\phi_i \in L^2(0, T; H^{-\frac{1}{2}}(\Gamma_i))$, $i = 2, 4, 7$ $\phi_i \in L^2(0, T; \mathbf{H}^{-\frac{1}{2}}(\Gamma_i))$, $i = 3, 5, 6$, $\alpha_{ij} \in L^\infty(0, T; L^\infty(\Gamma_5))$, $v_0 - U(x, 0) \in H$, where H is closure of $\mathbf{V}$ in $\mathbf{L}^2(\Omega)$, and $\Gamma_1 \neq \emptyset$.

Applying Theorems 3.1, 3.2 on $\Gamma_{ij}(i = 2, 3, 7)$ and taking $(v \cdot \nabla)v = \operatorname{rot} v \times v + \frac{1}{2}\operatorname{grad}|v|^2$ into account (cf. (3.46)), we get a variational formulation for problem (4.1), (4.2):

Problem 4.1 Find v such that

$$\begin{aligned} & v - U \in \mathbf{V}(Q), \\ & -\int_Q \left(v, \frac{\partial u}{\partial t}\right) dt + \int_0^T 2\mu(\mathcal{E}(v), \mathcal{E}(u))\, dt + \int_0^T \langle \operatorname{rot} v \times v, u\rangle\, dt \\ & \quad + \int_0^T 2\mu(k(x)v, u)_{\Gamma_2}\, dt + \int_0^T 2\mu(S\tilde{v}, \tilde{u})_{\Gamma_3}\, dt \\ & \quad + \int_0^T 2(\alpha(t, x)v_\tau, u)_{\Gamma_5}\, dt + \int_0^T \mu(k(x)v, u)_{\Gamma_7}\, dt \\ & = (v_0, u(0)) + \int_0^T \langle f, u\rangle\, dt + \int_0^T \sum_{i=2,4,7} \langle \phi_i, u_n\rangle_{\Gamma_i}\, dt + \int_0^T \sum_{i=3,5,6} \langle \phi_i, u\rangle_{\Gamma_i}\, dt \\ & \qquad \forall u \in \mathbf{\Lambda}(Q) \text{ with } u(x, T) = 0. \end{aligned} \tag{4.3}$$

Theorem 4.1 *Under Assumption 4.1 for any U, f and ϕ_i, $i = 2, \cdots, 7$, there exists a solution to Problem 4.1 such that*

$$\mathop{ess\,sup}_{t \in (0,T)} \|v\| \leq c.$$

For the proof of Theorem 4.1, we transform problem (4.3) into an equivalent one. Taking into account Assumption 4.1 and putting $v = z + U$, from Problem 4.1 we get an equivalent new problem:

Find $z \in \mathbf{V}(Q)$ such that

$$
\begin{aligned}
&-\int_Q\Big(z,\frac{\partial u}{\partial t}\Big)\,dt+\int_0^T 2\mu(\mathcal{E}(z),\mathcal{E}(u))\,dt+\int_0^T\langle\operatorname{rot} z\times z,u\rangle\,dt\\
&+\int_0^T(\operatorname{rot} z\times U+\operatorname{rot} U\times z,u)\;dt+\int_0^T 2\mu(k(x)z,u)_{\Gamma_2}\,dt\\
&+\int_0^T 2\mu(S\tilde z,\tilde u)_{\Gamma_3}\,dt+\int_0^T 2(\alpha(t,x)z,u)_{\Gamma_5}\,dt+\int_0^T\mu(k(x)z,u)_{\Gamma_7}\,dt\\
&=\int_0^T\Big(U,\frac{\partial u}{\partial t}\Big)\,dt-\int_0^T 2\mu(\mathcal{E}(U),\mathcal{E}(u))\,dt-\int_0^T(\operatorname{rot} U\times U,u)\,dt\\
&-\int_0^T 2\mu(k(x)U,u)_{\Gamma_2}\,dt-\int_0^T 2\mu(S\tilde U,\tilde u)_{\Gamma_3}\,dt\\
&-\int_0^T 2(\alpha(x)U_\tau,u)_{\Gamma_5}\,dt-\int_0^T\mu(k(x)U,u)_{\Gamma_7}+\int_0^T\langle f,u\rangle\,dt\\
&+\int_0^T\sum_{i=2,4,7}\langle\phi_i,u_n\rangle_{\Gamma_i}\,dt+\int_0^T\sum_{i=3,5,6}\langle\phi_i,u\rangle_{\Gamma_i}\,dt+(v_0,u(0))\\
&\qquad\qquad\forall u\in\mathbf{\Lambda}(Q)\text{ with }u(x,T)=0.
\end{aligned}\tag{4.4}
$$

In (4.4) let us make again a change of the unknown function by $w=e^{k_1t}z$ where k_1 is a constant to be determined in Lemma 4.1 later. Then we have

$$
\begin{aligned}
&-\int_Q z\frac{\partial u}{\partial t}\,dxdt=-\int_Q e^{-k_1t}w\frac{\partial u}{\partial t}\,dxdt=-\int_Q w\frac{\partial\hat u}{\partial t}\,dxdt-k_1\int_Q w\hat u\,dxdt,\\
&\int_Q U\frac{\partial u}{\partial t}\,dxdt=\int_Q Ue^{k_1t}e^{-k_1t}\frac{\partial u}{\partial t}\,dxdt=\int_Q Ue^{k_1t}\Big(\frac{\partial\hat u}{\partial t}+k_1\hat u\Big)\,dxdt,
\end{aligned}
$$

where $\hat u=e^{-k_1t}u$.

Substituting these into (4.4), we see that the problem to find a solution to (4.3) is equivalent to the following problem.

Find $w\in\mathbf{V}(Q)$ such that

$$
\begin{aligned}
&-\int_0^T \Big(w, \frac{\partial \hat{u}}{\partial t}\Big)\, dt + 2\mu \int_0^T (\mathcal{E}(w), \mathcal{E}(\hat{u}))\, dt + \int_0^T e^{-k_1 t} (\operatorname{rot} w \times w, \hat{u})\, dt \\
&+ \int_0^T \big(\operatorname{rot} w \times U + \operatorname{rot} U \times w, \hat{u}\big)\, dt + \int_0^T 2\mu (k(x) w, u)_{\Gamma_2}\, dt + \int_0^T 2\mu (S\tilde{w}, \tilde{\hat{u}})_{\Gamma_3}\, dt \\
&+ \int_0^T 2(\alpha(t,x) w, \hat{u})_{\Gamma_5}\, dt + \int_0^T \mu (k(x) w, \hat{u})_{\Gamma_7}\, dt - k_1 \int_0^T (w, \hat{u})\, dt \\
&= \int_Q \bar{U}\Big(\frac{\partial \hat{u}}{\partial t} + k_1 \hat{u}\Big)\, dx dt - 2\mu \int_0^T (\mathcal{E}(\bar{U}), \mathcal{E}(\hat{u}))\, dt - \int_0^T (\operatorname{rot} \bar{U} \times U, \hat{u})\, dt \\
&- \int_0^T 2\mu (k(x)\bar{U}, \hat{u})_{\Gamma_2}\, dt - \int_0^T 2\mu (S\tilde{\bar{U}}, \tilde{\hat{u}})_{\Gamma_3}\, dt \\
&- \int_0^T 2(\alpha(t,x)\bar{U}_\tau, \hat{u})_{\Gamma_5}\, dt - \int_0^T \mu (k(x) U, \hat{u})_{\Gamma_7} + \int_0^T \langle \bar{f}, \hat{u}\rangle\, dt \\
&+ \int_0^T \sum_{i=2,4,7} \langle \bar{\phi}_i, \hat{u}_n\rangle_{\Gamma_i}\, dt + \int_0^T \sum_{i=3,5,6} \langle \bar{\phi}_i, \hat{u}\rangle_{\Gamma_i}\, dt + (v_0, u(x,0)) \\
&\qquad\qquad \forall \hat{u} \in \mathbf{\Lambda}(Q), \text{ with } \hat{u}(x,T) = 0,
\end{aligned}
\tag{4.5}
$$

where $\bar{\gamma} = e^{k_1 t}\gamma$ for any γ.

Therefore, for the proof of Theorem 4.1 it suffices to prove the existence of a solution w to problem (4.5) and

$$
\operatorname*{ess\,sup}_{t \in (0,T)} \|w\| \le c. \tag{4.6}
$$

4.1.2 An Auxiliary Problem by Elliptic Regularization

First we will consider the following auxiliary problem by elliptic regularization.

For any positive integer m, we find the functions $w^m \in \mathbf{W}(Q)$ satisfying the following:

$$
\begin{aligned}
&\int_Q \left(\frac{1}{m}\frac{\partial w^m}{\partial t} - w^m\right)\cdot\frac{\partial u}{\partial t}\,dxdt + 2\mu\int_0^T (\mathcal{E}(w^m),\mathcal{E}(u))\,dt \\
&\quad + \int_0^T e^{-k_1 t}(\operatorname{rot} w^m \times w^m, u)\,dt + \int_0^T \left(\operatorname{rot} w^m \times U + \operatorname{rot} U \times w^m, u\right)\,dt \\
&\quad + \int_0^T 2\mu(k(x)w^m, u)_{\Gamma_2}\,dt + \int_0^T 2\mu(S\tilde{w}^m, \tilde{u})_{\Gamma_3}\,dt + \int_0^T 2(\alpha(t,x)w^m, u)_{\Gamma_5}\,dt \\
&\quad + \int_0^T \mu(k(x)w^m, u)_{\Gamma_7}\,dt - k_1\int_0^T (w^m, u)\,dt + (w^m(T), u(T)) \\
&= \int_Q \bar{U}\left(\frac{\partial u}{\partial t} + k_1 u\right)dxdt - 2\mu\int_0^T (\mathcal{E}(\bar{U}),\mathcal{E}(u))\,dt - \int_0^T (\operatorname{rot}\bar{U}\times U, u)_{\Omega(t)}\,dt \\
&\quad - \int_0^T 2\mu(k(x)\bar{U}, u)_{\Gamma_2}\,dt - \int_0^T 2\mu(S\tilde{\bar{U}}, \tilde{u})_{\Gamma_3}\,dt - \int_0^T 2(\alpha(t,x)\bar{U}_\tau, u)_{\Gamma_5}\,dt \\
&\quad - \int_0^T \mu(k(x)\bar{U}, u)_{\Gamma_7}\,dt + \int_0^T \langle \bar{f}, u\rangle\,dt + \int_0^T \sum_{i=2,4,7}\langle\bar{\phi}_i, u_n\rangle_{\Gamma_i}\,dt \\
&\quad + \int_0^T \sum_{i=3,5,6}\langle\bar{\phi}_i, u\rangle_{\Gamma_i}\,dt + (v_0, u(x,0)) \\
&\qquad\qquad\qquad\qquad \forall u \in \mathbf{W}(Q).
\end{aligned} \tag{4.7}
$$

For (4.7) we have the following result on the existence and uniqueness of the solution:

Lemma 4.1 *Under Assumptions 4.1, there exists a constant k_1 independent of m such that there exists a unique solution to problem* (4.7).

Proof Define an operator A_m from $\mathbf{W}(Q)$ into its dual space by

$$
\begin{aligned}
\langle A_m z, u\rangle &= \int_0^T \left(\frac{1}{m}\frac{\partial z}{\partial t}, \frac{\partial u}{\partial t}\right)dt - \int_0^T \left(z, \frac{\partial u}{\partial t}\right)dt \\
&\quad + 2\mu\int_0^T (\mathcal{E}(z),\mathcal{E}(u))\,dt + \int_0^T e^{-k_1 t}(\operatorname{rot} z \times z, u)\,dt \\
&\quad + \int_0^T (\operatorname{rot} z \times U + \operatorname{rot} U \times z, u)\,dt + \int_0^T 2\mu(k(x)z, u)_{\Gamma_2}\,dt \\
&\quad + \int_0^T 2\mu(S\tilde{z}, \tilde{u})_{\Gamma_3}\,dt + \int_0^T 2(\alpha(t,x)z, u)_{\Gamma_5}\,dt \\
&\quad + \int_0^T \mu(k(x)z, u)_{\Gamma_7}\,dt - k_1\int_0^T (z, u)\,dt + (z(T), u(T)) \\
&\qquad\qquad\qquad \forall z, u \in \mathbf{W}(Q).
\end{aligned} \tag{4.8}
$$

And also define an element $F \in \mathbf{W}(Q)^*$ by

$$\begin{aligned}\langle F, u\rangle = &\int_Q \bar{U}\Big(\frac{\partial u}{\partial t} + k_1 u\Big)\,dxdt - 2\mu\int_0^T (\mathcal{E}(\bar{U}), \mathcal{E}(u))\,dt \\ &- \int_0^T (\operatorname{rot}\bar{U}\times U, u)\,dt - \int_0^T 2\mu(k(x)\bar{U}, u)_{\Gamma_2}\,dt - \int_0^T 2\mu(S\tilde{\bar{U}}, \tilde{u})_{\Gamma_3}\,dt \\ &- \int_0^T 2(\alpha(t,x)\bar{U}_\tau, u)_{\Gamma_5}\,dt - \int_0^T \mu(k(x)\bar{U}, u)_{\Gamma_7}\,dt + \int_0^T \langle \bar{f}, u\rangle\,dt \\ &+ \int_0^T \sum_{i=2,4,7}\langle\bar{\phi}_i, u_n\rangle_{\Gamma_i}\,dt + \int_0^T \sum_{i=3,5,6}\langle\bar{\phi}_i, u\rangle_{\Gamma_i}\,dt + (v_0, u(x,0)) \\ &\qquad\qquad\qquad \forall u\in \mathbf{W}(Q).\end{aligned} \tag{4.9}$$

Now, let us consider the existence of a solution to the following problem:

$$A_m w^m = F, \tag{4.10}$$

which is equivalent to the existence of a solution to the auxiliary problem (4.7).

For all $z \in \mathbf{W}(Q)$ we have

$$\begin{aligned}&\langle A_m z, z\rangle = \\ &\int_0^T \Big(\frac{1}{m}\frac{\partial z}{\partial t}, \frac{\partial z}{\partial t}\Big)\,dt - \int_0^T \Big(z, \frac{\partial z}{\partial t}\Big)\,dt + 2\mu\int_0^T (\mathcal{E}(z), \mathcal{E}(z))\,dt \\ &+ \int_0^T (\operatorname{rot} z\times U, z)\,dt + \int_0^T 2\mu(k(x)z, z)_{\Gamma_2}\,dt + \int_0^T 2\mu(S\tilde{z}, \tilde{z})_{\Gamma_3}\,dt \\ &+ \int_0^T 2(\alpha(t,x)z, z)_{\Gamma_5}\,dt + \int_0^T \mu(k(x)z, z)_{\Gamma_7}\,dt - k_1\int_0^T \|z(t)\|^2\,dt + \|z(T)\|^2,\end{aligned} \tag{4.11}$$

where $(\operatorname{rot} z \times z, z) = 0$ and $(\operatorname{rot} U \times z, z) = 0$ were used.

Integrating by parts yields

$$-\int_Q z\frac{\partial z}{\partial t}\,dxdt = \frac{1}{2}\left(\|z(0)\|^2 - \|z(T)\|^2\right) \quad \text{for } z\in\mathbf{W}(Q). \tag{4.12}$$

By (4.11) and (4.12), for all $z \in \mathbf{W}(Q)$ we have

$$\begin{aligned}&\langle A_m z, z\rangle = \\ &\int_0^T \frac{1}{m}\Big\|\frac{\partial z}{\partial t}\Big\|^2\,dt + 2\mu\int_0^T \Sigma_{ij}\|\varepsilon_{ij}(z)\|^2\,dt + \int_0^T (\operatorname{rot} z\times U, z)\,dt \\ &+ \int_0^T 2\mu(k(x)z, z)_{\Gamma_2}\,dt + \int_0^T 2\mu(S\tilde{z}, \tilde{z})_{\Gamma_3}\,dt + \int_0^T 2(\alpha(t,x)z, z)_{\Gamma_5}\,dt \\ &+ \int_0^T \mu(k(x)z, z)_{\Gamma_7}\,dt - k_1\int_0^T \|z(t)\|^2\,dt + \frac{1}{2}\left(\|z(0)\|^2 + \|z(T)\|^2\right).\end{aligned} \tag{4.13}$$

By Korn's inequality,

$$2\int_0^T \Sigma_{ij}\|\varepsilon_{ij}(z)\|^2\,dt \geq c_1\int_0^T \|z\|^2_{\mathbf{H}^1(\Omega)}\,dt, \qquad c_1>0. \tag{4.14}$$

By Young's inequality and Assumption 4.1,

$$\left|\int_0^T (\operatorname{rot} z\times U, z)\,dt\right| \leq \frac{\mu c_1}{2}\int_0^T \|z\|^2_{\mathbf{H}^1(\Omega)}\,dt + c_2\int_0^T \|z\|^2\,dt. \tag{4.15}$$

On the other hand, by virtue of Remark 3.4 and Assumption 4.1 there exists a constant M such that

$$\|S(x)\|_\infty,\ \|k(x)\|_\infty,\ \|\alpha\|_{L^\infty(0,T;L^\infty(\Gamma_5))} \leq M.$$

Therefore, there exists a constant c_3 such that

$$\begin{aligned}\Bigg|\int_0^T 2(\alpha(t,x)z,z)_{\Gamma_5}\,dt &+ \int_0^T 2\mu(k(x)z,z)_{\Gamma_2}\,dt + \int_0^T 2\mu(S\tilde{z},\tilde{z})_{\Gamma_3}\,dt\\ &+\int_0^T \mu(k(x)z,z)_{\Gamma_7}\,dt\Bigg| \leq \int_0^T \left(\frac{\mu c_1}{4}\|z\|^2_{\mathbf{H}^1(\Omega)} + c_3\|z\|^2\right)dt \quad \forall z\in \mathbf{W}(Q)\end{aligned} \tag{4.16}$$

(see Theorem 1.27).

Taking $-k_1$ large enough in (4.13) independently of m, from (4.13)–(4.16) we have, therefore

$$\langle A_m z, z\rangle \geq c_4\left(\|z\|^2_{\mathbf{W}(Q)}\right), \ \exists c_4>0,\ \forall z\in\mathbf{W}(Q), \tag{4.17}$$

where c_4 depends on m.

Now, let us prove that if $z_k \rightharpoonup z$ in $\mathbf{W}(Q)$ as $k\to\infty$, then

$$\langle A_m z_k, u\rangle \to \langle A_m z, u\rangle \quad \forall u\in\mathbf{W}(Q). \tag{4.18}$$

First, let us prove that

$$\int_0^T e^{-k_1 t}(\operatorname{rot} z_k\times z_k, u)\,dt \to \int_0^T e^{-k_1 t}(\operatorname{rot} z\times z, u)\,dt \quad \forall u\in\mathbf{W}(Q) \quad \text{as } k\to\infty. \tag{4.19}$$

Indeed, straightforward calculation gives

$$\int_0^T e^{-k_1 t}\,(\operatorname{rot} z_k \times z_k, u)\,dt - \int_0^T e^{-k_1 t}\,(\operatorname{rot} z \times z, u)\,dt =$$
$$\int_0^T e^{-k_1 t}\,(\operatorname{rot} z_k \times (z_k - z), u)\,dt + \int_0^T e^{-k_1 t}\,(\operatorname{rot}(z_k - z) \times z), u)\,dt. \tag{4.20}$$

By the embedding of $H^1(Q)$ in $L^4(Q)$ we have $e^{-k_1 t} z u \in \mathbf{L}^2(Q)$ and $\operatorname{rot}(z_k - z) \rightharpoonup 0$ in $\mathbf{L}^2(Q)$ as $z_k \rightharpoonup z$ in $\mathbf{W}(Q)$. Thus, the second integral on the right-hand side of (4.20) converges to zero when $k \to \infty$. Let us consider the first integral on the right-hand side of (4.20). For any $\varepsilon \geq 0$ we can choose $u_\varepsilon \in \mathbf{\Lambda}(Q)$ such that $\|u - u_\varepsilon\|_{\mathbf{W}(Q)} \leq \varepsilon$. Then,

$$\int_0^T e^{-k_1 t}\,(\operatorname{rot} z_k \times (z_k - z), u)\,dt$$
$$= \int_0^T e^{-k_1 t}\,(\operatorname{rot} z_k \times (z_k - z), u_\varepsilon)\,dt + \int_0^T e^{-k_1 t}\,(\operatorname{rot} z_k \times (z_k - z), u - u_\varepsilon)\,dt. \tag{4.21}$$

Since $z_k \to z$ in $\mathbf{L}^2(Q)$ as $k \to \infty$,

$$\left|\int_0^T e^{-k_1 t}\,(\operatorname{rot} z_k \times (z_k - z), u_\varepsilon)\,dt\right| \leq C\|\nabla z_k\|_{\mathbf{L}^2(Q)}\|z_k - z\|_{\mathbf{L}^2(Q)}\|u_\varepsilon\|_{\mathbf{L}^\infty(Q)} \to 0 \quad \text{as } k \to \infty. \tag{4.22}$$

Also, since $\{z_k - z\}$ is bounded in $\mathbf{W}(Q)$, we have

$$\begin{aligned}\left|\int_0^T e^{-k_1 t}\,(\operatorname{rot} z_k \times (z_k - z), u - u_\varepsilon)\,dt\right| &\leq C\|\nabla z_k\|_{\mathbf{L}^2(Q)}\|z_k - z\|_{\mathbf{L}^4(Q)}\|u - u_\varepsilon\|_{\mathbf{L}^4(Q)} \\ &\leq C\|z_k\|_{\mathbf{W}(Q)}\|z_k - z\|_{\mathbf{W}(Q)}\|u - u_\varepsilon\|_{\mathbf{W}(Q)} \leq C\varepsilon.\end{aligned} \tag{4.23}$$

From (4.21)–(4.23), we see that the first integral on the right-hand side of (4.20) goes to zero when $k \to \infty$, and so we get (4.19).

It is easy to check that other terms in $\langle A_m z_k, u\rangle$ converge when $k \to \infty$. This fact together with (4.19) implies (4.18).

By (4.17) and (4.18), there exists a solution to (4.10) (see Theorem 1.43), and therefore the assertion is proved. □

4.1.3 Proof of the Existence of a Solution

Now, we prove Theorem 4.1. To do this we need to establish two lemmas first.

Lemma 4.2 *If $w^m \in \mathbf{W}(Q)$ are solutions to problem* (4.7) *with the k_1 in Lemma 4.1, then*

$$\int_Q \frac{1}{m}\frac{\partial w^m}{\partial t}\frac{\partial u}{\partial t}\,dxdt \to 0 \quad \forall u \in \mathbf{W}(Q) \quad \text{as } m \to \infty. \tag{4.24}$$

Proof By (4.10), (4.13)–(4.16), we have

$$\int_Q \frac{1}{m}\left|\frac{\partial w^m}{\partial t}\right|^2 dxdt + \frac{\mu c_1}{4}\int_0^T \|w^m\|^2_{\mathbf{H}^1(\Omega)}\,dt + \frac{1}{2}\left(\|w^m(x,0)\|^2 + \|w^m(x,T)\|^2\right) \le \left\langle F, w^m\right\rangle. \tag{4.25}$$

On the other hand,

$$\int_Q \bar{U}\frac{\partial w^m}{\partial t}\,dxdt = (\bar{U}, w^m(x,T)) - (\bar{U}, w^m(x,0)) - \int_Q \frac{\partial \bar{U}}{\partial t} w^m\,dxdt \quad \forall w^m \in \mathbf{W}(Q). \tag{4.26}$$

Taking (4.26) into account in (4.9) and applying Young's inequality to the right-hand side of (4.25), we have

$$\int_0^T \frac{1}{m}\left\|\frac{\partial w^m}{\partial t}\right\|^2 dt + \frac{\mu c_1}{8}\int_0^T \|w^m\|^2_{\mathbf{H}^1(\Omega)}\,dt + \frac{1}{8}\left(\|w^m(x,0)\|^2 + \|w^m(x,T)\|^2\right) \le c, \tag{4.27}$$

where c is independent of m and depends on $\bar{U}$, $\bar{\phi}_i$, v_0, $\bar{f}$, S, k and k_1.

Using $\int_Q \frac{1}{m}\left|\frac{\partial w^m}{\partial t}\right|^2 dxdt \le c$, we can get (4.24). Indeed, by Hölder's inequality

$$\left|\int_Q \frac{1}{m}\frac{\partial w^m}{\partial t}\frac{\partial u}{\partial t}\,dxdt\right| \le \frac{1}{\sqrt{m}}\left(\int_Q \left|\frac{1}{\sqrt{m}}\frac{\partial w^m}{\partial t}\right|^2 dxdt \cdot \int_Q \left|\frac{\partial u}{\partial t}\right|^2 dxdt\right)^{\frac{1}{2}},$$

which shows (4.24). □

Lemma 4.3 *If $w^m \in \mathbf{W}(Q)$ are solutions to problem* (4.7) *guaranteed by Lemma 4.1, then $\{w^m\}$ are relatively compact in $\mathbf{L}^2(Q)$.*

Proof Putting $\bar{w}^m(x,t) = 0$ on $(\Omega \times (-T, 2T)) \setminus Q$, let us make $\bar{w}^m(x,t)$ an extension of w^m. Thus, by virtue of (4.27) we get

$$\begin{aligned}
&\int_{\Omega\times(0,T)} \frac{1}{m}\left|\frac{\partial \bar{w}^m}{\partial t}\right|^2 dxdt \le c,\\
&\int_0^T \|\bar{w}^m\|^2_{\mathbf{H}^1(\Omega)}\, dt \le c,\\
&\|\bar{w}^m(x,0)\| \le c,\\
&\|\bar{w}^m(x,T)\| \le c.
\end{aligned}\tag{4.28}$$

For $0 < h < T$ let

$$w_h^m(x,t) = \frac{1}{h}\int_t^{t+h} \bar{w}^m(x,s)\,ds.$$

Then, $w_h^m|_Q \in W(Q)$ and

$$(w_h^m)_t := \frac{\partial w_h^m(x,t)}{\partial t} = \frac{1}{h}\left(\bar{w}^m(x,t+h) - \bar{w}^m(x,t)\right).$$

Replacing u by w_h^m in (4.7), we have

$$\begin{aligned}
&\int_0^T \frac{1}{m}\Big(\frac{\partial w^m(x,t)}{\partial t}, (w_h^m)_t\Big)\,dt - \int_0^T \left(w^m, (w_h^m)_t\right)dt + 2\mu\int_0^T \left(\mathcal{E}(w^m), \mathcal{E}(w_h^m)\right)dt\\
&+\int_0^T \left(e^{-k_1 t}\operatorname{rot} w^m \times w^m + \operatorname{rot} w^m \times U + \operatorname{rot} U \times w^m, w_h^m\right)dt\\
&+\int_0^T 2\mu\big(k(x)w^m, w_h^m\big)_{\Gamma_2}\,dt + \int_0^T 2\mu\big(S\tilde{w^m}, \tilde{w_h^m}\big)_{\Gamma_3}\,dt\\
&+\int_0^T 2\big(\alpha(t,x)w^m, w_h^m\big)_{\Gamma_5}\,dt + \int_0^T \mu\big(k(x)w^m, w_h^m\big)_{\Gamma_7}\,dt - k_1\int_0^T (w^m, w_h^m)\,dt\\
&= \int_0^T \langle F(t), w_h^m(t)\rangle\, dxdt,
\end{aligned}\tag{4.29}$$

where $w_h^m(T) = 0$ was used and F is the one in (4.9).

Assuming $\bar{w}(x,t) \in \mathbf{C}^1(\bar{\Omega}\times[0,T])$, we estimate

$$\frac{1}{m}\int\limits_{\Omega\times(0,T)} \left|\frac{1}{h}\left[\bar{w}(x,t+h) - \bar{w}(x,t)\right]\right|^2 dxdt.$$

Applying Hölder's inequality, we have

$$
\begin{aligned}
&\frac{1}{m}\int_{\Omega\times(0,T-h)}\frac{1}{h^2}\Big|\int_t^{t+h}\frac{\partial}{\partial s}\bar{w}(x,s)\,ds\Big|^2dxdt\\
&\qquad+\frac{1}{m}\int_{\Omega\times(T-h,T)}\frac{1}{h^2}\Big[\bar{w}(x,T)+\int_T^t\frac{\partial}{\partial s}\bar{w}(x,s)\,ds\Big]^2dxdt\\
&\quad\le\frac{1}{m}\int_{\Omega\times(0,T-h)}\frac{1}{h}\int_t^{t+h}\Big|\frac{\partial\bar{w}(x,s)}{\partial s}\Big|^2ds\,dxdt+\frac{1}{m}\frac{2}{h}\|\bar{w}(x,T)\|^2\\
&\qquad+\frac{1}{m}\frac{2}{h^2}\int_{\Omega\times(T-h,T)}\int_t^T\Big|\frac{\partial}{\partial s}\bar{w}(x,s)\Big|^2ds\cdot h\,dxdt \qquad (4.30)\\
&\quad\le\frac{1}{h}\frac{1}{m}\Big[(T-h)\int_{\Omega\times(0,T)}\Big|\frac{\partial}{\partial t}\bar{w}(x,t)\Big|^2dxdt+2\|\bar{w}(x,T)\|^2\\
&\qquad+2h\int_{\Omega\times(0,T)}\Big|\frac{\partial}{\partial t}\bar{w}(x,t)\Big|^2dxdt\Big]\\
&\quad\le\frac{1}{h}\frac{1}{m}\Big[(T+h)\int_{\Omega\times(0,T)}\Big|\frac{\partial}{\partial t}\bar{w}(x,t)\Big|^2dxdt+2\|\bar{w}(x,T)\|^2\Big].
\end{aligned}
$$

Since $\mathbf{C}^1(\bar{\Omega}\times[0,T])$ is dense in $\mathbf{H}^1(\Omega\times(0,T))$, by (4.28) and (4.30) for any $\bar{w}^m\in\mathbf{H}^1(\Omega\times(0,T))$ we have

$$
\Big\{\frac{1}{m}\int_{\Omega\times(0,T)}\Big|\frac{1}{h}\left[\bar{w}^m(x,t)-\bar{w}^m(x,t-h)\right]\Big|^2dxdt\Big\}^{\frac{1}{2}}\le\frac{c}{\sqrt{h}}. \tag{4.31}
$$

By Hölder's inequality, we get from (4.27) and (4.31)

$$
\begin{aligned}
&\Big|\int_Q\frac{1}{m}\frac{\partial w^m(x,t)}{\partial t}\cdot\frac{1}{h}\left[\bar{w}^m(x,t)-\bar{w}^m(x,t-h)\right]dxdt\Big|\\
&\quad\le\Big(\int_Q\frac{1}{m}\Big|\frac{\partial w^m}{\partial t}\Big|^2dxdt\Big)^{\frac{1}{2}}\Big(\frac{1}{m}\int_Q\Big|\frac{1}{h}\left[\bar{w}^m(x,t)-\bar{w}^m(x,t-h)\right]\Big|^2dxdt\Big)^{\frac{1}{2}}\\
&\quad\le\frac{c}{\sqrt{h}}.
\end{aligned} \tag{4.32}
$$

Using (4.28), we have

$$
\begin{aligned}
\Big|2\mu\int_0^T(\mathcal{E}(w^m),\mathcal{E}(w_h^m))\,dt\Big|&\le c\int_0^T\|w^m\|_{\mathbf{H}^1(\Omega)}\Big\|\frac{1}{h}\int_t^{t+h}\bar{w}^m(x,s)\,ds\Big\|_{\mathbf{H}^1(\Omega)}dt\\
&\le c\int_0^T\|w^m\|_{\mathbf{H}^1(\Omega)}\frac{1}{\sqrt{h}}\Big|\int_t^{t+h}\|\bar{w}^m(x,s)\|^2_{\mathbf{H}^1(\Omega)}ds\Big|^{\frac{1}{2}}dt\le c/\sqrt{h}.
\end{aligned} \tag{4.33}
$$

Now, let us estimate $\int_0^T\left(e^{-k_1t}\mathrm{rot}\,w^m\times w^m,w_h^m\right)dt$.
Since

$$\int_t^{t+h} \|w^m\|_{\mathbf{L}^3(\Omega)}\, ds \le \sqrt{h}\Big(\int_t^{t+h} \|w^m\|^2_{\mathbf{L}^3(\Omega)}\, ds\Big)^{\frac{1}{2}} \le c\sqrt{h}\|\bar{w}^m\|^2_{L^2(0,T;\mathbf{H}^1(\Omega))},$$

by Hölder's inequality and (4.28),

$$\begin{aligned}\Big|\int_0^T \Big(e^{-k_1 t}\mathrm{rot}\, w^m \times w^m, w_h^m\Big)\, dt\Big| &\le \frac{c}{h}\int_0^T \Big[\|\mathrm{rot}\, w^m\| \cdot \|w^m\|_{\mathbf{L}^6} \int_t^{t+h} \|w^m\|_{\mathbf{L}^3}\, ds\Big] dt\\ &\le \frac{c}{\sqrt{h}}\int_0^T \|w^m\|^2_{\mathbf{H}^1(\Omega)}\, dt \le \frac{c}{\sqrt{h}}.\end{aligned} \tag{4.34}$$

In the same way, we get

$$\begin{aligned}&\Big|\int_0^T \Big(\mathrm{rot}\, w^m \times U + \mathrm{rot}\, U \times w^m, w_h^m\Big)\, dt\Big| \le c/\sqrt{h},\\ &\Big| - k_1 \int_0^T (w^m, w_h^m)\, dt\Big| \le c/\sqrt{h}.\end{aligned} \tag{4.35}$$

Taking into account

$$\begin{aligned}\int_Q \bar{U}\frac{\partial w_h^m}{\partial t}\, dxdt &= \big(\bar{U}(T), w_h^m(T)\big) - \big(\bar{U}(0), w_h^m(0)\big) - \int_Q \frac{\partial \bar{U}}{\partial t} w_h^m\, dxdt\\ &= -\big(\bar{U}(0), w_h^m(0)\big) - \int_0^T \Big(\frac{\partial \bar{U}}{\partial t}, w_h^m\Big)\, dt,\end{aligned}$$

we have

$$\begin{aligned}\langle F, w_h^m\rangle &= -(\bar{U}(0), w_h^m(0)) - \int_0^T \Big(\frac{\partial \bar{U}}{\partial t}, w_h^m\Big)\, dt + \int_0^T (\bar{U}k_1 w_h^m)\, dt\\ &- 2\mu \int_0^T (\mathcal{E}(\bar{U}), \mathcal{E}(w_h^m))\, dt - \int_0^T (\mathrm{rot}\, \bar{U} \times U, w_h^m)\, dt - \int_0^T 2\mu(k(x)\bar{U}, w_h^m)_{\Gamma_2}\, dt\\ &- \int_0^T 2\mu(S\bar{\bar{U}}, \bar{w}_h^m)_{\Gamma_3}\, dt - \int_0^T 2(\alpha(t,x)\bar{U}, w_h^m)_{\Gamma_5}\, dt - \int_0^T \mu(k(x)\bar{U}, w_h^m)_{\Gamma_7}\, dt\\ &+ \int_0^T \langle \bar{f}, w_h^m\rangle\, dt + \int_0^T \sum_{i=2,4,7} \langle \bar{\phi}_i, w_h^m \cdot n\rangle_{\Gamma_i}\, dt + \int_0^T \sum_{i=3,5,6} \langle \bar{\phi}_i, w_h^m\rangle_{\Gamma_i}\, dt + (v_0, w_h^m(0)).\end{aligned}$$

Similarly, we have

$$|\langle F, w_h^m\rangle| \le \frac{c}{\sqrt{h}}. \tag{4.36}$$

Let us estimate

$$-\int_Q w^m \cdot \frac{1}{h}\big[\bar{w}^m(x, t+h) - \bar{w}^m(x, t)\big]\, dxdt.$$

Using $-ab = -\frac{1}{2}[(a+b)^2 - a^2 - b^2]$, we have

$$
\begin{aligned}
-\frac{1}{h}\int_0^T & \left(\bar{w}^m(t), \bar{w}^m(x,t+h) - \bar{w}^m(x,t)\right)_\Omega dt \\
&= \frac{1}{2h}\int_0^T \|w^m(x,t)\|^2\, dt - \frac{1}{2h}\int_0^T \|\bar{w}^m(x,t+h)\|^2\, dt \\
&\qquad + \frac{1}{2h}\int_0^T \|\bar{w}^m(x,t+h) - \bar{w}^m(x,t)\|^2\, dt \\
&= \frac{1}{2h}\int_0^T \|w^m(x,t)\|^2\, dt - \frac{1}{2h}\int_h^T \|\bar{w}^m(x,t)\|^2\, dt \\
&\qquad + \frac{1}{2h}\int_0^T \|\bar{w}^m(x,t+h) - \bar{w}^m(x,t)\|^2\, dt \\
&= \frac{1}{2h}\int_0^T \|w^m(x,t)\|^2\, dt - \frac{1}{2h}\int_h^T \|w^m(x,t)\|^2\, dt \\
&\qquad + \frac{1}{2h}\int_0^T \|\bar{w}^m(x,t+h) - \bar{w}^m(x,t)\|^2\, dt \\
&\geq \frac{1}{2h}\int_0^T \|\bar{w}^m(x,t+h) - \bar{w}^m(x,t)\|^2\, dt.
\end{aligned}
\tag{4.37}
$$

Formulas (4.29) and (4.32)–(4.37) imply

$$
\int_0^T \|\bar{w}^m(x,t+h) - \bar{w}^m(x,t)\|^2\, dt \leq c[\sqrt{h} + \omega(h)^{\frac{2}{3}}] \quad \text{for } T > h > 0.
$$

Therefore, by Theorem 1.38 we have the conclusion. □

Proof of Theorem 4.1.
Let $\{w^m(Q)\}$ be the sequence of solutions to (4.7) guaranteed by Lemma 4.1. By (4.27) $\{w^m(Q)\}$ is bounded in $\mathbf{V}(Q)$. Then by Lemma 4.3, we can choose its subsequence $\{w^k(Q)\}$ such that $w^k(Q) \to w \in \mathbf{V}(Q)$ in $\mathbf{L}^2(Q)$.

First, let us prove that for $u \in \mathbf{\Lambda}(Q)$

$$
\int_0^T e^{-k_1 t}\left(\operatorname{rot} w^k \times w^k, u\right) dt \to \int_0^T e^{-k_1 t}\left(\operatorname{rot} w \times w, u\right) dt \quad \text{as } k \to \infty.
\tag{4.38}
$$

We can write

$$
\begin{aligned}
&\int_0^T e^{-k_1 t}\left(\operatorname{rot} w^k \times w^k, u\right) dt-\int_0^T e^{-k_1 t}(\operatorname{rot} w \times w, u)\, dt \\
&=\int_0^T e^{-k_1 t}\left(\operatorname{rot} w^k \times (w^k-w), u\right) dt+\int_0^T e^{-k_1 t}\left(\operatorname{rot}(w^k-w) \times w, u\right) dt .
\end{aligned}
\tag{4.39}
$$

Since $\{\operatorname{rot} w^k\}$ is bounded in $\mathbf{L}^2(Q)$, $w^k \to w$ in $\mathbf{L}^2(Q)$ and $u \in \mathbf{L}^\infty(Q)$, the first integral on the right-hand side of (4.39) converges to zero as $k \to \infty$. Meanwhile, since $e^{-k_1 t} wu \in \mathbf{L}^2(Q)$ and $w_k \rightharpoonup w$ in $\mathbf{V}(Q)$, the second integral on the right-hand side of (4.39) converges to zero. Thus, (4.38) holds.

It is easy to verify the convergence of other terms in (4.7) when $k \to \infty$. Thus, passing to limit as $k \to \infty$ in (4.7) with k instead of m, by Lemma 4.2 we have (4.5). The estimate (4.6) can be obtained in the same way as in the proof of (4.2) in [1]. □

4.1.4 The Stokes Problem

Let us consider the Stokes problem

$$
\begin{cases}
\dfrac{\partial v}{\partial t}-\mu \Delta v+\nabla p=f, \quad \operatorname{div} v=0, \\
v(0)=v_0
\end{cases}
\tag{4.40}
$$

with boundary condition

$$
\begin{aligned}
&(1) \quad v|_{\Gamma_1}=h_1, \\
&(2) \quad v_\tau|_{\Gamma_2}=0, \ -p|_{\Gamma_2}=\phi_2, \\
&(3) \quad v_n|_{\Gamma_3}=0, \ \operatorname{rot} v \times n|_{\Gamma_3}=\phi_3/\mu, \\
&(4) \quad v_\tau|_{\Gamma_4}=h_4, \ (-p+2\mu\varepsilon_{nn}(v))|_{\Gamma_4}=\phi_4, \\
&(5) \quad v_n|_{\Gamma_5}=h_5, \ 2(\mu\varepsilon_{n\tau}(v)+\alpha v_\tau)|_{\Gamma_5}=\phi_5, \\
&(6) \quad (-pn+2\mu\varepsilon_n(v))|_{\Gamma_6}=\phi_6, \\
&(7) \quad v_\tau|_{\Gamma_7}=0, \ (-p+\mu\frac{\partial v}{\partial n}\cdot n)|_{\Gamma_7}=\phi_7.
\end{aligned}
\tag{4.41}
$$

Applying Theorems 3.1 and 3.2 on $\Gamma_{ij}(i=2,3,7)$ (and so using (3.41) and (3.42)), we get a variational formulation for problem (4.40), (4.41):

Problem 4.2 Find v such that

$$
\begin{aligned}
& v-U \in \mathbf{V}(Q),\\
& -\int_0^T \left(v, \frac{\partial u}{\partial t}\right) dt + 2\mu \int_0^T (\mathcal{E}(v), \mathcal{E}(u))\, dt + 2\mu \int_0^T (k(x)v, u)_{\Gamma_2}\, dt\\
&\qquad + 2\mu \int_0^T (S\tilde{v}, \tilde{u})_{\Gamma_3}\, dt + 2\int_0^T (\alpha(t,x)v_\tau, u)_{\Gamma_5}\, dt + \int_0^T \mu(k(x)v, u)_{\Gamma_7}\\
&\quad = (v_0, u(0)) + \int_0^T \langle f, u\rangle\, dt + \int_0^T \sum_{i=2,4,7} \langle \phi_i, u_n\rangle_{\Gamma_i}\, dt + \int_0^T \sum_{i=3,5,6} \langle \phi_i, u\rangle_{\Gamma_i}\, dt\\
&\qquad\qquad \forall u \in \mathbf{\Lambda}(Q) \text{ with } u(x,T)=0.
\end{aligned}
\tag{4.42}
$$

Theorem 4.2 *Under Assumption 4.1 there exists a unique solution to Problem 4.2 for the non-stationary Stokes system with mixed boundary condition* (4.41) *and the solution belongs to* $C([0,T]; \boldsymbol{L}^2(\Omega))$.

Proof Put $v = z + U$, $w = e^{k_1 t} z$, where k_1 is taken as in Lemma 4.1 under consideration of the fact that there is not any nonlinear term. Then, we get the new problem equivalent to Problem 4.2:

Find $w \in L^2(0,T; \mathbf{V})$ such that

$$
\begin{aligned}
& -\int_0^T \left(w, \frac{\partial \hat{u}}{\partial t}\right) dt + 2\mu \int_0^T (\mathcal{E}(w), \mathcal{E}(\hat{u}))\, dt + \int_0^T 2\mu(k(x)w, \hat{u})_{\Gamma_2}\, dt\\
&\quad + \int_0^T 2\mu(S\tilde{w}, \tilde{\hat{u}})_{\Gamma_3}\, dt + \int_0^T 2(\alpha(t,x)w, \hat{u})_{\Gamma_5}\, dt\\
&\quad + \int_0^T \mu(k(x)w, \hat{u})_{\Gamma_7}\, dt - k_1 \int_0^T (w, \hat{u})\, dt\\
& = \int_Q k_1 \bar{U}\hat{u}\, dxdt - \int_Q \frac{\partial \bar{U}}{\partial t} \hat{u}\, dxdt - 2\mu \int_0^T (\mathcal{E}(\bar{U}), \mathcal{E}(\hat{u}))\, dt\\
&\quad - \int_0^T 2\mu(k(x)\bar{U}, \hat{u})_{\Gamma_2}\, dt - \int_0^T 2\mu(S\tilde{\bar{U}}, \tilde{\hat{u}})_{\Gamma_3}\, dt - \int_0^T 2(\alpha(t,x)\bar{U}_\tau, \hat{u})_{\Gamma_5}\, dt\\
&\quad - \int_0^T \mu(k(x)\bar{U}, u)_{\Gamma_7} + \int_0^T \langle \bar{f}, \hat{u}\rangle\, dt + \int_0^T \sum_{i=2,4,7} \langle \bar{\phi}_i, \hat{u}_n\rangle_{\Gamma_i}\, dt\\
&\quad + \int_0^T \sum_{i=3,5,6} \langle \bar{\phi}_i, \hat{u}\rangle_{\Gamma_i}\, dt + \big(v_0 - U(x,0), u(x,0)\big) \quad \forall \hat{u} \in \mathbf{\Lambda}(Q),\ \ \hat{u}(x,T)=0,
\end{aligned}
\tag{4.43}
$$

where $\bar{\gamma} = e^{k_1 t}\gamma$ for γ.

Existence of a solution to (4.43) is proved in much the same way as Theorem 4.1 without applying the condition $U \in \mathbf{L}^\infty(Q)$ and Lemma 4.3, which are needed for the nonlinear term. To complete the proof, by Assumption 4.1 it is enough to prove that the solutions to (4.43) belong to $C([0,T]; H)$ and are unique.

When $w \in L^2(0,T; \mathbf{V})$ is a solution to (4.43), estimating

$$
\begin{aligned}
I \equiv \int_0^T \Big[& -2\mu(\mathcal{E}(w), \mathcal{E}(\hat{u})) - 2\mu(k(x)w, u)_{\Gamma_2} - 2\mu(S\tilde{w}, \tilde{\hat{u}})_{\Gamma_3} \\
& - 2(\alpha(t,x)w, \hat{u})_{\Gamma_5} - \mu(k(x)w, \hat{u})_{\Gamma_7} - k_1(w, \hat{u}) + k_1(\bar{U}, \hat{u}) - \left(\frac{\partial \bar{U}}{\partial t}, \hat{u}\right) \\
& - 2\mu(\mathcal{E}(\bar{U}), \mathcal{E}(\hat{u})) - 2\mu(k(x)\bar{U}, \hat{u})_{\Gamma_2} - 2\mu(S\tilde{\bar{U}}, \tilde{\hat{u}})_{\Gamma_3} - 2(\alpha(t,x)\bar{U}_\tau, \hat{u})_{\Gamma_5} \\
& - \mu(k(x)U, u)_{\Gamma_7} + \langle \bar{f}, \hat{u}\rangle + \sum_{i=2,4,7} \langle \bar{\phi}_i, \hat{u}_n\rangle_{\Gamma_i} + \sum_{i=3,5,6} \langle \bar{\phi}_i, \hat{u}\rangle_{\Gamma_i} \Big] dt \\
& \qquad \forall \hat{u} \in L^2(0,T; \mathbf{V}),
\end{aligned}
$$

we get

$$
|I| \le K \|\hat{u}\|_{L^2(0,T;\mathbf{V})} \quad \forall \hat{u} \in L^2(0,T; \mathbf{V}).
$$

This means that I is a continuous linear functional on $L^2(0,T; \mathbf{V})$. Thus, there exists a $\overline{F} \in L^2(0,T; \mathbf{V}^*)$ such that

$$
I = \int_0^T \langle \overline{F}, \hat{u}\rangle \, dt. \tag{4.44}
$$

Taking any $\phi \in \mathscr{D}(0,T)$ and $u \in \mathbf{V} \cap C^2(\overline{\Omega})$ and putting $\hat{u} = \phi u$, by definition of $w' \in \mathscr{D}^*(0,T; V^*)$ (see Definition 1.13), (4.43), (4.44) and Theorem 1.31 we have

$$
\begin{aligned}
\langle w'(\phi), u\rangle = -\Big\langle \int_0^T w\phi' \, dt, u \Big\rangle = -\int_0^T \langle w, \phi' u\rangle \, dt &= \int_0^T \langle \overline{F}, \phi u\rangle \, dt \\
&= \Big\langle \int_0^T \overline{F}\phi \, dt, u \Big\rangle,
\end{aligned}
$$

which means $w' = \overline{F} \in L^2(0,T; \mathbf{V}^*)$.

Therefore, $w \in C([0,T]; H)$ and

$$
\int_0^T \langle w, w'\rangle \, dt = \frac{1}{2}\big[\|w(T)\|^2 - \|w(0)\|^2\big]. \tag{4.45}
$$

Let w_1, w_2 be solutions to (4.43) corresponding to the same given data and $w = w_1 - w_2$. Then, by (4.43) we have

$$
\begin{aligned}
& \int_0^T \langle w', \hat{u}\rangle \, dt + 2\mu \int_0^T (\mathcal{E}(w), \mathcal{E}(\hat{u})) \, dt + \int_0^T 2\mu(k(x)w, u)_{\Gamma_2} \, dt \, dt \\
& \quad + \int_0^T 2\mu(S\tilde{w}, \tilde{\hat{u}})_{\Gamma_3} + \int_0^T 2(\alpha(t,x)w, \hat{u})_{\Gamma_5} \, dt \\
& \quad + \int_0^T \mu(k(x)w, \hat{u})_{\Gamma_7} \, dt - k_1 \int_0^T (w, \hat{u}) \, dt = 0 \qquad \text{for } \hat{u} = \phi \cdot u.
\end{aligned} \tag{4.46}
$$

Since the set $\{\hat{u} = \phi u : \phi \in \mathscr{D}(0, T), u \in \mathbf{V} \cap C^2(\overline{\Omega})\}$ is dense in $L^2(0, T; \mathbf{V})$ (see Remark 1.8), (4.46) is valid for $\hat{u} = w$. Thus, from (4.45) and (4.46) it follows that

$$\|w(T)\|^2 + \beta \int_0^T \|w\|_V^2 \, dt \le 0, \quad \beta > 0$$

(cf. (4.14), (4.16)), which shows that $w \equiv 0$, that is, the solution to (4.43) is unique. □

4.2 Existence and Uniqueness of Solutions: The Case of Static Pressure

In this section we study the Navier-Stokes problem

$$\begin{cases} \dfrac{\partial v}{\partial t} - \mu \Delta v + (v \cdot \nabla) v + \nabla p = f, \\ \operatorname{div} v = 0, \\ v(0) = v_0 \end{cases} \tag{4.47}$$

with boundary conditions including the static pressure. We are concerned with the Problems I and II, which are distinguished according to boundary conditions. Problem I is the one with the boundary conditions

$$\begin{aligned} &(1) \quad v|_{\Gamma_1} = h_1, \\ &(2) \quad v_\tau|_{\Gamma_2} = 0, \ \ -p|_{\Gamma_2} = \phi_2, \\ &(3) \quad v_n|_{\Gamma_3} = 0, \ \ \operatorname{rot} v \times n|_{\Gamma_3} = \phi_3/\mu, \\ &(4) \quad v_\tau|_{\Gamma_4} = h_4, \ \ (-p + 2\mu\varepsilon_{nn}(v))|_{\Gamma_4} = \phi_4, \\ &(5) \quad v_n|_{\Gamma_5} = h_5, \ \ 2(\mu\varepsilon_{n\tau}(v) + \alpha v_\tau)|_{\Gamma_5} = \phi_5, \\ &(6) \quad (-pn + 2\mu\varepsilon_n(v))|_{\Gamma_6} = \phi_6, \\ &(7) \quad v_\tau|_{\Gamma_7} = 0, \ \Big(-p + \mu \frac{\partial v}{\partial n} \cdot n\Big)\Big|_{\Gamma_7} = \phi_7, \end{aligned} \tag{4.48}$$

and Problem II is the one with the conditions

$$
\begin{aligned}
&(1)\quad v|_{\Gamma_1} = h_1,\\
&(2)\quad v_\tau|_{\Gamma_2} = 0,\ \ -p|_{\Gamma_2} = \phi_2,\\
&(3)\quad v_n|_{\Gamma_3} = 0,\ \ \operatorname{rot} v \times n|_{\Gamma_3} = \phi_3/\mu,\\
&(4)\quad v_\tau|_{\Gamma_4} = h_4,\ \ (-p + \mu\varepsilon_{nn}(v))|_{\Gamma_4} = \phi_4,\\
&(5)\quad v_n|_{\Gamma_5} = h_5,\ \ 2(\mu\varepsilon_{n\tau}(v) + \alpha v_\tau)|_{\Gamma_5} = \phi_5,\\
&(6)\quad \Big(-pn + \mu\frac{\partial v}{\partial n}\Big)\Big|_{\Gamma_7} = \phi_7
\end{aligned}
\tag{4.49}
$$

together $\Gamma_6 = \emptyset$. Condition (6) of (4.49) is "do nothing" condition, but (7) of (4.48) is rather different from "do nothing" condition (see Remark 3.7).

For Problems I and II stated above (for the corresponding perturbation problems in Sect. 4.2.3), respectively, conditions $\Gamma_{ij} \in C^{2,1}, i = 2, 3, 7$, and $\Gamma_{ij} \in C^{2,1}, i = 2, \cdots, 5$, are used since Theorems 3.1, 3.2 are applied for variational formulations for Problem I and Problem II, respectively, on $\Gamma_{ij} \in C^{2,1} (i = 2, 3, 7)$ and $\Gamma_{ij} \in C^{2,1} (i = 2, \cdots, 5)$.

For the proof of the existence of a unique solution to Problems I and II we use the Local inverse mapping theorem (see Theorem 1.7).

4.2.1 Existence and Uniqueness of Solutions to Problem I

We use the following notation:

$$
\mathbf{V} = \big\{u \in \mathbf{H}^1(\Omega) : \operatorname{div} u = 0,\ u|_{\Gamma_1} = 0,\ u_\tau|_{\Gamma_2\cup\Gamma_4\cup\Gamma_7} = 0,\ u_n|_{\Gamma_3\cup\Gamma_5} = 0\big\}
$$

and

$$
\mathbf{V}_{\Gamma 237}(\Omega) = \big\{u \in \mathbf{H}^1(\Omega) : \operatorname{div} u = 0,\ u_\tau|_{\Gamma_2\cup\Gamma_7} = 0,\ u_n|_{\Gamma_3} = 0\big\}.
$$

Denote by H the completion of $\mathbf{V}$ in the space $\mathbf{L}^2(\Omega)$. Throughout this section $\widetilde{\mathbf{V}} = \{u \in \mathbf{H}^1(\Omega) : \operatorname{div} u = 0\}$.

Let

$$
\begin{aligned}
&\mathscr{X} = \{w \in L^2(0, T; \mathbf{V});\ w' \in L^2(0, T; \mathbf{V}),\ w'' \in L^2(0, T; \mathbf{V}^*)\},\\
&\qquad \|w\|_{\mathscr{X}} = \|w\|_{L^2(0,T;\mathbf{V})} + \|w'\|_{L^2(0,T;\mathbf{V})} + \|w''\|_{L^2(0,T;\mathbf{V}^*)},\\
&\mathscr{Y} = \{w \in L^2(0, T; \mathbf{V}^*);\ w' \in L^2(0, T; \mathbf{V}^*)\},\\
&\qquad \|w\|_{\mathscr{Y}} = \|w\|_{L^2(0,T;\mathbf{V}^*)} + \|w'\|_{L^2(0,T;\mathbf{V}^*)},\\
&\mathscr{W} = \{w \in L^2(0, T; \widetilde{\mathbf{V}});\ w' \in L^2(0, T; \widetilde{\mathbf{V}}),\ w'' \in L^2(0, T; \widetilde{\mathbf{V}}^*)\},\\
&\qquad \|w\|_{\mathscr{W}} = \|w\|_{L^2(0,T;\widetilde{\mathbf{V}})} + \|w'\|_{L^2(0,T;\widetilde{\mathbf{V}})} + \|w''\|_{L^2(0,T;\widetilde{\mathbf{V}}^*)}.
\end{aligned}
$$

Here and in what follows w' means the derivative with respect to t of $w(t)$.

For Problem I, we will use the following assumptions:

Assumption 4.2 $f,\ f' \in L^2(0,T;\mathbf{V}^*),\ \phi_i,\ \phi_i' \in L^2(0,T;H^{-\frac{1}{2}}(\Gamma_i)),\ i=2,4,7,$ $\phi_i,\ \phi_i' \in L^2(0,T;\mathbf{H}^{-\frac{1}{2}}(\Gamma_i)), i=3,5,6,\ \alpha_{ij} \in L^\infty(\Gamma_5)$, where α_{ij} are components of the matrix α, and $\Gamma_1 \neq \emptyset$.

Assumption 4.3 There exists a function $U \in \mathscr{W}$ such that

$$\operatorname{div} U=0,\ U|_{\Gamma_1}=h_1,\ U_\tau|_{\Gamma_2\cup\Gamma_7}=0,\ U_n|_{\Gamma_3}=0,\ U_\tau|_{\Gamma_4}=h_4,\ U_n|_{\Gamma_5}=h_5.$$

Also, $U(0,x)-v_0 \in \mathbf{V}$.

Remark 4.1 For the particular situation such that $h_4, h_5=0,\ h_1(t,x) \in H^{\frac{1}{2}}_{00}(\Gamma_1)$, $\int_{\Gamma_1} h_1(t,x)\cdot n\,dx=0$ for every fixed t, and $h_1(t,x) \in C^2(0,T;H^{\frac{1}{2}}(\Gamma_1))$, applying the existence and estimate of a solution to the steady Stokes problem with non-homogeneous Dirichlet boundary condition (see Theorem IV.1.1 [2]), we can prove the existence of a function U above.

Taking (3.43) into account, we get the following variational formulation for Problem I (with boundary conditions (4.48)):

Problem 4.3 Find v such that

$$\begin{aligned}
&v-U \in L^2(0,T;\mathbf{V}),\\
&v(0)=v_0,\\
&\langle v',u\rangle+2\mu(\mathcal{E}(v),\mathcal{E}(u))+\langle (v\cdot\nabla)v,u\rangle+2\mu(k(x)v,u)_{\Gamma_2}\\
&\qquad+2\mu(S\tilde{v},\tilde{u})_{\Gamma_3}+2(\alpha(x)v_\tau,u)_{\Gamma_5}+\mu(k(x)v,u)_{\Gamma_7}\\
&\qquad=\langle f,u\rangle+\sum_{i=2,4,7}\langle\phi_i,u_n\rangle_{\Gamma_i}+\sum_{i=3,5,6}\langle\phi_i,u\rangle_{\Gamma_i}\quad\text{for all } u\in\mathbf{V}.
\end{aligned}\tag{4.50}$$

Using Assumption 4.3 and taking $v=\overline{z}+U$, we deduce the following problem equivalent to Problem 4.3:

Find $\overline{z}$ such that

$$\begin{aligned}
&\overline{z}\in L^2(0,T;\mathbf{V}),\\
&\overline{z}(0)=\overline{z}_0\equiv v_0-U(0)\in\mathbf{V},\\
&\langle\overline{z}',u\rangle+2\mu(\mathcal{E}(\overline{z}),\mathcal{E}(u))+\langle(\overline{z}\cdot\nabla)\overline{z},u\rangle+\langle(U\cdot\nabla)\overline{z},u\rangle+\langle(\overline{z}\cdot\nabla)U,u\rangle\\
&\qquad+2\mu(k(x)\overline{z},u)_{\Gamma_2}+2\mu(S\tilde{\overline{z}},\tilde{u})_{\Gamma_3}+2(\alpha(x)\overline{z},u)_{\Gamma_5}+\mu(k(x)\overline{z},u)_{\Gamma_7}\\
&=-(U',u)-2\mu(\mathcal{E}(U),\mathcal{E}(u))-\langle(U\cdot\nabla)U,u\rangle-2\mu(k(x)U,u)_{\Gamma_2}\\
&\qquad-2\mu(S\tilde{U},\tilde{u})_{\Gamma_3}-2(\alpha(x)U_\tau,u)_{\Gamma_5}-\mu(k(x)U,u)_{\Gamma_7}+\langle f,u\rangle\\
&\qquad+\sum_{i=2,4,7}\langle\phi_i,u_n\rangle_{\Gamma_i}+\sum_{i=3,5,6}\langle\phi_i,u\rangle_{\Gamma_i}\qquad\text{for all } u\in\mathbf{V}.
\end{aligned}\tag{4.51}$$

Now, define an operator $A_0 : \mathbf{V} \to \mathbf{V}^*$ by

$$\begin{aligned}\langle A_0 y, u\rangle =& 2\mu(\mathcal{E}(y), \mathcal{E}(u)) + 2\mu(k(x)y, u)_{\Gamma_2} + 2\mu(S\tilde{y}, \tilde{u})_{\Gamma_3} \\ &+ 2(\alpha(x)y, u)_{\Gamma_5} + \mu(k(x)y, u)_{\Gamma_7} \quad \text{for all } y, u \in \mathbf{V}.\end{aligned} \tag{4.52}$$

Lemma 4.4 $\exists \delta > 0, \exists k_0 \geq 0; \langle A_0 u, u\rangle \geq \delta \|u\|_{\mathbf{V}}^2 - k_0\|u\|^2$ *for all* $u \in \mathbf{V}$.

Proof By Korn's inequality we have

$$2\mu(\mathcal{E}(u), \mathcal{E}(u)) \geq \beta\|u\|_{\mathbf{V}}^2 \quad \exists \beta > 0, \text{ for all } u \in \mathbf{V}. \tag{4.53}$$

By Remark 3.4 and Assumption 4.2, there exists a constant M such that

$$\|S(x)\|_\infty, \ \|k(x)\|_\infty, \ \|\alpha(x)\|_\infty \leq M,$$

and so there exists a constant c_0 (depending on β) such that

$$\begin{aligned}&\left|2\mu(k(x)z, z)_{\Gamma_2} + 2\mu(S\tilde{z}, \tilde{z})_{\Gamma_3} + \mu(k(x)z, z)_{\Gamma_7} + 2(\alpha(x)y, u)_{\Gamma_5}\right| \\ &\qquad \leq \frac{\beta}{2}\|z\|_{\mathbf{H}^1(\Omega)}^2 + c_0\|z\|^2\, dt \quad \text{for all } z \in \mathbf{V}\end{aligned} \tag{4.54}$$

(see Theorem 1.27). Set $\delta = \frac{\beta}{2}$ and $k_0 = c_0$. Then, by (4.53) and (4.54) we come to the asserted conclusion. □

Remark 4.2 In the process of proving Lemma 4.4, we see that if $\Gamma_i = \emptyset, \ i = 2, 3, 7,$ or these are unions of pieces of planes (segments in case of 2-D) and $\Gamma_5 = \emptyset$ or $\alpha(x) = 0$, then we can take $k_0 = 0$.

When $k_0 > 0$, if k_0 is not small enough, then the operator defined by (4.52) is not positive, and so let us transform the unknown function to get a positive operator A in (4.56) below. Now, let k_0 be the constant in Lemma 4.4 and put $z = e^{-k_0 t}\overline{z}$. Then, since $e^{-k_0 t}\overline{z}' = z' + k_0 z$, we get the following problem equivalent to problem (4.51):

Find z such that

$$
\begin{aligned}
&z \in L^2(0,T;\mathbf{V}),\\
&z(0) = v_0 - U(0) \in \mathbf{V},\\
&\langle z'(t), u\rangle + 2\mu(\mathcal{E}(z(t)), \mathcal{E}(u)) + e^{k_0 t}\langle (z(t)\cdot\nabla)z(t), u\rangle + \langle (U(t)\cdot\nabla)z(t), u\rangle\\
&\qquad + \langle (z(t)\cdot\nabla)U(t), u\rangle + k_0(z(t), u) + 2\mu(k(x)z(t), u)_{\Gamma_2}\\
&\qquad + 2\mu(S\tilde{z}(t), \tilde{u})_{\Gamma_3} + 2(\alpha(x)z(t), u)_{\Gamma_5} + \mu(k(x)z(t), u)_{\Gamma_7}\\
&= e^{-k_0 t}\Big[- (U'(t), u) - 2\mu(\mathcal{E}(U(t)), \mathcal{E}(u)) - \langle (U(t)\cdot\nabla)U(t), u\rangle\\
&\qquad - 2\mu(k(x)U(t), u)_{\Gamma_2} - 2\mu(S\tilde{U}(t), \tilde{u})_{\Gamma_3} - 2(\alpha(x)U(t)_\tau, u)_{\Gamma_5}\\
&\qquad - \mu(k(x)U(t), u)_{\Gamma_7} + \langle f(t), u\rangle + \sum_{i=2,4,7}\langle \phi_i(t), u_n\rangle_{\Gamma_i}\\
&\qquad + \sum_{i=3,5,6}\langle \phi_i(t), u\rangle_{\Gamma_i}\Big] \quad \text{for all } u \in \mathbf{V}.
\end{aligned}
\tag{4.55}
$$

Define operators A, $A_U(t) : \mathbf{V} \to \mathbf{V}^*$, respectively, by

$$
\langle Av, u\rangle = \langle A_0 v, u\rangle + (k_0 v, u) \quad \text{for all } v, u \in \mathbf{V}, \tag{4.56}
$$

and

$$
\langle A_U(t)v, u\rangle = \langle (U(t,x)\cdot\nabla)v, u\rangle + \langle (v\cdot\nabla)U(t,x), u\rangle \quad \text{for all } v, u \in \mathbf{V}, \tag{4.57}
$$

where A_0 is the operator by (4.52) and k_0 is the one in Lemma 4.4. Since $U \in \mathscr{W}$, we have $U \in C([0,T]; \mathbf{H}^1(\Omega))$ and so such a definition makes sense. Then, the operator A is positive definite, and this fact is used in (4.67).

Define an operator $B(t) : \mathbf{V} \to \mathbf{V}^*$ and an element $F(t) \in V^*$, respectively, by

$$
\langle B(t)v, u\rangle = e^{k_0 t}\langle (v\cdot\nabla)v, u\rangle \quad \text{for all } v, u \in \mathbf{V}, \tag{4.58}
$$

and

$$
\begin{aligned}
\langle F(t), u\rangle = &e^{-k_0 t}\Big[- (U'(t), u) - 2\mu(\mathcal{E}(U(t)), \mathcal{E}(u)) - \langle (U(t)\cdot\nabla)U(t), u\rangle\\
&- 2\mu(k(x)U(t), u)_{\Gamma_2} - 2\mu(S\tilde{U}(t), \tilde{u})_{\Gamma_3} - 2(\alpha(x)U(t)_\tau, u)_{\Gamma_5}\\
&- \mu(k(x)U(t), u)_{\Gamma_7} + \langle f(t), u\rangle + \sum_{i=2,4,7}\langle \phi_i(t), u_n\rangle_{\Gamma_i} + \sum_{i=3,5,6}\langle \phi_i(t), u\rangle_{\Gamma_i}\Big]\\
&\text{for all } u \in \mathbf{V}.
\end{aligned}
\tag{4.59}
$$

Then, (4.55) is written as

$$
\begin{aligned}
&z \in L^2(0,T;\mathbf{V}),\\
&z(0) = v_0 - U(0) \in \mathbf{V},\\
&z'(t) + (A + A_U(t) + B(t))\,z(t) = F(t).
\end{aligned}
\tag{4.60}
$$

Now, define operators $L,\ \widetilde{A}_U,\ L_U,\ \widetilde{B} : \mathscr{X} \to \mathscr{Y}, C : \mathscr{X} \times \mathscr{X} \to \mathscr{Y}$ and an element $F \in \mathscr{Y}$ by

$$
\begin{aligned}
&\langle (Lz)(t), u\rangle = \langle z'(t), u\rangle + \langle Az(t), u\rangle \quad \text{for all } z \in \mathscr{X}, \text{for all } u \in \mathbf{V},\\
&\langle (\widetilde{A}_U z)(t), u\rangle = \langle A_U(t) z(t), u\rangle \quad \text{for all } z \in \mathscr{X}, \text{for all } u \in \mathbf{V},\\
&\langle (L_U z)(t), u\rangle = \langle z'(t), u\rangle + \langle (A + A_U(t)) z(t), u\rangle \quad \text{for all } z \in \mathscr{X}, \text{for all } u \in \mathbf{V},\\
&\langle (\widetilde{B} z)(t), u\rangle = \langle B(t) z(t), u\rangle \quad \text{for all } z \in \mathscr{X}, \text{for all } u \in \mathbf{V},\\
&\langle C(w,z)(t), u\rangle = e^{k_0 t}\langle (w(t)\cdot\nabla) z(t), u\rangle + e^{k_0 t}\langle (z(t)\cdot\nabla) w(t), u\rangle\\
&\qquad\qquad\qquad\qquad \text{for all } w, z \in \mathscr{X}, \text{for all } u \in \mathbf{V},\\
&(F)(t) = F(t).
\end{aligned}
\tag{4.61}
$$

For $\widetilde{A}_U$ and F, we have

Lemma 4.5 *C is a bilinear continuous operator such that $\mathscr{X} \times \mathscr{X} \to \mathscr{Y}$. Under Assumptions 4.2 and 4.3, $\widetilde{A}_U$ is a linear continuous operator such that $\mathscr{X} \to \mathscr{Y}$ and $F \in \mathscr{Y}$.*

Proof Obviously, C is bilinear. When $w \in \mathscr{X}$,

$$
w \in L^\infty(0,T;\mathbf{V}), \quad \|w\|_{L^\infty(0,T;\mathbf{V})} \le c\big[\|w\|_{L^2(0,T;\mathbf{V})} + \|w'\|_{L^2(0,T;\mathbf{V})}\big]
$$

and by virtue of Hölder's inequality and Sobolev's imbedding theorem (Theorem 1.20)

$$
\begin{aligned}
&\left|e^{k_0 t}\langle (w\cdot\nabla) z, u\rangle + e^{k_0 t}\langle (z\cdot\nabla) w, u\rangle\right|\\
&\qquad \le c(\|w\|_{\mathbf{L}^3}\|\nabla z\|_{\mathbf{L}^2}\|u\|_{\mathbf{L}^6} + \|z\|_{\mathbf{L}^3}\|\nabla w\|_{\mathbf{L}^2}\|u\|_{\mathbf{L}^6}) \le c\|w\|_{\mathbf{V}}\|z\|_{\mathbf{V}}\|u\|_{\mathbf{V}}\\
&\qquad\qquad\qquad \text{for all } w, z, u \in \mathbf{V}.
\end{aligned}
$$

Thus,

$$
\|C(w,z)\|_{L^2(0,T;\mathbf{V}^*)} \le c\|w\|_{L^\infty(0,T;\mathbf{V})}\|z\|_{L^2(0,T;\mathbf{V})} \le c\|w\|_{\mathscr{X}}\cdot\|z\|_{\mathscr{X}}. \tag{4.62}
$$

Also, since

$$
\begin{aligned}
|\langle C(w,z)'(t), u\rangle| = &e^{k_0 t}\big|k_0\langle (w\cdot\nabla) z, u\rangle + k_0\langle (z\cdot\nabla) w, u\rangle + \langle (w'\cdot\nabla) z, u\rangle\\
&+ \langle (w\cdot\nabla) z', u\rangle + \langle (z'\cdot\nabla) w, u\rangle + \langle (z\cdot\nabla) w', u\rangle\big|,
\end{aligned}
$$

taking (4.62) into account, we have

$$\begin{aligned}\|C(w,z)'\|_{L^2(0,T;\mathbf{V}^*)} &\le c\|C(w,z)\|_{L^2(0,T;\mathbf{V}^*)}\\ &\quad + c\big[\big(\|w'\|_{L^2(0,T;\mathbf{V})} + \|w\|_{L^\infty(0,T;\mathbf{V})}\big)\big(\|z'\|_{L^2(0,T;\mathbf{V})} + \|z\|_{L^\infty(0,T;\mathbf{V})}\big)\big]\\ &\le c\|w\|_{\mathscr{X}} \cdot \|z\|_{\mathscr{X}}.\end{aligned} \tag{4.63}$$

(4.62) and (4.63) imply

$$\|C(w,z)\|_{\mathscr{Y}} \le c\|w\|_{\mathscr{X}} \cdot \|z\|_{\mathscr{X}}. \tag{4.64}$$

By the same argument as above, we have

$$\|\widetilde{A}_U z\|_{\mathscr{Y}} \le c\|U\|_{\mathscr{W}} \cdot \|z\|_{\mathscr{X}}. \tag{4.65}$$

By Assumption 4.2, Remark 3.4 and the trace theorem, we can see that $F \in \mathscr{Y}$. □

Lemma 4.6 *The operator $\overline{L}$ defined by $\overline{L}z = (z'(0), Lz)$ for $z \in \mathscr{X}$ is a linear continuous and one-to-one operator from $\mathscr{X}$ onto $H \times \mathscr{Y}$.*

Proof The linearity of $\overline{L}$ is obvious. The fact $z \in \mathscr{X}$ implies that $z' \in C([0,T];H)$, $\|z'\|_{C([0,T];H)} \le c\|z\|_{\mathscr{X}}$, and so we see that a map $z \in \mathscr{X} \to z'(0) \in H$ is continuous.

Clearly, $\|z'\|_{\mathscr{Y}} \le c\|z\|_{\mathscr{X}}$. Also, by Assumption 4.2, Remark 3.4 and the trace theorem,

$$|\langle Av, u\rangle| \le c\|v\|_{\mathbf{V}} \cdot \|u\|_{\mathbf{V}} \quad \text{for all } v, u \in \mathbf{V}. \tag{4.66}$$

Equation (4.66) implies that the mapping $z \in \mathscr{X} \to Az \in \mathscr{Y}$ is continuous. Therefore, $\overline{L}$ is continuous.

Next, we show that $\overline{L}$ is a one-to-one and surjective operator from $\mathscr{X}$ onto $H \times \mathscr{Y}$.

First, let us prove that this operator is injective. For this, it is enough to prove that the inverse image of $(0_H, 0_{\mathscr{Y}}) \in H \times \mathscr{Y}$ by the operator $\overline{L}$ is $0_{\mathscr{X}}$. By Lemma 4.4 and (4.56), we can find $\delta > 0$ such that

$$\langle Av, v\rangle \ge \delta\|v\|^2_{\mathbf{V}} \quad \text{for all } v \in \mathbf{V}. \tag{4.67}$$

By (4.66) and (4.67), for any $q \in \mathbf{V}^*$ there exists a unique solution $y \in V$ to the following problem:

$$Ay = q. \tag{4.68}$$

Let $z \in \mathscr{X}$ be the inverse image of $(0_H, 0_{\mathscr{Y}}) \in H \times \mathscr{Y}$ by $\overline{L}$. Then since $z'(0) = 0_H$, putting $t = 0$ in the first equation of (4.61) we get

$$\langle Az(0), u\rangle = 0 \quad \text{for all } u \in \mathbf{V},$$

where $z(0) = z(0,x)$. This means that $z(0)$ is a unique solution to (4.68) for $q = 0_{\mathbf{V}^*}$, i.e. $z(0) = 0_{\mathbf{V}}$. Putting $w = z'$, we get $w(0) = z'(0) = 0_H$. Taking $Lz = 0$ into account and differentiating the first equation of (4.61), we have

$$\langle w'(t), u\rangle + \langle Aw(t), u\rangle = 0 \quad \text{for all } u \in \mathbf{V}. \tag{4.69}$$

The operator $\mathcal{A}$ defined by

$$(\mathcal{A}w)(t) = Aw(t) \quad \forall w \in L^2(0, T; \mathbf{V})$$

satisfies all conditions of Theorem 1.46. Thus, for problem (4.69) with an initial condition $w(0) \in H$ there exists a unique solution w such that $w \in L^2(0, T; \mathbf{V})$, $w' \in L^2(0, T; \mathbf{V}^*)$. Since $w(0) = 0_H$, we have $w = 0$, which means $z = 0_{\mathcal{X}}$ since $z(0) = 0_{\mathbf{V}}$.

Second, let us prove that $\overline{L}$ is surjective. Let $(w_0, g) \in H \times \mathcal{Y}$. Since $g \in \mathcal{Y}$, we have $g(0) \in \mathbf{V}^*$. Then, by (4.66) and (4.67), there exists a unique solution $z_0 \in \mathbf{V}$ to problem

$$Az_0 = g(0) - w_0. \tag{4.70}$$

Let us consider problem

$$\begin{cases} w' + Aw = g', \\ w(0) = w_0. \end{cases} \tag{4.71}$$

By Theorem 1.46, there exists a unique solution w such that $w \in L^2(0, T; \mathbf{V})$, $w' \in L^2(0, T; \mathbf{V}^*)$ to problem (4.71). Now, put

$$z = z_0 + \int_0^t w(s)\, ds, \tag{4.72}$$

where z_0 is the solution to (4.70). Then, $z' = w$ and $z \in \mathcal{X}$. Integrating both sides of the first one in (4.71) from 0 to t and using (4.72), we have

$$\langle w(t), u\rangle + \langle Az(t), u\rangle - [\langle w_0, u\rangle + \langle Az_0, u\rangle] = \langle g(t), u\rangle - \langle g(0), u\rangle \quad \text{for all } u \in \mathbf{V}. \tag{4.73}$$

Taking (4.70), (4.72) into account, from (4.73) we get

$$\langle z'(t), u\rangle + \langle Az(t), u\rangle = \langle g(t), u\rangle \quad \text{for all } u \in \mathbf{V}. \tag{4.74}$$

This means that $z \in \mathcal{X}$ defined by (4.72) is the inverse image of $(w_0, g) \in H \times \mathcal{Y}$ by the operator $\overline{L}$, i.e. $\overline{L}$ is surjective. □

Lemma 4.7 *Under Assumption 4.3, let $\|U(0, x)\|_{\tilde{\mathbf{V}}}$ be small enough. The operator $\overline{L}_U$ defined by $\overline{L}_U z = (z'(0), L_U z)$ for $z \in \mathcal{X}$ is a linear continuous one-to-one operator from $\mathcal{X}$ onto $H \times \mathcal{Y}$.*

Proof If $z \in \mathcal{X}$, then $z \in C([0, T]; \mathbf{V})$ and

$$\|z\|_{C([0,T];\mathbf{V})} \leq c\big[\|z\|_{L^2(0,T;\mathbf{V})} + \|z'\|_{L^2(0,T;\mathbf{V})}\big].$$

By virtue of this fact and Lemma 4.5, the operator $\overline{A}_U \in (\mathscr{X} \to H \times \mathscr{Y})$ defined by $\overline{A}_U z = (0_H, \widetilde{A}_U z)$ is continuous. Thus, the operator $\overline{L}_U$ defined on $\mathscr{X}$ is linear continuous.

As in Lemma 3.5 of [3] it is proved that the operator $\widetilde{A}_U \in (\mathscr{X} \to \mathscr{Y})$ is compact. Thus, $\overline{A}_U \in (\mathscr{X} \to H \times \mathscr{Y})$ is also compact. Since $\overline{L}_U = \overline{L} + \overline{A}_U$, by virtue of Theorem 1.8 and Lemma 4.6 we see that in order to prove that the operator $\overline{L}_U$ is one-to-one from $\mathscr{X}$ onto $H \times \mathscr{Y}$, it is enough to prove that $\overline{L}_U$ is one-to-one from $\mathscr{X}$ into $H \times \mathscr{Y}$. To prove the last fact it is enough to show that the inverse image of $(0_H, 0_{\mathscr{Y}})$ by $\overline{L}_U$ is $0_{\mathscr{X}}$. By Hölder's inequality and Sobolev's imbedding theorem

$$\left|\langle (U(t,x)\cdot\nabla)v, v\rangle + \langle (v\cdot\nabla)U(t,x), v\rangle\right| \le K_0\|v\|_{\mathbf{V}}\|U(t,x)\|_{\mathbf{H}^1}\|v\|_{\mathbf{V}}. \tag{4.75}$$

Thus, if $\|U(0,x)\|_{\widetilde{V}}$ is so small that $\|U(0,x)\|_{\mathbf{H}^1} \le \frac{\delta}{2K_0}$, then (4.66), (4.67) and (4.75) imply

$$\begin{aligned} &\left|\langle (A + A_U(0))v, u\rangle\right| \le c\|v\|_{\mathbf{V}}\cdot\|u\|_{\mathbf{V}} \quad \text{for all } v, u \in \mathbf{V}, \\ &\langle (A + A_U(0))v, v\rangle \ge \frac{\delta}{2}\|v\|_{\mathbf{V}}^2 \quad \text{for all } v \in \mathbf{V}. \end{aligned} \tag{4.76}$$

By (4.76), for any $q \in \mathbf{V}^*$ there exists a unique solution $y \in \mathbf{V}$ to

$$(A + A_U(0))y = q. \tag{4.77}$$

Let $z \in \mathscr{X}$ be the inverse image of $(0_H, 0_{\mathscr{Y}})$ by $\overline{L}$. Then $z'(0) = 0_H$, and putting $t = 0$ from the third one in (4.61) we get

$$\langle (A + A_U(0))z(0), u\rangle = 0 \quad \text{for all } u \in \mathbf{V},$$

where $z(0) = z(0,x)$. This means that $z(0)$ is the unique solution to (4.77) with $q = 0_{\mathbf{V}^*}$, i.e. $z(0) = 0_{\mathbf{V}}$. Therefore, $z \in \mathscr{X}$ satisfies

$$\begin{cases} z'(t) + (A + A_U(t))z(t) = 0, \\ z(0) = 0_{\mathbf{V}}. \end{cases} \tag{4.78}$$

Now, making duality pairing with $z(t)$ on both sides of

$$z'(t) + Az(t) = -A_U(t)z(t)$$

and taking (4.67) into account and using Gronwall's inequality, we can prove $z = 0_{\mathscr{X}}$ as in Lemma 3.8 of [3]. The proof is thus completed. □

Lemma 4.8 *Under Assumption 4.3 the operator T defined by $Tz = \big(z'(0), (L_U + \widetilde{B})z\big)$ for $z \in \mathscr{X}$ is continuously differentiable, $T(0_{\mathscr{X}}) = (0_H, 0_{\mathscr{Y}})$ and the Fréchet derivative of T at $0_{\mathscr{X}}$ is $\overline{L}_U$.*

Proof It is easy to verify that $T(0_{\mathscr{X}}) = (0_H, 0_{\mathscr{Y}})$. Since the operator L_U is linear, its Fréchet derivative is the same as itself. Therefore, if $\widetilde{B}$ is continuously differentiable, then so is T.

For any $w, z \in \mathscr{X}$,

$$\big(\widetilde{B}(w+z) - \widetilde{B}w\big)(t) = e^{k_0 t}\big(w(t)\cdot\nabla\big)z(t) + e^{k_0 t}\big(z(t)\cdot\nabla\big)w(t) + (\widetilde{B}z)(t).$$

By (4.64), we get

$$\lim_{\|z\|_{\mathscr{X}}\to 0} \frac{\|\widetilde{B}z\|_{\mathscr{Y}}}{\|z\|_{\mathscr{X}}} \le \lim_{\|z\|_{\mathscr{X}}\to 0} \frac{c\|z\|_{\mathscr{X}}^2}{\|z\|_{\mathscr{X}}} = 0.$$

Then, put

$$C(w,z)(t) \equiv e^{k_0 t}\big(w(t)\cdot\nabla\big)z(t) + e^{k_0 t}\big(z(t)\cdot\nabla\big)w(t) = (\widetilde{B}'_w z)(t).$$

By Lemma 4.5 $\widetilde{B}'_w : \mathscr{X} \to \mathscr{Y}$ is continuous, and it is the Fréchet derivative of $\widetilde{B}$ at w and also continuous with respect to w. Thus, T is continuously differentiable. Also from the formula above we can see that the Fréchet derivative of $\widetilde{B}$ at $0_{\mathscr{X}}$ is zero. Therefore, the Fréchet derivative of T at $0_{\mathscr{X}}$ is $\overline{L}_U$. □

Let us consider problem

$$\big(A + A_U(0) + B(0)\big)u = q. \tag{4.79}$$

Lemma 4.9 *Assume that* $\|U(0,x)\|_{\widetilde{\mathbf{V}}}$ *is small enough. If the norm of* $q \in V^*$ *is small enough, then there exists a unique solution to* (4.79) *in some* $\mathscr{O}_M(0_{\mathbf{V}})$.

Proof Since $\|U(0,x)\|_{\widetilde{\mathbf{V}}}$ is small enough, by (4.76), for any fixed $z \in \mathbf{V}$ there exists a unique solution to problem

$$\big(A + A_U(0)\big)w = q - B(0)z. \tag{4.80}$$

On the other hand,

$$|\langle B(0)w_1 - B(0)w_2, u\rangle| \le KM\|w_1 - w_2\|_{\mathbf{V}}\cdot\|u\|_{\mathbf{V}} \text{ for all } w_i \in \mathscr{O}_M(0_{\mathbf{V}}), \text{ for all } u \in \mathbf{V}. \tag{4.81}$$

Owing to (4.76) the solution w to (4.80) is estimated as follows:

$$\|w\|_{\mathbf{V}} \le \frac{2}{\delta}\big(\|q\|_{\mathbf{V}^*} + \|B(0)z\|_{\mathbf{V}^*}\big) \le \frac{2}{\delta}\big(\|q\|_{\mathbf{V}^*} + KM^2\big).$$

Thus, if $\|q\|_{\mathbf{V}^*}$ and M are small enough, then the operator $(z \to w)$ maps $\mathscr{O}_M(0_{\mathbf{V}})$ into itself and by (4.81) this operator is strictly contract. Therefore, in $\mathscr{O}_M(0_{\mathbf{V}})$ there exists a unique solution to (4.80). Thus, we come to the asserted conclusion. □

One of main results of this section is the following:

Theorem 4.3 *Suppose that Assumptions 4.2 and 4.3 hold. Assume that $\|U\|_{\mathscr{W}}$ and the norms of f, f', ϕ_i, ϕ_i' in the spaces where they belong to are small enough.*

If

$$w_0 \equiv F(0) - (A + A_U(0) + B(0))z_0 \in H \tag{4.82}$$

(compatibility condition at the initial time), where $z_0 = v_0 - U(0, \cdot)$, and $\|w_0\|_H$ is small enough, then there exists a unique solution to Problem 4.3 in the space $\mathscr{W}$.

Proof First, let us prove the existence of a solution.

If $\|U\|_{\mathscr{W}}$ and the norms of f, f', ϕ_i, ϕ_i' in the spaces that they belong to are small enough, then $\|F\|_{\mathscr{Y}}$ is also small enough. By virtue of Lemmas 4.7, 4.8 and the Local inverse mapping theorem (Theorem 1.7), for any $R_1 > 0$ small enough if $\|F\|_{\mathscr{Y}}$, R are small enough and $w_1 \in \mathscr{O}_R(0_H)$, then there exists a unique $z \in \mathscr{O}_{R_1}(0_{\mathscr{X}})$ such that

$$\begin{aligned} &z'(t) + \big(A + A_U(t) + B(t)\big)z(t) = F(t), \\ &z'(0) = w_1 \in \mathscr{O}_R(0_H). \end{aligned} \tag{4.83}$$

Putting $t = 0$ in (4.83), we get

$$F(0) - \big(A + A_U(0) + B(0)\big)z(0) = w_1 \in \mathscr{O}_R(0_H).$$

On the other hand, if $\|U\|_{\mathscr{W}}$ is small enough, then so is $\|U(0, x)\|_{\tilde{\mathbf{V}}}$. Thus, when $\|F(0) - w_1\|_{V^*}$ is small enough, by Lemma 4.9 there exists a unique solution $z_0 \in \mathscr{O}_{R_2}(0_V)$ for some $R_2 > 0$ to

$$\big(A + A_U(0) + B(0)\big)z_0 = F(0) - w_1. \tag{4.84}$$

Since $\|z(0)\|_V \le c\|z\|_{\mathscr{X}}$, we can choose R_1 such that $z(0) \in \mathscr{O}_{R_2}(0_V)$, and we have $z(0) = z_0$. Therefore, if $\|F\|_{\mathscr{Y}}$ is small enough, $F(0) - \big(A + A_U(0) + B(0)\big)z_0$ belongs to H and its norm is small enough, then $z \in \mathscr{X}$, the solution to (4.83), is a solution to problem

$$\begin{cases} z'(t) + \big(A + A_U(t) + B(t)\big)z(t) = F(t), \\ z(0) = z_0. \end{cases} \tag{4.85}$$

By definitions of $A, A_U(t), B(t), F$, the solution z of (4.85) is also a solution to (4.55) which is equivalent to (4.51). Thus, $e^{k_0 t}z + U \in \mathscr{W}$ is a solution to (4.50).

Second, let us prove uniqueness. We argue by contradiction.

Let v_1, v_2 be two solutions to (4.49) corresponding to the same data. Putting $\overline{w} = v_1 - v_2$, we have

$$\begin{aligned}
&\overline{w} \in L^2(0, T; \mathbf{V}),\\
&\overline{w}(0) = 0,\\
&\langle \overline{w}', u\rangle + 2\mu(\mathcal{E}(\overline{w}), \mathcal{E}(u)) + \langle (v_1 \cdot \nabla)\overline{w}, u\rangle + \langle (\overline{w} \cdot \nabla)v_2, u\rangle + 2\mu(k(x)\overline{w}, u)_{\Gamma_2}\\
&\qquad + 2\mu(S\tilde{\overline{w}}, \tilde{u})_{\Gamma_3} + 2(\alpha(x)\overline{w}, u)_{\Gamma_5} + \mu(k(x)\overline{w}, u)_{\Gamma_7} = 0 \quad \text{for all } u \in \mathbf{V}.
\end{aligned} \tag{4.86}$$

Putting $w = e^{-k_0 t}\overline{w}$, where k_0 is the constant in Lemma 4.4, we get $e^{-k_0 t}\overline{w}' = w' + k_0 w$. Then, we have

$$\begin{aligned}
&w \in L^2(0, T; \mathbf{V}),\\
&w(0) = 0,\\
&\langle w', u\rangle + 2\mu(\mathcal{E}(w), \mathcal{E}(u)) + \langle (v_1 \cdot \nabla)w, u\rangle + \langle (w \cdot \nabla)v_2, u\rangle\\
&\qquad + k_0(w, u) + 2\mu(k(x)w, u)_{\Gamma_2} + 2\mu(S\tilde{w}, \tilde{u})_{\Gamma_3}\\
&\qquad + 2(\alpha(x)w, u)_{\Gamma_5} + \mu(k(x)w, u)_{\Gamma_7} = 0 \quad \forall u \in \mathbf{V},
\end{aligned} \tag{4.87}$$

which is equivalent to (4.86). By Lemma 4.4,

$$\begin{aligned}
2\mu(\mathcal{E}(w), \mathcal{E}(w)) &+ k_0(w, w) + 2\mu(k(x)w, w)_{\Gamma_2} + 2\mu(S\tilde{w}, \tilde{w})_{\Gamma_3}\\
&+ 2(\alpha(x)w, w)_{\Gamma_5} + \mu(k(x)w, w)_{\Gamma_7} \geq \delta\|w\|^2_{\mathbf{V}}.
\end{aligned}$$

Taking this into account, we can prove $w = 0_{\mathscr{X}}$ as in Lemma 3.8 of [3]. Thus, uniqueness of solutions is proved, and so the proof is completed. □

Remark 4.3 Let us consider more precisely the condition that $F(0) - \big(A + A_U(0) + B(0)\big)z_0$ belongs to H and its norm is small enough. By (4.56)–(4.61), we have

$$\begin{aligned}
&\big\langle F(0) - \big(A + A_U(0) + B(0)\big)z_0, u\big\rangle =\\
&\Big[- (U'(0, x), u) - 2\mu(\mathcal{E}(U(0, x)), \mathcal{E}(u)) - \langle (U(0, x) \cdot \nabla)U(0, x), u\rangle\\
&\quad - 2\mu(k(x)U(0, x), u)_{\Gamma_2} - 2\mu(S\tilde{U}(0), \tilde{u})_{\Gamma_3} - 2(\alpha(x)U(0, x)_\tau, u)_{\Gamma_5}\\
&\quad - \mu(k(x)U(0, x), u)_{\Gamma_7} + \langle f(0), u\rangle + \sum_{i=2,4,7} \langle \phi_i(t), u_n\rangle_{\Gamma_i} + \sum_{i=3,5,6} \langle \phi_i(t), u\rangle_{\Gamma_i}\Big]\\
&- \big[2\mu(\mathcal{E}(z_0), \mathcal{E}(u)) + 2\mu(k(x)z_0, u)_{\Gamma_2} + 2\mu(S\tilde{z}_0, \tilde{u})_{\Gamma_3}\\
&\qquad\qquad + 2(\alpha(x)z_0, u)_{\Gamma_5} + \mu(k(x)z_0, u)_{\Gamma_7}\big]\\
&- \big[\langle (U(0, x) \cdot \nabla)z_0, u\rangle + \langle (z_0 \cdot \nabla)U(0, x), u\rangle\big] - \langle (z_0, \nabla)z_0, u\rangle \quad \text{for all } u \in \mathbf{V}.
\end{aligned} \tag{4.88}$$

Taking into account the fact that $U(0, x) + z_0 = v_0$, $U'(0, x) \in \mathbf{L}^2(\Omega)$ and its norm is small enough, from (4.88) we can see that the condition mentioned above is equivalent to the condition $\overline{w}_0 \in \mathscr{O}_R(0_H)$ for $R > 0$ small enough, where $\overline{w}_0$ is defined by

$$\langle \overline{w}_0, u\rangle \equiv \langle f(0), u\rangle + \sum_{i=2,4,7} \langle \phi_i(0,x), u_n\rangle_{\Gamma_i} + \sum_{i=3,5,6} \langle \phi_i(0,x), u\rangle_{\Gamma_i}$$
$$- \big[2\mu(\mathcal{E}(v_0), \mathcal{E}(u)) + 2\mu(k(x)v_0, u)_{\Gamma_2} + 2\mu(S\tilde{v}_0, \tilde{u})_{\Gamma_3} + 2(\alpha(x)v_0, u)_{\Gamma_5}$$
$$+ \mu(k(x)v_0, u)_{\Gamma_7} + \langle (v_0 \cdot \nabla)v_0, u\rangle + k_0(v_0, u)\big] \quad \text{for all } u \in \mathbf{V}. \tag{4.89}$$

Remark 4.4 If $\Gamma_i = \emptyset$, $i = 2, \cdots, 5, 7$, then the problem is reduced to the one in [4] where a local-in-time solution was studied. In this case $k_0 = 0$ (see Remark 4.1), and condition (4.82) is the same as (25) in [4]. And our condition for U is also the same as the one in [4].

4.2.2 Existence and Uniqueness of Solutions to Problem II

Let

$$\mathbf{V}_1 = \{u \in \mathbf{H}^1(\Omega) : \operatorname{div} u = 0,\ u|_{\Gamma_1} = 0,\ u_\tau|_{(\Gamma_2 \cup \Gamma_4)} = 0,\ u_n|_{(\Gamma_3 \cup \Gamma_5)} = 0\},$$
$$\mathbf{V}_{\Gamma 2-5} = \{u \in \mathbf{H}^1(\Omega) : \operatorname{div} u = 0,\ u_\tau|_{(\Gamma_2 \cup \Gamma_4)} = 0,\ u_n|_{(\Gamma_3 \cup \Gamma_5)} = 0\}.$$

Denote by H_1 the completion of $\mathbf{V}_1$ in the space $\mathbf{L}^2(\Omega)$.

By Theorems 3.1 and 3.2, for $v \in \mathbf{H}^2(\Omega) \cap \mathbf{V}_{\Gamma 2-5}(\Omega)$, $p \in H^1(\Omega)$ and $u \in \mathbf{V}_1$ we have

$$\begin{aligned} -\mu(\Delta v, u) &+ (\nabla p, u) \\ &= \mu(\nabla v, \nabla u) + \mu(k(x)v, u)_{\Gamma_2} + \mu(S\tilde{v}, \tilde{u})_{\Gamma_3} + \mu(k(x)v, u)_{\Gamma_7} \\ &\quad + (p, u\cdot n)_{\Gamma_2} - (\operatorname{rot} v \times u, u)_{\Gamma_3} - \big((-p + \mu\varepsilon_{nn}(v)), u\cdot n\big)_{\Gamma_4} \\ &\quad - 2\mu(\varepsilon_{n\tau}(v), u)_{\Gamma_5} - \Big(\big(-pn + \frac{\partial v}{\partial n}\big), u_n\Big)_{\Gamma_7}. \end{aligned} \tag{4.90}$$

Let

$$\mathscr{X}_1 = \{w \in L^2(0,T;\mathbf{V}_1);\ w' \in L^2(0,T;\mathbf{V}_1),\ w'' \in L^2(0,T;\mathbf{V}_1^*)\},$$
$$\|w\|_{\mathscr{X}_1} = \|w\|_{L^2(0,T;\mathbf{V}_1)} + \|w'\|_{L^2(0,T;\mathbf{V}_1)} + \|w''\|_{L^2(0,T;\mathbf{V}_1^*)},$$
$$\mathscr{Y}_1 = \{w \in L^2(0,T;\mathbf{V}_1^*);\ w' \in L^2(0,T;\mathbf{V}_1^*)\},$$
$$\|w\|_{\mathscr{Y}_1} = \|w\|_{L^2(0,T;\mathbf{V}_1^*)} + \|w'\|_{L^2(0,T;\mathbf{V}_1^*)}.$$

Unlike Problem I, for Problem II we do not require the condition $v_\tau|_{\Gamma_7} = 0$, and so instead of Assumptions 4.2 and 4.3, we use the following assumptions:

Assumption 4.4 Assumption 4.2 holds with $\phi_7, \phi_7' \in L^2(0,T;\mathbf{H}^{-\frac{1}{2}}(\Gamma_7))$ instead of $\phi_7, \phi_7' \in L^2(0,T;H^{-\frac{1}{2}}(\Gamma_7))$.

Assumption 4.5 There exists a function $U \in \mathscr{W}$ such that

$$\operatorname{div} U = 0,\ U|_{\Gamma_1} = h_1,\ U_\tau|_{\Gamma_2} = 0,\ U_n|_{\Gamma_3} = 0,\ U_\tau|_{\Gamma_4} = h_4,\ U_n|_{\Gamma_5} = h_5,$$

where $\mathscr{W}$ is the same as in the previous section. Also, $U(0, x) - v_0 \in \mathbf{V}_1$.

Applying (4.90), we get the following variational formulation for Problem II (with boundary conditions (4.49)):

Problem 4.4 Find v such that

$$\begin{aligned} &v - U \in L^2(0, T; \mathbf{V}_1), \\ &v(0) = v_0, \\ &\langle v', u\rangle + \mu(\nabla v, \nabla u) + \langle (v \cdot \nabla)v, u\rangle + \mu(k(x)v, u)_{\Gamma_2} \\ &\qquad + \mu(S\tilde{v}, \tilde{u})_{\Gamma_3} + 2(\alpha(x)v_\tau, u)_{\Gamma_5} - \mu(S\tilde{v}, \tilde{u})_{\Gamma_5} \\ &\qquad = \langle f, u\rangle + \sum_{i=2,4} \langle \phi_i, u_n\rangle_{\Gamma_i} + \sum_{i=3,5,7} \langle \phi_i, u\rangle_{\Gamma_i} \quad \text{for all } u \in \mathbf{V}_1. \end{aligned} \tag{4.91}$$

Taking into account Assumption 4.4 and putting $v = \overline{z} + U$, we get the following problem equivalent to Problem 4.4:
Find $\overline{z}$ such that

$$\begin{aligned} &\overline{z} \in L^2(0, T; \mathbf{V}_1), \\ &\overline{z}(0) \equiv v_0 - U(0) \in \mathbf{V}_1, \\ &\langle \overline{z}', u\rangle + \mu(\nabla \overline{z}, \nabla u) + \langle (\overline{z} \cdot \nabla)\overline{z}, u\rangle + \langle (U \cdot \nabla)\overline{z}, u\rangle + \langle (\overline{z} \cdot \nabla)U, u\rangle \\ &\qquad + \mu(k(x)\overline{z}, u)_{\Gamma_2} + \mu(S\tilde{\overline{z}}, \tilde{u})_{\Gamma_3} + 2(\alpha(x)\overline{z}, u)_{\Gamma_5} - \mu(S\tilde{\overline{z}}, \tilde{u})_{\Gamma_5} \\ &\quad = -\langle U', u\rangle - \mu(\nabla U, \nabla u) - \langle (U \cdot \nabla)U, u\rangle - \mu(k(x)U, u)_{\Gamma_2} - \mu(S\tilde{U}, \tilde{u})_{\Gamma_3} \\ &\qquad - 2(\alpha(x)U_\tau, u)_{\Gamma_5} + \mu(S\tilde{U}, \tilde{u})_{\Gamma_5} + \langle f, u\rangle + \sum_{i=2,4} \langle \phi_i, u_n\rangle_{\Gamma_i} + \sum_{i=3,5,7} \langle \phi_i, u\rangle_{\Gamma_i} \\ &\qquad\qquad \text{for all } u \in \mathbf{V}_1. \end{aligned} \tag{4.92}$$

Define an operator $A_{01} : \mathbf{V}_1 \to \mathbf{V}_1^*$ by

$$\begin{aligned} \langle A_{01} y, u\rangle = \mu(\nabla y, \nabla u) + \mu(k(x)y, u)_{\Gamma_2} + \mu(S\tilde{y}, \tilde{u})_{\Gamma_3} + 2(\alpha(x)y, u)_{\Gamma_5} - \mu(S\tilde{y}, \tilde{u})_{\Gamma_5} \\ \text{for all } y, u \in \mathbf{V}_1. \end{aligned} \tag{4.93}$$

By virtue of the same argument as in the proof of Lemma 4.4 we get

Lemma 4.10 $\exists \delta > 0,\ \exists k_1 \geq 0$:

$$\langle A_{01} u, u\rangle \geq \delta \|u\|^2_{\mathbf{V}_1} - k_1 \|u\|^2 \quad \textit{for all } u \in \mathbf{V}_1.$$

Putting $z = e^{-k_1 t}\overline{z}$, where k_1 is the constant in Lemma 4.10, and using the fact that $e^{-k_1 t}\overline{z}' = z' + k_1 z$, we get the following problem equivalent to (4.92):

Find z such that

$$\begin{aligned}
&z \in L^2(0, T; \mathbf{V}_1),\\
&z(0) = z_0 \equiv v_0 - U(0) \in \mathbf{V}_1,\\
&\langle z'(t), u\rangle + \mu(\nabla z, \nabla u) + e^{k_1 t}\langle (z(t)\cdot\nabla)z(t), u\rangle + \langle (U(t)\cdot\nabla)z(t), u\rangle\\
&\quad + \langle (z(t)\cdot\nabla)U(t), u\rangle + k_1(z(t), u) + \mu(k(x)z(t), u)_{\Gamma_2} + \mu(S\tilde{z}(t), \tilde{u})_{\Gamma_3}\\
&\quad + 2(\alpha(x)z(t), u)_{\Gamma_5} - \mu(S\tilde{z}(t), \tilde{u})_{\Gamma_5}\\
&= e^{-k_1 t}\Big[- (U'(t), u) - \mu(\nabla U, \nabla u) - \langle (U(t)\cdot\nabla)U(t), u\rangle\\
&\quad - \mu(k(x)U(t), u)_{\Gamma_2} - \mu(S\tilde{U}(t), \tilde{u})_{\Gamma_3} - 2(\alpha(x)U(t), u)_{\Gamma_5} - \mu(S\tilde{U}(t)_\tau, \tilde{u})_{\Gamma_5}\\
&\quad + \langle f(t), u\rangle + \sum_{i=2,4}\langle \phi_i(t), u_n\rangle_{\Gamma_i} + \sum_{i=3,5,7}\langle \phi_i(t), u\rangle_{\Gamma_i}\Big] \quad \text{for all } u \in \mathbf{V}_1.
\end{aligned} \tag{4.94}$$

Define operators A_1, $A_{1U}(t)$ by

$$\langle A_1 v, u\rangle = \langle A_{01} v, u\rangle + (k_1 v, u) \quad \text{for all } v, u \in \mathbf{V}_1, \tag{4.95}$$

$$\langle A_{1U}(t)v, u\rangle = \langle (U(t,x)\cdot\nabla)v, u\rangle + \langle (v\cdot\nabla)U(t,x), u\rangle \quad \text{for all } v, u \in \mathbf{V}_1, \tag{4.96}$$

where A_{01} is the one defined in (4.93). $U \in \mathscr{W}$ implies $U \in C([0, T]; \mathbf{H}^1(\Omega))$, and such definitions make sense. Also, define an operator $B_1(t) : \mathbf{V}_1 \to \mathbf{V}_1^*$ by

$$\langle B_1(t)v, u\rangle = e^{k_1 t}\langle (v\cdot\nabla)v, u\rangle \quad \text{for all } v, u \in \mathbf{V}_1. \tag{4.97}$$

Define an element $F_1 \in \mathscr{Y}_1$ by

$$\begin{aligned}
\langle F_1(t), u\rangle &= e^{-k_1 t}\Big[- \langle U'(t), u\rangle - \mu(\nabla U(t), \nabla u) - \langle (U(t)\cdot\nabla)U(t), u\rangle\\
&\quad - \mu(k(x)U(t), u)_{\Gamma_2} - \mu(S\tilde{U}(t), \tilde{u})_{\Gamma_3} - 2(\alpha(x)U(t)_\tau, u)_{\Gamma_5} + \mu(S\tilde{U}(t), \tilde{u})_{\Gamma_5}\\
&\quad + \langle f, u\rangle + \sum_{i=2,4}\langle \phi_i, u_n\rangle_{\Gamma_i} + \sum_{i=3,5,7}\langle \phi_i, u\rangle_{\Gamma_i}\Big] \quad \text{for all } u \in \mathbf{V}_1.
\end{aligned} \tag{4.98}$$

Now, in the same way as in the proof of Theorem 4.3 we can prove the following theorem which is one of the main results of this section.

Theorem 4.4 *Suppose that Assumptions 4.4 and 4.5 hold. Assume that $\|U\|_{\mathscr{W}}$ and the norms of f, f', ϕ_i, ϕ_i' in the spaces where they belong to are small enough.*

If (compatibility condition at the initial instance)

$$w_1 \equiv F_1(0) - (A_1 + A_{1U}(0) + B_1(0))z_0 \in H_1, \tag{4.99}$$

where $z_0 = v_0 - U(0,\cdot)$, *and* $\|w_1\|_{H_1}$ *is small enough, then in the space* $\mathscr{W}$ *there exists a unique solution to Problem 4.4.*

Remark 4.5 By the same argument as in Remark 4.3, we can see that the condition (4.99) is equivalent to the condition $\overline{w}_1 \in H_1$, where $\overline{w}_1 \in \mathbf{V}_1^*$ is defined by

$$\begin{aligned}\langle \overline{w}_1, u\rangle =&\langle f(0), u\rangle + \sum_{i=2,4} \langle \phi_i(0,x), u_n\rangle_{\Gamma_i} + \sum_{i=3,5,7} \langle \phi_i(0,x), u\rangle_{\Gamma_i}\\ &- \big[\mu(\nabla v_0, \nabla u) + \mu(k(x)v_0, u)_{\Gamma_2} + \mu(S\tilde{v}_0, \tilde{u})_{\Gamma_3} + 2(\alpha(x)v_0, u)_{\Gamma_5}\\ &- \mu(S\tilde{v}_0, \tilde{u})_{\Gamma_5} + \langle (v_0\cdot\nabla)v_0, u\rangle + k_1(v_0, u)\big] \quad \text{for all } u \in \mathbf{V}_1,\end{aligned} \tag{4.100}$$

with k_1 in Lemma 4.10.

Remark 4.6 If $U \equiv 0$ and $\Gamma_i = \emptyset, i = 2, \cdots, 5$, then problem (4.91) is reduced to one in [3]. In this case $k_1 = 0$ (see Remark 4.2). If $v_0 \in \mathbf{H}^{l/2}(\Omega)$, then $(v_0\cdot\nabla)v_0 \in \mathbf{L}^2(\Omega)$. Thus, the condition above for $\overline{w}_1$ being in H_1 is the same as one of conditions of Theorems 3.5–3.8 of [3], but we do not demand $v_0 \in \mathbf{H}^{r_0}(\Omega)$, $r_0 > \frac{l}{2}$.

4.2.3 Existence and Uniqueness of Solutions for Perturbed Data

In [3] it is proved that if a solution satisfying smoothness and a compatibility condition is given, then there exists a unique solution for small perturbed data satisfying the compatibility condition. In this section we get such results for the Problems I and II. In our results the conditions for a given solution is essentially the same as the one in [3], but the smoothness condition for the initial functions in the compatibility condition for small perturbed data is weaker than the one in [3] (see Remark 4.8).

Let $\widetilde{\mathbf{V}}^{r_0} = \{u \in \mathbf{H}^{r_0}(\Omega) : \operatorname{div} u = 0\}$, $r_0 > l/2$, and

$$\begin{aligned}\overline{\mathscr{W}} &= \left\{w \in L^2(0,T;\widetilde{\mathbf{V}});\, w' \in L^2(0,T;\widetilde{\mathbf{V}}),\, w'' \in L^2(0,T;\widetilde{\mathbf{V}}^*),\, w(0) \in \widetilde{\mathbf{V}}^{r_0}\right\},\\ \|w\|_{\overline{\mathscr{W}}} &= \|w\|_{L^2(0,T;\widetilde{\mathbf{V}})} + \|w'\|_{L^2(0,T;\widetilde{\mathbf{V}})} + \|w''\|_{L^2(0,T;\widetilde{\mathbf{V}}^*)} + \|w(0)\|_{\widetilde{\mathbf{V}}^{r_0}}.\end{aligned}$$

Let us first consider Problem I.

Let $W(x,t) \in \overline{\mathscr{W}}$ be a given solution to Problem I. Let v be the solution for the data perturbed except h_i, and put $v = \overline{z} + W$. Then, we get a problem for $\overline{z}$:

Find $\overline{z}$ such that

$$
\begin{aligned}
&\bar{z} \in L^2(0,T;\mathbf{V}),\\
&\bar{z}(0) = z_0 \equiv v_0 - W(0,x) \in \mathbf{V},\\
&\langle \bar{z}', u\rangle + 2\mu(\mathcal{E}(\bar{z}), \mathcal{E}(u)) + \langle (\bar{z}\cdot\nabla)\bar{z}, u\rangle + \langle (W\cdot\nabla)\bar{z}, u\rangle + \langle (\bar{z}\cdot\nabla)W, u\rangle\\
&\qquad + 2\mu(k(x)\bar{z}, u)_{\Gamma_2} + 2\mu(S\tilde{\bar{z}}, \tilde{u})_{\Gamma_3} + 2(\alpha(x)\bar{z}, u)_{\Gamma_5} + \mu(k(x)\bar{z}, u)_{\Gamma_7}\\
&\quad = \langle f, u\rangle + \sum_{i=2,4,7} \langle \phi_i, u_n\rangle_{\Gamma_i} + \sum_{i=3,5,6} \langle \phi_i, u\rangle_{\Gamma_i} \quad \text{for all } u \in \mathbf{V},
\end{aligned}
\tag{4.101}
$$

where z_0, f, ϕ_i are perturbations of corresponding data.

Remark 4.7 The proofs of this section are similar to the ones in the preceding subsections. The main difference is that unlike $U(0,x)$ in the preceding subsections we do not assume smallness of $W(0,x)$.

Define an operator $A_{02} : \mathbf{V} \to \mathbf{V}^*$ by

$$
\begin{aligned}
\langle A_{02}y, u\rangle =& 2\mu(\mathcal{E}(y), \mathcal{E}(u)) + 2\mu(k(x)y, u)_{\Gamma_2} + 2\mu(S\tilde{y}, \tilde{u})_{\Gamma_3}\\
&+ 2(\alpha(x)y, u)_{\Gamma_5} + \mu(k(x)y, u)_{\Gamma_7} + \langle (W(0,x)\cdot\nabla)y, u\rangle\\
&+ \langle (y\cdot\nabla)W(0,x), u\rangle \qquad \text{for all } y, u \in \mathbf{V}.
\end{aligned}
\tag{4.102}
$$

For A_{02} we have

Lemma 4.11 *There exists $\delta > 0$ and $k_2 \geq 0$ such that*

$$\langle A_{02}u, u\rangle \geq \delta\|u\|_{\mathbf{V}}^2 - k_2\|u\|^2 \quad \textit{for all } u \in \mathbf{V}.$$

Proof By Korn's inequality, there exists $\beta > 0$ such that

$$2\mu(\mathcal{E}(u), \mathcal{E}(u)) \geq \beta\|u\|_{\mathbf{V}}^2 \quad \text{for all } u \in \mathbf{V}. \tag{4.103}$$

By Remark 3.4, there exists a constant M such that

$$\|S(x)\|_\infty,\ \|k(x)\|_\infty,\ \|\alpha(x)\|_\infty \leq M.$$

Then, there exists a constant c_0 (depending on β) such that

$$
\begin{aligned}
&\left|2\mu(k(x)u, u)_{\Gamma_2} + 2\mu(S\tilde{u}, \tilde{u})_{\Gamma_3} + \mu(k(x)u, u)_{\Gamma_7} + 2(\alpha(x)u, u)_{\Gamma_5}\right|\\
&\qquad \leq \frac{\beta}{4}\|u\|_{\mathbf{H}^1(\Omega)}^2 + c_0\|u\|^2 \quad \text{for all } u \in \mathbf{V}
\end{aligned}
\tag{4.104}
$$

(see Theorem 1.27).

Let us estimate $\langle (W(0,x)\cdot\nabla)u, u\rangle + \langle (u\cdot\nabla)W(0,x), u\rangle$. Since $W(0,x) \in C(\overline{\Omega})$,

$$\left|\langle (W(0,x)\cdot\nabla)u, u\rangle\right| \leq \frac{\beta}{8}\|u\|_{\mathbf{H}^1(\Omega)}^2 + c_1\|u\|^2. \tag{4.105}$$

Taking $\operatorname{div} u = 0$ into account, we get

$$\begin{aligned}\big\langle (u\cdot\nabla)W(0,x),u\big\rangle &= \sum_j\int_\Omega u_j\frac{\partial W(0,x)}{\partial x_j}u\,dx\\ &= \int_{\partial\Omega}(u\cdot n)(W(0,x)\cdot u)\,d\Gamma - \sum_j\int_\Omega u_j\frac{\partial u}{\partial x_j}W(0,x)\,dx.\end{aligned}$$

Estimating the first term in the right-hand side of the equality above as in (4.104) and applying Hölder's inequality in the second term, we have

$$\big|\big\langle (u\cdot\nabla)W(0,x),u\big\rangle\big| \le \frac{\beta}{8}\|u\|^2_{\mathbf{H}^1(\Omega)} + c_2\|u\|^2. \tag{4.106}$$

Taking $\delta = \frac{\beta}{2}$, $k_2 = c_0 + c_1 + c_2$, from (4.103)–(4.106) we get the asserted conclusion. □

Put $z = e^{-k_2 t}\tilde{z}$, where k_2 is a constant in Lemma 4.11. Then $e^{-k_2 t}\tilde{z}' = z' + k_2 z$, and we have the following problem equivalent to (4.101):

Find z such that

$$\begin{aligned}
&z\in L^2(0,T;\mathbf{V}),\\
&z(0) = z_0 = v_0 - W(0)\in\mathbf{V},\\
&\langle z'(t),u\rangle + 2\mu(\mathcal{E}(z(t)),\mathcal{E}(u)) + e^{k_2 t}\langle (z(t)\cdot\nabla)z(t),u\rangle + \langle (W(t)\cdot\nabla)z(t),u\rangle\\
&\qquad + \langle (z(t)\cdot\nabla)W(t),u\rangle + k_2(z(t),u) + 2\mu(k(x)z(t),u)_{\Gamma_2} + 2\mu(S\tilde{z}(t),\tilde{u})_{\Gamma_3}\\
&\qquad + 2(\alpha(x)z(t),u)_{\Gamma_5} + \mu(k(x)z(t),u)_{\Gamma_7}\\
&\quad = e^{-k_2 t}\Big[\langle f(t),u\rangle + \sum_{i=2,4,7}\langle \phi_i(t),u_n\rangle_{\Gamma_i} + \sum_{i=3,5,6}\langle \phi_i(t),u\rangle_{\Gamma_i}\Big]\quad\text{for all } u\in\mathbf{V}.
\end{aligned} \tag{4.107}$$

Define operators $A_2, A_W(t):\mathbf{V}\to\mathbf{V}^*$, respectively, by

$$\begin{aligned}\langle A_2 y,u\rangle =& 2\mu(\mathcal{E}(y),\mathcal{E}(u)) + 2\mu(k(x)y,u)_{\Gamma_2} + 2\mu(S\tilde{y},\tilde{u})_{\Gamma_3}\\ &+ 2(\alpha(x)y,u)_{\Gamma_5} + \mu(k(x)y,u)_{\Gamma_7} + k_2(y,u)\quad\text{for all } y,u\in\mathbf{V},\end{aligned} \tag{4.108}$$

and

$$\langle A_W(t)v,u\rangle = \langle (W(t,x)\cdot\nabla)v,u\rangle + \langle (v\cdot\nabla)W(t,x),u\rangle\quad\text{for all } v,u\in\mathbf{V}, \tag{4.109}$$

where k_2 is the constant in Lemma 4.11. $W\in\overline{\mathscr{W}}$ implies $W\in C([0,T];\mathbf{H}^1(\Omega))$, and such definitions are well.

In the proof of Lemma 4.11 it is clear that

$$\langle A_2 u,u\rangle \ge \frac{3\beta}{4}\|u\|^2_{\mathbf{V}}. \tag{4.110}$$

Also, by Lemma 4.11 we have

$$\big\langle \big(A_2 + A_W(0)\big)u, u\big\rangle \geq \frac{\beta}{4}\|u\|^2_{\mathbf{V}}. \tag{4.111}$$

Define an operator $B_2(t) : \mathbf{V} \to \mathbf{V}^*$ by

$$\langle B_2(t)v, u\rangle = e^{k_2 t}\langle (v\cdot\nabla)v, u\rangle \quad \text{for all } v, u \in \mathbf{V}. \tag{4.112}$$

Define operators L_2, $\widetilde{A}_W$, L_{2W}, $\widetilde{B}_2 : \mathscr{X} \to \mathscr{Y}$, $C_2 : \mathscr{X} \times \mathscr{X} \to \mathscr{Y}$ and an element $F_2 \in \mathscr{Y}$ by

$$\begin{aligned}
&\langle (L_2 z)(t), u\rangle = \langle z'(t), u\rangle + \langle A_2 z(t), u\rangle \quad \forall z \in \mathscr{X},\ \forall u \in \mathbf{V},\\
&\langle (\widetilde{A}_W z)(t), u\rangle = \langle A_W(t) z(t), u\rangle \quad \forall z \in \mathscr{X},\ \forall u \in \mathbf{V},\\
&\langle (L_{2W} z)(t), u\rangle = \langle z'(t), u\rangle + \big\langle \big(A_2 + A_W(t)\big)z(t), u\big\rangle \quad \forall z \in \mathscr{X},\ \forall u \in \mathbf{V},\\
&\langle (\widetilde{B}_2 z)(t), u\rangle = \langle B_2(t) z(t), u\rangle \quad \forall z \in \mathscr{X},\ \forall u \in \mathbf{V},\\
&\langle C_2(w,z)(t), u\rangle = e^{k_2 t}\langle (w\cdot\nabla)z, u\rangle + e^{k_2 t}\langle (z\cdot\nabla)w, u\rangle \quad \forall z \in \mathscr{X},\ \forall u \in \mathbf{V},\\
&\langle (F_2)(t), u\rangle = e^{-k_2 t}\Big[\langle f(t), u\rangle + \sum_{i=2,4,7}\langle \phi_i(t), u_n\rangle_{\Gamma_i} + \sum_{i=3,5,6}\langle \phi_i(t), u\rangle_{\Gamma_i}\Big] \quad \forall u \in \mathbf{V}.
\end{aligned} \tag{4.113}$$

By the same argument as in Lemma 4.5 we get

Lemma 4.12 *C_2 is a bilinear continuous operator such that $\mathscr{X} \times \mathscr{X} \to \mathscr{Y}$. Under Assumption 4.2 $\widetilde{A}_W$ is a linear continuous operator such that $\mathscr{X} \to \mathscr{Y}$ and $F_2 \in \mathscr{Y}$.*

Using (4.110) instead of (4.67), as Lemma 4.6 we get

Lemma 4.13 *The operator $\overline{L}_2$ defined by $\overline{L}_2 z = (z'(0), L_2 z)$ for $z \in \mathscr{X}$ is a linear continuous one-to-one operator from $\mathscr{X}$ onto $H \times \mathscr{Y}$.*

Now, using (4.111) without assuming the fact that $\|W(0,x)\|_{\tilde{\mathbf{V}}}$ is small enough, as in Lemma 4.7 we prove the following:

Lemma 4.14 *The operator $\overline{L}_{2W}$ defined by $\overline{L}_{2W} z = (z'(0), L_{2W} z)$ for $z \in \mathscr{X}$ is a linear continuous one-to-one operator from $\mathscr{X}$ onto $H \times \mathscr{Y}$.*

Proof As in Lemma 3.5 of [3] the operator $\widetilde{A}_W : \mathscr{X} \to \mathscr{Y}$ is compact. Thus, $\overline{A}_W : \mathscr{X} \to H \times \mathscr{Y}$ defined by $\overline{A}_W z = \{0_H, \widetilde{A}_W z\}$ is also compact. Since $\overline{L}_{2W} = \overline{L}_2 + \overline{A}_W$, in order to get the asserted conclusion by virtue of Theorem 1.8 it suffices to prove that $\overline{L}_{2W}$ is one-to-one from $\mathscr{X}$ into $H \times \mathscr{Y}$.

To prove the last statement it is enough to show that the inverse image of $(0_H, 0_{\mathscr{Y}})$ by $\overline{L}_{2W}$ is $0_{\mathscr{X}}$. It is easy to verify that

$$\big|\big\langle \big(A_2 + A_W(0)\big)v, u\big\rangle\big| \leq c\|v\|_{\mathbf{V}}\cdot\|u\|_{\mathbf{V}} \quad \text{for all } v, u \in \mathbf{V}. \tag{4.114}$$

By (4.111) and (4.114), for any $q \in \mathbf{V}^*$ there exists a unique solution $y \in \mathbf{V}$ to

$$(A_2 + A_W(0))y = q. \tag{4.115}$$

Let $z \in \mathscr{X}$ be the inverse image of $(0_H, 0_{\mathscr{Y}})$ by $\overline{L}$. Then, $z'(0) = 0_H$, and putting $t = 0$ from the third one of (4.113) we get

$$\langle\!\langle (A_2 + A_W(0))z(0), u\rangle\!\rangle = 0 \quad \text{for all } u \in \mathbf{V},$$

where $z(0) = z(0, x)$. This means that $z(0)$ is the unique solution to (4.115) with $q = 0_{\mathbf{V}^*}$, i.e. $z(0) = 0_{\mathbf{V}}$. Therefore, $z \in \mathscr{X}$ satisfies

$$\begin{cases} z'(t) + \big(A_2 + A_W(t)\big)z(t) = 0, \\ z(0) = 0_{\mathbf{V}}. \end{cases} \tag{4.116}$$

Now, using (4.116) and Gronwall's inequality, as in Lemma 3.8 of [3] we can prove $z = 0_{\mathscr{X}}$. The proof is thus completed. □

By the same argument as in Lemma 4.8 we get

Lemma 4.15 *The operator T_2 defined by $T_2 z = \big(z'(0), (L_{2W} + \widetilde{B}_2)z\big)$ for $z \in \mathscr{X}$ is continuously differentiable, $T_2 0_{\mathscr{X}} = (0_H, 0_{\mathscr{Y}})$ and the Fréchet derivative of T_2 at $0_{\mathscr{X}}$ is $\overline{L}_{2W}$.*

Let us now turn to the following problem:

$$\big(A_2 + A_W(0) + B_2(0)\big)u = q. \tag{4.117}$$

Using (4.111) without assuming the fact that $\|W(0, x)\|_{\widetilde{\mathbf{V}}}$ is small enough, as in Lemma 4.9 we can prove

Lemma 4.16 *If the norm of $q \in \mathbf{V}^*$ is small enough, then there exists a unique solution to* (4.117) *in some $\mathscr{O}_M(0_{\mathbf{V}})$.*

Using Lemmas 4.12–4.15 and Theorem 1.7 (Local inverse mapping theorem), in the same way as in Theorem 4.3 we get

Theorem 4.5 *Suppose that Assumption 4.2 holds and the norms of f, f', ϕ_i, ϕ_i' in the spaces that they belong to are small enough.*

If

$$w_2 \equiv F_2(0) - (A_2 + A_{2W}(0) + B_2(0))z_0 \in H, \tag{4.118}$$

where $z_0 = v_0 - U(0, \cdot)$, and $\|w_2\|_H$ is small enough, then there exists a unique solution to (4.101) *in the space $\mathscr{W}$.*

Remark 4.8 By the same argument as in Remark 4.3, we can see that the condition (4.118) is equivalent to the condition $\overline{w}_2 \in H_1$, where $\overline{w}_2 \in \mathbf{V}_1^*$ is defined by

$$\begin{aligned}\langle \overline{w}_2, u\rangle =&\langle f(0), u\rangle + \sum_{i=2,4}\langle \phi_i(0,x), u_n\rangle_{\Gamma_i} + \sum_{i=3,5,7}\langle \phi_i(0,x), u\rangle_{\Gamma_i}\\ &- \big[2\mu(\mathcal{E}(z_0), \mathcal{E}(u)) + 2\mu(k(x)z_0, u)_{\Gamma_2} + 2\mu(S\tilde{z}_0, \tilde{u})_{\Gamma_3}\\ &+ 2(\alpha(x)z_0, u)_{\Gamma_5} + \mu(k(x)z_0, u)_{\Gamma_7} + \langle (W(0,x)\cdot\nabla)z_0, u\rangle\\ &+ \langle (z_0\cdot\nabla)W(0,x), u\rangle + k_2(z_0,u) + \langle (z_0\cdot\nabla)z_0, u\rangle\big] \quad \text{for all } u\in\mathbf{V}\end{aligned} \tag{4.119}$$

with k_2 in Lemma 4.11.

Now let us consider Problem II.

Let $W(x,t)\in\overline{\mathscr{W}}$ be a given solution to Problem II. Let v be the solution for the data perturbed except h_i and put $v = \overline{z} + W$. Then, we get a problem for $\overline{z}$ as follows: Find $\overline{z}$ such that

$$\begin{aligned}&\overline{z}\in L^2(0,T;\mathbf{V}_1),\\ &\overline{z}(0) = z_0 \equiv v_0 - W(0,x)\in\mathbf{V}_1,\\ &\langle \overline{z}', u\rangle + \mu(\nabla\overline{z}, \nabla u) + \langle (\overline{z}\cdot\nabla)\overline{z}, u\rangle + \langle (W\cdot\nabla)\overline{z}, u\rangle + \langle (\overline{z}\cdot\nabla)W, u\rangle\\ &\qquad + \mu(k(x)\overline{z}, u)_{\Gamma_2} + \mu(S\tilde{\overline{z}}, \tilde{u})_{\Gamma_3} + 2(\alpha(x)\overline{z}, u)_{\Gamma_5} - \mu(S\tilde{\overline{z}}, \tilde{u})_{\Gamma_5}\\ &\quad = \langle f, u\rangle + \sum_{i=2,4}\langle \phi_i, u_n\rangle_{\Gamma_i} + \sum_{i=3,5,7}\langle \phi_i, u\rangle_{\Gamma_i} \quad \text{for all } u\in\mathbf{V}_1,\end{aligned} \tag{4.120}$$

where z_0, f, ϕ_i are perturbations of corresponding data.

By the same argument as in Theorem 4.5 we have

Theorem 4.6 *Suppose that Assumption 4.2 holds and the norms of f, f', ϕ_i, ϕ_i' in the spaces that they belong to are small enough. Define an element $w_3\in\mathbf{V}_1^*$ by*

$$\begin{aligned}\langle w_3, u\rangle =&\langle f(0), u\rangle + \sum_{i=2,4}\langle \phi_i(0,x), u_n\rangle_{\Gamma_i} + \sum_{i=3,5,7}\langle \phi_i(0,x), u\rangle_{\Gamma_i}\\ &- \big[\mu(\nabla z_0, \nabla u) + \mu(k(x)z_0, u)_{\Gamma_2} + \mu(S\tilde{z}_0, \tilde{u})_{\Gamma_3} + 2(\alpha(x)z_0, u)_{\Gamma_5}\\ &- \mu(S\tilde{z}_0, \tilde{u})_{\Gamma_5} + \langle (W(0,x)\cdot\nabla)z_0, u\rangle + \langle (z_0\cdot\nabla)W(0,x), u\rangle\\ &+ k_3(z_0,u) + \langle (z_0\cdot\nabla)z_0, u\rangle\big] \quad \textit{for all } u\in\mathbf{V}_1,\end{aligned} \tag{4.121}$$

where k_3 is a constant determined as in Lemma 4.11.

If $w_3\in\mathscr{O}_R(0_{H_1})$ for $R>0$ small enough, then there exists a unique solution to (4.120) *in the space $\mathscr{W}$.*

Remark 4.9 If $\Gamma_i = \emptyset$, $i = 2,\cdots,5$, then problem (4.120) is reduced to the one in [3]. If $z_0\in\mathbf{H}^{l/2}(\Omega)$, then $(z_0\cdot\nabla)z_0$, $(W(0,x)\cdot\nabla)z_0$, $(z_0\cdot\nabla)W(0,x)\in\mathbf{L}^2(\Omega)$ and $k_3z_0\in\mathbf{L}^2(\Omega)$. Thus, the last four terms on the right-hand side of (4.121) do not give any effect to the condition for w_3 being in H_1, and so the conditions in Theorem 4.6 are the same as one of conditions of Theorems 3.5–3.8 of [3]. Thus, Theorem 4.6 guarantees existence of a unique solution under a condition weaker than the one in [3]. Note that taking $W(t,x)\equiv 0$ in Theorems 4.5 and 4.6, we can not get Theorems 4.3 and 4.4, since there $h_i\neq 0$.

Remark 4.10 Let $\mathbf{H}^k(\Omega) = (W^{k,2}(\Omega))^l$ be a Sobolev spaces on Ω with dimension l, V be a divergence-free subspace of $\mathbf{H}^1$ satisfying appropriate boundary conditions, H—the closure of V in $(L^2(\Omega))^l$, $V^{r_0}(\Omega) = V \cap \mathbf{H}^{r_0}(\Omega)$, where $r_0 > l/2$,

$$\mathscr{X} = \{w \in L^2(0, T; V); w' \in L^2(0, T; V), w'' \in L^2(0, T; V^*)\},$$

$$\mathscr{Y} = \{w \in L^2(0, T; V^*); w' \in L^2(0, T; V^*)\}$$

and $A : V \to V^*$—the Stokes operator. Concerned with the linearized problem of

$$\begin{cases} u'(t) + (A + B)u(t) = f(t), \\ u(0) = \varphi, \end{cases}$$

in [3] it is proved that a map $u \to \{u(0), Lu \equiv u' + Au\}$ is linear continuous one-to-one from $\mathcal{X} = \{u \in \mathscr{X} : u(0) \in V^{r_0}(\Omega)\}$ onto $\mathcal{Y} = \{[\varphi, h] : \varphi \in V^{r_0}(\Omega), h \in \mathscr{Y}, h(0) - A\varphi \in H\}$ and the compatibility condition

$$f(0) - A\varphi \in H, \quad \varphi \in V^{r_0}(\Omega)$$

for the existence of a unique solution to the nonlinear problem is obtained.

However, in Sect. 4.2 it is proved that for a modified operator A a map $u \to \{u'(0), Lu \equiv u' + Au\}$ is linear continuous one-to-one from $\mathscr{X}$ onto $H \times \mathscr{Y}$ (Lemma 4.6). Then, it is obtained the compatibility condition

$$f(0) - A\varphi - B(0)\varphi \in H$$

without $\varphi \in V^{r_0}(\Omega)$ for the existence of unique solution to a nonlinear problems. Since $B(0)\varphi = (\varphi \cdot \nabla)\varphi$, for $\varphi \in V^{l/2}(\Omega)$ we get $B(0)\varphi \in H$, and so from point of view of smoothness of the initial functions, the compatibility conditions in Sect. 4.2 are weaker than the one in [3].

4.3 Bibliographical Remarks

The content of Sect. 4.1 is the one of Sect. 5 of [5] and the content of Sect. 4.2 is taken from [6].

The Navier-Stokes equations on a domain decreasing in time with mixture of Dirichlet condition, the total pressure and vorticity conditions were studied in [1]. For a more general problem we refer to [7]. For the 2-D Navier-Stokes equations with mixture of the Dirichlet and pressure boundary conditions, existence of a unique solution for small data was proved in [8].

The Navier-Stokes equations with mixture of Dirichlet condition and stress condition were studied. In [4] under smoothness condition and a compatibility condition of data at the initial time the existence of a unique local-in-time solution to the 3-D Navier-Stokes equations was discussed. In [9] for the Navier-Stokes equations on the polyhedral domain with mixture of Dirichlet condition, Navier slip condition and stress condition a local-in-time solution was considered. Here smoothness of solutions to the corresponding steady Stokes problem was used essentially.

The Navier-Stokes equations with mixture of Dirichlet condition and the outflow boundary condition were studied. For 2-D Navier-Stokes equations a local-in-time solution in [10] and a solution for small data were obtained in [11]. Here also smoothness of solutions to the corresponding 2-D steady linear problem is important. In [3] it was proved that if under a compatibility condition at initial time there exists a unique solution, then so does for small perturbed data. This result shows that under the compatibility condition there exists a unique solution for small data. Smoothness of initial function in the compatibility condition of [3] is stronger than the one in [4, 12].

The Navier-Stokes equations with mixture of Dirichlet condition and the static pressure boundary condition were studied. In [8] for the 2-D Navier-Stokes problem existence of a unique solution for small data was established.

References

1. G. Łukaszewicz, On the Navier-Stokes equations in time dependent domains and with boundary conditions involving the pressure. J. Math. Sci. Univ. Tokyo **4**(3), 529–550 (1997)
2. G.P. Galdi, *An Introduction to the Mathematical Theory of the Navier-Stokes Equations* (Springer, Berlin, 2011)
3. P. Kučera, Basic properties of solution of the non-steady Navier-Stokes equations with mixed boundary conditions in a bounded domain. Ann. Univ. Ferrara **55**, 289–308 (2009)
4. P. Kučera, Z. Skalák, Local solutions to the Navier-Stokes equations with mixed boundary conditions. Acta Appl. Math. **54**, 275–288 (1998)
5. T. Kim, D. Cao, Some properties on the surfaces of vector fields and its application to the Stokes and Navier-Stokes problems with mixed boundary conditions. Nonlinear Anal. **113**, 94–114 (2015). Erratum, ibid **135**, 249–250 (2016)
6. T. Kim, D. Cao, Existence of solutions to the heat convection equations in a time-dependent domain with mixed boundary conditions. J. Math. Sci. Univ. Tokyo **22**, 531–568 (2015)
7. T. Kim, D. Cao, A non-steady system with friction boundary conditions for flow of heat-conducting incompressible viscous fluids. J. Math. Anal. Appl. **484**, 123676 (2020)
8. S. Marušić, On the Navier-Stokes system with pressure boundary condition. Ann. Univ. Ferrara **53**, 319–331 (2007)
9. M. Beneš, Mixed initial-boundary value problem for the three-dimensional Navier-Stokes equations in polyhedral domains. Discret. Contin. Dyn. Syst. Suppl. **02**, 135–144 (2011)
10. M. Beneš, P. Kučera, Non-steady Navier-Stokes equations with homogeneous mixed boundary conditions and arbitrarily large initial condition. Carpathian J. Math. **23**(1–2), 32–40 (2007)
11. M. Beneš, P. Kučera, Solutions to the Navier-Stokes equations with mixed boundary conditions in two-dimensional bounded domains. Math. Nachr. **289**(2–3), 194–212 (2016)
12. Z. Skalák, P. Kučera, An existence theorem for the Boussinesq equations with non-Dirichlet boundary conditions. Appl. Math. **45**(2), 81–98 (2000)

Chapter 5
The Steady Navier-Stokes System with Friction Boundary Conditions

In this chapter we are concerned with the steady Navier-Stokes systems with mixed boundary conditions which may include Tresca slip condition, leak boundary condition, one-sided leak boundary conditions, velocity, pressure, vorticity, stress and normal derivative of velocity together. Relying on the results in Sect. 3.1 and using the strain bilinear form, we embed all these boundary conditions into variational formulations of corresponding problems.

In Sect. 5.1 for every problem, which is distinguished according to boundary conditions, we first get the variational formulation which consists of five formulas with five unknown functions, that is, using velocity, tangent stress on slip surface, normal stress on leak surface, normal stresses on one-sided leak surfaces together as unknown functions. To show that such formulations are well defined, we note that when the solution is smooth enough, these variational formulations are equivalent to the original PDE problems. It is shown that if the boundary condition for pressure or stress on a portion of boundary where there is flux is given, then the pressure is determined uniquely. Then, we get variational inequalities with one unknown function (velocity) equivalent to the variational formulations with five unknown functions, by which the Navier-Stokes problem with 11 kinds of boundary conditions is reduced to one variational inequality.

In Sect. 5.2 we study three kinds of variational inequality obtained in Sect. 5.1.

In Sect. 5.3, using the results in Sect. 5.2, we study the existence, uniqueness and estimates of solutions to the Navier-Stokes problems with 11 kinds of boundary conditions. For the problem with boundary conditions involving the static pressure and stress, the existence of a unique solution is proved when the data are small enough. For the problem with boundary conditions involving the total pressure and total stress, the existence and estimate of solutions are proved without smallness of the data of problem.

T. Kim and D. Cao, *Equations of Motion for Incompressible Viscous Fluids*,
Advances in Mathematical Fluid Mechanics,
https://doi.org/10.1007/978-3-030-78659-5_5

From now on throughout this book the following notations are used. When $f \in H^{-1/2}(\Gamma_i)$, if $\langle f, w\rangle_{\Gamma_i} \geq 0$ (≤ 0) $\forall w \in C_0^\infty(\Gamma_i)$ with $w \geq 0$, then we denoted by $f \geq 0$ (≤ 0).

5.1 Variational Formulations of Problems

Let Ω be a bounded domain of $\mathbb{R}^l$, $l = 2, 3$. $\partial\Omega \in C^{0,1}$, $\partial\Omega = \cup_{i=1}^{11} \overline{\Gamma}_i, \Gamma_i \cap \Gamma_j = \emptyset$ for $i \neq j$, $\Gamma_i = \bigcup_j \Gamma_{ij}$, where Γ_{ij} are connected open subsets of $\partial\Omega$ and $\Gamma_{ij} \in C^2$ for $i = 2, 3, 7$ and $\Gamma_{ij} \in C^1$ for others (see Remark 5.2).

We are concerned with the Problems I and II for the Navier-Stokes equations

$$- \mu\Delta v + (v \cdot \nabla)v + \nabla p = f, \quad \operatorname{div} v = 0, \tag{5.1}$$

which are distinguished according to boundary conditions. Problem I is one with the following boundary conditions:

$$\begin{aligned}
&(1)\quad v|_{\Gamma_1} = h_1,\\
&(2)\quad v_\tau|_{\Gamma_2} = 0,\ \ -p|_{\Gamma_2} = \phi_2,\\
&(3)\quad v_n|_{\Gamma_3} = 0,\ \ \operatorname{rot} v \times n|_{\Gamma_3} = \phi_3/\mu,\\
&(4)\quad v_\tau|_{\Gamma_4} = h_4,\ \ (-p + 2\mu\varepsilon_{nn}(v))|_{\Gamma_4} = \phi_4,\\
&(5)\quad v_n|_{\Gamma_5} = h_5,\ \ 2(\mu\varepsilon_{n\tau}(v) + \alpha v_\tau)|_{\Gamma_5} = \phi_5,\ \ \alpha : \text{a matrix},\\
&(6)\quad (-pn + 2\mu\varepsilon_n(v))|_{\Gamma_6} = \phi_6,\\
&(7)\quad v_\tau|_{\Gamma_7} = 0,\ (-p + \mu\frac{\partial v}{\partial n} \cdot n)|_{\Gamma_7} = \phi_7,\\
&(8)\quad v_n = h_8,\ |\sigma_\tau(v)| \leq g_\tau,\ \sigma_\tau(v) \cdot v_\tau + g_\tau|v_\tau| = 0 \text{ on } \Gamma_8,\\
&(9)\quad v_\tau = h_9,\ |\sigma_n(v)| \leq g_n,\ \sigma_n(v)v_n + g_n|v_n| = 0 \text{ on } \Gamma_9,\\
&(10)\quad v_\tau = 0,\ v_n \geq 0,\quad \sigma_n(v) + g_{+n} \geq 0,\quad (\sigma_n(v) + g_{+n})v_n = 0 \text{ on } \Gamma_{10},\\
&(11)\quad v_\tau = 0,\ v_n \leq 0,\quad \sigma_n(v) - g_{-n} \leq 0,\quad (\sigma_n(v) - g_{-n})v_n = 0 \text{ on } \Gamma_{11},
\end{aligned} \tag{5.2}$$

and Problem II is one with the following boundary conditions:

$$
\begin{aligned}
&(1)\ \ v|_{\Gamma_1} = h_1,\\
&(2)\ \ v_\tau|_{\Gamma_2} = 0,\ \ -(p + 1/2|v|^2)|_{\Gamma_2} = \phi_2,\\
&(3)\ \ v_n|_{\Gamma_3} = 0,\ \ \operatorname{rot} v \times n|_{\Gamma_3} = \phi_3/\mu,\\
&(4)\ \ v_\tau|_{\Gamma_4} = h_4,\ \ (-p - 1/2|v|^2 + 2\mu\varepsilon_{nn}(v))|_{\Gamma_4} = \phi_4,\\
&(5)\ \ v_n|_{\Gamma_5} = h_5,\ \ 2(\mu\varepsilon_{n\tau}(v) + \alpha v_\tau)|_{\Gamma_5} = \phi_5,\ \ \alpha : \text{a matrix},\\
&(6)\ \ (-pn - 1/2|v|^2 n + 2\mu\varepsilon_n(v))|_{\Gamma_6} = \phi_6,\\
&(7)\ \ v_\tau|_{\Gamma_7} = 0,\ (-p - 1/2|v|^2 + \mu\frac{\partial v}{\partial n}\cdot n)|_{\Gamma_7} = \phi_7,\\
&(8)\ \ v_n = h_8,\ |\sigma^t_\tau(v)| \le g_\tau,\ \sigma^t_\tau(v)\cdot v_\tau + g_\tau|v_\tau| = 0 \text{ on } \Gamma_8,\\
&(9)\ \ v_\tau = h_9,\ |\sigma^t_n(v)| \le g_n,\ \sigma^t_n(v)v_n + g_n|v_n| = 0 \text{ on } \Gamma_9,\\
&(10)\ \ v_\tau = 0,\ \ v_n \ge 0,\ \ \ \sigma^t_n(v) + g_{+n} \ge 0,\ \ \ (\sigma^t_n(v) + g_{+n})v_n = 0 \text{ on } \Gamma_{10},\\
&(11)\ \ v_\tau = 0,\ v_n \le 0,\ \ \ \sigma^t_n(v) - g_{-n} \le 0,\ \ \ (\sigma^t_n(v) - g_{-n})v_n = 0 \text{ on } \Gamma_{11},
\end{aligned}
\tag{5.3}
$$

where and in what follows $\varepsilon_n(v) = \mathcal{E}(v)n, \varepsilon_{nn}(v) = (\mathcal{E}(v)n, n)_{\mathbb{R}^l}, \varepsilon_{n\tau}(v) = \mathcal{E}(v)n - \varepsilon_{nn}(v)n$ and h_i, ϕ_i, α_{kl} (components of matrix α) are given functions or vectors of functions. σ_n and σ^t_n are, respectively, the normal components of stress and total stress on a boundary surface, that is, $\sigma_n = \sigma\cdot n$ and $\sigma^t_n = \sigma^t\cdot n$. Also, $\sigma_\tau(v, p) = \sigma(v, p) - \sigma_n(v, p)n, \sigma^t_\tau(v, p) = \sigma^t(v, p) - \sigma^t_n(v, p)n$.

Throughout this book we always assume that $g_\tau \in L^2(\Gamma_8)$, $g_n \in L^2(\Gamma_9)$, $g_{+n} \in L^2(\Gamma_{10})$, $g_{-n} \in L^2(\Gamma_{11})$ and that $g_\tau > 0$, $g_n > 0$, $g_{+n} > 0$, $g_{-n} > 0$, for a.e. x of the portions of boundary.

For Problem II the static pressure p and the stress in the boundary conditions for Problem I are, respectively, changed with the total pressure and the total stress. Thus, as in Chaps. 3, 4 for convenience in what follows, the Navier-Stokes problems with boundary conditions (5.2) and (5.3) are, respectively, called *the case of static pressure* and *the case of total pressure*.

We also consider the Stokes equations

$$
-\mu\Delta v + \nabla p = f, \quad \operatorname{div} v = 0 \tag{5.4}
$$

with the boundary conditions (5.2), which is Problem III.

Assuming that $f \in \mathbf{L}^2(\Omega)$, $\phi_i \in L^2(\Gamma_i)$, $i = 2, 4, 7$, and $\phi_i \in \mathbf{L}^2(\Gamma_i)$, $i = 3$, 5, 6, formally we introduce the following:

Definition 5.1 A function $(v, p) \in \mathbf{H}^2(\Omega) \times H^1(\Omega)$ is called a solution to Problem I (or Problem II) if (5.1) holds in $\mathbf{L}^2(\Omega)$ and each of (5.2) (or (5.3)) holds in $L^2(\Gamma_i)$ or $\mathbf{L}^2(\Gamma_i)$. A function $(v, p) \in \mathbf{H}^2(\Omega) \times H^1(\Omega)$ is called a solution to Problem III if (5.4) holds in $\mathbf{L}^2(\Omega)$ and each of (5.2) holds in $L^2(\Gamma_i)$ or $\mathbf{L}^2(\Gamma_i)$.

Usually, it is difficult (or impossible) to find solutions in the sense of the definition above, and we will consider variational formulations of the problems.

5.1.1 Variational Formulation: The Case of Static Pressure

Let

$$\begin{aligned}
&\mathbf{V} = \left\{u \in \mathbf{H}^1(\Omega) : \operatorname{div} u = 0,\ u|_{\Gamma_1} = 0,\ u_\tau|_{(\cup_{i=2,4,7,9,10,11}\Gamma_i)} = 0,\ u_n|_{(\cup_{i=3,5,8}\Gamma_i)} = 0\right\},\\
&\mathbf{V}_{\Gamma 237}(\Omega) = \left\{u \in \mathbf{H}^1(\Omega) : \operatorname{div} u = 0,\ u_\tau|_{(\Gamma_2\cup\Gamma_7)} = 0,\ u_n|_{\Gamma_3} = 0\right\},\\
&K(\Omega) = \left\{u \in \mathbf{V} : u_n|_{\Gamma_{10}} \geq 0,\ u_n|_{\Gamma_{11}} \leq 0\right\}.
\end{aligned} \tag{5.5}$$

Taking into account (2.81) and applying Theorems 3.1 and 3.2 on $\Gamma_{ij}(i = 2, 3, 7)$, for $v \in \mathbf{H}^2(\Omega) \cap \mathbf{V}_{\Gamma 237}(\Omega)$ and $u \in \mathbf{V}$ we have

$$\begin{aligned}
-(\Delta v, u) &= 2(\mathcal{E}(v), \mathcal{E}(u)) - 2(\mathcal{E}(v)n, u)_{\cup_{i=2}^{11}\Gamma_i}\\
&= 2(\mathcal{E}(v), \mathcal{E}(u)) + 2(k(x)v, u)_{\Gamma_2} - (\operatorname{rot} v \times n, u)_{\Gamma_3} + 2(S\tilde{v}, \tilde{u})_{\Gamma_3}\\
&\quad - 2(\varepsilon_{nn}(v), u_n)_{\Gamma_4} - 2(\varepsilon_{n\tau}(v), u)_{\Gamma_5} - 2(\varepsilon_n(v), u)_{\Gamma_6}\\
&\quad - \left(\frac{\partial v}{\partial n}, u\right)_{\Gamma_7} + (k(x)v, u)_{\Gamma_7} - 2(\varepsilon_{n\tau}(v), u)_{\Gamma_8}\\
&\quad - 2(\varepsilon_{nn}(v), u_n)_{\Gamma_9} - 2(\varepsilon_{nn}(v), u_n)_{\Gamma_{10}} - 2(\varepsilon_{nn}(v), u_n)_{\Gamma_{11}}.
\end{aligned} \tag{5.6}$$

Also, for $p \in H^1(\Omega)$ and $u \in \mathbf{V}$ we have

$$(\nabla p, u) = (p, u_n)_{\cup_{i=2}^{11}\Gamma_i} = (p, u_n)_{\Gamma_2} + (p, u_n)_{\Gamma_4\cup\Gamma_7\cup\Gamma_9\cup\Gamma_{10}\cup\Gamma_{11}} + (pn, u)_{\Gamma_6}, \tag{5.7}$$

where $u_n\, |_{\Gamma_3\cup\Gamma_5\cup\Gamma_8}= 0$ was used.

We assume that the following holds:

Assumption 5.1 (1) There exists a function $U \in \mathbf{H}^1(\Omega)$ such that

$$\begin{aligned}
&\operatorname{div} U = 0, U|_{\Gamma_1} = h_1, U_\tau|_{(\Gamma_2\cup\Gamma_7)} = 0, U_n|_{\Gamma_3} = 0, U_\tau|_{\Gamma_4} = h_4,\\
&U_n|_{\Gamma_5} = h_5, U|_{\Gamma_8} = h_8 n,\ U|_{\Gamma_9} = h_9,\ U|_{\Gamma_{10}\cup\Gamma_{11}} = 0.
\end{aligned}$$

(2) $f \in \mathbf{V}^*$, $\phi_i \in H^{-\frac{1}{2}}(\Gamma_i), i = 2, 4, 7,$ $\phi_i \in \mathbf{H}^{-\frac{1}{2}}(\Gamma_i), i = 3, 5, 6,$ $\alpha_{ij} \in L^\infty(\Gamma_5)$, and $\Gamma_1 \neq \emptyset$.

(3) If Γ_i, where i is 10 or 11, is nonempty, then at least one of $\{\Gamma_j : j \in \{2, 4, 7, 9 - 11\}\backslash\{i\}\}$ is nonempty and there exists a diffeomorphism in C^1 between Γ_i and Γ_j.

Having in mind Assumption 5.1 and putting $v = w + U$, by (5.6) and (5.7) we can see that solutions (v, p) of the problem (5.1), (5.2) in the sense of Definition 5.1 satisfy the following:

$$\begin{cases} v - U = w \in K(\Omega), \\ 2\mu(\mathcal{E}(w), \mathcal{E}(u)) + \langle (w \cdot \nabla)w, u\rangle + \langle (U \cdot \nabla)w, u\rangle + \langle (w \cdot \nabla)U, u\rangle \\ \qquad + 2\mu(k(x)w, u)_{\Gamma_2} + 2\mu(S\tilde{w}, \tilde{u})_{\Gamma_3} + 2(\alpha(x)w, u)_{\Gamma_5} + \mu(k(x)w, u)_{\Gamma_7} \\ \qquad - 2\mu(\varepsilon_{n\tau}(w + U), u)_{\Gamma_8} + (p - 2\mu\varepsilon_{nn}(w + U), u_n)_{\Gamma_9 \cup \Gamma_{10} \cup \Gamma_{11}} \\ \quad = -2\mu(\mathcal{E}(U), \mathcal{E}(u)) - \langle (U \cdot \nabla)U, u\rangle - 2\mu(k(x)U, u)_{\Gamma_2} \\ \qquad - 2\mu(S\tilde{U}, \tilde{u})_{\Gamma_3} - 2(\alpha(x)U_\tau, u)_{\Gamma_5} - \mu(k(x)U, u)_{\Gamma_7} + \langle f, u\rangle \\ \qquad + \sum_{i=2,4,7} \langle \phi_i, u_n\rangle_{\Gamma_i} + \sum_{i=3,5,6} \langle \phi_i, u\rangle_{\Gamma_i} \qquad \forall u \in \mathbf{V}, \\ |\sigma_\tau(v)| \le g_\tau,\ \sigma_\tau(v) \cdot v_\tau + g_\tau |v_\tau| = 0 \quad \text{on } \Gamma_8, \\ |\sigma_n(v)| \le g_n,\ \sigma_n(v)v_n + g_n|v_n| = 0 \quad \text{on } \Gamma_9, \\ \sigma_n(v) + g_{+n} \ge 0, \quad (\sigma_n(v) + g_{+n})v_n = 0 \quad \text{on } \Gamma_{10}, \\ \sigma_n(v) - g_{-n} \le 0, \quad (\sigma_n(v) - g_{-n})v_n = 0 \quad \text{on } \Gamma_{11}. \end{cases} \tag{5.8}$$

Define $a_{01}(\cdot,\cdot), a_{11}(\cdot,\cdot,\cdot)$ and $F_1 \in V^*$ by

$$\begin{aligned} a_{01}(w, u) &= 2\mu(\mathcal{E}(w), \mathcal{E}(u)) + \langle (U \cdot \nabla)w, u\rangle + \langle (w \cdot \nabla)U, u\rangle \\ &\quad + 2\mu(k(x)w, u)_{\Gamma_2} + 2\mu(S\tilde{w}, \tilde{u})_{\Gamma_3} + 2(\alpha(x)w, u)_{\Gamma_5} \\ &\quad + \mu(k(x)w, u)_{\Gamma_7} \quad \forall w, u \in \mathbf{V}, \\ a_{11}(w, u, v) &= \langle (w \cdot \nabla)u, v\rangle \quad \forall w, u, v \in \mathbf{V}, \\ \langle F_1, u\rangle &= -2\mu(\mathcal{E}(U), \mathcal{E}(u)) - \langle (U \cdot \nabla)U, u\rangle - 2\mu(k(x)U, u)_{\Gamma_2} \\ &\quad - 2\mu(S\tilde{U}, \tilde{u})_{\Gamma_3} - 2(\alpha(x)U_\tau, u)_{\Gamma_5} - \mu(k(x)U, u)_{\Gamma_7} + \langle f, u\rangle \\ &\quad + \sum_{i=2,4,7} \langle \phi_i, u_n\rangle_{\Gamma_i} + \sum_{i=3,5,6} \langle \phi_i, u\rangle_{\Gamma_i} \quad \forall u \in \mathbf{V}. \end{aligned} \tag{5.9}$$

Then, taking into account

$$\sigma_\tau(v) = 2\mu\varepsilon_{n\tau}(v), \qquad \sigma_n(v) = -p + 2\mu\varepsilon_{nn}(v)$$

and (5.8), we introduce the following variational formulation for problem (5.1), (5.2).

Problem I-VE. Find $(v, \sigma_\tau, \sigma_n, \sigma_{+n}, \sigma_{-n}) \in \big(U + K(\Omega)\big) \times \mathbf{L}^2_\tau(\Gamma_8) \times L^2(\Gamma_9) \times H^{-1/2}(\Gamma_{10}) \times H^{-1/2}(\Gamma_{11})$ such that

$$\begin{cases} v - U = w \in K(\Omega), \\ a_{01}(w,u) + a_{11}(w,w,u) - (\sigma_\tau, u_\tau)_{\Gamma_8} - (\sigma_n, u_n)_{\Gamma_9} \\ \qquad - \langle \sigma_{+n}, u_n \rangle_{\Gamma_{10}} - \langle \sigma_{-n}, u_n \rangle_{\Gamma_{11}} = \langle F_1, u \rangle \quad \forall u \in \mathbf{V}, \\ |\sigma_\tau| \le g_\tau, \ \sigma_\tau \cdot v_\tau + g_\tau |v_\tau| = 0 \quad \text{on } \Gamma_8, \\ |\sigma_n| \le g_n, \ \sigma_n v_n + g_n |v_n| = 0 \quad \text{on } \Gamma_9, \\ \sigma_{+n} + g_{+n} \ge 0, \quad \langle \sigma_{+n} + g_{+n}, v_n \rangle_{\Gamma_{10}} = 0 \quad \text{on } \Gamma_{10}, \\ \sigma_{-n} - g_{-n} \le 0, \quad \langle \sigma_{-n} - g_{-n}, v_n \rangle_{\Gamma_{11}} = 0 \quad \text{on } \Gamma_{11}, \end{cases} \tag{5.10}$$

where $\mathbf{L}^2_\tau(\Gamma_8)$ is the subspace of $\mathbf{L}^2(\Gamma_8)$ consisting of functions such that[$(u, n)_{\mathbf{L}^2(\Gamma_8)} = 0$.

Remark 5.1 If $u \in H^1(\Omega)$ and $u|_{\partial\Omega} = 0$ on $O(\Gamma_i)\backslash\overline{\Gamma}_i$, where $O(\Gamma_i)$ is an open subset of $\partial\Omega$ such that $\overline{\Gamma}_i \subset O(\Gamma_i)$, then $u|_{\Gamma_i} \in H^{\frac{1}{2}}_{00}(\Gamma_i)$ (see Sect. 1.3.2). Since $H^{\frac{1}{2}}_{00}(\Gamma_i) \hookrightarrow H^{\frac{1}{2}}_0(\Gamma_i)$ and $H^{\frac{1}{2}}(\Gamma_i) = H^{\frac{1}{2}}_0(\Gamma_i)$ (see (1.11)),

$$H^{\frac{1}{2}}_{00}(\Gamma_i) \hookrightarrow H^{\frac{1}{2}}(\Gamma_i) \hookrightarrow H^{-\frac{1}{2}}(\Gamma_i) \hookrightarrow (H^{\frac{1}{2}}_{00}(\Gamma_i))^*.$$

Thus, under condition $u|_{\partial\Omega} = 0$ on $O(\Gamma_i)\backslash\overline{\Gamma}_i$, for $\phi_i \in (H^{\frac{1}{2}}_{00}(\Gamma_i))'$ a dual product $\langle \phi_i, u \rangle_{\Gamma_i}$ is well defined. However, if $\phi_i \in H^{-\frac{1}{2}}(\Gamma_i)$, then without knowing that $u|_{\partial\Omega} = 0$ on $O(\Gamma_i)\backslash\overline{\Gamma}_i$ the dual product $\langle \phi_i, u \rangle_{\Gamma_i}$ is meaningful. Therefore, under (2) of Assumption 5.1 the dual products on Γ_i in (5.8) is meaningful.

The following theorem shows that the variational formulation above is well defined.

Theorem 5.1 *Let Assumption 5.1 hold. If (v, p) is a solution in the sense of Definition 5.1 of the problem* (5.1), (5.2), *then $(v, \sigma_\tau|_{\Gamma_8}, \sigma_n|_{\Gamma_9}, \sigma_n|_{\Gamma_{10}}, \sigma_n|_{\Gamma_{11}})$ is a solution to Problem I-VE.*

Conversely, if $f \in \mathbf{L}^2(\Omega)$ and Problem I-VE has a smooth solution $(v, \sigma_\tau, \sigma_n, \sigma_+, \sigma_-)$ such that $v \in \mathbf{H}^2(\Omega)$, $\sigma_+ \in L^2(\Gamma_{10})$ and $\sigma_- \in L^2(\Gamma_{11})$, then there exists $p \in H^1(\Omega)$ such that (v, p) is a solution to the problem (5.1), (5.2). *Moreover, if at least one of the sets $\Gamma_i, i = 2, 4, 6, 7$, is nonempty, then p is unique.*

Proof It suffices to prove the conversion from Problem I-VE to the problem (5.1)–(5.2).

Let v be a solution smooth enough as above to Problem I-VE. From (5.9) and (5.10) we have

$$\begin{aligned} &2\mu(\mathcal{E}(v), \mathcal{E}(u)) + \langle (v \cdot \nabla)v, u \rangle + 2\mu(k(x)v, u)_{\Gamma_2} + 2\mu(S\tilde{v}, \tilde{u})_{\Gamma_3} \\ &\quad + 2(\alpha(x)v_\tau, u)_{\Gamma_5} + \mu(k(x)v, u)_{\Gamma_7} - (\sigma_\tau, u_\tau)_{\Gamma_8} - (\sigma_n, u_n)_{\Gamma_9} \\ &\quad - \langle \sigma_{+n}, u_n \rangle_{\Gamma_{10}} - \langle \sigma_{-n}, u_n \rangle_{\Gamma_{11}} - \sum_{i=2,4,7} \langle \phi_i, u_n \rangle_{\Gamma_i} - \sum_{i=3,5,6} \langle \phi_i, u \rangle_{\Gamma_i} \\ &= \langle f, u \rangle \quad \forall u \in \mathbf{V}. \end{aligned} \tag{5.11}$$

From (5.6) we get

$$\begin{aligned}2\mu(\mathcal{E}(v),\mathcal{E}(u)) = &-\mu(\Delta v,u) - 2\mu(k(x)v,u)_{\Gamma_2} + \mu(\operatorname{rot} v\times n,u)_{\Gamma_3} - 2\mu(S\tilde{v},\tilde{u})_{\Gamma_3}\\ &+ 2\mu(\varepsilon_{nn}(v),u\cdot n)_{\Gamma_4} + 2\mu(\varepsilon_{n\tau}(v),u)_{\Gamma_5} + 2\mu(\varepsilon_n(v),u)_{\Gamma_6}\\ &+\mu\left(\frac{\partial v}{\partial n},u\right)_{\Gamma_7} - \mu(k(x)v,u)_{\Gamma_7} + 2\mu(\varepsilon_{n\tau}(v),u)_{\Gamma_8} + 2\mu(\varepsilon_{nn}(v),u)_{\Gamma_9}\\ &+ 2\mu(\varepsilon_{nn}(v),u)_{\Gamma_{10}} + 2\mu(\varepsilon_{nn}(v),u)_{\Gamma_{11}}.\end{aligned} \tag{5.12}$$

Combining (5.11) with (5.12) yields

$$\begin{aligned}&(-\mu\Delta v + (v\cdot\nabla)v - f,u) + \mu(\operatorname{rot} v\times n,u)_{\Gamma_3} + 2\mu(\varepsilon_{nn}(v),u\cdot n)_{\Gamma_4}\\ &+ 2\mu(\varepsilon_{n\tau}(v),u)_{\Gamma_5} + 2(\alpha(x)v_\tau,u)_{\Gamma_5} + 2\mu(\varepsilon_n(v),u)_{\Gamma_6} + \mu\left(\frac{\partial v}{\partial n},u\right)_{\Gamma_7}\\ &+ 2\mu(\varepsilon_{n\tau}(v),u)_{\Gamma_8} + 2\mu(\varepsilon_{nn}(v),u_n)_{\Gamma_9} + 2\mu(\varepsilon_{nn}(v),u_n)_{\Gamma_{10}} + 2\mu(\varepsilon_{nn}(v),u_n)_{\Gamma_{11}}\\ &- (\sigma_\tau,u_\tau)_{\Gamma_8} - (\sigma_n,u_n)_{\Gamma_9} - \langle\sigma_{+n},u_n\rangle_{\Gamma_{10}} - \langle\sigma_{-n},u_n\rangle_{\Gamma_{11}}\\ &- \sum_{i=2,4,7}\langle\phi_i,u_n\rangle_{\Gamma_i} - \sum_{i=3,5,6}\langle\phi_i,u\rangle_{\Gamma_i} = 0.\end{aligned} \tag{5.13}$$

Taking any $u\in C_0^\infty(\Omega)$ with $\operatorname{div} u = 0$, we have

$$(-\mu\Delta v + (v\cdot\nabla)v - f,u) = 0,$$

and by Proposition 2.3 there exists a unique $P\in L^2(\Omega)$ such that $\int_\Omega P\,dx = 0$ and

$$-\mu\Delta v + (v\cdot\nabla)v - f = -\nabla P. \tag{5.14}$$

Since $v\in\mathbf{H}^2(\Omega)$, $f\in\mathbf{L}^2(\Omega)$ and $(v\cdot\nabla)v\in\mathbf{L}^2(\Omega)$, we have that $\nabla P\in\mathbf{L}^2(\Omega)$, which shows that $P\in H^1(\Omega)$.

Substituting (5.14) into (5.13), integrating by parts and taking into account (5.7), we have (see Remark 1.17)

$$\begin{aligned}&(-P-\phi_2,u_n)_{\Gamma_2} + \mu(\operatorname{rot} v\times n - \phi_3/\mu,u)_{\Gamma_3} + (-P + 2\mu\varepsilon_{nn}(v) - \phi_4,u_n)_{\Gamma_4}\\ &+ (2\mu\varepsilon_{n\tau}(v) + \alpha(x)v_\tau - \phi_5,u)_{\Gamma_5} + (-Pn + 2\mu\varepsilon_n(v) - \phi_6,u)_{\Gamma_6}\\ &+\Big(-P + \mu\frac{\partial v}{\partial n}\cdot n - \phi_7,u_n\Big)_{\Gamma_7} + (2\mu\varepsilon_{n\tau}(v) - \sigma_\tau,u)_{\Gamma_8}\\ &+ (-P + 2\mu\varepsilon_{nn}(v) - \sigma_n,u_n)_{\Gamma_9} + (-P + 2\mu\varepsilon_{nn}(v) - \sigma_{+n},u_n)_{\Gamma_{10}}\\ &+ (-P + 2\mu\varepsilon_{nn}(v) - \sigma_{-n},u_n)_{\Gamma_{11}} = 0,\end{aligned} \tag{5.15}$$

where $(v,u)_{\Gamma_5} = (v_\tau,u)_{\Gamma_5}$ and $(\mu\frac{\partial v}{\partial n},u)_{\Gamma_7} = (\mu\frac{\partial v}{\partial n}\cdot n,u_n)_{\Gamma_7}$ were used.

For every $i = 3, 5, 8$, respectively, let us take any $u\in\mathbf{V}$ such that $u_n|_{\Gamma_i} = 0$ and $u|_{(\partial\Omega\setminus\Gamma_i)} = 0$. The set of traces of such functions on Γ_i includes any tangent vector fields $\varphi\in\mathbf{H}_{00}^{1/2}(\Gamma_i)$ (see Sect. 2.3.2.1), and so from (5.15) we get

$$
\begin{aligned}
&\operatorname{rot} v \times n = \phi_3/\mu \quad \text{on } \Gamma_3,\\
&2\mu\varepsilon_{n\tau}(v) + \alpha(x)v_\tau - \phi_5 = 0 \quad \text{on } \Gamma_5,\\
&2\mu\varepsilon_{n\tau}(v) - \sigma_\tau = 0 \quad \text{on } \Gamma_8.
\end{aligned}
\tag{5.16}
$$

If for all $i = 2, 4, 6, 7, 9, 10, 11$, $\Gamma_i = \emptyset$, then putting $p = P + c$, where c is any constant, we get a solution (v, p) to problem (5.1), (5.2).

Assume that among $\Gamma_i, i = 2, 4, 6, 7, 9, 10, 11$, at least one is nonempty. For every $i = 2, 4, 7, 9, 10, 11$, respectively, let us take any $u \in \mathbf{V}$ such that $u_\tau|_{\Gamma_i} = 0$ and $u|_{(\partial\Omega\setminus\Gamma_i)} = 0$. The set of traces of such functions on Γ_i include any normal vector fields $\varphi \in \mathbf{H}^{1/2}_{00}(\Gamma_i)$ such that $\int_{\Gamma_i} \varphi\, dx = 0$ (see Theorem IV.1.1 of [1]), and so from (5.15) we have that for some constants c_i, $i = 2, 4, 7, 9, 10, 11$, respectively,

$$
\begin{aligned}
&-P - \phi_2 = c_2 \quad \text{on } \Gamma_2,\\
&-P + 2\mu\varepsilon_{nn}(v) - \phi_4 = c_4 \quad \text{on } \Gamma_4,\\
&-P + \mu\frac{\partial v}{\partial n}\cdot n - \phi_7 = c_7 \quad \text{on } \Gamma_7,\\
&-P + 2\mu\varepsilon_{nn}(v) - \sigma_n = c_9 \quad \text{on } \Gamma_9,\\
&-P + 2\mu\varepsilon_{nn}(v) - \sigma_{+n} = c_{10} \quad \text{on } \Gamma_{10},\\
&-P + 2\mu\varepsilon_{nn}(v) - \sigma_{-n} = c_{11} \quad \text{on } \Gamma_{11}.
\end{aligned}
\tag{5.17}
$$

(See the proof of (2.90).) Taking any $u \in \mathbf{V}$ such that $u|_{(\partial\Omega\setminus\Gamma_6)} = 0$ and arguing as in the proof of (2.90), we have from (5.15) that for a constant c_6

$$
-Pn + 2\mu\varepsilon_n(v) - \phi_6 = c_6 n \quad \text{on } \Gamma_6. \tag{5.18}
$$

Let us prove that indeed, all c_i are equal to one constant c. For example, assume that Γ_2 and Γ_4 are nonempty. Taking any $u \in \mathbf{V}$ such that $u|_{\partial\Omega} = 0$ on $\partial\Omega \setminus (\Gamma_2 \cup \Gamma_4)$, we get from (5.15)

$$
c_2 \int_{\Gamma_2} u_n\, dx + c_4 \int_{\Gamma_4} u_n\, dx = 0,
$$

which implies $c_2 = c_4 = c$ since $\int_{\Gamma_2} u_n\, dx = -\int_{\Gamma_4} u_n\, dx$. Thus, from (5.14), (5.17) and (5.18) we see that $p = P + c$ satisfies

$$
-\mu\Delta v + (v\cdot\nabla) + \nabla p = f
$$

and all the following boundary conditions:

$$\begin{aligned}
&-p=\phi_2 \quad \text{on } \Gamma_2,\\
&-p+2\mu\varepsilon_{nn}(v)=\phi_4 \quad \text{on } \Gamma_4,\\
&-pn+2\mu\varepsilon_n(v)=\phi_6 \quad \text{on } \Gamma_6,\\
&-p+\mu\frac{\partial v}{\partial n}\cdot n=\phi_7 \quad \text{on } \Gamma_7,\\
&-p+2\mu\varepsilon_{nn}(v)=\sigma_n \quad \text{on } \Gamma_9,\\
&-p+2\mu\varepsilon_{nn}(v)=\sigma_{+n} \quad \text{on } \Gamma_{10},\\
&-p+2\mu\varepsilon_{nn}(v)=\sigma_{-n} \quad \text{on } \Gamma_{11}.
\end{aligned} \tag{5.19}$$

By virtue of (5.10), (5.16) and (5.19), all conditions in (5.2) are satisfied. Therefore, (v, p) is a solution to problem (5.1), (5.2). If at least one of the sets Γ_i, $i = 2, 4, 6, 7$, is nonempty, then the equality with the given ϕ_i above on Γ_i holds, and p is unique. □

We will find a variational inequality equivalent to Problem I-VE.

Let $(v, \sigma_\tau, \sigma_n, \sigma_{+n}, \sigma_{-n})$ be a solution of Problem I-VE. Subtracting the second formula of (5.10) with $u = w$ from the second one of (5.10), we get

$$\begin{aligned}
&a_{01}(w, u-w)+a_{11}(w, w, u-w)-(\sigma_\tau, u_\tau-w_\tau)_{\Gamma_8}-(\sigma_n, u_n-w_n)_{\Gamma_9}\\
&\quad-\langle\sigma_{+n}, u_n-w_n\rangle_{\Gamma_{10}}-\langle\sigma_{-n}, u_n-w_n\rangle_{\Gamma_{11}}=\langle F_1, u-w\rangle \quad \forall u\in \mathbf{V}.
\end{aligned} \tag{5.20}$$

Define the functionals $\phi_\tau, \phi_n, \phi_+, \phi_-$, respectively, by

$$\begin{aligned}
&\phi_\tau(\eta)=\int_{\Gamma_8} g_\tau|\eta|\,dx \quad \forall\eta\in \mathbf{L}^2_\tau(\Gamma_8),\\
&\phi_n(\eta)=\int_{\Gamma_9} g_n|\eta|\,dx \quad \forall\eta\in L^2(\Gamma_9),\\
&\phi_+(\eta)=\int_{\Gamma_{10}} g_{+n}\eta\,dx \quad \forall\eta\in L^2(\Gamma_{10}),\\
&\phi_-(\eta)=-\int_{\Gamma_{11}} g_{-n}\eta\,dx \quad \forall\eta\in L^2(\Gamma_{11}).
\end{aligned} \tag{5.21}$$

Since if $u\in K(\Omega)$, then $u|_{\Gamma_8}\in \mathbf{L}^2_\tau(\Gamma_8)$, $u_n|_{\Gamma_9}\in L^2(\Gamma_9)$, $u_n|_{\Gamma_{10}}\in L^2(\Gamma_{10})$ and $u_n|_{\Gamma_{11}}\in L^2(\Gamma_{11})$, in what follows for convenience we use the notation

$$\begin{aligned}
&\phi_\tau(u)=\phi_\tau(u|_{\Gamma_8}),\\
&\phi_n(u)=\phi_n(u_n|_{\Gamma_9}),\\
&\phi_+(u)=\phi_+(u_n|_{\Gamma_{10}}),\\
&\phi_-(u)=\phi_-(u_n|_{\Gamma_{11}}) \quad \forall u\in K(\Omega).
\end{aligned}$$

Define a functional $\Phi:\mathbf{V}\to\overline{\mathbb{R}}$ by

$$\Phi(u) = \begin{cases} \phi_\tau(u) + \phi_n(u) + \phi_+(u) + \phi_-(u) & \forall u \in K(\Omega), \\ +\infty & \forall u \notin K(\Omega). \end{cases} \tag{5.22}$$

Then, Φ is proper, convex and weakly lower semi-continuous (cf. Example 1.2). Note $\Phi \geq 0$ since $u_n|_{\Gamma_{10}} \geq 0$ and $u_n|_{\Gamma_{11}} \leq 0$ for $u \in K(\Omega)$.

By Assumption 5.1, $w_\tau = v_\tau$ on Γ_8 and $w_n = v_n$ on $\cup_{i=9}^{11}\Gamma_i$. Taking into account the fact that $g_\tau|v_\tau| + \sigma_\tau \cdot v_\tau = 0$ and $|\sigma_\tau| \leq g_\tau$ on Γ_8, we have that

$$\begin{aligned} &\phi_\tau(u) - \phi_\tau(w) + (\sigma_\tau, u_\tau)_{\Gamma_8} - (\sigma_\tau, w_\tau)_{\Gamma_8} \\ &\quad = \int_{\Gamma_8} (g_\tau|u_\tau| + \sigma_\tau \cdot u_\tau)\, dx - \int_{\Gamma_8} (g_\tau|w_\tau| + \sigma_\tau \cdot w_\tau)\, dx \\ &\quad = \int_{\Gamma_8} (g_\tau|u_\tau| + \sigma_\tau \cdot u_\tau)\, dx - \int_{\Gamma_8} (g_\tau|v_\tau| + \sigma_\tau \cdot v_\tau)\, dx \geq 0 \quad \forall u \in K(\Omega). \end{aligned}$$

Taking into account the fact that $g_n|v_n| + \sigma_n \cdot v_n = 0$ and $|\sigma_n| \leq g_n$ on Γ_9, in the same way we have

$$\phi_n(u) - \phi_n(w) + (\sigma_n, u_n)_{\Gamma_9} - (\sigma_n, w_n)_{\Gamma_9} \geq 0. \tag{5.23}$$

Also,

$$\begin{aligned} &\phi_+(u) - \phi_+(w) + \langle \sigma_{+n}, u_n \rangle_{\Gamma_{10}} - \langle \sigma_{+n}, w_n \rangle_{\Gamma_{10}} \\ &= \langle g_{+n} + \sigma_{+n}, u_n \rangle_{\Gamma_{10}} - \langle g_{+n} + \sigma_{+n}, w_n \rangle_{\Gamma_{10}} \geq 0, \end{aligned} \tag{5.24}$$

where the facts that $u_n \geq 0$, $\sigma_{+n} + g_{+n} \geq 0$ and $\langle \sigma_{+n} + g_{+n}, v_n \rangle_{\Gamma_{10}} = 0$, $w_n = v_n$ on Γ_{10} were used. In the same way, we have

$$\phi_-(u) - \phi_-(w) + \langle \sigma_{-n}, u_n \rangle_{\Gamma_{11}} - \langle \sigma_{-n}, w_n \rangle_{\Gamma_{11}} \geq 0. \tag{5.25}$$

By virtue of (5.22)–(5.25), we have

$$\begin{aligned} \Phi(u) - \Phi(w) \geq &- (\sigma_\tau, u_\tau - w_\tau)_{\Gamma_8} - (\sigma_n, u_n - w_n)_{\Gamma_9} \\ &- \langle \sigma_{+n}, u_n - w_n \rangle_{\Gamma_{10}} - \langle \sigma_{-n}, u_n - w_n \rangle_{\Gamma_{11}} \quad \forall u \in \mathbf{V}. \end{aligned} \tag{5.26}$$

Therefore, we get from (5.20) and (5.26)

$$a_{01}(w, u - w) + a_{11}(w, w, u - w) + \Phi(u) - \Phi(w) \geq \langle F_1, u - w \rangle \quad \forall u \in \mathbf{V}. \tag{5.27}$$

Thus, we come to the following formulation associated with Problem I (the case of static pressure) by a variational inequality.

Problem I-VI Find $v = w + U$ such that

$$a_{01}(w, u - w) + a_{11}(w, w, u - w) + \Phi(u) - \Phi(w) \geq \langle F_1, u - w \rangle \quad \forall u \in \mathbf{V}, \tag{5.28}$$

where a_{01}, a_{11}, F_1 are the same as in (5.9), U is the same as in Assumption 5.1 and Φ is as in (5.22).

To prove equivalence of Problem I-VI and Problem I-VE we need first to establish the following result:

Lemma 5.1 *For $\psi \in C_0^\infty(\Gamma_i), i = 10, 11$, there exists a function $\overline{u} \in \mathbf{V}$ such that*

$$\overline{u}_n|_{\Gamma_i} = \psi, \quad \|\overline{u}\|_{\mathbf{V}} \le C_i \|\psi\|_{H^{1/2}(\Gamma_i)},$$

where C_i are independent of ψ.

Proof By (3) of Assumption 5.1 if $\Gamma_{10} \cup \Gamma_{11} \neq \emptyset$, then, for example, $\Gamma_2 \neq \emptyset$ and there exists a diffeomorphism $y = f_i(x) \in C^1$ from Γ_i onto Γ_2. Define $\varphi(y)$ at point $y \in \Gamma_2$ corresponding to point $x \in \Gamma_i$ by $\varphi(y) = \frac{1}{Df(x)}\psi(f_i^{-1}(y))$, where $Df_i(x)$ is Jacobian of the transformation f_i. Then,

$$\int_{\Gamma_2} \varphi(y)\,dy = \int_{\Gamma_i} \frac{1}{Df_i(x)}\psi(f_i^{-1}(y))Df_i(x)\,dx = \int_{\Gamma_i} \psi(x)\,dx, \tag{5.29}$$

and

$$\|\varphi(y)\|_{H^{\frac{1}{2}}(\Gamma_2)} \le \left\| \frac{1}{Df_i(x)} \right\|_{C(\overline{\Gamma}_i)} \|\psi(x)\|_{H^{\frac{1}{2}}(\Gamma_i)} \le c_i \|\psi(x)\|_{H^{\frac{1}{2}}(\Gamma_i)}. \tag{5.30}$$

When $\psi \in C_0^\infty(\Gamma_i)$, define a function $\overline{\phi} \in \mathbf{H}^{1/2}(\partial\Omega)$ on $\partial\Omega$ as follows:

$$\overline{\phi} \times n|_{\Gamma_2 \cup \Gamma_i} = 0, \ \overline{\phi}_n|_{\Gamma_2} = -\varphi, \ \overline{\phi}_n|_{\Gamma_{10}} = \psi, \ \overline{\phi}|_{(\cup_{i=1,3\sim 9,11}\Gamma_i)} = 0.$$

Thus, by (5.29) $\int_{\partial\Omega} \overline{\phi}_n\,ds = 0$. Then, there exists a solution $\overline{u} \in \mathbf{W}^{1,2}(\Omega)$ to the Stokes problem

$$\begin{cases} -\Delta u + \nabla p = 0, \\ \operatorname{div} u = 0, \\ u|_{\partial\Omega} = \overline{\phi} \end{cases}$$

and

$$\|\overline{u}\|_{\mathbf{V}} \le c\|\overline{\phi}\|_{\mathbf{H}^{1/2}(\partial\Omega)}.$$

(cf. Theorem IV.1.1 of [1]). Taking into account (5.30), we come to the asserted estimation with $C_i = 1 + c_i$. Thus $\overline{u}$ is the asserted function. □

Problem I-VE and Problem I-VI are equivalent in the following sense:

Theorem 5.2 *If $(v, \sigma_\tau, \sigma_n, \sigma_{+n}, \sigma_{-n})$ is a solution to Problem I-VE, then v is a solution to Problem I-VI.*

Conversely, if v is a solution to Problem I-VI, then there exist $\sigma_\tau, \sigma_n, \sigma_{+n}, \sigma_{-n}$ such that $(v, \sigma_\tau, \sigma_n, \sigma_{+n}, \sigma_{-n})$ is a solution to Problem I-VE.

Proof We already showed that if $(v, \sigma_\tau, \sigma_n, \sigma_{+n}, \sigma_{-n})$ is a solution to Problem I-VE, then v is a solution to Problem I-VI. Thus, it remains to prove the second assertion of theorem.

Since the functional Φ is proper, we have from (5.28)

$$v - U = w \in K(\Omega) \tag{5.31}$$

because if $w \notin K(\Omega)$, then the left hand side of (5.28) is $-\infty$, which is a contradiction to the fact that the right-hand side is finite.

Let $\psi \in \mathbf{V}_{8-11} \equiv \{u \in \mathbf{V} : u|_{\Gamma_8 \cup \Gamma_9 \cup \Gamma_{10} \cup \Gamma_{11}} = 0\}\ (\subset K(\Omega))$. Putting $u = w + \psi$, $u = w - \psi$ and taking into account

$$\begin{aligned}
\phi_\tau(w) &= \phi_\tau(w + \psi),\\
\phi_n(w) &= \phi_n(w + \psi),\\
\phi_+(w) &= \phi_+(w + \psi),\\
\phi_-(w) &= \phi_-(w + \psi),
\end{aligned}$$

we get from (5.22) and (5.28)

$$\begin{aligned}
&a_{01}(w, \psi) + a_{11}(w, w, \psi) \geq \langle F_1, \psi \rangle,\\
&a_{01}(w, -\psi) + a_{11}(w, w, -\psi) \geq \langle F_1, -\psi \rangle \quad \forall \psi \in \mathbf{V}_{8-11},
\end{aligned}$$

which imply

$$a_{01}(w, \psi) + a_{11}(w, w, \psi) = \langle F_1, \psi \rangle \quad \forall \psi \in \mathbf{V}_{8-11}. \tag{5.32}$$

When $u \in \mathbf{V}_{10-11} \equiv \{u \in \mathbf{V} : u|_{\Gamma_{10} \cup \Gamma_{11}} = 0\}\ (\subset K(\Omega))$, the set $\{(u|_{\Gamma_8}, u_n|_{\Gamma_9})\}$ is a subspace of $\mathbf{L}^2_\tau(\Gamma_8) \times L^2(\Gamma_9)$, where $u_n|_{\Gamma_9}$ is $u|_{\Gamma_9} \cdot n$.

Define a functional σ^* on the set by

$$\big\langle \sigma^*, (u|_{\Gamma_8}, u_n|_{\Gamma_9}) \big\rangle = a_{01}(w, u) + a_{11}(w, w, u) - \langle F_1, u \rangle \quad \forall u \in \mathbf{V}_{10-11}. \tag{5.33}$$

This functional is well defined. Because if $u, u_1 \in \mathbf{V}_{10-11}$ are such that $(u|_{\Gamma_8}, u|_{\Gamma_9}) = (u_1|_{\Gamma_8}, u_1|_{\Gamma_9})$, then since $u - u_1 \in \mathbf{V}_{8-11}$, by (5.32)

$$a_{01}(w, u - u_1) + a_{11}(w, w, u - u_1) - \langle F_1, u - u_1 \rangle = 0,$$

that is,

$$a_{01}(w, u) + a_{11}(w, w, u) - \langle F_1, u \rangle = a_{01}(w, u_1) + a_{11}(w, w, u_1) - \langle F_1, u_1 \rangle,$$

and so by (5.33)

$$\left\langle \sigma^*, (u|_{\Gamma_8}, u_n|_{\Gamma_9})\right\rangle = \left\langle \sigma^*, (u_1|_{\Gamma_8}, u_{1n}|_{\Gamma_9})\right\rangle.$$

This function is linear.

Putting $u = w + \psi$, where $\psi \in \mathbf{V}_{10-11}$, and taking into account

$$\phi_+(w+\psi) = \phi_+(w),$$
$$\phi_-(w+\psi) = \phi_-(w),$$

we have from (5.33) and (5.28)

$$\begin{aligned} -\left\langle \sigma^*, (\psi|_{\Gamma_8}, \psi_n|_{\Gamma_9})\right\rangle &= -\left[a_{01}(w,\psi) + a_{11}(w,w,\psi) - \langle F_1, \psi\rangle\right] \\ &\le \Phi(w+\psi) - \Phi(w) \\ &= \phi_\tau(w+\psi) - \phi_\tau(w) + \phi_n(w+\psi) - \phi_n(w) \\ &\le \int_{\Gamma_8} g_\tau |\psi|_{\Gamma_8}\, dx + \int_{\Gamma_9} g_n |\psi|_{\Gamma_9}\, dx \quad \forall \psi \in \mathbf{V}_{10-11}. \end{aligned} \tag{5.34}$$

Putting $u = w - \psi$, in the same way we have

$$\begin{aligned} \left\langle \sigma^*, (\psi|_{\Gamma_8}, \psi_n|_{\Gamma_9})\right\rangle &= \left[a_{01}(w,\psi) + a_{11}(w,w,\psi) - \langle F_1, \psi\rangle\right] \\ &\le \phi_\tau(w-\psi) - \phi_\tau(w) + \phi_n(w-\psi) - \phi_n(w) \\ &\le \int_{\Gamma_8} g_\tau |\psi|_{\Gamma_8}\, dx + \int_{\Gamma_9} g_n |\psi|_{\Gamma_9}\, dx \quad \forall \psi \in \mathbf{V}_{10-11}. \end{aligned} \tag{5.35}$$

By (5.34) and (5.35), we can see that σ^* is a bounded linear functional with a norm not greater than 1 on a subspace of $\mathbf{L}^1_{g_\tau}(\Gamma_8) \times L^1_{g_n}(\Gamma_9)$, where $\mathbf{L}^1_{g_\tau}(\Gamma_8)$, $L^1_{g_n}(\Gamma_9)$ are, respectively, the spaces of functions integrable with weights g_τ, g_n on Γ_8 and Γ_9. By the Hahn-Banach theorem the functional can be extended as a functional on $\mathbf{L}^1_{g_\tau}(\Gamma_8) \times L^1_{g_n}(\Gamma_9)$ norms of which is not greater than 1. Therefore, there exist elements $\sigma_\tau \in \mathbf{L}^\infty_{\frac{1}{g_\tau}}(\Gamma_8)$ with $\|\sigma_\tau\|_{\mathbf{L}^\infty_{\frac{1}{g_\tau}}(\Gamma_8)} \le 1$ and $\sigma_n \in L^\infty_{\frac{1}{g_n}}(\Gamma_9)$ with $\|\sigma_n\|_{L^\infty_{\frac{1}{g_n}}(\Gamma_9)} \le 1$, which imply

$$|\sigma_\tau| \le g_\tau, \quad |\sigma_n| \le g_n \tag{5.36}$$

and

$$\left\langle \sigma^*, (u|_{\Gamma_8}, u_n|_{\Gamma_9})\right\rangle = \left(\sigma_\tau, u|_{\Gamma_8}\right)_{\Gamma_8} + \left(\sigma_n, u_n|_{\Gamma_9}\right)_{\Gamma_9} \quad \forall u \in \mathbf{V}_{10-11}. \tag{5.37}$$

When $u \in \mathbf{V}$, the set $\{(u_n|_{\Gamma_{10}}, u_n|_{\Gamma_{11}})\}$ is a subspace of $H^{\frac{1}{2}}(\Gamma_{10}) \times H^{\frac{1}{2}}(\Gamma_{11})$. Define a functional σ_1^* on the set $\mathbf{V}$ by

$$\langle\sigma_1^*, (u_n|_{\Gamma_{10}}, u_n|_{\Gamma_{11}})\rangle = \\ a_{01}(w,u) + a_{11}(w,w,u) - (\sigma_\tau, u|_{\Gamma_8})_{\Gamma_8} - (\sigma_n, u|_{\Gamma_9})_{\Gamma_9} - \langle F_1, u\rangle \quad \forall u \in \mathbf{V}. \tag{5.38}$$

This functional is also well defined. Because if $u, u^1 \in \mathbf{V}$ are such that $(u_{\Gamma_{10}}, u|_{\Gamma_{11}}) = (u^1|_{\Gamma_{10}}, u^1|_{\Gamma_{11}})$, then since $u - u^1 \in \mathbf{V}_{10-11}(\Omega)$, by (5.33) and (5.37)

$$\begin{aligned} &a_{01}(w, u-u^1) + a_{11}(w,w,u-u^1) - \left(\sigma_\tau, (u-u^1)|_{\Gamma_8}\right)_{\Gamma_8} \\ &\quad - \left(\sigma_n, (u-u^1)|_{\Gamma_9}\right)_{\Gamma_9} - \langle F_1, u-u^1\rangle \\ &= \left\langle\sigma^*, ((u-u^1)|_{\Gamma_8}, (u-u^1)|_{\Gamma_9})\right\rangle - \left(\sigma_\tau, (u-u^1)|_{\Gamma_8}\right)_{\Gamma_8} - \left(\sigma_n, (u-u^1)|_{\Gamma_9}\right)_{\Gamma_9} = 0, \end{aligned}$$

and so by (5.38)

$$\langle\sigma_1^*, (u_n|_{\Gamma_{10}}, u_n|_{\Gamma_{11}})\rangle = \langle\sigma_1^*, (u_n^1|_{\Gamma_{10}}, u_n^1|_{\Gamma_{11}})\rangle.$$

The functional σ_1^* is linear. Let us prove next its continuity.

Let $\overline{u}$ be the function corresponding to $\psi \in C_0^\infty(\Gamma_{10})$ by Lemma 5.1. Then, by Lemma 5.1 we have from (5.38)

$$\begin{aligned} &|\langle\sigma_1^*, (\psi, 0)\rangle| \\ &\le C\left[\|w\|_{\mathbf{V}}\|\overline{u}\|_{\mathbf{V}} + \|w\|_{\mathbf{V}}^2\|\overline{u}\|_{\mathbf{V}} + (\|\sigma_\tau\|_{L_2(\Gamma_8)} + \|\sigma_\tau\|_{L_2(\Gamma_9)})\|\overline{u}\| + \|F_1\|_{\mathbf{V}^*}\|\overline{u}\|_{\mathbf{V}}\right] \\ &\le C\left[\|w\|_{\mathbf{V}}\| + \|w\|_{\mathbf{V}}^2 + (\|\sigma_\tau\|_{L_2(\Gamma_8)} + \|\sigma_\tau\|_{L_2(\Gamma_9)}) + \|F_1\|_{\mathbf{V}^*}\right] \cdot \|\psi\|_{H^{\frac{1}{2}}(\Gamma_{10})}. \end{aligned} \tag{5.39}$$

Also assuming that $\overline{u}$ is the function corresponding to $\psi \in C_0^\infty(\Gamma_{11})$ by Lemma 5.1, we have

$$\begin{aligned} &|\langle\sigma_1^*, (0, \psi)\rangle| \\ &\le C\left[\|w\|_{\mathbf{V}}\|\overline{u}\|_{\mathbf{V}} + \|w\|_{\mathbf{V}}^2\|\overline{u}\|_{\mathbf{V}} + (\|\sigma_\tau\|_{L_2(\Gamma_8)} + \|\sigma_\tau\|_{L_2(\Gamma_9)})\|\overline{u}\| + \|F_1\|_{\mathbf{V}^*}\|\overline{u}\|_{\mathbf{V}}\right] \\ &\le C\left[\|w\|_{\mathbf{V}} + \|w\|_{\mathbf{V}}^2 + (\|\sigma_\tau\|_{L_2(\Gamma_8)} + \|\sigma_\tau\|_{L_2(\Gamma_9)}) + \|F_1\|_{\mathbf{V}^*}\right] \cdot \|\psi\|_{H^{\frac{1}{2}}(\Gamma_{11})}. \end{aligned} \tag{5.40}$$

Since $H_0^{1/2}(\Gamma_i) = H^{\frac{1}{2}}(\Gamma_i)$, $i = 10, 11$, (cf. (1.11)), (5.39) and (5.40) show that the functional σ_1^* is continuous on the subspace of $H^{\frac{1}{2}}(\Gamma_{10}) \times H^{\frac{1}{2}}(\Gamma_{11})$ mentioned above. Thus, by the Hahn-Banach theorem the functional is extended as a functional on $H^{\frac{1}{2}}(\Gamma_{10}) \times H^{\frac{1}{2}}(\Gamma_{11})$.

Therefore, there exists an element $(\sigma_{+n}, \sigma_{-n}) \in H^{-1/2}(\Gamma_{10}) \times H^{-1/2}(\Gamma_{11})$ such that

$$\langle\sigma_1^*, (u|_{\Gamma_{10}}, u|_{\Gamma_{11}})\rangle = \langle\sigma_{+n}, u|_{\Gamma_{10}}\rangle_{\Gamma_{10}} + \langle\sigma_{-n}, u|_{\Gamma_{11}}\rangle_{\Gamma_{11}} \quad \forall u \in \mathbf{V}. \tag{5.41}$$

When $\psi \ge 0$ is such that $\psi \in C_0^\infty(\Gamma_{10})$, let $\overline{u} \in K(\Omega)$ be the function asserted in Lemma 5.1. Putting $u = w + \overline{u}$, by (5.28) we have

$$a_{01}(w,\overline{u})+a_{11}(w,w,\overline{u})+\Phi(w+\overline{u})-\Phi(w)-\langle F_1,\overline{u}\rangle\geq 0. \tag{5.42}$$

On the other hand, by (5.38), (5.41) and property of $\overline{u}$,

$$a_{01}(w,\overline{u})+a_{11}(w,w,\overline{u})-\langle F_1,\overline{u}\rangle=\langle\sigma_{+n},\psi\rangle_{\Gamma_{10}}$$

and so from (5.42) we have that

$$\langle\sigma_{+n},\psi\rangle_{\Gamma_{10}}+\Phi(w+\overline{u})-\Phi(w)\geq 0. \tag{5.43}$$

By (5.21), (5.22) and the property of $\overline{u}$, we get

$$\Phi(w+\overline{u})-\Phi(w)=\langle g_{+n},\psi\rangle_{\Gamma_{10}},$$

which combined with (5.43) gives

$$\langle\sigma_{+n},\psi\rangle_{\Gamma_{10}}+(g_{+n},\psi)_{\Gamma_{10}}\geq 0,$$

that is,

$$\sigma_{+n}+g_{+n}\geq 0. \tag{5.44}$$

When $\psi\leq 0$ is such that $\psi\in C_0^\infty(\Gamma_{11})$, let $\overline{u}\in K(\Omega)$ be the function asserted in Lemma 5.1. Then, in the same way we have that

$$\langle\sigma_{-n},-\psi\rangle_{\Gamma_{11}}-(g_{-n},-\psi)_{\Gamma_{11}}\geq 0,$$

that is,

$$\sigma_{-n}-g_{-n}\leq 0. \tag{5.45}$$

From (5.38) and (5.41), we have

$$\begin{aligned}a_{01}(w,u)+a_{11}(w,w,u)-(\sigma_\tau,u_\tau)_{\Gamma_8}-(\sigma_n,u)_{\Gamma_9}-\langle\sigma_{+n},u\rangle_{\Gamma_{10}}-\langle\sigma_{-n},u\rangle_{\Gamma_{11}}\\=\langle F_1,u\rangle\quad\forall u\in\mathbf{V}.\end{aligned} \tag{5.46}$$

Putting $u=0$ in (5.28) and taking into account (5.46) with $u=w$, we have

$$\begin{aligned}(\sigma_\tau,w)_{\Gamma_8}+(\sigma_n,w)_{\Gamma_9}+\langle\sigma_{+n},w_n\rangle_{\Gamma_{10}}+\langle\sigma_{-n},w_n\rangle_{\Gamma_{11}}\\+\phi_\tau(w)+\phi_n(w)+\phi_+(w)+\phi_-(w)\leq 0,\end{aligned}$$

that is,

$$\begin{aligned}\int_{\Gamma_8}(\sigma_\tau w_\tau+g_\tau|w_\tau|)\,ds+\int_{\Gamma_9}(\sigma_n w_n+g_n|w_n|)\,ds\\+\langle\sigma_{+n}+g_{+n},w_n\rangle_{\Gamma_{10}}+\langle\sigma_{-n}-g_{-n},w_n\rangle_{\Gamma_{11}}\leq 0.\end{aligned} \tag{5.47}$$

Since on Γ_8, Γ_9, Γ_{10} and Γ_{11}, respectively, $w_\tau = v_\tau$, $w_n = v_n$, $w_n = v_n \geq 0$ and $w_n = v_n \leq 0$, taking into account (5.36), (5.44), (5.45), by (5.47) we have

$$\begin{aligned} &\sigma_\tau v_\tau + g_\tau |v_\tau| = 0, \\ &\sigma_n v_n + g_n |v_n| = 0, \\ &\langle \sigma_{+n} + g_{+n}, v_n \rangle = 0, \\ &\langle \sigma_{-n} - g_{-n}, v_n \rangle = 0. \end{aligned} \tag{5.48}$$

Therefore, by virtue of (5.31), (5.36), (5.44)–(5.46) and (5.48), we come to the conclusion. □

5.1.2 Variational Formulation: The Case of Total Pressure

Taking $(v \cdot \nabla)v = \operatorname{rot} v \times v + \frac{1}{2}\operatorname{grad}|v|^2$ into account and putting $v = w + U$, by (5.6), (5.7) and Assumption 5.1 we can see that solutions (v, p) of the problem (5.1), (5.3) in the sense of Definition 5.1 satisfy the following:

$$\begin{cases} v - U = w \in K(\Omega), \\ 2\mu(\mathcal{E}(w), \mathcal{E}(u)) + \langle \operatorname{rot} w \times w, u \rangle + \langle \operatorname{rot} U \times w, u \rangle \\ \qquad + \langle \operatorname{rot} w \times U, u \rangle + 2\mu(k(x)w, u)_{\Gamma_2} + 2\mu(S\tilde{w}, \tilde{u})_{\Gamma_3} \\ \qquad + 2(\alpha(x)w, u)_{\Gamma_5} + \mu(k(x)w, u)_{\Gamma_7} - 2\mu(\varepsilon_{n\tau}(w + U), u)_{\Gamma_8} \\ \qquad + \left(p + \dfrac{1}{2}|v|^2 - 2\mu\varepsilon_{nn}(w + U), u_n\right)_{\Gamma_9 \cup \Gamma_{10} \cup \Gamma_{11}} \\ \quad = -2\mu(\mathcal{E}(U), \mathcal{E}(u)) - \langle \operatorname{rot} U \times U, u \rangle - 2\mu(k(x)U, u)_{\Gamma_2} \\ \qquad - 2\mu(S\tilde{U}, \tilde{u})_{\Gamma_3} - 2(\alpha(x)U_\tau, u)_{\Gamma_5} - \mu(k(x)U, u)_{\Gamma_7} \\ \qquad + \langle f, u \rangle + \sum\limits_{i=2,4,7} \langle \phi_i, u_n \rangle_{\Gamma_i} + \sum\limits_{i=3,5,6} \langle \phi_i, u \rangle_{\Gamma_i} \\ \qquad\qquad\qquad\qquad \forall u \in \mathbf{V}, \\ |\sigma_\tau^t(v)| \leq g_\tau,\ \sigma_\tau^t \cdot v_\tau + g_\tau |v_\tau| = 0 \quad \text{on } \Gamma_8, \\ |\sigma_n^t(v)| \leq g_n,\ \sigma_n^t(v) v_n + g_n |v_n| = 0 \quad \text{on } \Gamma_9, \\ \sigma_n^t(v) + g_{+n} \geq 0, \quad (\sigma_n^t(v) + g_{+n}) v_n = 0 \quad \text{on } \Gamma_{10}, \\ \sigma_n^t(v) - g_{-n} \leq 0, \quad (\sigma_n^t(v) - g_{-n}) v_n = 0 \quad \text{on } \Gamma_{11}. \end{cases} \tag{5.49}$$

Define $a_{02}(\cdot, \cdot)$, $a_{12}(\cdot, \cdot, \cdot)$ and $F_2 \in \mathbf{V}^*$, respectively, by

$$
\begin{aligned}
a_{02}(w,u) &= 2\mu(\mathcal{E}(w),\mathcal{E}(u)) + \langle \operatorname{rot} U \times w, u\rangle + \langle \operatorname{rot} w \times U, u\rangle \\
&\quad + 2\mu(k(x)w,u)_{\Gamma_2} + 2\mu(S\tilde{w},\tilde{u})_{\Gamma_3} + 2(\alpha(x)w,u)_{\Gamma_5} \\
&\quad + \mu(k(x)w,u)_{\Gamma_7} \qquad \forall w,u \in \mathbf{V}, \\
a_{12}(w,u,v) &= \langle \operatorname{rot} w \times u, v\rangle \qquad \forall w,u,v \in \mathbf{V}, \\
\langle F_2, u\rangle &= -2\mu(\mathcal{E}(U),\mathcal{E}(u)) - \langle \operatorname{rot} U \times U, u\rangle - 2\mu(k(x)U,u)_{\Gamma_2} \\
&\quad - 2\mu(S\tilde{U},\tilde{u})_{\Gamma_3} - 2(\alpha(x)U_\tau,u)_{\Gamma_5} - \mu(k(x)U,u)_{\Gamma_7} \\
&\quad + \langle f,u\rangle + \sum_{i=2,4,7} \langle \phi_i, u_n\rangle_{\Gamma_i} + \sum_{i=3,5,6} \langle \phi_i, u\rangle_{\Gamma_i} \qquad \forall u \in \mathbf{V}.
\end{aligned}
\tag{5.50}
$$

Then, taking into account

$$
\begin{aligned}
\sigma^t_\tau(v) &= 2\mu\varepsilon_{n\tau}(v), \\
\sigma^t_n(v) &= -(p + \frac{1}{2}|v|^2) + 2\mu\varepsilon_{nn}(v)
\end{aligned}
$$

and (5.49), we introduce the following variational formulation for the problem (5.1), (5.3).

Problem II-VE. Find $(v, \sigma^t_\tau, \sigma^t_n, \sigma^t_{+n}, \sigma^t_{-n}) \in \big(U + K(\Omega)\big) \times \mathbf{L}^2_\tau(\Gamma_8) \times L^2(\Gamma_9) \times H^{-\frac{1}{2}}(\Gamma_{10}) \times H^{-\frac{1}{2}}(\Gamma_{11})$ such that

$$
\begin{cases}
v - U = w \in K(\Omega), \\
a_{02}(w,u) + a_{12}(w,w,u) - (\sigma^t_\tau, u_\tau)_{\Gamma_8} - (\sigma^t_n, u_n)_{\Gamma_9} \\
\qquad - \big\langle \sigma^t_{+n}, u_n\big\rangle_{\Gamma_{10}} - \big\langle \sigma^t_{-n}, u_n\big\rangle_{\Gamma_{11}} = \langle F_2, u\rangle \quad \forall u \in \mathbf{V}, \\
|\sigma^t_\tau| \le g_\tau, \ \sigma^t_\tau \cdot v_\tau + g_\tau|v_\tau| = 0 \quad \text{on } \Gamma_8, \\
|\sigma^t_n| \le g_n, \ \sigma^t_n v_n + g_n|v_n| = 0 \quad \text{on } \Gamma_9, \\
\sigma^t_{+n} + g_{+n} \ge 0, \quad \big\langle \sigma^t_{+n} + g_{+n}, v_n\big\rangle_{\Gamma_{10}} = 0 \quad \text{on } \Gamma_{10}, \\
\sigma^t_{-n} - g_{-n} \le 0, \quad \big\langle \sigma^t_{-n} - g_{-n}, v_n\big\rangle_{\Gamma_{11}} = 0 \quad \text{on } \Gamma_{11}.
\end{cases}
\tag{5.51}
$$

In the same way as in Theorem 5.1 we have

Theorem 5.3 *Let Assumption 5.1 hold. If (v, p) is a solution in the sense of Definition 5.1 of the problem (5.1), (5.3), then $(v, \sigma^t_\tau|_{\Gamma_8}, \sigma^t_n|_{\Gamma_9}, \sigma^t_n|_{\Gamma_{10}}, \sigma^t_n|_{\Gamma_{11}})$ is a solution to Problem II-VE.*

Conversely, if $f \in \mathbf{L}^2(\Omega)$ and Problem II-VE has a smooth solution $(v, \sigma^t_\tau, \sigma^t_n, \sigma^t_+, \sigma^t_-)$ such that $v \in \mathbf{H}^2(\Omega)$, $\sigma^t_+ \in L^2(\Gamma_{10})$ and $\sigma^t_- \in L^2(\Gamma_{11})$, then there exists $p \in H^1(\Omega)$ such that (v, p) is a solution to the problem (5.1), (5.3). Moreover, if at least one of the sets Γ_i, $i = 2, 4, 6, 7$, is nonempty, then p is unique.

Then, in the same way as in Problem I we get Problem II-VI (the case of total pressure) formulated by a variational inequality and can prove that the problem is equivalent to Problem II-VE.

Problem II-VI. Find $v = w + U$ such that

$$a_{02}(w, u-w) + a_{12}(w, w, u-w) + \Phi(u) - \Phi(w) \geq \langle F_2, u-w\rangle \quad \forall u \in \mathbf{V}, \tag{5.52}$$

where a_{02}, a_{12}, F_2 are in (5.50) and Φ is defined by (5.22).

Theorem 5.4 *If $(v, \sigma_\tau^t, \sigma_n^t, \sigma_{+n}^t, \sigma_{-n}^t)$ is a solution to Problem II-VE, then v is a solution to Problem II-VI.*

Conversely, if v is a solution to Problem II-VI, then there exist $\sigma_\tau^t, \sigma_n^t, \sigma_{+n}^t, \sigma_{-n}^t$ such that $(v, \sigma_\tau^t, \sigma_n^t, \sigma_{+n}^t, \sigma_{-n}^t)$ is a solution to Problem II-VE.

5.1.3 Variational Formulation: The Stokes Problem

In the same way as in Problem I we get the following equivalent formulations of Problem III for the Stokes equations with boundary condition (5.2).

Problem III-VE. Find $(v, \sigma_\tau, \sigma_n, \sigma_{+n}, \sigma_{-n}) \in \big(U + K(\Omega)\big) \times \mathbf{L}^2_\tau(\Gamma_8) \times L^2(\Gamma_9) \times H^{-\frac{1}{2}}(\Gamma_{10}) \times H^{-\frac{1}{2}}(\Gamma_{11})$ such that

$$\begin{cases} v - U = w \in K(\Omega), \\ a_{03}(w, u) - (\sigma_\tau, u_\tau)_{\Gamma_8} - (\sigma_n, u_n)_{\Gamma_9} - \langle \sigma_{+n}, u_n\rangle_{\Gamma_{10}} - \langle \sigma_{-n}, u_n\rangle_{\Gamma_{11}} = \langle F_3, u\rangle \\ \qquad\qquad\qquad\qquad\qquad\qquad\qquad\qquad\qquad \forall u \in \mathbf{V}, \\ |\sigma_\tau| \leq g_\tau, \ \sigma_\tau \cdot v_\tau + g_\tau |v_\tau| = 0 \quad \text{on } \Gamma_8, \\ |\sigma_n| \leq g_n, \ \sigma_n v_n + g_n |v_n| = 0 \quad \text{on } \Gamma_9, \\ \sigma_{+n} + g_{+n} \geq 0, \quad \langle \sigma_{+n} + g_{+n}, v_n\rangle = 0 \quad \text{on } \Gamma_{10}, \\ \sigma_{-n} - g_{-n} \leq 0, \quad \langle \sigma_{-n} - g_{-n}, v_n\rangle = 0 \quad \text{on } \Gamma_{11}, \end{cases} \tag{5.53}$$

where

$$\begin{aligned} a_{03}(w, u) &= 2\mu(\mathcal{E}(w), \mathcal{E}(u)) + 2\mu(k(x)w, u)_{\Gamma_2} \\ &\quad + 2\mu(S\tilde{w}, \tilde{u})_{\Gamma_3} + 2(\alpha(x)w, u)_{\Gamma_5} + \mu(k(x)w, u)_{\Gamma_7} \quad \forall w, u \in \mathbf{V}, \\ \langle F_3, u\rangle &= -2\mu(\mathcal{E}(U), \mathcal{E}(u)) - 2\mu(k(x)U, u)_{\Gamma_2} - 2\mu(S\tilde{U}, \tilde{u})_{\Gamma_3} - 2(\alpha(x)U_\tau, u)_{\Gamma_5} \\ &\quad - \mu(k(x)U, u)_{\Gamma_7} + \langle f, u\rangle + \sum_{i=2,4,7} \langle \phi_i, u_n\rangle_{\Gamma_i} + \sum_{i=3,5,6} \langle \phi_i, u\rangle_{\Gamma_i} \quad \forall u \in \mathbf{V}. \end{aligned} \tag{5.54}$$

Problem III-VI Find v such that

$$\begin{aligned} & v - U = w \in K(\Omega), \\ & a_{03}(w, u-w) + \Phi(u) - \Phi(w) \geq \langle F_3, u-w\rangle \quad \forall u \in \mathbf{V}, \end{aligned} \tag{5.55}$$

where the functionals Φ is defined by (5.22).

Remark 5.2 Using the strain bilinear form and applying Theorems 3.1 and 3.2 on $\Gamma_{ij}, i = 2, 3, 7$, we embedded all the boundary conditions into variational formulations of problems. For this, we used condition $\Gamma_{ij} \in C^2, i = 2, 3, 7$. The condition $\Gamma_{ij} \in C^1$ is for the existence of a diffeomorphism in C^1 between Γ_i and Γ_j in (3) of Assumption 5.1.

5.2 Solutions to Variational Inequalities

In this section we study some variational inequalities obtained in Sect. 5.1. We denote by $X \hookrightarrow\hookrightarrow Y$ compact imbedding of a space X into Y.

Theorem 5.5 *Let X, X_1 be real separable Hilbert spaces such that $X \hookrightarrow\hookrightarrow X_1$, and X^* be dual space of X. Assume the followings:*

(1) $\Phi : X \to [0, +\infty]$ is a proper lower semi-continuous convex functional such that $\Phi(0_X) = 0$.

(2) $a_0(\cdot, \cdot) \in (X \times X \to \mathbb{R})$ is a bilinear form such that

$$|a_0(u, v)| \leq K\|u\|_X\|v\|_X \quad \forall u, v \in X,$$
$$a_0(u, u) \geq \alpha\|u\|_X^2 \quad \exists \alpha > 0, \forall u \in X.$$

(3) $a_1(\cdot, \cdot, \cdot) \in (X_1 \times X \times X \to \mathbb{R})$ is a triple linear functional such that

$$a_1(w, u, u) = 0 \quad \forall w \in X_1, \forall u \in X,$$
$$|a_1(w, u, v)| \leq K\|w\|_{X_1}\|u\|_X\|v\|_X, \quad \forall w \in X_1, \forall u, v \in X.$$

Then for each $f \in X^*$ there exists a solution to the variational inequality

$$a_0(v, u - v) + a_1(v, v, u - v) + \Phi(u) - \Phi(v) \geq \langle f, u - v\rangle \quad \forall u \in X \tag{5.56}$$

and all solutions v satisfy the estimate

$$\|v\|_X \leq \frac{1}{\alpha}\|f\|_{X^*}. \tag{5.57}$$

If moreover

$$\frac{Kc}{\alpha^2}\|f\|_{X^*} < 1, \tag{5.58}$$

then solution is unique, where c is the constant in $\|\cdot\|_{X_1} \leq c\|\cdot\|_X$.

Proof Fix $w \in X_1$. Let us consider the following variational inequality:

$$a_0(v, u - v) + a_1(w, v, u - v) + \Phi(u) - \Phi(v) \geq \langle f, u - v\rangle \quad \forall u \in X, \tag{5.59}$$

where $f \in X^*$. From existing results there exists a unique solution to (5.59) (see Theorem 10.5 of [2]). Let v_1, v_2 be the solutions corresponding to f_1, f_2 instead of f. Then, under consideration of condition (2) it is easy to verify that

$$\|v_1 - v_2\|_X \le \frac{1}{\alpha}\|f_1 - f_2\|_{X^*}. \tag{5.60}$$

Now, let us consider the operator which maps w to the solution v of (5.59)

$$T \in (X_1 \to X) : w \to T(w) = v.$$

Taking into account condition (1), we can easily verify that the solution corresponding to $f = 0_{X^*}$ is 0_X. Thus, we have from (5.60)

$$\|v\|_X \le \frac{1}{\alpha}\|f\|_{X^*} \quad \forall w \in X_1. \tag{5.61}$$

Note that this estimate is independent of w.

Denote by v_1 and v_2, respectively, the solutions to (5.59) corresponding to w_1 and w_2. Then

$$\begin{aligned} a_0(v_1, u - v_1) + a_1(w_1, v_1, u - v_1) + \Phi(u) - \Phi(v_1) \ge \langle f, u - v_1\rangle \quad \forall u \in X, \\ a_0(v_2, u - v_2) + a_1(w_2, v_2, u - v_2) + \Phi(u) - \Phi(v_2) \ge \langle f, u - v_2\rangle \quad \forall u \in X. \end{aligned} \tag{5.62}$$

Putting $u = v_2$ and $u = v_1$, respectively, in the first formula and the second one of (5.62), and adding two formulas, we get

$$a_0(v_1 - v_2, v_2 - v_1) + a_1(w_1, v_1, v_2 - v_1) + a_1(w_2, v_2, v_1 - v_2) \ge 0. \tag{5.63}$$

From (5.63), conditions (2), (3) of Theorem and (5.61), we get

$$\begin{aligned} \|v_2 - v_1\|_X^2 &\le \frac{1}{\alpha}|a_1(w_1, v_1, v_2 - v_1) - a_1(w_2, v_1, v_2 - v_1) \\ &\qquad + a_1(w_2, v_1, v_2 - v_1) - a_1(w_2, v_2, v_2 - v_1)| \\ &\le \frac{1}{\alpha}|a_1(w_1 - w_2, v_1, v_2 - v_1)| + \frac{1}{\alpha}|a_1(w_2, v_2 - v_1, v_2 - v_1)| \\ &\le \frac{K}{\alpha}\|w_1 - w_2\|_{X_1}\|v_1\|_X\|v_2 - v_1\|_X \\ &\le \frac{K\|f\|_{X^*}}{\alpha^2}\|w_1 - w_2\|_{X_1}\|v_2 - v_1\|_X \quad \forall w_1, w_2 \in X_1, \end{aligned}$$

which implies

$$\|v_2 - v_1\|_X \le \frac{K\|f\|_{X^*}}{\alpha^2}\|w_1 - w_2\|_{X_1} \quad \forall w_1, w_2 \in X_1. \tag{5.64}$$

By (5.61), (5.64) and Schauder fixed-point theorem (Theorem 1.6), there exists a solution to (5.56). And any solution is a fixed point of operator T, and by (5.61) all solutions satisfy the estimate (5.57).

If (5.58) holds, then the operator $T : w \in X \to v \in X$ is contract, and so we come to the last conclusion. □

Let us next study variational inequalities when condition (3) of the above theorem is weakened.

Theorem 5.6 *Let X be a real separable Hilbert space. Assume the followings:*

(1) Condition (1) of Theorem 5.5 holds.

(2) Condition (2) of Theorem 5.5 holds.

(3) $a_1(\cdot, \cdot, \cdot) \in (X \times X \times X \to \mathbb{R})$ is a triple linear functional such that

$$|a_1(w, u, v)| \leq K\|w\|_X\|u\|_X\|v\|_X, \quad \forall w, u, v \in X.$$

If f is small enough, then in $\mathscr{O}_M(0_X)$, where M is determined in (5.73), there exists a unique solution to the variational inequality

$$a_0(v, u - v) + a_1(v, v, u - v) + \Phi(u) - \Phi(v) \geq \langle f, u - v\rangle \quad \forall u \in X. \tag{5.65}$$

Proof Fix $w \in X$. Let us consider a variational inequality

$$a_0(v, u - v) + a_1(w, w, u - v) + \Phi(u) - \Phi(v) \geq \langle f, u - v\rangle \quad \forall u \in X, \tag{5.66}$$

where $f \in X^*$. Defining an element $a_1(w) \in X^*$ by

$$\langle a_1(w), u\rangle = a_1(w, w, u) \quad \forall u \in X,$$

by condition (3) we have

$$\|a_1(w)\|_{X^*} \leq K\|w\|_X^2 \quad \forall w \in X. \tag{5.67}$$

Then, (5.66) can be rewritten as follows:

$$a_0(v, u - v) + \Phi(u) - \Phi(v) \geq \langle f - a_1(w), u - v\rangle \quad \forall u \in X. \tag{5.68}$$

By the same argument as in the proof of Theorem 5.5, there exists a unique solution v_w to (5.68) and

$$\|v_w\| \leq \frac{1}{\alpha}(\|f\|_{X^*} + \|a_1(w)\|_{X^*}) \leq \frac{1}{\alpha}(\|f\|_{X^*} + K\|w\|_X^2), \tag{5.69}$$

where (5.67) was used and α is the one of Theorem 5.5.

Now, let us consider the operator which maps w to the solution of (5.68)

$$T \in (X \to X) : w \to T(w) = v.$$

Denote by v_1 and v_2, respectively, the solutions to (5.66) corresponding to $w_1, w_2 \in \mathcal{O}_M(0_X)$, where M is to be determined below. Then

$$\|v_1 - v_2\|_X \le \frac{1}{\alpha}\|a_1(w_1) - a_1(w_2)\|_{X^*}. \tag{5.70}$$

By condition (3)

$$\|a_1(w_1) - a_1(w_2)\|_{X^*} \le K\left(\|w_2 - w_1\|_X\|w_2\|_X + \|w_1\|_X\|w_1 - w_2\|_X\right). \tag{5.71}$$

Thus, by (5.70) and (5.71),

$$\begin{aligned}\|v_1 - v_2\|_X &\le \frac{K}{\alpha}\left(\|w_2 - w_1\|_X\|w_2\|_X + \|w_1\|_X\|w_1 - w_2\|_X\right)\\ &\le \frac{2KM}{\alpha}\|w_2 - w_1\|_X \quad \forall w_1, w_2 \in \mathcal{O}_M(0_X).\end{aligned} \tag{5.72}$$

Therefore, if M is taken such that

$$\begin{cases} M = \dfrac{1}{\alpha}(\|f\|_{X^*} + KM^2),\\ \dfrac{2KM}{\alpha} < 1 \end{cases} \tag{5.73}$$

(If α is large and $\|f\|_{X^*}$ is small enough, then such a choice is possible.), then by (5.69) and (5.72) the operator T on $\mathcal{O}_M(0_X)$ is contract, and so there exists a unique solution to (5.65). □

Theorem 5.7 *Let X be a real separable Hilbert space and X^* be its dual space. Assume that*

(1) $\Phi : X \to \mathbb{R}$ is a finite weak continuous convex functional, $\Phi_\varepsilon : X \to \mathbb{R}$ is convex such that

$$\Phi_\varepsilon(v) \to \Phi(v) \quad \text{uniformly on } X \text{ as } \varepsilon \to 0,$$

Gâteaux derivative $D\Phi_\varepsilon \equiv A_\varepsilon \in (X \to X^*)$ is weak continuousand $A_\varepsilon(0_X) = 0_{X^{l*}}$.

(2) $a(\cdot,\cdot,\cdot) \in (X \times X \times X \to \mathbb{R})$ is a form such that

$$\begin{aligned}&\text{when } w \in X, (u, v) \to a(w; u, v) \text{ is bilinear on } X \times X,\\ &\quad a(v, v, v) \ge \alpha\|v\|_X^2 \;\; \exists\alpha > 0, \forall v \in X \text{ and}\\ &\text{when } v_m \rightharpoonup v \text{ in } X,\\ &\quad a(v_m, v_m, u) \to a(v, v, u)\ \forall u \in X \text{ and } \liminf_{m\to\infty} a(v_m, v_m, v_m) \ge a(v, v, v).\end{aligned}$$

Then, for $f \in X^*$ there exists a solution to a variational inequality

$$a(v, v, u - v) + \Phi(u) - \Phi(v) \geq \langle f, u - v \rangle \quad \forall u \in X \tag{5.74}$$

satisfying the estimate

$$\|v\|_X \leq \frac{1}{\alpha} \|f\|_{X^*}. \tag{5.75}$$

Proof First let us prove the existence of a solution to a variational equation

$$a(v, v, u) + \langle A_\varepsilon(v), u \rangle = \langle f, u \rangle \quad \forall u \in X. \tag{5.76}$$

We will rely on Theorem 1.43. Let $\{w_n\}$ be a basis of X and denote by X_m the subspace of X spanned by $w_1, \cdots, w_m$.

We will find $v_m = \sum_{i=1}^m \mu_i w_i \in X_m$ satisfying

$$a(v_m, v_m, u) + \langle A_\varepsilon(v_m), u \rangle = \langle f, u \rangle \quad \forall u \in X_m. \tag{5.77}$$

Define $F_m \in (X_m \to X_m)$ by

$$(F_m(v), w_i) = a(v, v, w_i) + \langle A_\varepsilon(v), w_i \rangle - \langle f, w_i \rangle, \quad 1 \leq i \leq m. \tag{5.78}$$

Since Gâteaux derivative of finite convex functional is monotone (see Remark 1.16) and $A_\varepsilon(0_X) = 0_{X^*}$, we get

$$\langle A_\varepsilon(u) - A_\varepsilon(0_X), u - 0_X \rangle = \langle A_\varepsilon(u), u \rangle \geq 0 \quad \forall u \in X.$$

Thus,

$$a(u, u, u) + \langle A_\varepsilon(u), u \rangle \geq \alpha \|u\|_X^2 \quad \forall u \in X. \tag{5.79}$$

Combining (5.78) with (5.79) yields

$$(F_m(v), v) \geq (\alpha \|v\|_X - \|f\|_{X^*}) \|v\|_X \quad \forall v \in X_m. \tag{5.80}$$

Therefore,

$$(F_m(v), v) \geq 0 \quad \forall v \in X \text{ with } \|v\|_X = \frac{\|f\|_{X^*}}{\alpha}.$$

And by virtue of assumption (2), F_m is continuous in X_m. Thus, there exists a solution $v_{\varepsilon m}$ to problem (5.77). By (5.80), we have that

$$0 = (F_m(v_{\varepsilon m}), v_{\varepsilon m}) \geq (\alpha \|v_{\varepsilon m}\|_X - \|f\|_{X^*}) \|v_{\varepsilon m}\|_X$$

for all solution $v_{\varepsilon m}$ to (5.77), which implies

$$\|v_{\varepsilon m}\|_X \le \frac{1}{\alpha}\|f\|_{X^*}. \tag{5.81}$$

Note this estimation is independent of ε, m. Thus, from $\{v_{\varepsilon m}\}$ we can extract a subsequence $\{v_{\varepsilon m_p}\}$ such that

$$v_{\varepsilon m_p} \rightharpoonup v_\varepsilon \quad \text{in } X \quad \text{as } p \to +\infty.$$

By the assumptions of theorem,

$$a(v_{\varepsilon m_p}, v_{\varepsilon m_p}, u) + \langle A_\varepsilon(v_{\varepsilon m_p}), u\rangle \to a(v_\varepsilon, v_\varepsilon, u) + \langle A_\varepsilon(v_\varepsilon), u\rangle \quad \forall u \in X. \tag{5.82}$$

By virtue of (5.77), (5.82) and (5.81), we see that v_ε is a solution to (5.76) and satisfies

$$\|v_\varepsilon\|_X \le \frac{1}{\alpha}\|f\|_{X^*}. \tag{5.83}$$

Subtracting the following two formulas which are got from (5.76)

$$a(v_\varepsilon, v_\varepsilon, u) + \langle A_\varepsilon(v_\varepsilon), u\rangle = \langle f, u\rangle \quad \forall u \in X,$$
$$a(v_\varepsilon, v_\varepsilon, v_\varepsilon) + \langle A_\varepsilon(v_\varepsilon), v_\varepsilon\rangle = \langle f, v_\varepsilon\rangle$$

and taking into account that

$$\Phi_\varepsilon(u) - \Phi_\varepsilon(v_\varepsilon) \ge \langle A_\varepsilon(v_\varepsilon), u - v_\varepsilon\rangle$$

which is due to convexity of Φ_ε, we come to the following inequality:

$$a(v_\varepsilon, v_\varepsilon, u - v_\varepsilon) + \Phi_\varepsilon(u) - \Phi_\varepsilon(v_\varepsilon) \ge \langle f, u - v_\varepsilon\rangle \quad \forall u \in X. \tag{5.84}$$

By (5.83) we can choose $\{v_{\varepsilon_k}\}$ such that

$$v_{\varepsilon_k} \rightharpoonup v^* \quad \text{in } X \quad \text{as } \varepsilon_k \to 0. \tag{5.85}$$

By virtue of assumption (1),

$$|\Phi_{\varepsilon_k}(v_{\varepsilon_k}) - \Phi(v^*)| \le |\Phi_{\varepsilon_k}(v_{\varepsilon_k}) - \Phi(v_{\varepsilon_k})| + |\Phi(v_{\varepsilon_k}) - \Phi(v^*)| \to 0 \quad \text{as } \varepsilon_k \to 0,$$

and so

$$\Phi_{\varepsilon_k}(v_{\varepsilon_k}) \to \Phi(v^*) \quad \text{as } \varepsilon_k \to 0. \tag{5.86}$$

Also

$$\Phi_{\varepsilon_k}(u) \to \Phi(u) \quad \forall u \in X \quad \text{as } \varepsilon_k \to 0. \tag{5.87}$$

By virtue of assumption (2),

$$\begin{aligned} &a(v_{\varepsilon_k}, v_{\varepsilon_k}, u) \to a(v^*, v^*, u) \quad \forall u \in X \quad \text{as } \varepsilon_k \to 0, \\ &\liminf_{k\to\infty} a(v_{\varepsilon_k}, v_{\varepsilon_k}, v_{\varepsilon_k}) \geq a(v^*, v^*, v^*). \end{aligned} \tag{5.88}$$

Taking into account (5.86)–(5.88), we get from (5.84)

$$a(v^*, v^*, u - v^*) + \Phi(u) - \Phi(v^*) \geq \langle f, u - v^* \rangle \quad \forall u \in X.$$

By (5.83) we have

$$\|v^*\|_X \leq \frac{1}{\alpha} \|f\|_{X^*}. \tag{5.89}$$

Therefore, we finish our proof. □

Remark 5.3 The estimate of solutions in Theorem 5.5 is for all solutions of the problem, but the one in Theorem 5.7 is for the solution whose existence is guaranteed by the theorem.

5.3 Existence and Uniqueness of Solutions to the Steady Navier-Stokes Problems

In this section, on the basis of the results in Sect. 5.2, we study the problems formulated in Sect. 5.1. Our first result is the following:

Theorem 5.8 *Let Assumption 5.1 hold, the surfaces Γ_{2j}, Γ_{3j}, Γ_{7j} be convex (see Definition 3.2), α positive and $\|U\|_{\mathbf{H}^1(\Omega)}$ small enough. Then, when f and ϕ_i, $i = 2, \cdots, 7$, are small enough, in a neighborhood of U in $\mathbf{H}^1(\Omega)$ there exists a unique solution to Problem I-VI (for the steady Navier-Stokes problem of the case of static pressure).*

Proof Define a functional Φ by (5.22). Trace operator is continuous and sum of convex functions is also convex. Thus, the functional satisfies condition (1) of Theorem 5.6.

Set $w = v - U$, where U is the function in Assumption 5.1. Let $a_{01}(\cdot, \cdot)$, $a_{11}(\cdot, \cdot, \cdot)$ and $F_1 \in \mathbf{V}^*$ be the same as in (5.9).

By Korn's inequality

$$2\mu(\mathcal{E}(w), \mathcal{E}(w)) \geq \delta \|w\|_{\mathbf{V}}^2, \quad \delta > 0. \tag{5.90}$$

On the other hand, applying Hölder's inequality, for $w \in \mathbf{V}$ we have

$$|\langle (U \cdot \nabla)w, w\rangle + \langle (w \cdot \nabla)U, w\rangle| \leq \gamma \|w\|_{\mathbf{V}}^2 \cdot \|U\|_{\mathbf{H}^1(\Omega)}. \tag{5.91}$$

Therefore, if $\delta - \gamma \|U\|_{\mathbf{H}^1(\Omega)} = \beta_1 > 0$, then by (5.90), (5.91), Assumption 5.1 and Lemma 3.1 we have

$$a_{01}(u,u) \ge \beta_1 \|u\|_{\mathbf{V}}^2 \quad \forall u \in \mathbf{V}. \tag{5.92}$$

It is easy to verify that

$$|a_{01}(u,v)| \le c\|u\|_{\mathbf{V}}\|v\|_{\mathbf{V}} \quad \forall u, v \in \mathbf{V}. \tag{5.93}$$

By (5.92) and (5.93), $a_0(u,v)$ satisfies condition (2) of Theorem 5.6.

By Hölder's inequality we can see

$$|a_{11}(w,u,v)| \le c\|w\|_{\mathbf{V}}\|u\|_{\mathbf{V}}\|v\|_{\mathbf{V}} \quad \forall w, u, v \in \mathbf{V}. \tag{5.94}$$

which means that $a_{11}(w,u,v)$ satisfies condition (3) of Theorem 5.6.

Also

$$\|F_1\|_{\mathbf{V}^*} \le M_1\Big(\|U\|_{\mathbf{H}^1} + \|U\|_{\mathbf{H}^1}^2 + \|f\|_{\mathbf{V}^*} + \sum_{i=2,4,7}\|\phi_i\|_{H^{-\frac{1}{2}}(\Gamma_i)} + \sum_{i=3,5,6}\|\phi_i\|_{\mathbf{H}^{-\frac{1}{2}}(\Gamma_i)}\Big), \tag{5.95}$$

where M_1 depends on mean curvature of Γ_7, shape operator of Γ_3, μ and α.

By Theorem 5.6, if $\|U\|_{\mathbf{H}^1}$, $\|f\|_{\mathbf{V}^*}$, $\|\phi_i\|_{H^{-\frac{1}{2}}(\Gamma_i)}$, $i = 2, 4, 7$, and $\|\phi_i\|_{\mathbf{H}^{-\frac{1}{2}}(\Gamma_i)}$, $i = 3, 5, 6$, are small enough, then there exists a unique solution $w \in K(\Omega)$ to

$$a_{01}(w, u-w) + a_{11}(w, w, u-w) + \Phi(u) - \Phi(w) \ge \langle F_1, u-w\rangle \quad \forall u \in K(\Omega). \tag{5.96}$$

Since $v = w + U$ is solution, we come to the asserted conclusion. □

Theorem 5.9 *Let Assumption 5.1 hold, the surfaces Γ_{2j}, Γ_{3j}, Γ_{7j} be convex, α positive and $\|U\|_{\mathbf{H}^1(\Omega)}$ small enough. Then, for any f ϕ_i, $i = 2 \sim 7$, there exists a solution v to Problem II-VI (the steady Navier-Stokes problem for the case of total pressure) in a neighborhood of U in $\mathbf{H}^1(\Omega)$ and all solutions satisfy*

$$\|v - U\|_{\mathbf{H}^1} \le \frac{M_1}{\delta - \gamma\|U\|_{\mathbf{H}^1}}\Big(\|U\|_{\mathbf{H}^1} + \|U\|_{\mathbf{H}^1}^2 + \|f\|_{\mathbf{V}^*} + \sum_{i=2,4,7}\|\phi_i\|_{H^{-\frac{1}{2}}(\Gamma_i)} + \sum_{i=3,5,6}\|\phi_i\|_{\mathbf{H}^{-\frac{1}{2}}(\Gamma_i)}\Big), \tag{5.97}$$

where δ, γ, M_1 are as in (5.98), (5.99), (5.107).

If $\|U\|_{\mathbf{H}^1}$, $\|f\|_{\mathbf{V}^*}$, $\|\phi_i\|_{H^{-\frac{1}{2}}(\Gamma_i)}$, $i = 2, 4, 7$, and $\|\phi_i\|_{\mathbf{H}^{-\frac{1}{2}}(\Gamma_i)}$, $i = 3, 5, 6$, are small enough, then the solution is unique.

Proof Define a functional $\Phi(u)$ by (5.22). Then, Φ satisfies condition (1) of Theorem 5.5.

Let $a_{02}(\cdot,\cdot)$, $a_{12}(\cdot,\cdot,\cdot)$ and $F_2 \in \mathbf{V}^*$ be as in (5.50).

By Korn's inequality

$$2\mu(\mathcal{E}(w), \mathcal{E}(w)) \geq \delta \|w\|_{\mathbf{V}}^2, \quad \delta > 0. \tag{5.98}$$

On the other hand, for any $w \in \mathbf{V}$ we have

$$\begin{aligned} &\langle \operatorname{rot} U \times w, w\rangle = 0, \\ &|\langle \operatorname{rot} w \times U, w\rangle| \leq \gamma \|w\|_{\mathbf{V}}^2 \cdot \|U\|_{\mathbf{H}^1(\Omega)}. \end{aligned} \tag{5.99}$$

Therefore, if $\delta - \gamma \|U\|_{\mathbf{H}^1(\Omega)} = \beta_1 > 0$, then by (5.98), (5.99), Assumption 5.1 and Lemma 3.1 we have

$$a_{02}(u, u) \geq \beta_1 \|u\|_{\mathbf{V}}^2 \quad \forall u \in \mathbf{V}. \tag{5.100}$$

It is easy to verify

$$|a_{02}(u, v)| \leq c\|u\|_{\mathbf{V}} \|v\|_{\mathbf{V}} \quad \forall u, v \in \mathbf{V}. \tag{5.101}$$

Then, (5.100) and (5.101) imply that $a_{02}(u, v)$ satisfies condition (2) of Theorem 5.5.

By a property of mixed product of vectors,

$$a_{12}(w, u, u) = \langle \operatorname{rot} w \times u, u\rangle = 0 \quad \forall w \in \mathbf{V}^{\frac{2}{3}}(\Omega), \forall u \in \mathbf{V}, \tag{5.102}$$

where

$$\mathbf{V}^{\frac{2}{3}}(\Omega) = \left\{u \in \mathbf{H}^{\frac{2}{3}}(\Omega) : \operatorname{div} u = 0,\ u|_{\Gamma_1} = 0,\ u_\tau|_{(\cup_{i=2,4,7,9}\Gamma_i)} = 0,\ u \cdot n|_{(\cup_{i=3,5,8}\Gamma_i)} = 0\right\}.$$

On the other hand, by density argument we get

$$a_{12}(w, u, v) = \langle \operatorname{rot} w \times u, v\rangle = -\langle \operatorname{rot} w, v \times u\rangle. \tag{5.103}$$

When $u, v \in \mathbf{V}$, $v \times u \in \mathbf{H}^{\frac{1}{2}}(\Omega)$ and

$$\|v \times u\|_{\mathbf{H}^{\frac{1}{3}}(\Omega)} \leq c_1 \|v \times u\|_{\mathbf{H}^{\frac{1}{2}}(\Omega)} \leq c\|v\|_{\mathbf{V}} \|u\|_{\mathbf{V}} \tag{5.104}$$

(cf. Theorem 1.23). Also, if $w \in \mathbf{V}^{\frac{2}{3}}(\Omega)$, then $\operatorname{rot} w \in \mathbf{H}^{-\frac{1}{3}}(\Omega)$ and

$$\|\operatorname{rot} w\|_{\mathbf{H}^{-\frac{1}{3}}(\Omega)} \leq c\|w\|_{\mathbf{H}^{\frac{2}{3}}(\Omega)} \tag{5.105}$$

(see Theorem 1.24). Since $H_0^{\frac{1}{3}}(\Omega) = H^{\frac{1}{3}}(\Omega)$ (see Theorem 1.15), by (5.103)–(5.105) we get

$$|a_{12}(w, u, v)| \leq K \|w\|_{\mathbf{V}^{\frac{2}{3}}(\Omega)} \|u\|_{\mathbf{V}} \|v\|_{\mathbf{V}} \quad \forall w \in \mathbf{V}^{\frac{2}{3}}(\Omega) \quad \forall u, v \in \mathbf{V}. \tag{5.106}$$

Since $\mathbf{V} \hookrightarrow\hookrightarrow \mathbf{V}^{\frac{2}{3}}(\Omega)$, setting $X = \mathbf{V}$, $X_1 = \mathbf{V}^{\frac{2}{3}}(\Omega)$, we can see that by (5.102) and (5.106) $a_{11}(w, u, v)$ satisfies condition (3) of Theorem 5.5.

Also, we have

$$\|F_2\|_{\mathbf{V}^*} \le M_1\Big(\|U\|_{\mathbf{H}^1} + \|U\|^2_{\mathbf{H}^1} + \|f\|_{\mathbf{V}^*} + \sum_{i=2,4,7} \|\phi_i\|_{H^{-\frac{1}{2}}(\Gamma_i)} + \sum_{i=3,5,6} \|\phi_i\|_{\mathbf{H}^{-\frac{1}{2}}(\Gamma_i)}\Big), \tag{5.107}$$

where M_1 depends on the mean curvature, shape operator, μ and α.

Therefore, by Theorem 5.5, we have the existence and estimate of solutions to

$$a_{02}(w, u-w) + a_{12}(w, w, u-w) + \Phi(u) - \Phi(w) \ge \langle F_2, u-w\rangle \quad \forall u \in \mathbf{V}.$$

Since $v = w + U$ is a solution to the given problem, we have the existence of solutions and the estimate (5.97).

If $\|U\|_{\mathbf{H}^1}$, $\|f\|_{\mathbf{V}^*}$, $\|\phi_i\|_{H^{-\frac{1}{2}}(\Gamma_i)}$, $i = 2, 4, 7$, and $\|\phi_i\|_{\mathbf{H}^{-\frac{1}{2}}(\Gamma_i)}$, $i = 3, 5, 6$, are small enough, then the solution is unique. □

Let us consider now a special case of the Navier-Stokes problem with boundary condition (5.2) in which there is no flux across boundary except Γ_1, Γ_8.

Theorem 5.10 *Let Assumption 5.1 hold, $\Gamma_i = \emptyset$ $(i = 2, 4, 6, 7, 9, 10, 11)$, the surfaces Γ_{3j} be convex, α positive and $\|U\|_{\mathbf{H}^1(\Omega)}$ small enough. Then, for any f and ϕ_i, $i = 3, 5$, there exists a solution v to Problem I-VI (the steady Navier-Stokes problem for the case of static pressure) and all solutions satisfy*

$$\|v - U\|_{\mathbf{H}^1} \le \frac{M_1}{\delta - \gamma\|U\|_{\mathbf{H}^1}}\Big(\|U\|^2_{\mathbf{H}^1} + \|f\|_{\mathbf{V}^*} + \sum_{i=3,5} \|\phi_i\|_{\mathbf{H}^{-\frac{1}{2}}(\Gamma_i)}\Big), \tag{5.108}$$

where δ, γ and M_1 are as in (5.90), (5.91) *and* (5.95), *respectively.*

In addition, if $\|f\|_{\mathbf{V}^*}$, $\|\phi_i\|_{\mathbf{H}^{-\frac{1}{2}}(\Gamma_i)}$, $i = 3, 5$, are small enough, then the solution is unique.

Proof Define a functional $\Phi(u) = \phi_\tau(u)$ by (5.21) and (5.22). Then, the functional satisfies condition (1) of Theorem 5.6.

Let $w = v - U$, U be a function in Assumption 5.1 and $a_{01}(\cdot,\cdot)$, $a_{11}(\cdot,\cdot,\cdot)$ and $F_1 \in \mathbf{V}^*$ be as in (5.9).

We can see that condition (2) in Theorem 5.5 is satisfied (see proof of Theorem 5.8).

By the condition of theorem,

$$a_{11}(w, u, u) = \langle (w \cdot \nabla)u, u\rangle = 0 \quad \forall w \in \mathbf{V}^{\frac{2}{3}}(\Omega), \forall u \in \mathbf{V}. \tag{5.109}$$

By Hölder's inequality we can see

$$|a_{11}(w, u, v)| \le K\|w\|_{\mathbf{V}^{\frac{2}{3}}(\Omega)}\|u\|_{\mathbf{V}}\|v\|_{\mathbf{V}} \quad \forall w \in \mathbf{V}^{\frac{2}{3}}(\Omega), \forall u, v \in \mathbf{V}. \tag{5.110}$$

By (5.109) and (5.110), $a_{11}(w, u, v)$ satisfies condition (3) of Theorem 5.5.
Applying Theorem 5.5 to

$$a_{01}(w, u - w) + a_{11}(w, w, u - w) + \Phi(u) - \Phi(w) \geq \langle F_1, u - w\rangle \quad \forall u \in K(\Omega),$$

we come to the asserted conclusion. □

Remark 5.4 Assumption $\Gamma_i = \emptyset, i = 2, 4, 6, 7, 9 - 11$, is only used to get (5.109).

Applying Theorem 5.7, next we revisit the problem concerned within Theorem 5.10. We will get a generalization of methods used in papers based on smooth approximation of functional in variational inequalities (see [3]).

Lemma 5.2 *Let X, Y be reflexive Banach spaces, an operator $i : X \to Y$ be completely linear continuous, $j : Y \to \mathbb{R}$ be convex and Gâteaux derivative $Dj(y) = a(y)$ for $y \in Y$. Then, $\Phi(v) \equiv j(iv) : X \to \mathbb{R}$ is convex, $D\Phi(v) \equiv A(v) = i^*a(iv)$, where i^* is the operator adjoint to i, and $A : X \to X^*$ is weak continuous.*

Proof It is easy to verify the convexity of Φ.

$$\begin{aligned}\langle A(v), u\rangle_X &= \lim_{t\to 0} \frac{\Phi(v + tu) - \Phi(u)}{t} = \lim_{t\to 0} \frac{j(i(v + tu)) - j(iu)}{t}\\ &= \langle a(iv), iu\rangle_Y = \langle i^*a(iv), u\rangle_X \qquad \forall v, u \in X,\end{aligned}$$

which means $A(v) = i^*a(iv)$.

Let $v_n \rightharpoonup v$ in X. Since gradient of a finite convex functional is monotone and demi-continuous (see Remark 1.15) and $iv_n \to iv$ in Y,

$$\langle A(v_n), u\rangle_X = \langle i^*a(iv_n), u\rangle_X = \langle a(iv_n), iu\rangle_Y \to \langle a(iv), iu\rangle_Y = \langle i^*a(iv), u\rangle_X \qquad \forall u \in X,$$

that is, $D\Phi = A : X \to X^*$ is weak continuous. □

Theorem 5.11 *Let Assumption 5.1 hold, $\Gamma_i = \emptyset$ $(i = 2, 4, 6, 7, 9, 10, 11)$, the surfaces Γ_{3j} be convex, α positive and $\|U\|_{\mathbf{H}^1(\Omega)}$ small enough. Then, for any f and ϕ_i, $i = 3, 5$, there exists a solution v to Problem I-VI (the steady Navier-Stokes problem for the case of static pressure) and the solution satisfies the estimate* (5.108).

Proof Define an operator $i : \mathbf{V} \to \mathbf{L}^2_\tau(\Gamma_8)$ by $iu = u|_{\Gamma_8}$ and a functional $\Phi : \mathbf{V} \to \mathbb{R}$ by $\Phi(v) \equiv \phi_\tau(iv)$, where ϕ_τ is as in (5.21). Since the trace operator $\mathbf{V} \to H^{\frac{1}{2}}(\partial\Omega)$ is continuous and $H^{\frac{1}{2}}(\partial\Omega) \hookrightarrow\hookrightarrow L^2(\partial\Omega)$, the operator i is compact, and by Lemma 5.2 $\Phi : \mathbf{V} \to \mathbb{R}$ is weak continuous and convex.

Define a functional $\Phi_\varepsilon : \mathbf{V} \to \mathbb{R}$ by

$$\begin{aligned}
&\Phi_\varepsilon(v) = \phi_{\tau\varepsilon}(iv),\\
&\phi_{\tau\varepsilon}(\eta) = \int_{\Gamma_8} g_\tau \rho_\varepsilon(\eta)\, ds,\\
&\rho_\varepsilon(\eta) = \begin{cases} |\eta| - \varepsilon/2 & |\eta| > \varepsilon,\\ |\eta|^2/2\varepsilon & |\eta| \le \varepsilon. \end{cases}
\end{aligned} \tag{5.111}$$

Since

$$|\phi_{\tau\varepsilon}(\eta) - \phi_\tau(\eta)| \le \frac{\varepsilon}{2}|g_\tau| \quad \forall \eta \in \mathbf{L}^2_\tau(\Gamma_8)$$

(cf. Lemma 2.1 of [3]), we have

$$|\Phi_\varepsilon(v) - \Phi(v)| \le \frac{\varepsilon}{2}|g_\tau| \quad \forall v \in \mathbf{V}. \tag{5.112}$$

Also, $\phi_{\tau\varepsilon}$ is convex, and so its Gâteaux derivative is demi-continuous. Thus, by Lemma 5.2 $D\Phi_\varepsilon \equiv A_\varepsilon \in (\mathbf{V} \to \mathbf{V}^*)$ is weak continuous. By this fact together (5.112), condition (1) of Theorem 5.7 is satisfied.

Under the assumption of theorem $a_{01}(\cdot,\cdot), a_{11}(\cdot,\cdot,\cdot)$ and $F_1 \in \mathbf{V}^*$ of (5.9) are as follows:

$$\begin{aligned}
a_{01}(u,v) &= 2\mu(\mathcal{E}(u),\mathcal{E}(v)) + \langle (U\cdot\nabla)u, v\rangle + \langle (u\cdot\nabla)U, v\rangle\\
&\quad + 2\mu(S\tilde{u},\tilde{v})_{\Gamma_3} + 2(\alpha(x)u, v)_{\Gamma_5} \quad \forall u, v \in \mathbf{V},\\
a_{11}(w,u,v) &= \langle (w\cdot\nabla)u, v\rangle \quad \forall w, u, v \in \mathbf{V},\\
\langle F_1, u\rangle &= -2\mu(\mathcal{E}(U),\mathcal{E}(u)) - \langle (U\cdot\nabla)U, u\rangle - 2\mu(S\tilde{U},\tilde{u})_{\Gamma_3}\\
&\quad - 2(\alpha(x)U_\tau, u)_{\Gamma_5} + \langle f, u\rangle + \sum_{i=3,5}\langle \phi_i, u\rangle_{\Gamma_i} \quad \forall u \in \mathbf{V}.
\end{aligned} \tag{5.113}$$

By Korn's inequality,

$$2\mu(\mathcal{E}(u),\mathcal{E}(u)) \ge \delta\|u\|^2_{\mathbf{V}}, \quad \delta > 0. \tag{5.114}$$

On the other hand, for any $u \in \mathbf{V}$ we have

$$|\langle (U\cdot\nabla)u, u\rangle + \langle (u\cdot\nabla)U, u\rangle| \le \gamma\|u\|^2_{\mathbf{V}} \cdot \|U\|_{\mathbf{H}^1(\Omega)}. \tag{5.115}$$

Therefore, if $\delta - \gamma\|U\|_{\mathbf{H}^1(\Omega)} = \beta_1 > 0$, then by (5.114), (5.115), Assumption 5.1 and Lemma 3.1 we have

$$a_{01}(u,u) \ge \beta_1\|u\|^2_{\mathbf{V}} \quad \forall u \in \mathbf{V}. \tag{5.116}$$

Under condition $\Gamma_i = \emptyset, i = 2, 4, 6, 7, 9, 10, 11$, it is easy to verify that

$$a_{11}(v,v,v) = 0 \quad \forall v \in \mathbf{V}. \tag{5.117}$$

Let

$$a(w, u, v) = a_{01}(u, v) + a_{11}(w, u, v).$$

Then, by (5.116) and (5.117) we have

$$a(v, v, v) \geq \beta_1 \|u\|_{\mathbf{V}}^2 \quad \forall v \in \mathbf{V}. \tag{5.118}$$

Let us prove that when $v_m \rightharpoonup v$ in $\mathbf{V}$, there exists a subsequence $\{v_{m_p}\}$ such that

$$a(v_{m_p}, v_{m_p}, u) \to a(v; v, u) \quad \forall u \in \mathbf{V} \quad \text{as } p \to \infty. \tag{5.119}$$

To this end, first let us prove that when $v_m \rightharpoonup v$ in $\mathbf{V}$, there exists a subsequence $\{v_{m_p}\}$ such that

$$a_{01}(v_{m_p}, u) \to a_{01}(v, u) \quad \forall u \in \mathbf{V} \quad \text{as } p \to \infty. \tag{5.120}$$

Since $U_i u_j \in L^2(\Omega), i, j = 1, 2, 3$, and $\partial_i v_m \rightharpoonup \partial_i v$ in $L^2(\Omega)$, we have

$$\langle (U \cdot \nabla) v_m, u\rangle \to \langle (U \cdot \nabla) v, u\rangle \quad \text{as } m \to \infty. \tag{5.121}$$

By Hölder's inequality,

$$|\langle ((v_m - v) \cdot \nabla) U, u\rangle| \leq c\|v_m - v\|_{\mathbf{L}^3(\Omega)} \|\nabla U\|_{\mathbf{L}^2(\Omega)} \|u\|_{\mathbf{L}^6(\Omega)}.$$

Since $H^1(\Omega) \hookrightarrow\hookrightarrow L^3(\Omega)$, we can choose a subsequence $\{v_{m_p}\}$ such that $v_{mp} \to v$ in $\mathbf{L}^3(\Omega)$. Then, we have

$$\langle (v_{m_p} \cdot \nabla) U, u\rangle \to \langle (v \cdot \nabla) U, u\rangle \quad \text{as } m_p \to \infty. \tag{5.122}$$

It is easy to verify the convergence of other terms. Thus, using (5.121) and (5.122), we have (5.120).

Using Hölder's inequality and $a_{11}(v, u, w) = -a_{11}(v, w, u)$, we have

$$\begin{aligned}
&|a_{11}(v_m, v_m, u) - a_{11}(v, v, u)| \\
&\leq |a_{11}(v_m, v_m, u) - a_{11}(v, v_m, u)| + |a_{11}(v, v_m, u) - a_{11}(v, v, u)| \\
&\leq c\big(\|v_m - v\|_{\mathbf{L}^3(\Omega)} \|\nabla v_m\|_{\mathbf{L}^2(\Omega)} \|u\|_{\mathbf{L}^6(\Omega)} + \|v\|_{\mathbf{L}^6(\Omega)} \|\nabla u\|_{\mathbf{L}^2(\Omega)} \|v_m - v\|_{\mathbf{L}^3(\Omega)}\big) \quad \forall u \in \mathbf{V}.
\end{aligned}$$

Thus, we have

$$a_{11}(v_{m_p}, v_{m_p}, u) \to a_{11}(v, v, u) \quad \forall u \in \mathbf{V} \quad \text{as } m_p \to \infty. \tag{5.123}$$

Combining (5.120) and (5.123) yields (5.119).

Let us prove that

$$\liminf_{m\to\infty} a(v_{m_p}, v_{m_p}, v_{m_p}) \geq a(v, v, v). \tag{5.124}$$

By lower semi-continuity of norm

$$\liminf_{m\to\infty} 2\mu(\mathcal{E}(v_m), \mathcal{E}(v_m)) \geq 2\mu(\mathcal{E}(v), \mathcal{E}(v)) \quad \text{as } v_m \rightharpoonup v \text{ in } \mathbf{V}. \tag{5.125}$$

It is easy to prove that

$$2\mu(S\tilde{v}_m, \tilde{u})_{\Gamma_3} + 2(\alpha(x)v_m, u)_{\Gamma_5} \to 2\mu(S\tilde{v}, \tilde{u})_{\Gamma_3} + 2(\alpha(x)v, u)_{\Gamma_5} \quad \forall u \in \mathbf{V}. \tag{5.126}$$

Using Hölder's inequality and $a_{11}(v, v_m, u) = -a_{11}(v, u, v_m)$, we have

$$\begin{aligned}
&|a_{11}(v_m, v_m, v_m) - a_{11}(v, v, v)| \\
&\quad \leq |a_{11}(v_m, v_m, v_m) - a_{11}(v, v_m, v_m)| + |a_{11}(v, v_m, v_m) - a_{11}(v, v_m, v)| \\
&\qquad + |a_{11}(v, v_m, v) - a_{11}(v, v, v)| \\
&\quad \leq c\big(\|v_m - v\|_{\mathbf{L}^3(\Omega)}\|\nabla v_m\|_{\mathbf{L}^2(\Omega)}\|v_m\|_{\mathbf{L}^6(\Omega)} + \|v\|_{\mathbf{L}^6(\Omega)}\|\nabla v_m\|_{\mathbf{L}^2(\Omega)}\|v_m - v\|_{\mathbf{L}^3(\Omega)} \\
&\qquad + \|v\|_{\mathbf{L}^6(\Omega)}\|\nabla v\|_{\mathbf{L}^2(\Omega)}\|v_m - v\|_{\mathbf{L}^3(\Omega)}\big),
\end{aligned}$$

which implies

$$a_{11}(v_{m_p}, v_{m_p}, v_{m_p}) \to a_{11}(v, v, v) \quad \text{as } m_p \to \infty. \tag{5.127}$$

From (5.124)–(5.127), we have (5.124).

By virtue of (5.118), (5.119) and (5.124), condition (2) of Theorem 5.7 is satisfied. Therefore, by Theorem 5.7 we have the existence and estimate of a solution $w \in \mathbf{V}$ to

$$a_{01}(w, u - w) + a_{11}(w, w, u - w) + \phi_\tau(u) - \phi_\tau(w) \geq \langle F_1, u - w\rangle \quad \forall u \in \mathbf{V}. \tag{5.128}$$

Since $v = w + U$ is a solution, we come to the asserted conclusion. □

Remark 5.5 The estimate of solution of Theorem 5.11 may not be true for all solutions, and so Theorem 5.11 is weaker than Theorem 5.10.

Let us consider Problem III for the Stokes system.

Theorem 5.12 *Let Assumption 5.1 hold, the surfaces Γ_{2j}, Γ_{3j}, Γ_{7j} be convex and α positive. Then, for any f ϕ_i, $i = 2, \cdots, 7$, there exists a unique solution v to Problem III-VI for the steady Stokes problem with mixed boundary condition* (5.2) *and the following estimate holds:*

$$\|v - U\|_{\mathbf{H}^1} \leq \frac{M_1}{\delta}\Big(\|U\|_{\mathbf{H}^1} + \|f\|_{\mathbf{V}^*} + \sum_{i=2,4,7} \|\phi_i\|_{H^{-\frac{1}{2}}(\Gamma_i)} + \sum_{i=3,5,6} \|\phi_i\|_{\mathbf{H}^{-\frac{1}{2}}(\Gamma_i)}\Big), \tag{5.129}$$

where δ and M_1 are, respectively, as in (5.98) *and* (5.107) *(for F_3 instead of F_2).*

If v_1, v_2 are solutions, respectively, to Problem-III-VI with $g_{\tau 1}, g_{n1}, g_{+n1}, g_{-n1}, f_1, h_i^1, \phi_i^1$ and $g_{\tau 2}, g_{n2}, g_{+n2}, g_{-n2}, f_2, h_i^1, \phi_i^2$, then

$$\begin{aligned}\|v_1 - v_2\|_{\mathbf{H}^1} \leq &\frac{M_1}{\delta}\Big(\|U_1 - U_2\|_{\mathbf{H}^1} + \|f_1 - f_2\|_{\mathbf{V}^*} + \|g_{\tau 1} - g_{\tau 2}\|_{\mathbf{L}^2_\tau(\Gamma_8)} \\ &+ \|g_{n1} - g_{n2}\|_{L^2(\Gamma_9)} + \|g_{+n1} - g_{+n2}\|_{L^2(\Gamma_{10})} + \|g_{-n1} - g_{-n2}\|_{L^2(\Gamma_{11})} \\ &+ \sum_{i=2,4,7} \|\phi_i^1 - \phi_i^2\|_{H^{-\frac{1}{2}}(\Gamma_i)} + \sum_{i=3,5,6} \|\phi_i^1 - \phi_i^2\|_{\mathbf{H}^{-\frac{1}{2}}(\Gamma_i)}\Big) + \|U_1 - U_2\|_{\mathbf{H}^1},\end{aligned} \tag{5.130}$$

where U_j, $j = 1, 2$, are the functions in Assumption 5.1 with h_i^j instead h_i.

Proof By an argument similar to the proof of Theorem 5.6 we can apply the well known result for variational inequality (see Theorem 10.5 of [2])

$$a_{03}(w, u - w) + \Phi(u) - \Phi(w) \geq \langle F_3, u - w\rangle \quad \forall u \in X, \tag{5.131}$$

where $\Phi(u)$ is defined by (5.22) and $a_{03}(v, u)$, F_3 are as in (5.54).

Thus, we have the existence of a unique solution and estimate (5.129).

If $v_1 = w_1 + U_1$, $v_2 = w_2 + U_2$ are solutions corresponding to the given data, we get

$$\begin{aligned}&a_{03}(w_1, u - w_1) + \Phi_1(u) - \Phi_1(w_1) \geq \langle F_3^1, u - w_1\rangle, \\ &a_{03}(w_2, u - w_2) + \Phi_2(u) - \Phi_2(w_2) \geq \langle F_3^2, u - w_2\rangle \quad \forall u \in \mathbf{V},\end{aligned} \tag{5.132}$$

where $\Phi_j(u)$, F_3^j, $j = 1, 2$, are the one corresponding to $U_j, g_{\tau j}, g_{nj}, g_{+nj}, g_{-nj}, f_j, h_i^j, \phi_i^j$. Putting $u = w_2$, $u = w_1$, respectively, in the first and second inequality in (5.132) and adding those obtained, we have

$$\begin{aligned}a_{03}(w_1 - w_2, w_2 - w_1) + \Phi_1(w_2) - \Phi_1(w_1) + \Phi_2(w_1) - \Phi_2(w_2) \\ \geq \langle F_3^1 - F_3^2, w_2 - w_1\rangle.\end{aligned} \tag{5.133}$$

By Korn's inequality and Lemma 3.1 we have

$$a_{03}(w_1 - w_2, w_1 - w_2) \geq \delta\|w_1 - w_2\|^2_{\mathbf{V}}. \tag{5.134}$$

From (5.133) and (5.134) we have

$$\|w_1 - w_2\|^2_{\mathbf{V}} \leq \frac{1}{\delta}\Big(|\langle F_3^1 - F_3^2, w_2 - w_1\rangle| + |\Phi_1(w_2) - \Phi_1(w_1) + \Phi_2(w_1) - \Phi_2(w_2)|\Big). \tag{5.135}$$

Since $w_1, w_2 \in K(\Omega)$,

$$
\begin{aligned}
\Phi_1(w_2)-\Phi_1(w_1)=&\int_{\Gamma_8} g_{\tau 1}(|w_{2\tau}|-|w_{1\tau}|)\,ds+\int_{\Gamma_9} g_{n1}(|w_{2n}|-|w_{1n}|)\,ds\\
&+\int_{\Gamma_{10}} g_{+n1}(w_{2n}-w_{1n})\,ds-\int_{\Gamma_{11}} g_{-n1}(w_{2n}-w_{1n})\,ds,\\
\Phi_2(w_2)-\Phi_2(w_1)=&\int_{\Gamma_8} g_{\tau 2}(|w_{2\tau}|-|w_{1\tau}|)\,ds+\int_{\Gamma_9} g_{n2}(|w_{2n}|-|w_{1n}|)\,ds\\
&+\int_{\Gamma_{10}} g_{+n2}(w_{2n}-w_{1n})\,ds-\int_{\Gamma_{11}} g_{-n2}(w_{2n}-w_{1n})\,ds.
\end{aligned}
\tag{5.136}
$$

Subtracting two formulas of (5.136), we have

$$
\begin{aligned}
&|\Phi_1(w_2)-\Phi_1(w_1)+\Phi_2(w_1)-\Phi_2(w_2)|\\
&\quad\le \|g_{\tau 1}-g_{\tau 2}\|_{\mathbf{L}^2_\tau(\Gamma_8)}\|w_{2\tau}-w_{1\tau}\|_{\mathbf{L}^2_\tau(\Gamma_8)}+\|g_{n1}-g_{n2}\|_{L^2(\Gamma_9)}\|w_{2n}-w_{1n}\|_{L^2(\Gamma_9)}\\
&\quad+\|g_{+n1}-g_{+n2}\|_{L^2(\Gamma_{10})}\|w_{n2}-w_{n1}\|_{L^2(\Gamma_{10})}+\|g_{-n1}-g_{-n2}\|_{L^2(\Gamma_{11})}\|w_{2n}-w_{1n}\|_{L^2(\Gamma_{11})}\\
&\quad\le M\big(\|g_{\tau 1}-g_{\tau 2}\|_{\mathbf{L}^2_\tau(\Gamma_8)}+\|g_{n1}-g_{n2}\|_{L^2(\Gamma_9)}+\|g_{+n1}-g_{+n2}\|_{L^2(\Gamma_{10})}\\
&\qquad\quad+\|g_{-n1}-g_{-n2}\|_{L^2(\Gamma_{11})}\big)\|w_2-w_1\|_{\mathbf{V}}.
\end{aligned}
\tag{5.137}
$$

By (5.135) and (5.137), we have

$$
\begin{aligned}
\|w_1-w_2\|_{\mathbf{V}}\le \frac{M}{\delta}\Big(&\|F_3^1-F_3^2\|_{\mathbf{V}^*}+\|g_{\tau 1}-g_{\tau 2}\|_{\mathbf{L}^2_\tau(\Gamma_8)}+\|g_{n1}-g_{n2}\|_{L^2(\Gamma_9)}\\
&+\|g_{+n1}-g_{+n2}\|_{L^2(\Gamma_{10})}+\|g_{-n1}-g_{-n2}\|_{L^2(\Gamma_{11})}\Big),
\end{aligned}
$$

from which we get (5.130). □

Remark 5.6 The estimates of solutions to the problems (5.97), (5.108) and (5.129) are independent of the thresholds $g_\tau, g_n, g_{+n}, g_{-n}$. (See (8) of [4] and (25) of [3].)

5.4 Bibliographical Remarks

The content of Chap. 5 is taken from [5].

Till now, for the Stokes and Navier-Stokes problems with friction type boundary conditions rather simple cases were studied. More clearly, problems with the Dirichlet boundary condition on a portion of boundary and either Tresca slip condition or the leak condition on the other portion have been dealt with.

In [6] the existence of solutions to the steady Stokes and Navier-stokes equations with the homogeneous Dirichlet boundary condition on a portion of boundary and leak or threshold slip boundary condition on the other portion was studied. Also, [7–9] concerned with the steady or non-steady Stokes equations with the homogeneous Dirichlet boundary condition and leak boundary condition.

When a portion of boundary with Dirichlet boundary condition and another moving portion where nonlinear slip occurs are separated, the existence, uniqueness and continuous dependence on the data were studied for the steady Stokes equations in [9]. In [10] when a portion of boundary with Dirichlet boundary condition and another portion with slip condition are separated, the existence of strong solution to the steady Stokes equations was established. In [11] when a portion with homogeneous Dirichlet boundary condition and another portion with nonlinear boundary condition are separated, for the steady Stokes equations a relation between a regularized problem and the original problem, the regularity of solution were discussed.

In [12] for the steady Navier-Stokes equations, the existence, uniqueness and continuous dependence on the data were considered when a portion of boundary with Dirichlet boundary condition and another moving portion where nonlinear slip occurs are separated. In [13] the existence and uniqueness of a local solution to the steady Navier-Stokes problem with homogeneous Dirichlet boundary condition and one of friction boundary conditions was investigated. In [4] the existence and uniqueness of solution to the steady rotating Navier-Stokes equations are studied when boundary consists of a portion with homogeneous Dirichlet boundary condition and other portions with a threshold slip. In [3] under similar boundary condition the steady Navier-Stokes problem is studied. With exception of [13] in all above-mentioned papers dealing the Navier-Stokes problem with friction boundary conditions, ones approximated the functionals in the considering variational inequalities with smooth one resulting to study of operator equation and it's convergence.

Numerical solution methods were studied for the Stokes and Navier-Stokes problems with friction boundary conditions. For the 2-D steady Stokes problems we refer to [14–18] and for the 3-D steady Stokes problems see [19]. For the 2-D steady Navier-Stokes problem we refer to [20–23].

References

1. G.P. Galdi, *An Introduction to the Mathematical Theory of the Navier-Stokes Equations* (Springer, Berlin, 2011)
2. C. Baiocchi, A. Capelo, *Variational and Quasivariational Inequalities: Applications to Free-Boundary Problems* (Wiley, Chichester, 1984) (Russian, 1988)
3. Y. Li, K. Li, Existence of the solution to stationary Navier-Stokes equations with nonlinear slip boundary conditions. J. Math. Anal. Appl. **381**, 1–9 (2011)
4. R. An, K. Li, Variational inequality for the rotating Navier-Stokes equations with subdifferential boundary conditions. Comput. Math. Appl. **55**, 581–587 (2008)
5. T. Kim, D. Cao, The steady Navier-Stokes and Stokes systems with mixed boundary conditions including one-sided leaks and pressure. Methods Appl. Anal. **23**, 329–364 (2016)
6. H. Fujita, A mathematical analysis of motions of viscous incompressible fluid under leak or slip boundary conditions. RIMS Kokyuroku **888**, 199–216 (1994)
7. H. Fujita, Non-stationary Stokes flows under leak boundary conditions of friction type. J. Comput. Math. **19**, 1–8 (2001)
8. H. Fujita, A coherent analysis of Stokes flows under boundary conditions of friction type. J. Comput. Appl. Math. **149**, 57–69 (2002)

9. C.L. Roux, Steady Stokes flows with threshold slip boundary conditions. Math. Model. Methods Appl. Sci. **15**, 1141–1168 (2005)
10. N. Saito, On the Stokes equations with the leak and slip boundary conditions of friction type: regularity of solutions. Publ. RIMS (Kyoto University) **40**, 345–383 (2004)
11. N. Saito, H. Fujita, Regularity of solutions to the Stokes equation under a certain nonlinear boundary condition. Lect. Notes Pure Appl. Math. **223**, 73–86 (2001)
12. C.L. Roux, A. Tani, Steady solutions of the Navier-Stokes equations with threshold slip boundary conditions. Math. Methods Appl. Sci. **30**, 595–624 (2007)
13. F. Saidi, On the Navier-Stokes equations with the slip boundary conditions of friction type: regularity of solution. Math. Model. Anal. **12**, 389–398 (2007)
14. M. Ayadi, M.K. Gdoura, T. Sassi, Mixed formulation for Stokes problem with Tresca friction. C. R. Acad. Sci. Paris Ser. I **348**, 1069–1072 (2010)
15. T. Kashiwabara, On a finite element approximation of the Stokes problem under a leak boundary condition of friction type. Jpn. J. Ind. Appl. Math. **30**, 227–261 (2013)
16. Y. Li, K. Li, Penalty finite element method for Stokes problem with nonlinear slip boundary conditions. Appl. Math. Comput. **204**, 216–226 (2008)
17. Y. Li, K. Li, Locally stabilized finite element method for Stokes problem with nonlinear slip boundary conditions. J. Comput. Math. **28**, 826–836 (2010)
18. Y. Li, K. Li, Uzawa iteration method for Stokes type variational inequality of the second kind. Acta Math. Appl. Sin. Engl. Ser. **27**, 303–316 (2011)
19. T. Kashiwabara, Finite element method for Stokes equations under leak boundary condition of friction type. SIAM J. Numer. Anal. **52**, 2448–2469 (2013)
20. R. An, Comparisons of Stokes/Oseen/Newton iteration methods for Navier-Stokes equations with friction boundary conditions. Appl. Math. Model. **38**, 5535–5544 (2014)
21. Y. Li, R. An, Penalty finite element method for Navier-Stokes equations with nonlinear slip boundary conditions. Int. J. Numer. Methods Fluids **69**, 550–566 (2012)
22. Y. Li, R. An, Two-level pressure projection finite element methods for Navier-Stokes equations with nonlinear slip boundary conditions. Appl. Numer. Math. **61**, 285–297 (2011)
23. Y. Li, R. An, Two-level variational multiscale finite element methods for Navier-Stokes type variational inequality problem. J. Comput. Appl. Math. **290**, 656–669 (2015)

Chapter 6
The Non-steady Navier-Stokes System with Friction Boundary Conditions

In this chapter we are concerned with the non-steady Navier-Stokes and Stokes problems corresponding to the steady problems in Chap. 5. In Sect. 6.1 relying on the results of Sect. 3.1, we embed all boundary conditions to variational formulations. We get variational inequalities with one unknown which are equivalent to the original PDE problems for the smooth solutions. In Sect. 6.2 we study the existence and uniqueness of solutions to the variational inequalities obtained in Sect. 6.1. In Sect. 6.3 using the results of Sect. 6.2, we get the existence, uniqueness and estimates of solutions to the Navier-Stokes and Stokes problems with the boundary conditions.

For the problem with boundary conditions involving the total pressure and total stress, the existence of a solution without restriction of data of the problem is proved. For the problem with boundary conditions involving the static pressure and stress, under a compatibility condition at the initial time for the small data it is proved that there exists a unique solution on the given interval of time.

6.1 Variational Formulations of Problems

Let Ω be a bounded domain of $\mathbb{R}^l$, $l = 2, 3$. $\partial\Omega \in C^{0,1}$, $\partial\Omega = \cup_{i=1}^{11}\overline{\Gamma}_i$, $\Gamma_i \cap \Gamma_j = \emptyset$ for $i \neq j$, $\Gamma_i = \bigcup_j \Gamma_{ij}$, where Γ_{ij} are connected open subsets of $\partial\Omega$ and $\Gamma_{ij} \in C^{2,1}$ for $i = 2, 3, 7$ and $\Gamma_{ij} \in C^1$ for others. Also, let $Q = \Omega \times (0, T)$, $\Sigma_i = \Gamma_i \times (0, T)$, $0 < T < \infty$.

T. Kim and D. Cao, *Equations of Motion for Incompressible Viscous Fluids*,
Advances in Mathematical Fluid Mechanics,
https://doi.org/10.1007/978-3-030-78659-5_6

Remark 6.1 In Chap. 5 for the boundary of Ω it is assumed that $\Gamma_{ij} \in C^2, i = 2, 3, 7$, and those are convex in Theorems 5.8–5.12, whereas in this chapter it is assumed that $\Gamma_{ij} \in C^{2,1}, i = 2, 3, 7$, and convexity of $\Gamma_{ij}, i = 2, 3, 7$, is not assumed in the following theorems. In the following chapters for the steady problems the assumptions as in Chap. 5 are used, whereas for the non-steady problems the assumptions as in this chapter are used (cf. Remark 5.2).

We are concerned with Problems I and II for the non-steady Navier-Stokes problem

$$\begin{cases} \dfrac{\partial v}{\partial t} - \mu \Delta v + (v \cdot \nabla) v + \nabla p = f, \\ \operatorname{div} v = 0, \\ v(0) = v_0, \end{cases} \tag{6.1}$$

which are distinguished according to boundary conditions. Problem I is the one with the boundary conditions (5.3) (the case of total pressure) and Problem II is the one with the boundary conditions (5.2) (the case of static pressure).

We also consider the Stokes problem

$$\begin{cases} \dfrac{\partial v}{\partial t} - \mu \Delta v + \nabla p = f, \\ \operatorname{div} v = 0, \\ v(0) = v_0 \end{cases} \tag{6.2}$$

with the boundary conditions (5.2), which is Problem III.

Assuming that $f \in L^2(0, T; \mathbf{L}^2(\Omega))$, $\phi_i \in L^2(0, T; L^2(\Gamma_i))$, $i = 2, 4, 7$, and $\phi_i \in L^2(0, T; \mathbf{L}^2(\Gamma_i))$, $i = 3, 5, 6$, formally we introduce the following:

Definition 6.1 A function $(v, p) \in L^2(0, T; \mathbf{H}^2(\Omega)) \times L^1(0, T; H^1(\Omega))$ is called a solution to Problem I (or Problem II) if the first two equations of (6.1) hold in $L^1(0, T; \mathbf{L}^2(\Omega))$, $v(0) = v_0$ in $\mathbf{L}^2(\Omega)$ and each of (5.3) (or (5.2)) holds in $L^1(0, T; L^2(\Gamma_i))$ or $L^1(0, T; \mathbf{L}^2(\Gamma_i))$.

A function $(v, p) \in L^2(0, T; \mathbf{H}^2(\Omega)) \times L^2(0, T; H^1(\Omega))$ is called a solution to Problem III if the first two equations of (6.2) hold in $L^2(0, T; \mathbf{L}^2(\Omega))$, $v(0) = v_0$ in $\mathbf{L}^2(\Omega)$ and each of (5.2) holds in $L^2(0, T; L^2(\Gamma_i))$ or $L^2(0, T; \mathbf{L}^2(\Gamma_i))$.

Remark 6.2 Under the conditions for (v, p) and f, we have that $v \in C([0, T]; \mathbf{L}^2(\Omega))$ (see Lemma 1.1 of Chap. III of [1]), and so the condition that $v(0) = v_0$ in $\mathbf{L}^2(\Omega)$ is meaningful.

6.1.1 Variational Formulation: The Case of Total Pressure

Let $\mathbf{V}$, $K(\Omega)$ be the same as in (5.5) and

$$H: \text{the closer of } \mathbf{V} \text{ in } \mathbf{L}^2(\Omega),$$
$$\mathbf{V}(Q) = L^2(0,T;\mathbf{V})$$
$$\mathbf{V}_{\Gamma 237}(\Omega) = \{u \in \mathbf{H}^1(\Omega) : \operatorname{div} u = 0,\ u_\tau|_{(\Gamma_2 \cup \Gamma_7)} = 0,\ u_n|_{\Gamma_3} = 0\},$$
$$K(Q) = \{u \in L^2(0,T;\mathbf{V}) : u' \in L^1(0,T;\mathbf{V}^*);\ u_n|_{\Gamma_{10}} \geq 0,\ u_n|_{\Gamma_{11}} \leq 0\},$$
$$\mathbf{\Lambda}(Q) = \{u \in \mathbf{C}^2(\bar{Q}) : \operatorname{div} u = 0, u\ |_{\Sigma_1} = 0,\ u_\tau\ |_{(\Sigma_2 \cup \Sigma_4 \cup \Sigma_7)} = 0, u_n\ |_{(\Sigma_3 \cup \Sigma_5)} = 0\}.$$

We will use the following assumption:

Assumption 6.1 The followings hold:
(1) There exists a function $U \in W^{1,2}(0,T;\mathbf{H}^1) \cap \mathbf{L}^\infty(Q)$ such that

$$\operatorname{div} U = 0,\ U\ |_{\Sigma_1} = h_1,\ U_\tau\ |_{(\Sigma_2 \cup \Sigma_7)} = 0,\ U_n\ |_{\Sigma_3} = 0,\ U_\tau\ |_{\Sigma_4} = h_4,$$
$$U_n\ |_{\Sigma_5} = h_5,\ U\ |_{\Sigma_8} = h_8 n,\ U\ |_{\Sigma_9} = h_9,\ U\ |_{\Sigma_{10} \cup \Sigma_{11}} = 0,$$

where $\Sigma_i = \Gamma_i \times (0,T)$.
(2) $f \in L^2(0,T;\mathbf{V}^*)$, $\phi_i \in L^2(0,T;H^{-\frac{1}{2}}(\Gamma_i)), i = 2,4,7,$ $\phi_i \in L^2(0,T;\mathbf{H}^{-\frac{1}{2}}(\Gamma_i)), i = 3,5,6$, $\alpha_{ij} \in L^\infty(\Gamma_5)$, $v_0 - U(x,0) \in H$, and $\Gamma_1 \neq \emptyset$.
(3) of Assumption 5.1 holds.

Remark 6.3 On the Γ_{10} (Γ_{11}) outflow (inflow) only is possible. Thus, to guarantee $div\, u = 0$ we assume that if $\Gamma_{10} \neq \emptyset$ ($\Gamma_{11} \neq \emptyset$), then for example $\Gamma_2 \neq \emptyset$. In Sect. 5.1 for proof of equivalence of variational formulations between variational inequalities, (3) of Assumption 5.1 was used via Lemma 5.1 to get existence of σ_{+n}, σ_{-n} such that $\sigma_{+n} + g_{+n} \geq 0, \sigma_{-n} - g_{-n} \leq 0$ from variational inequalities, which is also necessary for Theorems 6.2 and 6.4.

Taking $(v \cdot \nabla)v = \operatorname{rot} v \times v + \frac{1}{2}\operatorname{grad}|v|^2$ into account and putting $v = w + U$, by (5.6), (5.7) and Assumption 6.1 we can see that solutions (v, p) of the problem (6.1), (5.3) in the sense of Definition 6.1 satisfy the following:

$$
\begin{cases}
v - U = w \in K(Q), \;\; v_0 - U(x,0) = w(0), \\
\left(\dfrac{\partial w}{\partial t}, u\right) + 2\mu(\mathcal{E}(w), \mathcal{E}(u)) + \langle \operatorname{rot} w \times w, u\rangle + \langle \operatorname{rot} U \times w, u\rangle \\
\qquad + \langle \operatorname{rot} w \times U, u\rangle + 2\mu(k(x)w, u)_{\Gamma_2} + 2\mu(S\tilde{w}, \tilde{u})_{\Gamma_3} \\
\qquad + 2(\alpha(x)w, u)_{\Gamma_5} + \mu(k(x)w, u)_{\Gamma_7} - 2\mu(\varepsilon_{n\tau}(w+U), u)_{\Gamma_8} \\
\qquad + \left(p + \dfrac{1}{2}|v|^2 - 2\mu\varepsilon_{nn}(w+U), u_n\right)_{\Gamma_9\cup\Gamma_{10}\cup\Gamma_{11}} \\
\quad = -\left(\dfrac{\partial U}{\partial t}, u\right) - 2\mu(\mathcal{E}(U), \mathcal{E}(u)) - \langle \operatorname{rot} U \times U, u\rangle - 2\mu(k(x)U, u)_{\Gamma_2} \\
\qquad - 2\mu(S\tilde{U}, \tilde{u})_{\Gamma_3} - 2(\alpha(x)U_\tau, u)_{\Gamma_5} - \mu(k(x)U, u)_{\Gamma_7} \\
\qquad + \langle f, u\rangle + \sum_{i=2,4,7} \langle \phi_i, u_n\rangle_{\Gamma_i} + \sum_{i=3,5,6} \langle \phi_i, u\rangle_{\Gamma_i} \qquad \forall u \in \mathbf{V}, \\
|\sigma_\tau^t(v)| \le g_\tau, \; \sigma_\tau^t \cdot v_\tau + g_\tau |v_\tau| = 0 \;\; \text{on } \Gamma_8, \\
|\sigma_n^t(v)| \le g_n, \; \sigma_n^t(v) v_n + g_n |v_n| = 0 \;\; \text{on } \Gamma_9, \\
\sigma_n^t(v) + g_{+n} \ge 0, \;\; (\sigma_n^t(v) + g_{+n}) v_n = 0 \;\; \text{on } \Gamma_{10}, \\
\sigma_n^t(v) - g_{-n} \le 0, \;\; (\sigma_n^t(v) - g_{-n}) v_n = 0 \;\; \text{on } \Gamma_{11}.
\end{cases}
\tag{6.3}
$$

Define $a_{01}(t,\cdot,\cdot)$, $a_{11}(\cdot,\cdot,\cdot)$ and $F_1(t) \in \mathbf{V}^*$ by

$$
\begin{aligned}
a_{01}(t, w, u) &= 2\mu(\mathcal{E}(w), \mathcal{E}(u)) + \langle \operatorname{rot} U(t) \times w, u\rangle + \langle \operatorname{rot} w \times U(t), u\rangle \\
&\quad + 2\mu(k(x)w, u)_{\Gamma_2} + 2\mu(S\tilde{w}, \tilde{u})_{\Gamma_3} + 2(\alpha(x)w, u)_{\Gamma_5} \\
&\quad + \mu(k(x)w, u)_{\Gamma_7} \quad \forall w, u \in \mathbf{V}, \\
a_{11}(w, u, v) &= \langle \operatorname{rot} w \times u, v\rangle \quad \forall w, u, v \in \mathbf{V}, \\
\langle F_1(t), u\rangle &= -\left\langle \frac{\partial U(t)}{\partial t}, u\right\rangle - 2\mu(\mathcal{E}(U(t)), \mathcal{E}(u)) - \langle \operatorname{rot} U(t) \times U(t), u\rangle \\
&\quad - 2\mu(k(x)U(t), u)_{\Gamma_2} - 2\mu(S\tilde{U(t)}, \tilde{u})_{\Gamma_3} \\
&\quad - 2(\alpha(x)U(t)_\tau, u)_{\Gamma_5} - \mu(k(x)U(t), u)_{\Gamma_7} + \langle f(t), u\rangle \\
&\quad + \sum_{i=2,4,7} \langle \phi_i(t), u_n\rangle_{\Gamma_i} + \sum_{i=3,5,6} \langle \phi_i(t), u\rangle_{\Gamma_i} \qquad \forall u \in \mathbf{V}.
\end{aligned}
\tag{6.4}
$$

Then, taking into account

$$
\sigma_\tau^t(v) = 2\mu\varepsilon_{n\tau}(v), \quad \sigma_n^t(v) = -(p + \frac{1}{2}|v|^2) + 2\mu\varepsilon_{nn}(v)
$$

and (6.3), we introduce the following variational formulation for problem (6.1), (5.3).

Problem I-VE. Find $v{=}U{+}w \in (U + K(Q))$ and $(\sigma_\tau^t, \sigma_n^t, \sigma_{+n}^t, \sigma_{-n}^t) \in \mathbf{L}_\tau^2(\Gamma_8) \times L^2(\Gamma_9) \times H^{-\frac{1}{2}}(\Gamma_{10}) \times H^{-\frac{1}{2}}(\Gamma_{11})$ in a.e. $t \in (0,T)$ such that $w(0) = v_0 - U(0)$ and

$$
\begin{cases}
\left\langle \dfrac{\partial w}{\partial t}, u \right\rangle + a_{01}(t, w, u) + a_{11}(w, w, u) - (\sigma_\tau^t, u_\tau)_{\Gamma_8} - (\sigma_n^t, u_n)_{\Gamma_9} \\
\qquad - \left\langle \sigma_{+n}^t, u_n \right\rangle_{\Gamma_{10}} - \left\langle \sigma_{-n}^t, u_n \right\rangle_{\Gamma_{11}} = \langle F_1, u \rangle \\
\qquad\qquad\qquad \forall u \in L^2(0, T; \mathbf{V}), \\
|\sigma_\tau^t| \le g_\tau,\ \sigma_\tau^t \cdot v_\tau + g_\tau |v_\tau| = 0 \quad \text{on } \Gamma_8, \\
|\sigma_n^t| \le g_n,\ \sigma_n^t v_n + g_n |v_n| = 0 \quad \text{on } \Gamma_9, \\
\sigma_{+n}^t + g_{+n} \ge 0,\quad \left\langle \sigma_{+n}^t + g_{+n}, v_n \right\rangle_{\Gamma_{10}} = 0 \quad \text{on } \Gamma_{10}, \\
\sigma_{-n}^t - g_{-n} \le 0,\quad \left\langle \sigma_{-n}^t - g_{-n}, v_n \right\rangle_{\Gamma_{11}} = 0 \quad \text{on } \Gamma_{11},
\end{cases}
\tag{6.5}
$$

where and in what follows $\mathbf{L}_\tau^2(\Gamma_8)$ is the subspace of $\mathbf{L}^2(\Gamma_8)$ consisting of functions such that $(u, n)_{\mathbf{L}^2(\Gamma_8)} = 0$.

Remark 6.4 Under (2) of Assumption 6.1 the dual products on Γ_i in (6.4) and (6.5) is meaningful. For more detail we refer to Remark 5.1.

In the same way as Theorem 5.1, we have

Theorem 6.1 *Let Assumption 6.1 hold. If* (v, p) *is a solution in the sense of Definition 6.1 of the problem* (6.1), (5.3), *then* $(v, \sigma_\tau^t|_{\Gamma_8}, \sigma_n^t|_{\Gamma_9}, \sigma_n^t|_{\Gamma_{10}}, \sigma_n^t|_{\Gamma_{11}})$ *is a solution to Problem I-VE.*

Conversely, if $f \in L^2(0, T; \mathbf{L}^2(\Omega))$ *and Problem I-VE has a smooth solution* $(v, \sigma_\tau^t, \sigma_n^t, \sigma_+^t, \sigma_-^t)$ *such that* $v \in L^2(0, T; \mathbf{H}^2(\Omega))$, $\sigma_\tau^t \in L^2(0, T; \mathbf{L}^2(\Gamma_8))$, $\sigma_n^t \in L^2(0, T; L^2(\Gamma_9))$, $\sigma_+^t \in L^2(0, T; L^2(\Gamma_{10}))$ *and* $\sigma_-^t \in L^2(0, T; L^2(\Gamma_{11}))$, *then there exists* $p \in L^1(0, T; H^1(\Omega))$ *such that* (v, p) *is a solution to the problem* (6.1), (5.3). *Moreover, if at least one of the sets* $\Gamma_i, i = 2, 4, 6, 7$, *is nonempty, then* p *is unique.*

We will find a variational inequality corresponding to Problem I-VE.

Let $(v, \sigma_\tau, \sigma_n, \sigma_{+n}, \sigma_{-n})$ be a solution of Problem I-VE. Subtracting the first formula of (6.5) with $u = w$ from the first formula of (6.5), we get

$$
\begin{aligned}
&\left\langle \frac{\partial w}{\partial t}, u - w \right\rangle + a_{01}(t, w, u - w) + a_{11}(w, w, u - w) - (\sigma_\tau^t, u_\tau - w_\tau)_{\Gamma_8} \\
&\quad - (\sigma_n^t, u_n - w_n)_{\Gamma_9} - \left\langle \sigma_{+n}^t, u_n - w_n \right\rangle_{\Gamma_{10}} - \left\langle \sigma_{-n}^t, u_n - w_n \right\rangle_{\Gamma_{11}} \\
&\quad = \langle F_1, u - w \rangle.
\end{aligned}
\tag{6.6}
$$

Let $\Phi : \mathbf{V} \to \overline{\mathbb{R}}$ be the functional defined by (5.22). Define a functional $\Psi(u)$ by

$$
\Psi(u) = \begin{cases} \displaystyle\int_0^T \Phi(u(t))\, dt & if\ \ \Phi(u(t)) \in L^1(0, T), \\ +\infty & otherwise. \end{cases}
\tag{6.7}
$$

In the same way as in Problem I of Chap. 5, we get from (6.6)

$$\begin{aligned}\left\langle\frac{\partial w}{\partial t}, u-w\right\rangle+a_{01}(t, w, u-w)+a_{11}(w, w, u-w)+\Phi(u)-\Phi(w) \\ \geq\langle F_1, u-w\rangle .\end{aligned} \tag{6.8}$$

If $u-w \in \mathbf{\Lambda}(Q)$, then

$$\begin{aligned}\int_0^T\left\langle\frac{\partial w}{\partial t}, u-w\right\rangle dt &= \int_0^T\left\langle\frac{\partial(w-u)}{\partial t}, w-u\right\rangle dt+\int_0^T\left\langle\frac{\partial u}{\partial t}, u-w\right\rangle dt \\ &= \int_0^T\left\langle\frac{\partial u}{\partial t}, u-w\right\rangle dt+\frac{1}{2}(\|w_0-u(0)\|^2-\|w(T)-u(T)\|^2).\end{aligned} \tag{6.9}$$

Define operators $A_1(t): \mathbf{V} \rightarrow \mathbf{V}^*$ and $B_1: \mathbf{V} \times \mathbf{V} \rightarrow \mathbf{V}^*$ by

$$\begin{aligned}&\langle A_1(t) w, u\rangle=a_{01}(t, w, u) \quad \forall w, u \in \mathbf{V}, \\ &\langle B_1(w, u), v\rangle=a_{11}(w, u, v) \quad \forall w, u, v \in \mathbf{V}.\end{aligned} \tag{6.10}$$

If $w \in L^2(0, T, \mathbf{V}) \cap L^{\infty}(0, T ; \mathbf{L}^2(\Omega))$, then $w \in L^4(0, T, \mathbf{L}^3(\Omega))$ (see (1.21)) and $B_1(w, w) \in L^{\frac{4}{3}}(0, T, \mathbf{V}^*)$. Thus, when $u \in L^4(0, T ; \mathbf{V})$, the integral $\int_0^T\langle B_1(w, w), u-w\rangle dt$ is well defined since $\langle B_1(w, w), w\rangle=0$.

Thus, neglecting $\frac{1}{2}(\|w(T)-u(T)\|^2)$, from (6.8) and (6.9) we come to a variational inequality corresponding to Problem I (the case of total pressure).

Problem I-VI. Find v such that $v-U \equiv w \in L^2(0, T ; \mathbf{V}) \cap L^{\infty}(0, T ; H)$ and

$$\begin{aligned}\int_0^T\langle u'+A_1(t) w(t)+B_1(w(t), w(t))-F_1(t), u(t)-w(t)\rangle dt+\Psi(u)-\Psi(w) \\ \geq-\frac{1}{2}\|w_0-u(0)\|^2 \quad \forall u \in L^4(0, T ; \mathbf{V}) \text{ with } u' \in L^2(0, T ; \mathbf{V}^*).\end{aligned} \tag{6.11}$$

If the solutions to Problem I-VI is smooth as much as $v-U \equiv w \in L^2(0, T ; \mathbf{V})$, $w' \in L^2(0, T ; \mathbf{V}^*)$, then we can see that the solutions satisfy

$$\begin{aligned}\int_0^T\langle w'(t)+A_1(t) w(t)+B_1(w(t), w(t))-F_1(t), u(t)-w(t)\rangle dt+\Psi(u)-\Psi(w) \geq 0 \\ \forall u \in L^4(0, T ; \mathbf{V}).\end{aligned} \tag{6.12}$$

Especially, for the cases of 2-D and small data we will study such solutions to Problem I.

Thus, we come to another formulation associated with Problem I by a variational inequality.

Problem I-VI'. Find v such that $v-U \equiv w \in L^2(0, T ; \mathbf{V})$, $w' \in L^2(0, T ; \mathbf{V}^*)$, $w(0, x)=v_0-U(0, x)$ and (6.12) is satisfied.

Remark 6.5 (Remark, p. 114 of [2]) The inequality (6.12) is equivalent to

$$\big\langle w'(t) + A_1(t)w(t) + B_1(w(t), w(t)) - F_1(t), u - w(t)\big\rangle + \Phi(u) - \Phi(w(t)) \geq 0 \quad \text{for a.e. } t \in [0, T], \quad \forall u \in K(\Omega). \tag{6.13}$$

Further, the function w in (6.12) also satisfies the inequality (6.11).

In (6.13) putting $\langle F, u - w\rangle = \big\langle -\frac{\partial w}{\partial t} + F_1, u - w\big\rangle$, by the argument in the proof of Theorem 5.2 we can obtain existence of $(\sigma^t_\tau, \sigma^t_n, \sigma^t_{+n}, \sigma^t_{-n}) \in \mathbf{L}^2_\tau(\Gamma_8) \times L^2(\Gamma_9) \times H^{-1/2}(\Gamma_{10}) \times H^{-1/2}(\Gamma_{11})$ at a.e. $t \in (0, T)$ such that $(v, \sigma^t_\tau, \sigma^t_n, \sigma^t_{+n}, \sigma^t_{-n})$ is a solution to Problem I-VE. Thus, we have

Theorem 6.2 *If $(v, \sigma^t_\tau, \sigma^t_n, \sigma^t_{+n}, \sigma^t_{-n})$ is a solution to Problem I-VE such that $v' - U' \in L^2(0, T; \mathbf{V}^*)$, then v is a solution to Problem I-VI'.*

Conversely, if v is a solution to Problem I-VI', then there exist $\sigma^t_\tau, \sigma^t_n, \sigma^t_{+n}, \sigma^t_{-n}$ for a.e. $t \in (0, T)$ such that $(v, \sigma^t_\tau, \sigma^t_n, \sigma^t_{+n}, \sigma^t_{-n})$ is a solution to Problem I-VE.

6.1.2 Variational Formulation: The Case of Static Pressure

In the same way as in Problem I, we will get the variational formulations of Problem II for the Navier-Stokes equations with boundary condition (5.2).

Having in mind Assumption 6.1 and putting $v = w + U$, by (5.6) and (5.7) we can see that solutions (v, p) of problem (6.1), (5.2) in the sense of Definition 6.1 satisfy the following:

$$\begin{cases} v - U = w \in K(Q), \\ \Big(\dfrac{\partial w}{\partial t}, u\Big) + 2\mu(\mathcal{E}(w), \mathcal{E}(u)) + \langle (w \cdot \nabla)w, u\rangle + \langle (U \cdot \nabla)w, u\rangle \\ \qquad + \langle (w \cdot \nabla)U, u\rangle + 2\mu(k(x)w, u)_{\Gamma_2} + 2\mu(S\tilde{w}, \tilde{u})_{\Gamma_3} \\ \qquad + 2(\alpha(x)w, u)_{\Gamma_5} + \mu(k(x)w, u)_{\Gamma_7} - 2\mu(\varepsilon_{n\tau}(w + U), u)_{\Gamma_8} \\ \qquad + (p - 2\mu\varepsilon_{nn}(w + U), u_n)_{\Gamma_9 \cup \Gamma_{10} \cup \Gamma_{11}} \\ \quad = -\Big(\dfrac{\partial U}{\partial t}, u\Big) - 2\mu(\mathcal{E}(U), \mathcal{E}(u)) - \langle (U \cdot \nabla)U, u\rangle - 2\mu(k(x)U, u)_{\Gamma_2} \\ \qquad - 2\mu(S\tilde{U}, \tilde{u})_{\Gamma_3} - 2(\alpha(x)U_\tau, u)_{\Gamma_5} - \mu(k(x)U, u)_{\Gamma_7} \\ \qquad + \langle f, u\rangle + \displaystyle\sum_{i=2,4,7} \langle \phi_i, u_n\rangle_{\Gamma_i} + \sum_{i=3,5,6} \langle \phi_i, u\rangle_{\Gamma_i} \qquad \forall u \in \mathbf{V}, \\ |\sigma_\tau(v)| \leq g_\tau, \ \sigma_\tau(v) \cdot v_\tau + g_\tau |v_\tau| = 0 \ \text{ on } \Gamma_8, \\ |\sigma_n(v)| \leq g_n, \ \sigma_n(v) v_n + g_n |v_n| = 0 \ \text{ on } \Gamma_9, \\ \sigma_n(v) + g_{+n} \geq 0, \quad (\sigma_n(v) + g_{+n}) v_n = 0 \ \text{ on } \Gamma_{10}, \\ \sigma_n(v) - g_{-n} \leq 0, \quad (\sigma_n(v) - g_{-n}) v_n = 0 \ \text{ on } \Gamma_{11}. \end{cases} \tag{6.14}$$

Define $a_{02}(t,\cdot,\cdot), a_{12}(\cdot,\cdot,\cdot)$ and $F_2(t) \in \mathbf{V}^*$ by

$$
\begin{aligned}
a_{02}(t, w, u) &= 2\mu(\mathcal{E}(w), \mathcal{E}(u)) + \langle (U(t)\cdot\nabla)w, u\rangle + \langle (w\cdot\nabla)U(t), u\rangle \\
&\quad + 2\mu(k(x)w, u)_{\Gamma_2} + 2\mu(S\tilde{w}, \tilde{u})_{\Gamma_3} + 2(\alpha(x)w, u)_{\Gamma_5} \\
&\quad + \mu(k(x)w, u)_{\Gamma_7} \quad \forall w, u \in \mathbf{V}, \\
a_{12}(w, u, v) &= \langle (w\cdot\nabla)u, v\rangle \quad \forall w, u, v \in \mathbf{V}, \\
\langle F_2(t), u\rangle &= -\Big\langle \frac{\partial U(t)}{\partial t}, u\Big\rangle - 2\mu(\mathcal{E}(U(t)), \mathcal{E}(u)) - \langle (U(t)\cdot\nabla)U(t), u\rangle \qquad (6.15)\\
&\quad - 2\mu(k(x)U(t), u)_{\Gamma_2} - 2\mu(S\tilde{U(t)}, \tilde{u})_{\Gamma_3} \\
&\quad - 2(\alpha(x)U(t)_\tau, u)_{\Gamma_5} - \mu(k(x)U(t), u)_{\Gamma_7} + \langle f(t), u\rangle \\
&\quad + \sum_{i=2,4,7} \langle \phi_i(t), u_n\rangle_{\Gamma_i} + \sum_{i=3,5,6} \langle \phi_i(t), u\rangle_{\Gamma_i} \quad \forall u \in \mathbf{V}.
\end{aligned}
$$

Then, taking into account

$$\sigma_\tau(v) = 2\mu\varepsilon_{n\tau}(v), \quad \sigma_n(v) = -p + 2\mu\varepsilon_{nn}(v)$$

and (6.14), we introduce the following variational formulation for problem (6.1), (5.2).

Problem II-VE. Find $v = U + w \in \big(U + K(Q)\big)$ and $(\sigma_\tau, \sigma_n, \sigma_{+n}, \sigma_{-n}) \in \mathbf{L}^2_\tau(\Gamma_8) \times L^2(\Gamma_9) \times H^{-1/2}(\Gamma_{10}) \times H^{-1/2}(\Gamma_{11})$ in a.e. $t \in (0, T)$ such that

$$w(0) = v_0 - U(0)$$

and

$$
\begin{cases}
\Big\langle \dfrac{\partial w}{\partial t}, u\Big\rangle + a_{02}(t, w, u) + a_{12}(w, w, u) - (\sigma_\tau, u_\tau)_{\Gamma_8} - (\sigma_n, u_n)_{\Gamma_9} \\
\qquad - \langle \sigma_{+n}, u_n\rangle_{\Gamma_{10}} - \langle \sigma_{-n}, u_n\rangle_{\Gamma_{11}} = \langle F_2, u\rangle \quad \forall u \in L^2(0, T; \mathbf{V}), \\
|\sigma_\tau| \le g_\tau, \ \sigma_\tau \cdot v_\tau + g_\tau |v_\tau| = 0 \quad \text{on } \Gamma_8, \\
|\sigma_n| \le g_n, \ \sigma_n v_n + g_n |v_n| = 0 \quad \text{on } \Gamma_9, \\
\sigma_{+n} + g_{+n} \ge 0, \quad \langle \sigma_{+n} + g_{+n}, v_n\rangle_{\Gamma_{10}} = 0 \quad \text{on } \Gamma_{10}, \\
\sigma_{-n} - g_{-n} \le 0, \quad \langle \sigma_{-n} - g_{-n}, v_n\rangle_{\Gamma_{11}} = 0 \quad \text{on } \Gamma_{11}.
\end{cases}
\qquad (6.16)
$$

In the same way as in Theorem 5.1 we have

Theorem 6.3 *Let Assumption 6.1 hold. If (v, p) is a solution in the sense of Definition 6.1 of the problem* (6.1), (5.2), *then $(v, \sigma_\tau|_{\Gamma_8}, \sigma_n|_{\Gamma_9}, \sigma_n|_{\Gamma_{10}}, \sigma_n|_{\Gamma_{11}})$ is a solution to Problem II-VE.*

Conversely, if $f \in L^2(0, T; \mathbf{L}^2(\Omega))$ *and Problem II-VE has a smooth solution* $(v, \sigma_\tau, \sigma_n, \sigma_+, \sigma_-)$ *such that* $v \in L^2(0, T; \mathbf{H}^2(\Omega))$, $\sigma_\tau \in L^2(0, T; \mathbf{L}^2(\Gamma_8))$, $\sigma_n \in L^2(0, T; L^2(\Gamma_9))$, $\sigma_+ \in L^2(0, T; L^2(\Gamma_{10}))$ *and* $\sigma_- \in L^2(0, T; L^2(\Gamma_{11}))$, *then there exists* $p \in L^1(0, T; H^1(\Omega))$ *such that* (v, p) *is a solution to the problem* (6.1), (5.2). *Moreover, if at least one of the sets* $\Gamma_i, i = 2, 4, 6, 7$, *is nonempty, then* p *is unique.*

We will find a variational inequality corresponding to Problem II-VE.
Define operators $A_2(t) : \mathbf{V} \to \mathbf{V}^*$ and $B_2 : \mathbf{V} \times \mathbf{V} \to \mathbf{V}^*$ by

$$\begin{aligned} &\langle A_2(t)w, u\rangle = a_{02}(t, w, u) \quad \forall w, u \in \mathbf{V}, \\ &\langle B_2(w, u), v\rangle = a_{12}(w, u, v) \quad \forall w, u, v \in \mathbf{V}. \end{aligned} \tag{6.17}$$

Unlike Problem I, in this problem the property $\langle B_2(w, w), w\rangle = 0$ fails and for $w \in L^2(0, T; \mathbf{V})$ integral $\int_0^T \langle B_2(w, w), w\rangle\, ds$ is not meaningful. Thus, we will find more smooth solution $w \in L^4(0, T; \mathbf{V})$, such that $w' \in L^2(0, T; \mathbf{V}^*)$.

Thus, as Problem I-VI' we come to the following variational inequality.

Problem II-VI. (the case of static pressure) Find v such that $w \equiv v - U \in L^4(0, T; \mathbf{V})$, $\frac{\partial w}{\partial t} \in L^2(0, T; \mathbf{V}^*)$, $w(0) = v_0 - U(0, x)$ and

$$\begin{aligned} \int_0^T \langle \frac{\partial w}{\partial t} + A_2(t)w(t) + B_2(w(t), w(t)) - F_2(t), u(t) - w(t)\rangle\, dt + \Psi(u) - \Psi(w) \geq 0 \\ \forall u \in L^4(0, T; \mathbf{V}), \end{aligned} \tag{6.18}$$

where Ψ is defined by (6.7).

As Theorem 6.2 we have the following theorem:

Theorem 6.4 *If* $(v, \sigma_\tau, \sigma_n, \sigma_{+n}, \sigma_{-n})$ *is a solution to Problem II-VE such that* $v' - U' \in L^2(0, T; \mathbf{V}^*)$, *then* v *is a solution to Problem II-VI.*

Conversely, if v *is a solution to Problem II-VI, then there exist* $\sigma_\tau, \sigma_n, \sigma_{+n}, \sigma_{-n}$ *for a.e.* $t \in (0, T)$ *such that* $(v, \sigma_\tau, \sigma_n, \sigma_{+n}, \sigma_{-n})$ *is a solution to Problem II-VE.*

6.1.3 Variational Formulation: The Stokes Problem

In the same way as Problem I we get the equivalent formulations of Problem III for the Stokes equation with boundary condition (5.2).

Problem III-VE. Find $v = U + w \in \big(U + K(Q)\big)$ and $(\sigma_\tau, \sigma_n, \sigma_{+n}, \sigma_{-n}) \in \mathbf{L}^2_\tau(\Gamma_8) \times L^2(\Gamma_9) \times H^{-\frac{1}{2}}(\Gamma_{10}) \times H^{-\frac{1}{2}}(\Gamma_{11})$ in a.e. t such that $w(0) = v_0 - U(0)$ and

$$
\begin{cases}
\left\langle \dfrac{\partial w}{\partial t}, u\right\rangle + a_{03}(w,u) - (\sigma_\tau, u_\tau)_{\Gamma_8} - (\sigma_n, u_n)_{\Gamma_9} - \langle \sigma_{+n}, u_n\rangle_{\Gamma_{10}} - \langle \sigma_{-n}, u_n\rangle_{\Gamma_{11}} \\
\qquad = \langle F_3, u\rangle \quad \forall u \in L^2(0,T;\mathbf{V}), \\
|\sigma_\tau| \le g_\tau, \ \sigma_\tau \cdot v_\tau + g_\tau |v_\tau| = 0 \quad \text{on } \Gamma_8, \\
|\sigma_n| \le g_n, \ \sigma_n v_n + g_n |v_n| = 0 \quad \text{on } \Gamma_9, \\
\sigma_{+n} + g_{+n} \ge 0, \quad \langle \sigma_{+n} + g_{+n}, v_n\rangle = 0 \quad \text{on } \Gamma_{10}, \\
\sigma_{-n} - g_{-n} \le 0, \quad \langle \sigma_{-n} - g_{-n}, v_n\rangle = 0 \quad \text{on } \Gamma_{11},
\end{cases}
\tag{6.19}
$$

where

$$
\begin{aligned}
a_{03}(w,u) &= 2\mu(\mathcal{E}(w), \mathcal{E}(u)) + 2\mu(k(x)w, u)_{\Gamma_2} + 2\mu(S\tilde{w}, \tilde{u})_{\Gamma_3} \\
&\quad + 2(\alpha(x)w, u)_{\Gamma_5} + \mu(k(x)w, u)_{\Gamma_7} \quad \forall w, u \in \mathbf{V}, \\
\langle F_3, u\rangle &= -\left\langle \frac{\partial U}{\partial t}, u\right\rangle - 2\mu(\mathcal{E}(U), \mathcal{E}(u)) - 2\mu(k(x)U, u)_{\Gamma_2} \\
&\quad - 2\mu(S\tilde{U}, \tilde{u})_{\Gamma_3} - 2(\alpha(x)U_\tau, u)_{\Gamma_5} - \mu(k(x)U, u)_{\Gamma_7} + \langle f, u\rangle \\
&\quad + \sum_{i=2,4,7} \langle \phi_i, u_n\rangle_{\Gamma_i} + \sum_{i=3,5,6} \langle \phi_i, u\rangle_{\Gamma_i} \quad \forall u \in \mathbf{V}.
\end{aligned}
\tag{6.20}
$$

Problem III-VI. Find v such that $w \equiv v - U \in L^2(0,T;\mathbf{V})$, $\frac{\partial w}{\partial t} \in L^2(0,T;\mathbf{V}^*)$, $w(0) = v_0 - U(0)$ and

$$
\begin{aligned}
\int_0^T \langle \frac{\partial w}{\partial t} + A_3 w(t) - F_3(t), u(t) - w(t)\rangle\, dt + \Psi(u) - \Psi(w) \ge 0 \\
\forall u \in L^2(0,T;\mathbf{V}),
\end{aligned}
\tag{6.21}
$$

where $A_3 : \mathbf{V} \to \mathbf{V}^*$ is defined by

$$
\langle A_3 w, u\rangle = a_{03}(w,u) \quad \forall w, u \in \mathbf{V}. \tag{6.22}
$$

As Theorem 6.2 we have

Theorem 6.5 *If $(v, \sigma_\tau, \sigma_n, \sigma_{+n}, \sigma_{-n})$ is a solution to Problem III-VE such that $v' - U' \in L^2(0,T;\mathbf{V}^*)$, then v is a solution to Problem III-VI. Conversely, if v is a solution to Problem III-VI, then there exist $\sigma_\tau, \sigma_n, \sigma_{+n}, \sigma_{-n}$ at a.e. $t \in (0,T)$ such that $(v, \sigma_\tau, \sigma_n, \sigma_{+n}, \sigma_{-n})$ is a solution to Problem III-VE.*

6.2 The Existence and Uniqueness of Solutions to Variational Inequalities

In this section we study the variational inequalities obtained in Sect. 6.1.

Let us start with the existence of a solution $w \in L^2(0, T; \mathbf{V})$ to the following problem:

$$\int_0^T \langle u'(t) + A(t)w(t) + B(w(t), w(t)) - F(t), u(t) - w(t)\rangle\, dt + \Psi(u) - \Psi(w)$$
$$\geq -\frac{1}{2}\|w_0 - u(0)\|^2 \quad \forall u \in L^4(0, T; \mathbf{V}) \text{ with } u' \in L^2(0, T; \mathbf{V}^*). \tag{6.23}$$

Theorem 6.6 *Assume that*
(1) For any $u, v \in \mathbf{V}$, $\langle A(t)u, v\rangle$ is continuous with respect to t and

$$\begin{aligned}
&\langle A(t)u, u\rangle \geq c_1\|u\|_{\mathbf{V}}^2 - c_2\|u\|^2 \quad \exists c_1 > 0,\ c_2 \geq 0,\ \forall u \in \mathbf{V},\\
&|\langle A(t)u, v\rangle| \leq c_3\|u\|_{\mathbf{V}}\|v\|_{\mathbf{V}} \quad \exists c_3 > 0,\ \forall u, v \in \mathbf{V},\\
&\liminf_{m\to\infty} \langle A(t)u_m, u_m\rangle \geq \langle A(t)u, u\rangle \quad \textit{as } u_m \rightharpoonup u \textit{ in } V;
\end{aligned}$$

(2) $B(u, v)$ is bilinear continuous from $\mathbf{V} \times \mathbf{V}$ into $\mathbf{V}^*$ and

$$\begin{aligned}
&|\langle B(w, v), z\rangle| \leq K\|w\|_{\mathbf{H}^1}\|v\|^{\frac{1}{2}}\|v\|_{\mathbf{H}^1}^{\frac{1}{2}}\|z\|_{\mathbf{H}^1},\\
&\langle B(w, v), v\rangle = 0;
\end{aligned}$$

(3) $F \in L^2(0, T; \mathbf{V}^*)$;
(4) The functional $\Phi : \mathbf{V} \to \overline{\mathbb{R}}$ is proper, convex, weakly lower semi-continuous and

$$\Phi(u) \geq 0 \quad \forall u \in \mathbf{V}, \quad \Phi(0_{\mathbf{V}}) = 0.$$

Then, for any initial function $w_0 \in H$ there exists a solution to (6.23) *satisfying*

$$\begin{aligned}
&\|w(t)\|^2 \leq K_1(\|w_0\|^2 + \|F\|_{L^2(0,T;\mathbf{V}^*)}^2),\\
&\|w\|_{L^2(0,T;V)}^2 \leq K_1(\|w_0\|^2 + \|F\|_{L^2(0,T;\mathbf{V}^*)}^2),
\end{aligned} \tag{6.24}$$

where K_1 is independent of Φ.

Proof For every $\varepsilon > 0$, let Φ_ε be the Moreau-Yosida approximation of Φ and $\nabla\Phi_\varepsilon$ be Fréchet derivative of Φ_ε (see Sect. 1.6).

Let $\{v_j, j = 1, 2, \cdots\}$ be a basis of the space $\mathbf{V}$. We find a solution $w_m = \sum_{j=1}^m g_{jm}(t)v_j$ to problem

$$\begin{cases}
\langle w_m'(t), v_j\rangle + \langle A(t)w_m(t), v_j\rangle + \langle B(w_m(t), w_m(t)), v_j\rangle + \langle \nabla\Phi_\varepsilon(w_m(t)), v_j\rangle\\
\qquad\qquad = \langle F(t), v_j\rangle,\\
w_m(0) = w_{0m},
\end{cases} \tag{6.25}$$

where $w'_m(t) = \frac{d}{dt}w_m(t)$ and w_{0m} is such that $w_{0m} = \sum_{i=1}^m k_j v_j \to w_0$ in H as $m \to \infty$. The solutions to (6.25) depend on ε, but for convenience of notation here and in what follows we use subindex m instead of subindex m, ε. For a t_m there exists an absolute continuous function $g_{jm}(t)$ on $[0, t_m)$ and by the first a priori estimate of (6.30) continuation of the function beyond t_m is possible. Thus, we can see that $t_m = T$.

Multiplying (6.25) by g_{jm} and summing over $j = 1, \cdots, m$, yield

$$\frac{1}{2}\frac{d\|w_m(t)\|^2}{dt} + \langle A(t)w_m(t) + B(w_m(t), w_m(t)) + \nabla\Phi_\varepsilon(w_m(t)), w_m(t)\rangle = \langle F(t), w_m(t)\rangle . \tag{6.26}$$

By (1.37) and condition (4) we know that $\Phi_\varepsilon(0_{\mathbf{V}}) = 0$. Also, since Φ_ε is convex, continuous and Fréchet differentiable, we have

$$\Phi_\varepsilon(v) - \Phi_\varepsilon(w_m(t)) \geq \langle \nabla\Phi_\varepsilon(w_m(t)), v - w_m(t)\rangle \quad \forall v \in \mathbf{V}, \tag{6.27}$$

and so by $\Phi_\varepsilon(0_{\mathbf{V}}) = 0$,

$$0 \leq \Phi_\varepsilon(w_m(t)) \leq \langle \nabla\Phi_\varepsilon(w_m(t)), w_m(t)\rangle . \tag{6.28}$$

By virtue of conditions (1), (2) and (6.28), we have from (6.26)

$$\frac{1}{2}\frac{d\|w_m(t)\|^2}{dt} + c_1\|w_m(t)\|_{\mathbf{V}}^2 + \Phi_\varepsilon(w_m(t)) \leq \langle F(t), w_m(t)\rangle + c_2\|w_m(t)\|^2,$$

$$\frac{d}{dt}\|w_m(t)\|^2 + 2c_1\|w_m(t)\|_{\mathbf{V}}^2 + 2\Phi_\varepsilon(w_m(t)) \leq \frac{1}{c_1}\|F\|_{\mathbf{V}^*}^2 + c_1\|w_m(t)\|_{\mathbf{V}}^2 + 2c_2\|w_m(t)\|^2,$$

and

$$\frac{d}{dt}\|w_m(t)\|^2 + c_1\|w_m(t)\|_{\mathbf{V}}^2 + 2\Phi_\varepsilon(w_m(t)) \leq \frac{1}{c_1}\|F\|_{\mathbf{V}^*}^2 + 2c_2\|w_m(t)\|^2. \tag{6.29}$$

We have from (6.29)

$$\begin{aligned}
\|w_m(t)\|^2 &\leq (\|w_0\|^2 + \frac{1}{c_1}\|F\|_{L^2(0,T;\mathbf{V}^*)}^2)e^{2c_2T} \leq C_1(\|w_0\|^2 + \|F\|_{L^2(0,T;\mathbf{V}^*)}^2),\\
\|w_m(t)\|_{L^2(0,T;\mathbf{V})}^2 &\leq \frac{2c_2}{c_1}(\|w_0\|^2 + \frac{1}{c_1}\|F\|_{L^2(0,T;\mathbf{V}^*)}^2)e^{2c_2T} + \frac{1}{c_1^2}\|F\|_{L^2(0,T;\mathbf{V}^*)}^2\\
&\leq C_2(\|w_0\|^2 + \|F\|_{L^2(0,T;\mathbf{V}^*)}^2)
\end{aligned} \tag{6.30}$$

and

$$\int_0^T \Phi_\varepsilon(w_m(t))\,dt \leq C_3, \tag{6.31}$$

where C_i, $i = 1, 2, 3$, are independent of m, ε and Φ. From (1.36), (1.37) and (6.31) we have

$$\int_0^T \|w_m(t) - J_\varepsilon(w_m(t))\|_{\mathbf{V}}^2 \, dt \le c_7 \varepsilon \tag{6.32}$$

with c_7 independent of m and ε.

Multiplying (6.25) by $\big(g_{jm}(t) - g_{jm}(s)\big)$, $s \in (0, T)$, and summing over $j = 1, \cdots, m$, we have

$$\begin{aligned}
\frac{1}{2}\frac{d\|w_m(t) - w_m(s)\|^2}{dt} &+ \langle A(t)w_m(t) + B(w_m(t), w_m(t)) - F(t), w_m(t) - w_m(s)\rangle \\
&= \langle \nabla \Phi_\varepsilon(w_m(t)), w_m(s) - w_m(t)\rangle \\
&\le \Phi_\varepsilon(w_m(s)) - \Phi_\varepsilon(w_m(t)) \\
&\le \Phi_\varepsilon(w_m(s)).
\end{aligned}$$

By conditions (1) and (2), from the above it follows that

$$\begin{aligned}
\frac{d\|w_m(t) - w_m(s)\|^2}{dt} \le &\Phi_\varepsilon(w_m(s)) + \langle A(t)w_m(t), w_m(s)\rangle + \langle B(w_m(t), w_m(t)), w_m(s)\rangle \\
&+ \langle F(t), w_m(t) - w_m(s)\rangle + c_2\|w_m(t)\|^2.
\end{aligned} \tag{6.33}$$

Let us integrate every terms in (6.33) first with respect to t from s to $s + h$ and then with respect to s from 0 to T, where $w_m(t) = 0$ when $t \in (T, T + h)$. We get

$$\int_0^T \int_s^{s+h} \frac{d\|w_m(t) - w_m(s)\|^2}{dt} \, dtds = \int_0^T \|w_m(s + h) - w_m(s)\|^2 \, ds. \tag{6.34}$$

By (6.31),

$$\int_0^T \int_s^{s+h} \Phi_\varepsilon(w_m(s)) \, dtds \le h \int_0^T \Phi_\varepsilon(w_m(s)) \, ds \le c_6 h. \tag{6.35}$$

By conditions (1) and (6.30) we have

$$\int_0^T \int_s^{s+h} \langle A(t)w_m(t), w_m(s)\rangle \, dtds \le c_8\sqrt{h} \int_0^T \|w_m(s)\|_{\mathbf{V}}^2 \, ds. \tag{6.36}$$

By conditions (2) and (6.30) we have

$$\begin{aligned}
&\int_0^T \int_s^{s+h} \langle B(w_m(t), w_m(t))w_m(s)\rangle \, dtds \\
&\quad \le K \int_0^T \int_s^{s+h} \|w_m(t)\|_{\mathbf{V}}^{\frac{3}{2}} \|w_m(t)\|^{\frac{1}{2}} \|w_m(s)\|_{\mathbf{V}} \, dtds \le c_9 h^{\frac{1}{4}}.
\end{aligned} \tag{6.37}$$

Using (6.30) we have

$$\begin{aligned}&\int_0^T \int_s^{s+h} \langle F(t), w_m(t)\rangle \, dtds \le \int_0^T |\langle F(t), w_m(t)\rangle| \Big(\int_{t-h}^t ds\Big)dt \le Kh,\\ &\int_0^T \int_s^{s+h} \langle F(t), -w_m(s)\rangle \, dtds \le K \int_0^T \|w_m(s)\|_{\mathbf{V}} \int_s^{s+h} \|F(t)\|_{\mathbf{V}^*} \, dtds \le c_9\sqrt{h}.\end{aligned} \tag{6.38}$$

By virtue of (6.34)–(6.38), uniformly with respect to m and ε

$$\int_0^T \|w_m(s+h) - w_m(s)\|^2 \, ds \le O(h^{\frac{1}{4}}), \tag{6.39}$$

which implies that the set $\{w_m\}$ is relatively compact in $L^2(0, T; H)$ (see Theorem 1.38). Therefore, by (6.30) and (6.39), there exists w and a subsequence $\{w_{m_k}\}$ such that

$$\begin{aligned} w_{m_k} &\overset{*}{\rightharpoonup} w \quad \text{in } L^\infty(0, T; H),\\ w_{m_k} &\to w \quad \text{in } L^2(0, T; H),\\ w_{m_k} &\rightharpoonup w \quad \text{in } L^2(0, T; \mathbf{V})\end{aligned} \tag{6.40}$$

when $m_k \to \infty$ and $\varepsilon \to 0$ (note that for convenience we used subindex m instead of m, ε).

On the other hand, putting $v = \sum_{j=1}^M k_j(t)v_j$, where $k_j(t) \in C^1[0, T]$ and M-positive integer, multiply (6.25) by $k_j(t)$ and sum for $j = 1, \cdots, M$. Then, multiply (6.25) by $g_{jm}(t)$ and sum for $j = 1, \cdots, m$. Substituting the resulting equations, we have

$$\begin{aligned}&\big\langle w_m'(t) + A(t)w_m(t) + B(w_m(t), w_m(t)) + \nabla\Phi_\varepsilon(w_m), v(t) - w_m(t)\big\rangle\\ &\qquad = \langle F(t), v(t) - w_m(t)\rangle .\end{aligned} \tag{6.41}$$

Since

$$\begin{aligned}&\int_0^T \big\langle w_m'(t), v(t) - w_m(t)\big\rangle \, dt =\\ &\quad \int_0^T \big\langle v'(t), v(t) - w_m(t)\big\rangle \, dt - \frac{1}{2}\|w_m(T) - v(T)\|^2 + \frac{1}{2}\|w_m(0) - v(0)\|^2,\end{aligned}$$

we have

$$\begin{aligned}&\int_0^T \big\langle v'(t) + A(t)w_m(t) + B(w_m(t), w_m(t)) + \nabla\Phi_\varepsilon(w_m(t)) - F(t), v(t) - w_m(t)\big\rangle \, dt\\ &\qquad = -\frac{1}{2}\|w_m(0) - v(0)\|^2 + \frac{1}{2}\|w_m(T) - v(T)\|^2.\end{aligned} \tag{6.42}$$

Taking into account (6.27), we have from (6.42)

$$\begin{aligned}\int_0^T &\big\langle v'(t) + A(t)w_m(t) + B(w_m(t), w_m(t)) - F(t), v(t) - w_m(t)\big\rangle \, dt \\ &+ \int_0^T \big(\Phi_\varepsilon(v(t)) - \Phi_\varepsilon(w_m(t))\big) \, dt \geq -\frac{1}{2}\|w_m(0) - v(0)\|^2.\end{aligned} \tag{6.43}$$

Since $\Phi(v) \geq \Phi_\varepsilon(v)$ and $\Phi(J_\varepsilon w_m(t)) \leq \Phi_\varepsilon(w_m(t))$ (see (1.37)), we have from (6.43)

$$\begin{aligned}\int_0^T &\big\langle v'(t) + A(t)w_m(t) + B(w_m(t), w_m(t)) - F(t), v(t) - w_m(t)\big\rangle \, dt \\ &+ \Psi(v) - \int_0^T \Phi(J_\varepsilon w_m(t)) \, dt \geq -\frac{1}{2}\|w_m(0) - v(0)\|^2.\end{aligned} \tag{6.44}$$

By (6.32) it follows that $J_\varepsilon w_{m_k} \rightharpoonup w$ in $L^2(0, T; \mathbf{V})$ as $m_k \to \infty$ and $\varepsilon \to 0$, and

$$\liminf_{m_k \to \infty, \varepsilon \to 0} \int_0^T \Phi(J_\varepsilon w_{m_k}(t)) \, dt \geq \int_0^T \Phi(w(t)) \, dt = \Psi(w). \tag{6.45}$$

It is not difficult to prove that

$$\int_0^T \big\langle B(w_{m_k}(t), w_{m_k}(t)), v(t)\big\rangle \, dt \to \int_0^T \langle B(w(t), w(t)), v(t)\rangle \, dt \tag{6.46}$$

as $m_k \to \infty$ and $\varepsilon \to 0$. We give a proof below for convenience of the readers.

$$\begin{aligned}\int_0^T &(\big\langle B(w_{m_k}(t), w_{m_k}(t)), v(t)\big\rangle - \langle B(w(t), w(t)), v(t)\rangle) \, dt \\ &= \int_0^T \big\langle B(w_{m_k}(t), w_{m_k}(t) - w(t)), v(t)\big\rangle \, dt + \int_0^T \big\langle B(w_{m_k}(t) - w(t), w(t)), v(t)\big\rangle \, dt \\ &\equiv I_1 + I_2.\end{aligned} \tag{6.47}$$

By condition (2), Hölder's inequality with exponents 2, 4, 4 and (6.40) we have

$$\begin{aligned}|I_1| &\leq K \sup_{t\in[0,T]} \|v(t)\|_{\mathbf{H}^1} \int_0^T \|w_{m_k}(t)\|_{\mathbf{H}^1} \|w_{m_k}(t) - w(t)\|_{\mathbf{H}^1}^{\frac{1}{2}} \|w_{m_k}(t) - w(t)\|^{\frac{1}{2}} \, dt \\ &\leq K \sup_{t\in[0,T]} \|v(t)\|_{\mathbf{H}^1} \|w_{m_k}\|_{L^2(0,T;\mathbf{H}^1)} \|w_{m_k} - w\|_{L^2(0,T;\mathbf{H}^1)}^{\frac{1}{2}} \|w_{m_k} - w\|_{L^2(0,T;\mathbf{L}^2)}^{\frac{1}{2}} \to 0.\end{aligned} \tag{6.48}$$

By condition (2),

$$\Big|\int_0^T \langle B(z(t), w(t)), v(t)\rangle\, dt\Big| \le K \sup_{t\in[0,T]} \|v(t)\|_{\mathbf{H}^1} \|z(t)\|_{L^2(0,T;\mathbf{H}^1)} \|w\|^{\frac12}_{L^2(0,T;\mathbf{H}^1)} \|w\|^{\frac12}_{L^2(0,T;\mathbf{L}^2)}$$
$$\le K_1 \|z(t)\|_{L^2(0,T;\mathbf{H}^1)},$$

which means that

$$z(t) \in L^2(0, T; \mathbf{V}) \to \int_0^T \langle B(z(t), w(t)), v(t)\rangle\, dt$$

is a continuous linear on $L^2(0, T; \mathbf{V})$, that is, there exists a $f \in L^2(0, T; \mathbf{V})^*$ such that

$$\int_0^T \langle B(z(t), w(t)), v(t)\rangle\, dt = \langle z(t), f(t)\rangle$$

and so by (6.40) it follows that

$$|I_2| \to 0 \quad \text{as } m_k \to \infty \text{ and } \varepsilon \to 0. \tag{6.49}$$

By (6.47)–(6.49) we get (6.46).

Since $\liminf_{m_k\to\infty,\varepsilon\to 0} \big\langle A(t)w_{m_k}(t), w_{m_k}(t)\big\rangle \ge \langle A(t)w(t), w(t)\rangle$, by (6.45) and (6.46) we have from (6.44)

$$\int_0^T \big\langle v'(t) + A(t)w(t) + B(w(t), w(t)) - F(t), v(t) - w(t)\big\rangle\, dt + \Psi(v) - \Psi(w)$$
$$\ge -\frac{1}{2}\|w(0) - v(0)\|^2. \tag{6.50}$$

On the other hand, $\|B(w(t), w(t))\|_{\mathbf{V}^*} \le K\|w(t)\|^{\frac32}_{\mathbf{V}}\|w(t)\|^{\frac12}$ (cf. condition(2)) and $w(t) \in L^\infty(0, T; \mathbf{L}^2(\Omega))$, and we can get that $B(w(t), w(t)) \in L^{\frac43}(0, T; \mathbf{V}^*)$. Therefore, taking into account the fact that the set $\{v = \sum_{i=1}^{M} k_j(t)v_j; k_j(t) \in C^1[0, T]\}$ is dense in $\{L^4(0, T; \mathbf{V}) : u' \in L^2(0, T; \mathbf{V}^*)\}$ (see Remark 1.8), we see that (6.50) is valid for all $v \in \{L^4(0, T; \mathbf{V}) : u' \in L^2(0, T; \mathbf{V}^*)\}$.

From (6.30) we get estimates (6.24). □

Remark 6.6 The operator $A_1(t)$ in Problem I-VI is not symmetric and depends on time t. Thus, unlike Theorem 1 of [2] we can not use eigenfunctions of a symmetric operator as the basis for Galërkin method.

Next, let us study the following problem:

$$\begin{cases} \displaystyle\int_0^T \big\langle w'(t) + A(t)w(t) + B(w(t), w(t)) - F(t), u(t) - w(t)\big\rangle\, dt + \Psi(u) - \Psi(w) \ge 0 \\ \qquad\qquad\qquad\qquad \forall u \in L^4(0, T; \mathbf{V}), \\ w(0) = w_0. \end{cases} \tag{6.51}$$

Condition (2) for $B(w, w)$ in the following theorem is for 2-D problem corresponding to Problem I (the case of total pressure).

Theorem 6.7 *Assume that*
(1) Let $l = 2$. For any $u, v \in \mathbf{V}$, $\langle A(t)u, v\rangle$ and $\langle A'(t)u, v\rangle$ are continuous with respect to t and

$$\begin{aligned} &\langle A(t)u, u\rangle \geq c_1\|u\|_{\mathbf{V}}^2 - c_2\|u\|^2 \ \ \exists c_1 > 0,\ c_2 \geq 0,\ \forall u \in \mathbf{V}, \\ &|\langle A(t)u, v\rangle| \leq c_3\|u\|_{\mathbf{V}}\|v\|_{\mathbf{V}} \ \ \exists c_3 > 0,\ \forall u, v \in \mathbf{V}, \\ &|\langle A'(t)u, v\rangle| \leq c_4\|u\|_{\mathbf{V}}\|v\|_{\mathbf{V}} \ \ \exists c_4 \geq 0,\ \forall u, v \in \mathbf{V}; \end{aligned} \tag{6.52}$$

(2) $B(u, v)$ is bilinear continuous from $\mathbf{V} \times \mathbf{V}$ into $\mathbf{V}^*$ and

$$\begin{aligned} &|\langle B(w, v), z\rangle| \leq K\|w\|_{\mathbf{H}^1}\|v\|_{\mathbf{L}^4}\|z\|^{\frac{1}{3}}\|z\|_{\mathbf{H}^1}^{\frac{2}{3}}, \\ &\langle B(w, v), v\rangle = 0; \end{aligned}$$

(3) $F, F' \in L^2(0, T; \mathbf{V}^*)$;
(4) The functional $\Phi : \mathbf{V} \to \overline{\mathbb{R}}$ is proper, weakly lower semi-continuous and

$$\Phi(u) \geq 0 \ \ \forall u \in \mathbf{V}, \ \ \Phi(0_{\mathbf{V}}) = 0;$$

(5) $w_0 \in \mathbf{V}$, $\Phi(w_0) = 0$ and $(A(0)w_0 + B(w_0, w_0) - F(0)) \in H$ (a compatibility condition at initial time).

Then, there exists a unique solution w to (6.51) *such that*

$$w \in C([0, T]; \mathbf{V}), \quad w' \in L^2(0, T; \mathbf{V}) \cap L^\infty(0, T; H), \quad w(0) = w_0, \tag{6.53}$$

and the estimates (6.24) *and the following hold:*

$$\|w'(t)\|^2 \leq K_1, \quad \|w'\|_{L^2(0,T;\mathbf{V})}^2 \leq K_1, \tag{6.54}$$

where K_1 depends on $\|A(0)w_0 + B(w_0, w_0) - F(0)\|$, $\|w_0\|$, $\|F_1\|_{L^2(0,T;\mathbf{V}^*)}^2$, $\|F_1'\|_{L^2(0,T;\mathbf{V}^*)}^2$ and is independent of Φ.

Proof Let $\{v_j, j = 1, 2, \cdots\}$ be a basis of the space $\mathbf{V}$. Without loss of generality, we assume that $v_1 = w_0$. We find a solution $w_m = \sum_{i=1}^m g_{im}(t)v_i$ to problem

$$\begin{cases} \langle w_m'(t), v_j\rangle + \langle A(t)w_m(t), v_j\rangle + \langle B(w_m(t), w_m(t)), v_j\rangle + \langle \nabla\Phi_\varepsilon(w_m(t)), v_j\rangle = \langle F, v_j\rangle, \\ w_m(0) = w_0. \end{cases} \tag{6.55}$$

By the same argument as in the proof of Theorem 6.6, there exists continuously differentiable function g_m on the interval $[0, T]$. Also, as (6.30) in the proof of Theorem 6.6 we have

$$\begin{aligned} &\|w_m(t)\|^2 \leq C_1(\|w_0\|^2 + \|F\|^2_{L^2(0,T;\mathbf{V}^*)}), \\ &\|w_m\|^2_{L^2(0,T;\mathbf{V})} \leq C_2(\|w_0\|^2 + \|F\|^2_{L^2(0,T;\mathbf{V}^*)}), \end{aligned} \tag{6.56}$$

where C_1, C_2 are independent of m, ε.

We are going to derive a priori estimates for functions w'_m. In the same way as in the proof of Theorem 6.6 we see that there exists a solution $g_{im}(t)$ on the interval $[0, T]$. Owing to $F' \in L^2(0, T; \mathbf{V}^*)$, the function $\langle F, v_j\rangle$ is absolute continuous with respect to t. Therefore, by the Lipschitz continuity of $\nabla\Phi_\varepsilon$ and continuity of $\langle A'(t)u, v\rangle$ with respect to t, $g'_{im}(t)$ is in fact absolute continuous.

Setting $t = 0$ in (6.55), multiplying by $g'_{im}(0)$ and summing over $i = 1, \cdots, m$, we get

$$\begin{aligned} \|w'_m(0)\|^2 + \langle A(0)w_m(0), w'_m(0)\rangle + \langle B(w_m(0), w_m(0)), w'_m(0)\rangle + \langle\nabla\Phi_\varepsilon(w_0), w'_m(0)\rangle \\ = \langle F(0), w'_m(0)\rangle. \end{aligned} \tag{6.57}$$

By conditions (4) and (5), for any $u \in \mathbf{V}$ we have $\Phi(u) \geq \Phi(w_0)$, and so $\nabla\Phi_\varepsilon(w_0) = 0$. Thus, we have from (6.57)

$$\|w'_m(0)\| \leq \|A(0)w_0 + B(w_0, w_0) - F(0)\|. \tag{6.58}$$

Differentiating (6.55) yields

$$\begin{aligned} \langle w''_m(t), v_j\rangle + \langle A'(t)w_m(t), v_j\rangle + \langle A(t)w'_m(t), v_j\rangle + \langle B(w'_m(t), w_m(t)), v_j\rangle \\ + \langle B(w_m(t), w'_m(t)), v_j\rangle + \langle(\nabla\Phi_\varepsilon(w_m))', v_j\rangle = \langle F', v_j\rangle, \end{aligned} \tag{6.59}$$

where $(\nabla\Phi_\varepsilon(w_m))' = \frac{d}{dt}(\nabla\Phi_\varepsilon(w_m(t)))$. Multiplying (6.59) by $g'_{im}(t)$ and summing for i, we have

$$\begin{aligned} \langle w''_m(t), w'_m(t)\rangle + \langle A'(t)w_m(t), w'_m(t)\rangle + \langle A(t)w'_m(t), w'_m(t)\rangle \\ + \langle B(w'_m(t), w_m(t)), w'_m(t)\rangle + \langle(\nabla\Phi_\varepsilon(w_m))', w'_m(t)\rangle = \langle F', w'_m(t)\rangle, \end{aligned} \tag{6.60}$$

where the fact that $\langle B(w_m, w'_m), w'_m\rangle = 0$ was used. Using monotonicity of $\nabla\Phi_\varepsilon$, we know that $\langle(\nabla\Phi_\varepsilon(w_m))', w'_m\rangle \geq 0$ (see (1.38)). By condition (2), (6.56) and Young's inequality with exponents 6/5, 6, we get

$$
\begin{aligned}
2|\langle B(w'_m, w_m), w'_m\rangle| &\le 2K\|w'_m\|_{\mathbf{V}}\|w_m\|_{\mathbf{L}^4}\|w'_m\|^{\frac{1}{3}}\|w'_m\|^{\frac{2}{3}}_{\mathbf{V}} \\
&\le 2K\|w'_m\|^{5/3}_{\mathbf{V}}\|w_m\|_{\mathbf{L}^4}\|w'_m\|^{\frac{1}{3}} \\
&\le \frac{c_1}{4}\|w'_m\|^2_{\mathbf{V}} + K_1\|w_m\|^6_{\mathbf{L}^4}\|w'_m\|^2.
\end{aligned}
$$

Thus, we have from (6.60)

$$
\begin{aligned}
&\frac{d}{dt}\|w'_m(t)\|^2 + c_1\|w'_m\|^2_{\mathbf{V}} + c_1\|w'_m\|^2_{\mathbf{V}} - c_1\|w'_m\|^2_{\mathbf{V}} - \frac{c_4}{c_1}\|w_m\|^2_{\mathbf{V}} \\
&\le \frac{c_1}{4}\|w'_m\|^2_{\mathbf{V}} + \frac{4}{c_1}\|F'_1\|^2_{\mathbf{V}^*} + 2c_2\|w'_m\|^2 + \frac{c_1}{4}\|w'_m\|^2_{\mathbf{V}} + K_2\|w'_m\|^2\|w_m\|^6_{\mathbf{L}^4},
\end{aligned}
$$

and so

$$
\frac{d}{dt}\|w'_m(t)\|^2 + \frac{c_1}{2}\|w'_m\|^2_{\mathbf{V}} \le \frac{c_4}{c_1}\|w_m\|^2_{\mathbf{V}} + \frac{4}{c_1}\|F'\|^2_{\mathbf{V}^*} + (2c_2 + K_2\|w_m\|^6_{\mathbf{L}^4})\|w'_m\|^2. \tag{6.61}
$$

Integrating (6.61) yields

$$
\begin{aligned}
\|w'_m(t)\|^2 + \int_0^t \frac{c_1}{2}\|w'_m(s)\|^2_{\mathbf{V}}\,ds \le{}& \|w'_m(0)\|^2 + \frac{c_4}{c_1}\|w_m\|^2_{L^2(0,T;\mathbf{V})} + \frac{4}{c_1}\|F'_1\|^2_{L^2(0,T;\mathbf{V}^*)} \\
&+ \int_0^t \left(2c_2 + K_2\|w_m(s)\|^6_{\mathbf{L}^4}\right)\|w'_m(s)\|^2\,ds.
\end{aligned} \tag{6.62}
$$

By (6.56),

$$
w_m \in L^2(0, T; \mathbf{V}) \cap L^\infty(0, T; \mathbf{L}^2(\Omega))
$$

and $\{w_m\}$ is bounded in the space above. Since $\mathbf{V} \hookrightarrow \mathbf{L}^q(\Omega)$ for any $1 < q < \infty$ when $l = 2$, by virtue of Theorems 1.12 and 1.33 we have that

$$
w_m \in L^6(0, T; \mathbf{L}^4(\Omega))
$$

and $\{w_m\}$ is bounded in $L^6(0, T; \mathbf{L}^4(\Omega))$. Taking into account (6.56), (6.58) and Gronwall's inequality (see Theorem 1.51), we have from (6.62)

$$
\|w'_m(t)\|^2 \le C_4 e^{\int_0^t \left(2c_2 + K_2\|w_m(s)\|^6_{\mathbf{L}^4}\right)ds}, \tag{6.63}
$$

where C_4 depends on $\|Aw_0 + B(w_0, w_0) - F(0)\|$, $\|w_0\|$, $\|F_1\|^2_{L^2(0,T;\mathbf{V}^*)}$, $\|F'_1\|^2_{L^2(0,T;\mathbf{V}^*)}$ and is independent of ε, m, Φ.

From (6.56), (6.62) and (6.63) we have

$$
\int_0^T \|w'_m(s)\|^2_{\mathbf{V}}\,ds \le C_5, \tag{6.64}
$$

where C_5 is similar to C_4.

Therefore, by (6.56), (6.63) and (6.64)

$$\begin{aligned}
&\|w_m(t)\| \le \text{const} \quad \forall t \in [0, T],\ \forall m,\ \forall \varepsilon > 0;\\
&\|w_m\|_{L^2(0,T;\mathbf{V})} \le \text{const} \quad \forall m,\ \forall \varepsilon > 0;\\
&\|w_m'(t)\| \le \text{const} \quad \forall t \in [0, T],\ \forall m,\ \forall \varepsilon > 0;\\
&\|w_m'\|_{L^2(0,T;\mathbf{V})} \le \text{const} \quad \forall m,\ \forall \varepsilon > 0.
\end{aligned} \tag{6.65}$$

Extracting a subsequence and passing to limit in (6.55) as $m \to \infty$, and then letting $\varepsilon \to 0$ and using (6.27), (6.45), similarly to the proof of Theorem 6.6 (from (6.41)) we can prove the existence of a solution $w \in L^2(0, T; \mathbf{V})$ to (6.51) satisfying that $w' \in L^2(0, T; \mathbf{V}) \cap L^\infty(0, T; H)$.

From (6.56), (6.63) and (6.64), we get the estimates (6.24) and (6.54).

Let us prove uniqueness next.

Let w_1, w_2 be two solutions asserted in Theorem 6.7. Then, taking into account (6.13) and Remark 6.5, we have from (6.51)

$$\begin{aligned}
&\big\langle w_1'(t) + A(t)w_1(t) + B(w_1(t), w_1(t)) - F(t), w_2(t) - w_1(t)\big\rangle\\
&\qquad\qquad + \Phi(w_2(t)) - \Phi(w_1(t)) \ge 0,\\
&\big\langle w_2'(t) + A(t)w_2(t) + B(w_2(t), w_2(t)) - F(t), w_1(t) - w_2(t)\big\rangle\\
&\qquad\qquad + \Phi(w_1(t)) - \Phi(w_2(t)) \ge 0,
\end{aligned}$$

which imply

$$\begin{aligned}
&\big\langle w_1'(t) - w_2'(t), w_1(t) - w_2(t)\big\rangle + \langle A(t)(w_1(t) - w_2(t)), w_1(t) - w_2(t)\rangle \le\\
&\qquad | \langle B(w_1(t), w_1(t)) - B(w_2(t), w_2(t)), w_1(t) - w_2(t)\rangle |.
\end{aligned} \tag{6.66}$$

By virtue of conditions (1) and (2) we have

$$\begin{aligned}
&\frac{1}{2}\frac{\partial \|(w_1(t) - w_2(t)\|}{\partial t} + \langle A(t)(w_1(t) - w_2(t)), w_1(t) - w_2(t)\rangle\\
&\qquad \le \langle |B(w_1(t), w_1(t)) - B(w_2(t), w_2(t)), w_1(t) - w_2(t)\rangle|,
\end{aligned}$$

$$\begin{aligned}
&\frac{1}{2}\frac{\partial \|w_1(t) - w_2(t)\|^2}{\partial t} + c_1\|w_1(t) - w_2(t)\|_{\mathbf{V}}^2\\
&\quad \le c_2\|w_1(t) - w_2(t)\|^2 + \|w_2(t)\|_{\mathbf{L}^4}\|w_1(t) - w_2\|_{\mathbf{V}}^{\frac{5}{3}}\|w_1(t) - w_2\|^{\frac{1}{3}}\\
&\quad \le c_2\|(w_1(t) - w_2(t)\|^2 + c_1\|w_1(t) - w_2\|_{\mathbf{V}}^2 + K\|w_2(t)\|_{\mathbf{L}^4}^6\|w_1(t) - w_2\|^2.
\end{aligned} \tag{6.67}$$

We have from (6.67)

$$\|(w_1(t) - w_2(t)\|^2 \le \text{const} \int_0^t (c_2 + K\|w_2(s)\|_{\mathbf{L}^4}^6)\|w_1(s) - w_2(s)\|^2\, ds. \tag{6.68}$$

Since $w_2 \in C([0,T];\mathbf{V})$, by Gronwall's inequality we have that $w_1(t) = w_2(t)$ for all $t \in [0,T]$. □

The following theorem is for small data in (6.51) corresponding to Problem I (the case of total pressure).

Theorem 6.8 *Assume that*
(1) For any $u, v \in \mathbf{V}$, $\langle A(t)u, v\rangle$ and $\langle A'(t)u, v\rangle$ are continuous with respect to t and

$$\begin{aligned}
&\langle A(t)u, u\rangle \geq c_1\|u\|_{\mathbf{V}}^2 - c_2\|u\|^2, \quad \exists c_1 > 0,\ c_2 \geq 0,\ \ \forall u \in \mathbf{V},\\
&|\langle A(t)u, v\rangle| \leq c_3\|u\|_{\mathbf{V}}\|v\|_{\mathbf{V}} \quad \exists c_3 > 0,\ \forall u, v \in \mathbf{V},\\
&|\langle A'(t)u, v\rangle| \leq c_4\|u\|_{\mathbf{V}}\|v\|_{\mathbf{V}} \quad \exists c_4 > 0,\ \forall u, v \in \mathbf{V};
\end{aligned}$$

(2) $B(u,v)$ is bilinear continuous from $\mathbf{V} \times \mathbf{V}$ into $\mathbf{V}^*$ and

$$\begin{aligned}
&\langle B(w, v), u\rangle \leq c_5\|w\|_V\|v\|_V\|u\|^{\frac{1}{2}}\|u\|_V^{\frac{1}{2}},\\
&\langle B(w, v), v\rangle = 0;
\end{aligned}$$

(3) $F, F' \in L^2(0,T;\mathbf{V}^*)$;
(4) The functional $\Phi : \mathbf{V} \to \overline{\mathbb{R}}$ is proper, weakly lower semi-continuous and

$$\Phi(u) \geq 0 \ \ \forall u \in \mathbf{V}, \quad \Phi(0_{\mathbf{V}}) = 0;$$

(5) $w_0 \in \mathbf{V}$, $\Phi(w_0) = 0$ and $(A(0)w_0 + B(w_0, w_0) - F(0)) \in H$ (a compatibility condition at the initial time).

If $\|w_0\|_{\mathbf{V}}$, $\|(A(0)w_0 + B(w_0, w_0) - F(0))\|_H$ and norms of F, F' in the spaces where they belong to are small enough, then there exists a unique solution w to (6.51) *such that*

$$w \in C([0,T];\mathbf{V}), \quad w' \in L^2(0,T;\mathbf{V}) \cap L^\infty(0,T;H), \tag{6.69}$$

and (6.24), (6.54) *hold.*

Proof First, let us prove the existence of the asserted solution.

Let $\{v_j, j = 1, 2, \cdots\}$ be a basis of the space $\mathbf{V}$. Without loss of generality, we assume that $v_1 = w_0$. We find a solution $w_m = \sum_{i=1}^m g_{im}(t)v_i$ to problem

$$\begin{cases}
\langle w_m'(t), v_j\rangle + \langle A(t)w_m(t), v_j\rangle + \langle B(w_m, w_m), v_j\rangle + \langle \nabla\Phi_\varepsilon(w_m), v_j\rangle = \langle F, v_j\rangle,\\
w_m(0) = w_0.
\end{cases} \tag{6.70}$$

As (6.30) in the proof of Theorem 6.6 we get

$$\begin{aligned}
&\|w_m(t)\|^2 \leq C_1(\|w_0\|^2 + \|F\|^2_{L^2(0,T;\mathbf{V}^*)}),\\
&\|w_m\|^2_{L^2(0,T;\mathbf{V})} \leq C_2(\|w_0\|^2 + \|F\|^2_{L^2(0,T;\mathbf{V}^*)}),
\end{aligned} \tag{6.71}$$

where C_1, C_2 are independent of m and ε.

For a $\tilde{t}_m$ there exists a differentiable function $g_{im}(t)$ on $[0, \tilde{t}_m]$ such that $g'_{im}(t)$ is absolute continuous (cf. proof of Theorem 6.7).

Multiplying (6.70) by $g_{im}(t)$ and summing over $i = 1, \cdots, m$, yield

$$\begin{cases} \left\langle w'_m(t), w_m(t)\right\rangle + \langle A(t)w_m(t), w_m(t)\rangle + \langle B(w_m(t), w_m(t)), w_m(t)\rangle \\ \qquad + \langle \nabla\Phi_\varepsilon(w_m), w_m(t)\rangle = \langle F, w_m(t)\rangle, \\ w_m(0) = w_0. \end{cases} \tag{6.72}$$

We are going to derive a priori estimates for w'_m.

Setting $t = 0$ in (6.70), multiplying the resulting equation by $g'_{im}(0)$ and adding on $i = 1, \cdots, m$, we get

$$\begin{aligned} &\|w'_m(0)\|^2 + \left\langle A(0)w_m(0), w'_m(0)\right\rangle + \left\langle B(w_m(0), w_m(0)), w'_m(0)\right\rangle \\ &\qquad + \left\langle \nabla\Phi_\varepsilon(w_0), w'_m(0)\right\rangle = \left\langle F(0), w'_m(0)\right\rangle. \end{aligned} \tag{6.73}$$

By conditions (4), (5) for any $u \in \mathbf{V}$, $\Phi(u) \geq \Phi(w_0)$, and so $\nabla\Phi_\varepsilon(w_0) = 0$. Thus, we have from (6.73)

$$\|w'_m(0)\| \leq \|A(0)w_0 + B(w_0, w_0) - F(0)\|. \tag{6.74}$$

On the other hand, taking into account (6.28), we have from (6.72)

$$c_1\|w_m(t)\|^2_{\mathbf{V}} \leq \|F\|_{\mathbf{V}^*}\|w_m(t)\|_{\mathbf{V}} + c_2\|w_m(t)\|^2 - (w_m(t), w'_m(t)),$$

and so

$$c_1\|w_m(t)\|_{\mathbf{V}} \leq \|F\|_{L^\infty(0,T;\mathbf{V}^*)} + \alpha_1\|w'_m(t)\| + c_2\alpha_1\|w_m(t)\|, \tag{6.75}$$

where and in what follows of this section α_1 is the one in $\|\cdot\| \leq \alpha_1\|\cdot\|_{\mathbf{V}}$.

Differentiating (6.70) yields

$$\begin{aligned} &\left\langle w''_m(t), v_j\right\rangle + \left\langle A(t)'w_m(t), v_j\right\rangle + \left\langle A(t)w'_m(t), v_j\right\rangle + \left\langle B(w'_m(t), w_m(t)), v_j\right\rangle \\ &\qquad + \left\langle B(w_m(t), w'_m(t)), v_j\right\rangle + \left\langle (\nabla\Phi_\varepsilon(w_m))', v_j\right\rangle = \left\langle F', v_j\right\rangle. \end{aligned} \tag{6.76}$$

Multiplying (6.76) by $g'_{im}(t)$, summing for i and observing the fact that $\langle(\nabla\Phi_\varepsilon(w_m))', w'_m\rangle \geq 0$ due to monotonicity of $\nabla\Phi_\varepsilon$ (see (1.38)), by conditions (2), (3) we have

$$\begin{aligned} &\frac{d}{dt}\|w'_m(t)\|^2 + (c_1 - 2c_5\sqrt{\alpha_1}\|w_m(t)\|_{\mathbf{V}})\|w'_m\|^2_{\mathbf{V}} + c_1\|w'_m\|^2_{\mathbf{V}} - \frac{c_1}{2}\|w'_m\|^2_{\mathbf{V}} - \frac{2c_4^2}{c_1}\|w_m\|^2_{\mathbf{V}} \\ &\qquad \leq \frac{c_1}{2}\|w'_m\|^2_{\mathbf{V}} + \frac{2}{c_1}\|F'\|^2_{\mathbf{V}^*} + 2c_2\|w'_m\|^2, \end{aligned}$$

where $|\langle B(w'_m, w_m), w'_m\rangle| \le c_5\sqrt{\alpha_1}\|w_m\|_{\mathbf{V}}\|w'_m\|^2_{\mathbf{V}}$, $\langle B(w_m, w'_m), w'_m\rangle = 0$ and

$$2|\langle A(t)' w_m(t), w'_m(t)\rangle| \le \frac{c_1}{2}\|w'_m\|^2_{\mathbf{V}} + \frac{2c_4^2}{c_1}\|w_m\|^2_{\mathbf{V}}$$

were used. Thus,

$$\frac{d}{dt}\|w'_m(t)\|^2 + (c_1 - 2c_5\sqrt{\alpha_1}\|w_m(t)\|_{\mathbf{V}})\|w'_m\|^2_{\mathbf{V}} \le \frac{2}{c_1}\|F'\|^2_{\mathbf{V}^*} + 2c_2\|w'_m\|^2 + \frac{2c_4^2}{c_1}\|w_m\|^2_{\mathbf{V}}. \tag{6.77}$$

If $\|w_m(0)\|_{\mathbf{V}} = \|w_0\|_{\mathbf{V}} < \frac{c_1}{4c_5\sqrt{\alpha_1}}$, then

$$c_1 - 2c_5\sqrt{\alpha_1}\|w_0\|_{\mathbf{V}} \ge \frac{c_1}{2}, \tag{6.78}$$

and there exists a t_m such that $c_1 - 2c_5\sqrt{\alpha_1}\|w_m(t)\|_{\mathbf{V}} > 0$ for $t \in [0, t_m]$. Therefore, we have from (6.74) and (6.77) that

$$\begin{aligned}\|w'_m(t)\|^2 \le &\|Aw_0 + B(w_0, w_0) - F(0)\|^2\\ &+ \frac{2}{c_1}\int_0^T \|F'\|^2_{\mathbf{V}^*}\,dt + \frac{2c_4^2}{c_1}\int_0^T \|w_m(s)\|^2_{\mathbf{V}}\,ds + 2c_2\int_0^t \|w'_m(s)\|^2\,ds\end{aligned}$$

for $t \in [0, t_m]$, from which by Gronwall's inequality and (6.71) it follows that

$$\begin{aligned}\|w'_m(t)\|^2 &\le \Big(\|Aw_0 + B(w_0, w_0) - F(0)\|^2 + \frac{2c_4^2}{c_1}\|w_m\|^2_{L^2(0,T;\mathbf{V})} + \frac{2}{c_1}\|F'\|^2_{L^2(0,T;\mathbf{V}^*)}\Big)e^{2c_2T}\\ &\le \Big(\|Aw_0 + B(w_0, w_0) - F(0)\|^2 + \frac{2c_4^2 C_2}{c_1}(\|w_0\|^2 + \|F\|^2_{L^2(0,T;\mathbf{V}^*)})\\ &\quad + \frac{2}{c_1}\|F'\|^2_{L^2(0,T;\mathbf{V}^*)}\Big)e^{2c_2T}.\end{aligned} \tag{6.79}$$

Using (6.71) and (6.79), we have from (6.75)

$$\begin{aligned}\|w_m(t)\|_{\mathbf{V}} &\le \frac{1}{c_1}(\|F\|_{L^\infty(0,T;\mathbf{V}^*)} + (\alpha_1\|w'_m(t)\| + c_2\alpha_1\|w_m(t)\|))\\ &\le \frac{1}{c_1}\Big(\|F\|_{L^\infty(0,T;\mathbf{V}^*)} + (\alpha_1\|Aw_0 + B(w_0, w_0) - F(0)\| + \alpha_1\sqrt{\frac{2}{c_1}}\|F'\|_{L^2(0,T;\mathbf{V}^*)})e^{c_2T}\\ &\quad + (c_2\alpha_1\sqrt{C_1} + \alpha_1\sqrt{\frac{2c_4^2C_2}{c_1}}e^{c_2T})(\|w_0\| + \|F\|_{L^2(0,T;\mathbf{V}^*)})\Big).\end{aligned} \tag{6.80}$$

Putting

$$\beta \equiv \frac{1}{c_1}\Big(\|F\|_{L^\infty(0,T;\mathbf{V}^*)} + \big(\alpha_1\|Aw_0 + B(w_0,w_0) - F(0)\| + \alpha_1\sqrt{\frac{2}{c_1}}\|F'\|_{L^2(0,T;\mathbf{V}^*)}\big)e^{c_2T}$$
$$+ \big(c_2\alpha_1\sqrt{C_1} + \alpha_1\sqrt{\frac{2c_4^2C_2}{c_1}e^{c_2T}}\big)(\|w_0\| + \|F\|_{L^2(0,T;\mathbf{V}^*)})\Big), \tag{6.81}$$

we have from (6.80)

$$\|w_m(t)\|_{\mathbf{V}} \le \beta \quad \forall t \in [0, t_m). \tag{6.82}$$

By the assumption of theorem, we can assume that β is small so that

$$c_1 - 2c_5\sqrt{\alpha_1}\beta \ge \frac{c_1}{2}. \tag{6.83}$$

Now, let us prove that if $\|w(0)\|_{\mathbf{V}} \le \frac{c_1}{4c_5\sqrt{\alpha_1}}$ and β satisfies (6.83), then for any m

$$c_1 - 2c_5\sqrt{\alpha_1}\|w_m(t)\|_{\mathbf{V}} > 0 \quad \forall t \in [0, T]. \tag{6.84}$$

To this end, let us assume that there exists an m and $\bar{t}_m(\le T)$ such that

$$\begin{aligned} &c_1 - 2c_5\sqrt{\alpha_1}\|w_m(t)\|_{\mathbf{V}} > 0 \quad \forall t \in [0, \bar{t}_m), \\ &c_1 - 2c_5\sqrt{\alpha_1}\|w_m(\bar{t}_m)\|_{\mathbf{V}} = 0. \end{aligned} \tag{6.85}$$

Then, by (6.82) and (6.83), we have

$$c_1 - 2c_5\sqrt{\alpha_1}\|w_m(t)\|_{\mathbf{V}} > \frac{c_1}{2} \quad \forall t \in [0, \bar{t}_m), \tag{6.86}$$

which is a contradiction to the second formula of (6.85), and so we get (6.84). From (6.83) we have (6.82) with $t_m = T$.

Therefore, by (6.79), (6.83), (6.77) and (6.86) we have that

$$\begin{aligned} &\|w_m(t)\|_{\mathbf{V}} \le \text{const} \quad \forall t \in [0, T], \forall m, \forall \varepsilon > 0, \\ &\|w_m'(t)\| \le \text{const} \quad \forall t \in [0, T], \forall m, \forall \varepsilon > 0, \\ &\|w_m'\|_{L^2(0,T;\mathbf{V})} \le \text{const} \quad \forall m, \forall \varepsilon > 0. \end{aligned} \tag{6.87}$$

Extracting a subsequence and passing to limit in (6.70) as $m \to \infty$, and then $\varepsilon \to 0$ and using (6.27), (6.45), in a similar way as in the proof of Theorem 6.6 (from (6.41)) we can prove the existence of a solution w to (6.51) such that $w' \in L^2(0, T; \mathbf{V}) \cap L^\infty(0, T; H)$, and so $w \in C([0, T]; \mathbf{V})$.

By the same argument as in Theorem 6.7 we get the estimates for solutions.

It only remains to prove the uniqueness of solution to the problem.

Let w_1, w_2 be two solutions asserted in Theorem 6.7. Then, as (6.67) we have from (6.51)

$$
\begin{aligned}
&\frac{1}{2}\frac{\partial\|w_1(t)-w_2(t)\|^2}{\partial t}+c_1\|w_1(t)-w_2(t)\|_{\mathbf{V}}^2\\
&\quad\le c_2\|w_1(t)-w_2(t)\|^2+\|w_2(t)\|_{\mathbf{V}}\|w_1(t)-w_2\|_{\mathbf{V}}^{\frac{3}{2}}\|w_1(t)-w_2\|^{\frac{1}{2}}\\
&\quad\le c_2\|w_1(t)-w_2(t)\|^2+c_1\|w_1(t)-w_2\|_{\mathbf{V}}^2+K\|w_2(t)\|_{\mathbf{V}}^4\|w_1(t)-w_2\|^2.
\end{aligned}
\tag{6.88}
$$

Equation (6.88) implies

$$
\|w_1(t)-w_2(t)\|^2\le \text{const}\int_0^t(c_2+K\|w_2(s)\|_{\mathbf{V}}^4)\|w_1(s)-w_2(s)\|^2\,ds,
$$

which by Gronwall's inequality deduces $w_1(t)=w_2(t)$ for all $t\in[0,T]$. □

Remark 6.7 Following Theorem 3.7 of [1], to study existence of a unique solution to a variational inequality under smallness condition of data in Theorem 2 of [2] the author assumed that $F\in L^\infty(0,T;\mathbf{V}^*)\cap L^1(0,T;H),\ F'\in L^1(0,T;H)$.

The right-hand side of (6.15) for the definition of F_2 includes some boundary integrals:

$$
\begin{aligned}
&-2\mu(k(x)U,u)_{\Gamma_2}-2\mu(S\tilde{U},\tilde{u})_{\Gamma_3}-2(\alpha(x)U_\tau,u)_{\Gamma_5}\\
&\qquad-\mu(k(x)U,u)_{\Gamma_7}+\sum_{i=2,4,7}\langle\phi_i,u_n\rangle_{\Gamma_i}+\sum_{i=3,5,6}\langle\phi_i,u\rangle_{\Gamma_i}\quad\forall u\in\mathbf{V}.
\end{aligned}
$$

For fixed t let us assume that $\phi_i(t)\in H^{\frac{1}{2}}(\Gamma_i),\ i=2,4,7$. Due to $u_\tau|_{\Gamma_i}=0$,

$$
u\cdot n\in H^{-\frac{1}{2}}(\Gamma_i),\quad \|u\cdot n\|_{H^{-\frac{1}{2}}(\Gamma_i)}\le K\|u\|_H
$$

(see Proposition 2.1), and so

$$
|\langle\phi_i,u_n\rangle_{\Gamma_i}|\le\|\phi_i\|_{H^{\frac{1}{2}}(\Gamma_i)}\|u_n\|_{H^{-\frac{1}{2}}(\Gamma_i)}\le K\|\phi_i\|_{H^{\frac{1}{2}}(\Gamma_i)}\|u\|_H\le K_1\|u\|_H.
$$

Then, there exist elements $f_i\in H$ such that $\langle\phi_i,u_n\rangle_{\Gamma_i}=(f_i,u)$. Thus, for $\phi_i\in L^2(0,T;H^{\frac{1}{2}}(\Gamma_i))$, $i=2,4,7$, there exist $f_i\in L^2(0,T;H)$ such that

$$
\sum_{i=2,4,7}\langle\phi_i,u_n\rangle_{\Gamma_i}=\sum_{i=2,4,7}(f_i,u)_H\quad\forall u\in\mathbf{V}.
$$

But, for other boundary integrals we can not find such functions belonging to $L^2(0,T;H)$.

Therefore, the assumption such that $F \in L^1(0, T; H)$, $F' \in L^1(0, T; H)$ is not suitable for our problems with mixed boundary conditions, and in Theorems 6.8 and 6.9 we assume that $F \in L^2(0, T; \mathbf{V}^*)$, $F' \in L^2(0, T; \mathbf{V}^*)$.

The following theorem is for small data in (6.51) corresponding to Problem II (the case of static pressure).

Theorem 6.9 *Assume that*

(1) For any $u, v \in \mathbf{V}$, $\langle A(t)u, v\rangle$, $\langle A'(t)u, v\rangle$ *are continuous with respect to* t *and*

$$\begin{aligned}
&\langle A(t)u, u\rangle \geq c_1\|u\|^2_{\mathbf{V}} - c_2\|u\|^2 \quad \exists c_1 > 0,\ c_2 \geq 0,\ \forall u \in \mathbf{V},\\
&|\langle A(t)u, v\rangle| \leq c_3\|u\|_{\mathbf{V}}\|v\|_{\mathbf{V}} \quad \exists c_3 > 0,\ \forall u, v \in \mathbf{V},\\
&|\langle A'(t)u, v\rangle| \leq c_1\alpha_2\|u\|_{\mathbf{V}}\|v\|_{\mathbf{V}} \quad \exists \alpha_2 (0 \leq \alpha_2 < 1),\ \forall u, v \in \mathbf{V};
\end{aligned}$$

(2) $B(u, v)$ *is bilinear continuous from* $\mathbf{V} \times \mathbf{V}$ *into* $\mathbf{V}^*$ *and*

$$\langle B(w, v), u\rangle \leq \frac{c_5}{\sqrt{\alpha_1}}\|w\|_V\|v\|_V\|u\|^{\frac{1}{2}}\|u\|^{\frac{1}{2}}_{\mathbf{V}};$$

(3) $F, F' \in L^2(0, T; \mathbf{V}^*)$;
(4) The functional $\Phi : \mathbf{V} \to \overline{\mathbb{R}}$ *is proper lower weak semi-continuous and*

$$\Phi(u) \geq 0 \quad \forall u \in \mathbf{V}, \quad \Phi(0_{\mathbf{V}}) = 0;$$

(5) $w_0 \in \mathbf{V}$, $\Phi(w_0) = 0$ *and* $(A(0)w_0 + B(w_0, w_0) - F(0)) \in H$ *(a compatibility condition at the initial time).*

If $\|w_0\|_{\mathbf{V}}$, $\|(A(0)w_0 + B(w_0, w_0) - F(0))\|_H$ *and norms of* F, F' *in the spaces that they belong to are small enough, then there exists a unique solution* v *to problem* (6.51) *such that*

$$w \in C([0, T]; \mathbf{V}), \quad w' \in L^2(0, T; \mathbf{V}) \cap L^\infty(0, T; H). \tag{6.89}$$

Proof First, let us prove the existence of the asserted solution.

Let $\{v_j, j = 1, 2, \cdots\}$ be a basis of $\mathbf{V}$. Without loss of generality, we assume that $v_1 = w_0$. We find a solution $w_m = \sum_{i=1}^m g_{im}(t)v_i$ to problem

$$\begin{cases}
\langle w_m'(t), v_j\rangle + \langle A(t)w_m(t), v_j\rangle + \langle B(w_m, w_m), v_j\rangle + \langle \nabla\Phi_\varepsilon(w_m), v_j\rangle = \langle F, v_j\rangle,\\
w_m(0) = w_0,
\end{cases} \tag{6.90}$$

which gives us an ordinary differential system for $g_{im}(t)$, $i = 1, \cdots, m$. In the same way as in Theorem 6.7 we see that for a $\tilde{t}_m > 0$ there exists a solution $g_{im}(t)$ on the interval $[0, \tilde{t}_m]$ and $g_{im}'(t)$ is absolute continuous. If $\|w_m(t)\|$ is bounded, then $g_{im}(t)$

can be extended over $\tilde{t}_m$. Under smallness of data we will find estimate for $\|w_m(t)\|$ below, by which we can say that $\tilde{t}_m = T$.

Multiplying (6.90) by $g_{im}(t)$ and adding for $i = 1, \cdots, m$, yield

$$\begin{cases} \left\langle w'_m(t), w_m(t)\right\rangle + \langle A(t)w_m(t), w_m(t)\rangle + \langle B(w_m, w_m), w_m(t)\rangle + \langle \nabla\Phi_\varepsilon(w_m), w_m(t)\rangle \\ \qquad\qquad = \langle F, w_m(t)\rangle, \\ w_m(0) = w_0. \end{cases} \tag{6.91}$$

We will find a priori estimates for $\|w_m(t)\|^2 + \|w'_m(t)\|^2$.
By virtue of (6.28), we have from (6.91)

$$\frac{d}{dt}\|w_m(t)\|^2 + 2c_1\|w_m(t)\|^2_{\mathbf{V}} - 2c_5\|w_m(t)\|^3_{\mathbf{V}} \\ \leq \frac{1}{c_1(1-\alpha_2)}\|F\|^2_{\mathbf{V}^*} + c_1(1-\alpha_2)\|w_m(t)\|^2_{\mathbf{V}} + 2c_2\|w_m(t)\|^2,$$

and so

$$\frac{d}{dt}\|w_m(t)\|^2 + \left(c_1 - 2c_5\|w_m(t)\|_{\mathbf{V}}\right)\|w_m(t)\|^2_{\mathbf{V}} + c_1\alpha_2\|w_m(t)\|^2_{\mathbf{V}} \\ \leq \frac{1}{c_1(1-\alpha_2)}\|F_2\|^2_{\mathbf{V}^*} + 2c_2\|w_m(t)\|^2. \tag{6.92}$$

Setting $t = 0$ in (6.90), multiplying the resulting equation by $g'_{im}(0)$ and adding for $i = 1, \cdots, m$, we get

$$\|w'_m(0)\|^2 + \left\langle A(0)w_m(0), w'_m(0)\right\rangle + \left\langle B(w_m(0), w_m(0)), w'_m(0)\right\rangle + \left\langle \nabla\Phi_\varepsilon(w_0), w'_m(0)\right\rangle \\ = \left\langle F(0), w'_m(0)\right\rangle. \tag{6.93}$$

By conditions (4), (5) we have that for any $u \in \mathbf{V}$, $\Phi(u) \geq \Phi(w_0)$, and so $\nabla\Phi_\varepsilon(w_0) = 0$. Thus, we have from (6.91)

$$\|w'_m(0)\| \leq \|A(0)w_0 + B(w_0, w_0) - F(0)\|. \tag{6.94}$$

On the other hand, taking into account (6.28), we have from (6.91)

$$c_1\|w_m(t)\|^2_{\mathbf{V}} \leq \|F(t)\|_{\mathbf{V}^*}\|w_m(t)\|_{\mathbf{V}} + c_5\|w_m(t)\|^3_{\mathbf{V}} + c_2\|w_m(t)\|^2 - (w_m(t), w'_m(t)),$$

and so

$$c_1\|w_m(t)\|_{\mathbf{V}} \leq \|F(t)\|_{\mathbf{V}^*} + c_5\|w_m(t)\|^2_{\mathbf{V}} + (\alpha_1\|w'_m(t)\| + c_2\alpha_1\|w_m(t)\|), \tag{6.95}$$

where α_1 is in $\|\cdot\| \leq \alpha_1\|\cdot\|_{\mathbf{V}}$.

Differentiating (6.90) yields

$$\begin{aligned}\big\langle w_m''(t), v_j\big\rangle + \big\langle A'(t)w_m(t), v_j\big\rangle + \big\langle A(t)w_m'(t), v_j\big\rangle + \big\langle B(w_m'(t), w_m(t)), v_j\big\rangle \\ + \big\langle B(w_m(t), w_m'(t)), v_j\big\rangle + \big\langle (\nabla\Phi_\varepsilon(w_m))', v_j\big\rangle = \big\langle F', v_j\big\rangle.\end{aligned} \tag{6.96}$$

Multiplying (6.96) by $g_{im}'(t)$, summing for i and observing the fact that $\langle(\nabla\Phi_\varepsilon(w_m))', w_m'\rangle \geq 0$ (see (1.38)), by conditions (2), (3) we have

$$\begin{aligned}&\frac{d}{dt}\|w_m'(t)\|^2 + (c_1 - 4c_5\|w_m(t)\|_{\mathbf{V}})\|w_m'(t)\|_{\mathbf{V}}^2 + c_1\|w_m'(t)\|_{\mathbf{V}}^2 \\ &\qquad - c_1\alpha_2\|w_m'(t)\|_{\mathbf{V}}^2 - c_1\alpha_2\|w_m(t)\|_{\mathbf{V}}^2 \\ &\qquad \leq c_1(1-\alpha_2)\|w_m'(t)\|_{\mathbf{V}}^2 + \frac{1}{c_1(1-\alpha_2)}\|F'(t)\|_{\mathbf{V}^*}^2 + 2c_2\|w_m'(t)\|^2,\end{aligned}$$

where $|\big\langle B(w_m', w_m), w_m'\big\rangle + \big\langle B(w_m, w_m'), w_m'\big\rangle| \leq 2c_5\|w_m\|_{\mathbf{V}}\|w_m'\|_{\mathbf{V}}^2$ was used. Thus,

$$\begin{aligned}&\frac{d}{dt}\|w_m'(t)\|^2 + (c_1 - 4c_5\|w_m(t)\|_{\mathbf{V}})\|w_m'(t)\|_{\mathbf{V}}^2 - c_1\alpha_2\|w_m(t)\|_{\mathbf{V}}^2 \\ &\qquad \leq \frac{1}{c_1(1-\alpha_2)}\|F'(t)\|_{\mathbf{V}^*}^2 + 2c_2\|w_m'(t)\|^2.\end{aligned} \tag{6.97}$$

Adding (6.92) and (6.97) yields

$$\begin{aligned}&\frac{d}{dt}\left(\|w_m(t)\|^2 + \|w_m'(t)\|^2\right) + (c_1 - 4c_5\|w_m(t)\|_{\mathbf{V}})\left(\|w_m(t)\|_{\mathbf{V}}^2 + \|w_m'(t)\|_{\mathbf{V}}^2\right) \\ &\qquad \leq \frac{1}{c_1(1-\alpha_2)}\left(\|F(t)\|_{\mathbf{V}^*}^2 + \|F'(t)\|_{\mathbf{V}^*}^2\right) + 2c_2\left(\|w_m(t)\|^2 + \|w_m'(t)\|^2\right).\end{aligned} \tag{6.98}$$

If $\|w_m(0)\|_{\mathbf{V}} = \|w_0\|_{\mathbf{V}} < \frac{c_1}{4c_5}$, then there exists a t_m such that $c_1 - 4c_5\|w_m(t)\|_{\mathbf{V}} \geq 0$ for $t \in [0, t_m]$. Therefore, we from (6.94) and (6.98) have

$$\begin{aligned}\left(\|w_m(t)\|^2 + \|w_m'(t)\|^2\right) &\leq \left(\alpha_1\|w_0\|^2 + \|Aw_0 + B(w_0, w_0) - F(0)\|^2\right) \\ &\quad + \frac{1}{c_1(1-\alpha_2)}\int_0^T \left(\|F\|_{\mathbf{V}^*}^2 + \|F'\|_{\mathbf{V}^*}^2\right) dt \\ &\quad + 2c_2\int_0^t \left(\|w_m(s)\|^2 + \|w_m'(s)\|^2\right) ds\end{aligned}$$

for $t \in [0, t_m]$, from which by Gronwall's inequality we get

$$\begin{aligned}\left(\|w_m(t)\|^2 + \|w_m'(t)\|^2\right) \leq &\Big(\alpha_1\|w_0\|_{\mathbf{V}}^2 + \|Aw_0 + B(w_0, w_0) - F(0)\|^2 \\ &+ \frac{1}{c_1(1-\alpha_2)}\left(\|F\|_{L^2(0,T;\mathbf{V}^*)}^2 + \|F'\|_{L^2(0,T;\mathbf{V}^*)}^2\right)\Big)e^{2c_2T}.\end{aligned} \tag{6.99}$$

Using the estimate, we will obtain a quadratic inequality satisfied by $\|w_m(t)\|_{\mathbf{V}}$. Put

$$\beta \equiv \|F\|_{L^\infty(0,T;\mathbf{V}^*)} + \max\{\alpha_1, c_2\alpha_1\}\sqrt{2}\Big(\alpha_1\|w_0\|_{\mathbf{V}}^2 + \|Aw_0 + B(w_0,w_0) - F(0)\|^2 + \frac{1}{c_1(1-\alpha_2)}\big(\|F\|^2_{L^2(0,T;\mathbf{V}^*)} + \|F'\|^2_{L^2(0,T;\mathbf{V}^*)}\big)\Big)^{\frac{1}{2}} e^{c_2T}. \tag{6.100}$$

Taking into account

$$\left(\|w_m(t)\| + \|w_m'(t)\|\right) \le \sqrt{2}\left(\|w_m(t)\|^2 + \|w_m'(t)\|^2\right)^{\frac{1}{2}},$$

we have from (6.99)

$$\left(\|w_m(t)\| + \|w_m'(t)\|\right) \le \sqrt{2}\Big(\alpha_1\|w_0\|_{\mathbf{V}}^2 + \|Aw_0 + B(w_0,w_0) - F(0)\|^2 + \frac{1}{c_1(1-\alpha_2)}\big(\|F\|^2_{L^2(0,T;\mathbf{V}^*)} + \|F'\|^2_{L^2(0,T;\mathbf{V}^*)}\big)\Big)^{1/2} e^{c_2T}. \tag{6.101}$$

Taking into account (6.101) and (6.100), we have from (6.95) a quadratic inequality for $\|w_m(t)\|_{\mathbf{V}}$

$$0 \le \beta - c_1\|w_m(t)\|_{\mathbf{V}} + c_5\|w_m(t)\|^2_{\mathbf{V}} \quad \forall t \in [0, t_m], \tag{6.102}$$

which is the one we want.

By the assumption of theorem, we can assume that β is small so that

$$c_1^2 - 4c_5\beta > \frac{4c_1^2}{9}. \tag{6.103}$$

Now, let us prove that if $\|w(0)\|_{\mathbf{V}} \le \frac{c_1-\sqrt{c_1^2-4c_5\beta}}{2c_5} (< \frac{c_1}{6c_5})$, then for any m, ε

$$c_1 - 4c_5\|w_m(t)\|_{\mathbf{V}} > \frac{c_1}{4} \quad \forall t \in [0, T]. \tag{6.104}$$

Taking into account

$$c_1 - 4c_5\|w_m(0)\|_{\mathbf{V}} > \frac{c_1}{3},$$

let us assume that there exists an m and $\bar{t}_m (\le T)$ such that

$$c_1 - 4c_5\|w_m(t)\|_{\mathbf{V}} > \frac{c_1}{4} \quad \forall t \in [0, \bar{t}_m) \quad \text{and} \quad c_1 - 4c_5\|w_m(\bar{t}_m)\|_{\mathbf{V}} = \frac{c_1}{4}. \tag{6.105}$$

Setting $y = \|w_m(t)\|_{\mathbf{V}}$ in (6.102), we get

$$0 \le \beta - c_1 y + c_5 y^2 \quad \forall t \in [0, \bar{t}_m]. \tag{6.106}$$

By virtue of (6.103), there exist two real roots of quadratic polynomial $\beta - c_1 y + c_5 y^2$

$$y_1 = \frac{c_1 - \sqrt{c_1^2 - 4c_5\beta}}{2c_5} \quad \text{and} \quad y_2 = \frac{c_1 + \sqrt{c_1^2 - 4c_5\beta}}{2c_5},$$

and on the intervals $[0, y_1]$ and $[y_2, +\infty)$ (6.106) holds. Thus, by the continuity of $\|w_m(t)\|_{\mathbf{V}}$ with respect to t we have from $\|w(0)\|_{\mathbf{V}} \in [0, y_1]$ that $\|w_m(t)\|_{\mathbf{V}} \in [0, y_1]\ \forall t \in [0, \bar{t}_m)$, that is,

$$\|w_m(t)\|_{\mathbf{V}} \le \frac{c_1 - \sqrt{c_1^2 - 4c_5\beta}}{2c_5} < \frac{c_1}{6c_5} \quad \forall t \in [0, \bar{t}_m].$$

Thus,

$$c_1 - 4c_5\|w_m(t)\|_{\mathbf{V}} > \frac{c_1}{3} \quad \forall t \in [0, \bar{t}_m], \tag{6.107}$$

which is a contradiction to (6.105), and so we get (6.104).

Therefore, by (6.98), (6.99) and (6.104) we have

$$\begin{aligned}
&\|w_m(t)\| \le \text{const} \quad \forall t \in [0, T],\ \forall m,\ \forall \varepsilon > 0;\\
&\|w_m\|_{L^2(0,T;\mathbf{V})} \le \text{const} \quad \forall m,\ \forall \varepsilon > 0;\\
&\|w_m'(t)\| \le \text{const} \quad \forall t \in [0, T],\ \forall m,\ \forall \varepsilon > 0;\\
&\|w_m'\|_{L^2(0,T;\mathbf{V})} \le \text{const} \quad \forall m,\ \forall \varepsilon > 0.
\end{aligned} \tag{6.108}$$

Extracting a subsequence, passing to limit in (6.41) as $m \to \infty$ and then letting $\varepsilon \to 0$ and using (6.27) and (6.45), by a way similar to the proof of Theorem 6.6 we can prove the existence of a solution w to (6.51) such that $w' \in L^2(0, T; \mathbf{V}) \cap L^\infty(0, T; H)$, and so $w \in C([0, T]; \mathbf{V})$.

As in Theorem 6.7 we can prove uniqueness. □

Remark 6.8 In Sect. 4.2 the non-steady Navier-Stokes equations with boundary condition (5.2) with $\Gamma_i = \emptyset, i = 8, 9, 10, 11$, is considered. For the proof of existence of a unique solution, the following property of the linear problem obtained by neglecting the nonlinear term $B(w, w)$ from

$$\begin{cases} w'(t) + Aw + B(w, w) = f(t),\\ u(0) = \varphi, \end{cases}$$

was essential.

Property: (Lemma 4.6) When $\mathscr{X} = \{w \in L_2(0, T; \mathbf{V});\ w' \in L_2(0, T; \mathbf{V}),\ w'' \in L_2(0, T; \mathbf{V}^*)\}$ and $\mathscr{Y} = \{w \in L_2(0, T; \mathbf{V}^*);\ w' \in L_2(0, T; \mathbf{V}^*)\}$, a map $u \to \{u'(0), Lu \equiv u' + Au\}$ is linear continuous one-to-one from $\mathscr{X}$ onto $H \times \mathscr{Y}$.

But, in the case of Problem II in this chapter, owing to the boundary conditions on $\Gamma_i(\neq \emptyset), i = 8, 9, 10, 11$, neglecting the nonlinear term $B(w, w)$, we only get a nonlinear problem

$$\begin{cases} \left\langle \dfrac{\partial w}{\partial t}, u - w \right\rangle + \langle Aw, u - w\rangle + \Phi(u) - \Phi(w) \geq \langle f, u - w\rangle \\ \qquad\qquad\qquad\qquad \text{for a.e. } t \in (0, T) \ \forall u \in \mathbf{V}, \\ w(0) = w_0 \in \mathbf{V}, \end{cases}$$

which seems not to have the property above (cf. Proposition 6.1), and so we can not use the method in Sect. 4.2.

6.3 Solutions to the Non-steady Navier-Stokes Problems

6.3.1 Existence of a Solution: The Case of Total Pressure

Let us study Problem I-VI.

Theorem 6.10 *Under Assumption 6.1 there exists a solution to Problem I-VI (the non-steady Navier-Stokes problem for the case of total pressure) for any f, ϕ_i, $i = 2 \sim 7$ and initial function v_0 such that $v_0 - U(0) \in H$. Moreover the following estimate for the solution holds:*

$$\|v(t) - U(t)\| \leq K \quad \forall t \in [0, T], \quad \|v - U\|_{L^2(0,T;V)} \leq K, \tag{6.109}$$

where

$$\begin{aligned} K =&\Big(\|v_0 - U(0)\| + \Big\|\frac{\partial U}{\partial t}\Big\|_{L^2(0,T;\mathbf{V}^*)} + \|U\|_{L^2(0,T;\mathbf{H}^1)} + \|U\|_{L^2(0,T;\mathbf{H}^1)}\|U\|_{L^\infty(0,T;\mathbf{H}^1)} \\ &+ \|f\|_{L^2(0,T;\mathbf{V}^*)} + \sum_{i=2,4,7} \|\phi_i\|_{L^2(0,T;H^{-\frac{1}{2}}(\Gamma_i))} + \sum_{i=3,5,6} \|\phi_i\|_{L^2(0,T;\mathbf{H}^{-\frac{1}{2}}(\Gamma_i))}\Big). \end{aligned}$$

Proof By Assumption 6.1 we can prove that $F_1 \in L^2(0, T; \mathbf{V}^*)$ and $F_1' \in L^2(0, T; \mathbf{V}^*)$.

Define an operator $A_1(t) : \mathbf{V} \to \mathbf{V}^*$ by

$$\langle A_1(t)v, u\rangle = a_{01}(t, v, u) \quad \forall v, u \in \mathbf{V}, \tag{6.110}$$

where $a_{01}(t, v, u)$ is the one in (6.4). Then, $\langle A_1(t)v, u\rangle$ is continuous with respect to t.

By Korn's inequality,

$$2\Sigma_{ij}\|\varepsilon_{ij}(z)\|^2 \geq C_1\|z\|^2_{\mathbf{H}^1(\Omega)}, \quad C_1 > 0. \tag{6.111}$$

Since $\langle \text{rot}\, U(t) \times w, w\rangle = 0$, by Young's inequality and Assumption 6.1, we have

$$|\langle \text{rot}\, U(t) \times w, w\rangle + \langle \text{rot}\, w \times U(t), w\rangle| \leq \frac{\mu C_1}{4}\|w\|^2_{\mathbf{H}^1(\Omega)} + C_2\|w\|^2, \quad C_2 > 0 \tag{6.112}$$

for any $t \in [0, T]$. On the other hand, by virtue of Remark 3.4 and Assumption 6.1 there exists a constant M such that

$$\|S(x)\|_\infty,\ \|k(x)\|_\infty,\ \|\alpha\|_{L_\infty(\Gamma_5)} \leq M.$$

Therefore, there exists C_3 such that

$$\left|\mu(k(x)z, z)_{\Gamma_2} + 2\mu(S\tilde{z}, \tilde{z})_{\Gamma_3} + (\alpha(x)z, z)_{\Gamma_5} + \mu(k(x)z, z)_{\Gamma_7}\right| \leq \frac{\mu C_1}{4}\|z\|^2_{\mathbf{H}^1(\Omega)} + C_3\|z\|^2 \quad \forall z \in \mathbf{V} \tag{6.113}$$

(see Theorem 1.27). Setting $c_2 = C_2 + C_3$, we have therefore from (6.110)–(6.113)

$$\langle A_1(t)z, z\rangle \geq c_1\|z\|^2_{\mathbf{V}} - c_2\|z\|^2, \quad c_1 = \frac{\mu C_1}{2} > 0, \quad \forall z \in \mathbf{V}. \tag{6.114}$$

By Hölder's inequality and Assumption 6.1, we have

$$|\langle \text{rot}\, U(t) \times u, v\rangle + \langle \text{rot}\, u \times U(t), v\rangle| \leq K\|u\|_{\mathbf{V}}\|v\|_{\mathbf{V}} \quad \forall u, v \in \mathbf{V} \tag{6.115}$$

for any $t \in [0, T]$. Also, we can prove that

$$|\langle A_1(t)u, v\rangle| \leq K\|u\|_{\mathbf{V}}\|v\|_{\mathbf{V}} \quad \forall u, v \in \mathbf{V}. \tag{6.116}$$

Let us prove that

$$\liminf_{m\to\infty}\langle A_1(t)u_m, u_m\rangle \geq \langle A_1(t)u, u\rangle \quad \text{as } u_m \rightharpoonup u \text{ in } \mathbf{V}. \tag{6.117}$$

By (6.4) we have

$$\begin{aligned}\langle A_1(t)u_m, u_m\rangle &\equiv a_{01}(t, u_m, u_m)\\ &= 2\mu(\mathcal{E}(u_m), \mathcal{E}(u_m)) + \langle \text{rot}\, U(t) \times u_m, u_m\rangle + \langle \text{rot}\, u_m \times U(t), u_m\rangle\\ &\quad + 2\mu(k(x)u_m, u_m)_{\Gamma_2} + 2\mu(S\tilde{u}_m, \tilde{u}_m)_{\Gamma_3} + 2(\alpha(x)u_m, u_m)_{\Gamma_5} + \mu(k(x)u_m, u_m)_{\Gamma_7}.\end{aligned} \tag{6.118}$$

Taking into account

$$\liminf_{m\to\infty}(\mathcal{E}(u_m), \mathcal{E}(u_m)) \geq (\mathcal{E}(u), \mathcal{E}(u)) \quad \text{as } u_m \rightharpoonup u \text{ in } \mathbf{V}$$

(see (1.1)), from (6.118) we get (6.117).

Define a bilinear operator $B_1 : \mathbf{V} \times \mathbf{V} \to \mathbf{V}^*$ by

$$\langle B_1(z, v), u\rangle = a_{11}(z, v, u) \quad \forall z, v, u \in \mathbf{V}, \tag{6.119}$$

where $a_{11}(t, v, u)$ is the one in (6.4). Then, by a property of mixed production of vectors,

$$\langle B_1(w, v), v\rangle = \langle \mathrm{rot}\, w \times v, v\rangle = 0. \tag{6.120}$$

Using Hölder's inequality, imbedding theorems and interpolation inequality $\|v\|_{\mathbf{L}^3} \le K\|v\|_{\mathbf{L}^2}^{\frac{1}{2}}\|v\|_{\mathbf{L}^6}^{\frac{1}{2}}$, we have

$$\begin{aligned} |\langle B_1(w, v), z\rangle| &= |\langle \mathrm{rot}\, w \times v, z\rangle| \le K\|\mathrm{rot}\, w\|_{\mathbf{L}^2}\|v\|_{\mathbf{L}^3}\|z\|_{\mathbf{L}^6} \\ &\le K\|w\|_{\mathbf{V}}\|v\|^{\frac{1}{2}}\|v\|_{\mathbf{V}}^{\frac{1}{2}}\|z\|_{\mathbf{V}}. \end{aligned} \tag{6.121}$$

By the definition of the functional Φ, we know $\Phi(u) \ge 0$ and $\Phi(0_{\mathbf{V}}) = 0$.
Let F_1 be the one in (6.4). Then, by Assumption 6.1 we can prove

$$\begin{aligned} \|F_1\|_{L^2(0,T;\mathbf{V}^*)} \le & M_1\Big(\|\frac{\partial U}{\partial t}\|_{L^2(0,T;\mathbf{V}^*)} + \|U\|_{L^2(0,T;\mathbf{H}^1)} + \|U\|_{L^2(0,T;\mathbf{H}^1)}\|U\|_{L^\infty(0,T;\mathbf{H}^1)} \\ & + \|f\|_{L^2(0,T;\mathbf{V}^*)} + \sum_{i=2,4,7}\|\phi_i\|_{L^2(0,T;H^{-\frac{1}{2}}(\Gamma_i))} + \sum_{i=3,5,6}\|\phi_i\|_{L^2(0,T;\mathbf{H}^{-\frac{1}{2}}(\Gamma_i))}\Big). \end{aligned} \tag{6.122}$$

By virtue of (6.114), (6.116) and (6.120)–(6.122), we can verify that all conditions of Theorem 6.6 are satisfied, and we come to the conclusion. □

Theorem 6.11 *Assume that*

(1) In addition to Assumption 6.1, $\frac{\partial}{\partial t}U \in W^{1,2}(0, T; \mathbf{H}^1(\Omega))$*;*
(2) $f, f' \in L^2(0, T; \mathbf{V}^*)$, $\phi_i \in W^{1,2}(0, T; H^{-\frac{1}{2}}(\Gamma_i)), i = 2, 4, 7,$ *and* $\phi_i \in W^{1,2}(0, T; \mathbf{H}^{-\frac{1}{2}}(\Gamma_i)), i = 3, 5, 6$*;*
(3) $w_0 \equiv v_0 - U \in \mathbf{V}$, $w_0|_{\cup_{i=8-11}\Gamma_i} = 0$ *and*

$$(A_1(0)w_0 + B_1(w_0, w_0) - F_1(0)) \in H, \tag{6.123}$$

where F_1 *is the one in* (6.4) *and* A_1, B_1 *are in* (6.110), (6.119), *respectively.*
If $l = 2$*, then there exists a unique solution to Problem I-VI' satisfying* (6.109) *and*

$$v - U \equiv w \in C([0, T]; \mathbf{V}), \quad w' \in L^2(0, T; \mathbf{V}) \cap L^\infty(0, T; H). \tag{6.124}$$

In the case of $l = 3$*, if* $\|w_0\|_V$, $\|A_1w_0 + B_1(w_0, w_0) - F_1(0)\|$, $\|U\|_{W^{1,2}(0,T;\mathbf{H}^1)}$, $\|\frac{\partial^2}{\partial t^2}U\|_{L^2(0,T;(\mathbf{H}^1)^*)}$ *and norms of* f, f', ϕ_i *in the spaces of condition (2) are small* *enough, then there exists a unique solution* v *to Problem I-VI' satisfying* (6.124) *and* (6.109).

Proof We know that the first two of condition (6.52) of Theorem 6.7 are satisfied (see (6.114), (6.116)). By (6.110), we know that

$$\left\langle A_1'(t)v, u\right\rangle = \frac{\partial a_{01}(t, v, u)}{\partial t} = \left(\operatorname{rot}\frac{\partial U(t)}{\partial t} \times v, u\right) + \left(\operatorname{rot} v \times \frac{\partial U(t)}{\partial t}, u\right),$$

which show that $\langle A_1'(t)v, u\rangle$ is Lipschitz continuous with respect to t. Also,

$$|\langle A_1'(t)v, u\rangle| \leq \left\|\operatorname{rot}\frac{\partial U(t)}{\partial t}\right\|_{\mathbf{L}^2}\|v\|_{\mathbf{V}}\|u\|_{\mathbf{V}} + \|\operatorname{rot} v\|_{\mathbf{L}^2}\left\|\frac{\partial U(t)}{\partial t}\right\|_{\mathbf{H}^1}\|u\|_{\mathbf{V}}$$

and the third one of condition (6.52) of Theorem 6.7 is satisfied since

$$\left\|\frac{\partial}{\partial t}U(t)\right\|_{C([0,T];\mathbf{H}^1)} \leq K\left\|\frac{\partial}{\partial t}U(t)\right\|_{W^{1,2}(0,T;\mathbf{H}^1)}.$$

By Hölder's inequality and interpolation inequality, we have

$$\begin{aligned} |\langle B_1(w, v), z\rangle| &= |\langle \operatorname{rot} w \times v, z\rangle| \leq K\|\operatorname{rot} w\|_{\mathbf{L}^2}\|v\|_{\mathbf{L}^4}\|z\|_{\mathbf{L}^4} \\ &\leq K\|w\|_{\mathbf{V}}\|v\|_{\mathbf{L}^4}\|z\|^{\frac{1}{3}}\|z\|^{\frac{2}{3}}_{\mathbf{V}} \end{aligned} \tag{6.125}$$

for $l = 2$ and

$$\begin{aligned} |\langle B_1(w, v), z\rangle| &= |\langle \operatorname{rot} w \times v, z\rangle| \leq K\|\operatorname{rot} w\|_{\mathbf{L}^2}\|v\|_{\mathbf{L}^6}\|z\|_{\mathbf{L}^3} \\ &\leq K\|w\|_{\mathbf{V}}\|v\|_{\mathbf{V}}\|z\|^{\frac{1}{2}}\|z\|^{\frac{1}{2}}_{\mathbf{V}} \end{aligned} \tag{6.126}$$

for $l = 3$. Since

$$\begin{aligned} \langle F_1, u\rangle = &-\left(\frac{\partial U}{\partial t}, u\right) - 2\mu(\mathcal{E}(U), \mathcal{E}(u)) - \langle \operatorname{rot} U \times U, u\rangle \\ &- 2\mu(k(x)U, u)_{\Gamma_2} - 2\mu(S\tilde{U}, \tilde{u})_{\Gamma_3} - 2(\alpha(x)U_\tau, u)_{\Gamma_5} \\ &- \mu(k(x)U, u)_{\Gamma_7} + \langle f, u\rangle + \sum_{i=2,4,7}\langle \phi_i, u_n\rangle_{\Gamma_i} + \sum_{i=3,5,6}\langle \phi_i, u\rangle_{\Gamma_i}, \end{aligned}$$

by the conditions we have

$$F_1, F_1' \in L^2(0, T; \mathbf{V}^*).$$

It is easy to verify that condition (4) of Theorem 6.7 is satisfied. By the condition $w_0|_{\cup_{i=8-11}\Gamma_i} = 0$ and (5.22), we know $\Phi(w_0) = 0$, which shows that condition (5) of Theorem 6.7 is satisfied.

Thus, in the case of 2-D by Theorem 6.7 we come to the conclusion.

If $\|\frac{\partial U}{\partial t}\|_{L^2(0,T;(\mathbf{H}^1)^*)}$, $\|U\|_{L^2(0,T;\mathbf{H}^1)}$ are small and $U \in C(0, T; \mathbf{H}^1)$, then under smallness conditions of the theorem for f, ϕ_i we can know that $\|F_1\|_{L^2(0,T;\mathbf{V}^*)}$ is small. Also if $\|\frac{\partial^2 U}{\partial t^2}\|_{L^2(0,T;(\mathbf{H}^1)^*)}$, $\|\frac{\partial}{\partial t}U\|_{L^2(0,T;\mathbf{H}^1)}$ are small and $U \in C(0, T; \mathbf{H}^1)$, then under smallness conditions of the theorem for f, ϕ_i we can know that $\|F_1'\|_{L^2(0,T;\mathbf{V}^*)}$ is small. Thus, if $\|U\|_{W^{1,2}(0,T;\mathbf{H}^1)}$, $\|\frac{\partial^2}{\partial t^2}U\|_{L^2(0,T;(\mathbf{H}^1)^*)}$ and norms of f, f', ϕ_i in the spaces of condition (2) are small enough, then norms of F_1, F_1' in the space $L^2(0, T; \mathbf{V}^*)$ are small. Therefore, in case of 3-D by Theorem 6.8 we come to the conclusion. □

Remark 6.9 From (6.109) we know that the estimates for solutions are independent of the thresholds of friction conditions $g_\tau, g_n, g_{+n}, g_{-n}$.

6.3.2 Existence of a Unique Solution: The Case of Static Pressure

Let us study now Problem II-VI.

Theorem 6.12 *Assume that*

(1) Assumption 6.1 and $\frac{dU}{dt} \in W^{1,2}(0, T; \mathbf{H}^1(\Omega))$ *hold;*

(2) $f, f' \in L^2(0, T; \mathbf{V}^*)$, $\phi_i \in W^{1,2}(0, T; H^{-\frac{1}{2}}(\Gamma_i)), i = 2, 4, 7,$ *and* $\phi_i \in W^{1,2}(0, T; \mathbf{H}^{-\frac{1}{2}}(\Gamma_i)), i = 3, 5, 6;$

(3) $w_0 \equiv v_0 - U \in \mathbf{V}$ *and* $w_0|_{\cup_{i=8-11}\Gamma_i} = 0$ *and*

$$(A_2(0)w_0 + B_2(w_0, w_0) - F_2(0)) \in H, \tag{6.127}$$

where A_2, B_2 *and* F_2 *are the one, respectively,* (6.17) *and* (6.15).

If $\|w_0\|_{\mathbf{V}}$, $\|(A_2(0)w_0 + B_2(w_0, w_0) - F_2(0))\|$, $\|U\|_{W^{1,2}(0,T;\mathbf{H}^1)}$, $\|U'\|_{W^{1,2}(0,T;\mathbf{H}^1)}$ *and norms of* f, f', ϕ_i *in the spaces that they belong to are small enough, then there exists a unique solution* v *to Problem II-VI such that*

$$v - U \equiv w \in C([0, T]; \mathbf{V}), \quad w' \in L^2(0, T; \mathbf{V}) \cap L^\infty(0, T; H). \tag{6.128}$$

Proof First, let us prove the existence of the desired solution.

The operator $A_2(t) : \mathbf{V} \to \mathbf{V}^*$ was defined in (6.17) by

$$\langle A_2(t)v, u\rangle = a_{02}(t, v, u) \quad \forall v, u \in \mathbf{V}. \tag{6.129}$$

Then, $\langle A_2(t)v, u\rangle$ is continuous with respect to t. Note the operator $A_2(t)$ is not symmetric. By Korn's inequality

$$2\Sigma_{ij}\|\varepsilon_{ij}(z)\|^2 \geq C_1\|z\|^2_{\mathbf{H}^1(\Omega)}, \quad C_1 > 0. \tag{6.130}$$

If $\|U\|_{C([0,T;\mathbf{H}^1)}$ is small enough, then

$$|\langle (U(t)\cdot\nabla)w,u\rangle+\langle (w\cdot\nabla)U(t),u\rangle|\le\frac{\mu C_1}{4}\|w\|_{\mathbf{H}^1(\Omega)}\|u\|_{\mathbf{H}^1(\Omega)}. \tag{6.131}$$

By (6.130), (6.131) and (6.113), we have

$$\langle A_2(t)z,z\rangle\ge c_1\|z\|_{\mathbf{V}}^2-c_2\|z\|^2\quad c_1=\frac{\mu C_1}{2}>0,\,c_2>0\quad\forall z\in\mathbf{V}. \tag{6.132}$$

Using (6.130), we can prove that

$$|\langle A_2(t)u,v\rangle|\le c_3\|u\|_{\mathbf{V}}\|v\|_{\mathbf{V}}\quad\forall u,v\in\mathbf{V}. \tag{6.133}$$

By (6.129), we know

$$\big\langle A_2'(t)w,u\big\rangle=\Big\langle\Big(\frac{\partial U(t)}{\partial t}\cdot\nabla\Big)w,u\Big\rangle+\Big\langle(w\cdot\nabla)\frac{\partial U(t)}{\partial t},u\Big\rangle,$$

which shows that $\langle A_2'(t)w,u\rangle$ is continuous with respect to t.

Also, $\|U'(t)\|_{C([0,T;\mathbf{H}^1)}$ is small enough, then

$$\begin{aligned}|\big\langle A_2'(t)w,u\big\rangle|=&\Big|\Big\langle\Big(\frac{\partial U(t)}{\partial t}\cdot\nabla\Big)w,u\Big\rangle+\Big\langle(w\cdot\nabla)\frac{\partial U(t)}{\partial t},u\Big\rangle\Big|\\ \le&\, c_1\alpha\|w\|_{\mathbf{V}}\|u\|_{\mathbf{V}},\quad\alpha<1.\end{aligned} \tag{6.134}$$

The bilinear operator $B_2:\mathbf{V}\times\mathbf{V}\to\mathbf{V}^*$ was defined in (6.17) by

$$\langle B_2(w,v),u\rangle=a_{12}(z,v,u)\quad\forall w,v,u\in\mathbf{V}. \tag{6.135}$$

Then, by virtue of Hölder's inequality, imbedding theorems and interpolation inequality $\|u\|_{\mathbf{L}^3}\le K\|u\|^{\frac12}\|u\|_{\mathbf{L}^6}^{\frac12}$ we have

$$\begin{aligned}|\langle B_2(w,v),u\rangle|=&|\langle(w\cdot\nabla v),u\rangle|\le\|w\|_{\mathbf{L}^6}\|\nabla v\|_{\mathbf{L}^2}\|u\|_{\mathbf{L}^3}\\ \le&\, K\|w\|_{\mathbf{V}}\|v\|_{\mathbf{V}}\|u\|^{\frac12}\|u\|_{\mathbf{V}}^{\frac12}.\end{aligned} \tag{6.136}$$

Since

$$\begin{aligned}\langle F_2,u\rangle=&-\Big(\frac{\partial U}{\partial t},u\Big)-2\mu(\mathcal{E}(U),\mathcal{E}(u))-\langle\operatorname{rot}U\times U,u\rangle\\ &-2\mu(k(x)U,u)_{\Gamma_2}-2\mu(S\tilde U,\tilde u)_{\Gamma_3}-2(\alpha(x)U,u)_{\Gamma_5}\\ &-\mu(k(x)U,u)_{\Gamma_7}+\langle f,u\rangle+\sum_{i=2,4,7}\langle\phi_i,u_n\rangle_{\Gamma_i}+\sum_{i=3,5,6}\langle\phi_i,u\rangle_{\Gamma_i},\end{aligned}$$

by Conditions (1), (2) we have

$$F_2,\ F_2' \in L^2(0, T; \mathbf{V}^*). \tag{6.137}$$

As in the proof of Theorem 6.11 we know that under conditions of the theorem $\|F_2\|_{L^2(0,T;\mathbf{V}^*)}$ and $\|F_2'\|_{L^2(0,T;\mathbf{V}^*)}$ are also small enough.

As in Theorem 6.11, we can verify that other conditions of Theorem 6.9 are satisfied as well. Therefore, by Theorem 6.9 we come to the asserted conclusion. □

6.3.3 Existence of a Unique Solution: The Stokes Problem

To study Problem III-VI, we use the following known result. To state it we need to introduce a problem first.

Let A be a linear continuous and symmetric operator from $\mathbf{V}$ to $\mathbf{V}^*$ satisfying the coercivity condition

$$\langle Au, u\rangle + \alpha\|u\| \geq \omega\|u\|_{\mathbf{V}}^2 \quad \omega > 0,\ \alpha \in \mathbb{R}.$$

We find $w \in L^2(0, T; \mathbf{V}) \cap C([0, T]; H) \cap W^{1,2}(0, T; \mathbf{V}^*)$ such that

$$\begin{cases} \left\langle \dfrac{\partial w}{\partial t}, u - w\right\rangle + \langle Aw, u - w\rangle + \Phi(u) - \Phi(w) \geq \langle f, u - w\rangle \\ \qquad\qquad\qquad\qquad \text{for a.e. } t \in (0, T)\ \forall u \in \mathbf{V}, \\ w(0) = w_0 \in \mathbf{V}, \end{cases} \tag{6.138}$$

where $f \in L^2(0, T; \mathbf{V}^*)$.

Proposition 6.1 *(Theorem 5.1 of [3]) Let $f \in L^2(0, T; \mathbf{V}^*)$ and $w_0 \in \mathbf{V}$ be such that*

$$\{Aw_0 + \partial\Phi(w_0) - f(0)\} \cap H \neq \emptyset. \tag{6.139}$$

Then, problem (6.138) *has a unique solution $w \in W^{1,2}([0, T]; \mathbf{V}) \cap W^{1,\infty}([0, T]; H)$ and the map $(w_0, f) \to w$ is Lipschitz from $H \times L^2(0, T; \mathbf{V}^*)$ to $C([0, T]; H) \cap L^2(0, T; \mathbf{V})$. If $f \in W^{1,2}([0, T]; \mathbf{V}^*)$ and $\Phi(w_0) < \infty$, then problem* (6.138) *has a unique solution $w \in W^{1,2}([0, T]; H) \cap C_w([0, T]; \mathbf{V})$.*

Theorem 6.13 *Assume that*

(1) $U \in L^2(0, T; \mathbf{H}^1)$, $\frac{\partial U}{\partial t} \in L^2(0, T; (\mathbf{H}^1)^*)$;

(2) $f \in L^2(0, T; \mathbf{V}^*)$, $\phi_i \in L^2(0, T; H^{\frac{1}{2}}(\Gamma_i)), i = 2, 4, 7,$ and $\phi_i \in L_2(0, T; \mathbf{H}^{-\frac{1}{2}}(\Gamma_i)), i = 3, 5, 6$;

(3) $w_0 \equiv v_0 - U \in \mathbf{V}$ and $w_0|_{\cup_{i=8-11}\Gamma_i} = 0$ and

$$(A_3 w_0 - F_3(0)) \in H, \tag{6.140}$$

where A_3 and F_3 are, respectively, as in (6.22) *and* (6.20).

Then, there exists a unique solution v to Problem III-VI for the non-steady Stokes problem with mixed boundary condition (5.2) *and*

$$\begin{aligned}
&\|v - U\|_{C([0,T];\mathbf{L}^2)\cap L^2(0,T;\mathbf{H}^1)}\\
&\quad\le K\Big(\|w_0\| + \|U\|_{L^2(0,T;\mathbf{H}^1)} + \Big\|\frac{\partial U}{\partial t}\Big\|_{L^2(0,T;(\mathbf{H}^1)^*)} + \|f\|_{L^2(0,T;\mathbf{V}^*)}\\
&\qquad + \sum_{i=2,4,7}\|\phi_i\|_{L^2(0,T;H^{-\frac{1}{2}}(\Gamma_i)} + \sum_{i=3,5,6}\|\phi_i\|_{L^2(0,T;\mathbf{H}^{-\frac{1}{2}}(\Gamma_i)}\Big).
\end{aligned} \tag{6.141}$$

If v_1, v_2 are solutions, respectively, to Problem-III-VI corresponding to data $v_{01}, g_{\tau 1}, g_{n1}, g_{+n1}, g_{-n1}, f_1, h_i^1, \phi_i^1$ and $v_{02}, g_{\tau 2}, g_{n2}, g_{+n2}, g_{-n2}, f_2, h_i^2, \phi_i^2$, then

$$\begin{aligned}
&\|v_1 - v_2\|_{C([0,T];\mathbf{L}^2)\cap L^2(0,T;\mathbf{H}^1)}\\
&\quad\le K\Big(\|v_{01} - v_{02}\|_{\mathbf{V}} + \|U_1 - U_2\|_{C([0,T];\mathbf{L}^2)} + \|U_1 - U_2\|_{L^2(0,T;\mathbf{H}^1)}\\
&\quad + \Big\|\frac{\partial U_1 - U_2}{\partial t}\Big\|_{L^2(0,T;(\mathbf{H}^1)^*)} + \|f_1 - f_2\|_{L^2(0,T;\mathbf{V}^*)} + \|g_{\tau 1} - g_{\tau 2}\|_{\mathbf{L}^2_\tau(\Gamma_8)}\\
&\quad + \|g_{n1} - g_{n2}\|_{L^2(\Gamma_9)} + \|g_{+n1} - g_{+n2}\|_{L^2(\Gamma_{10})} + \|g_{-n1} - g_{-n2}\|_{L^2(\Gamma_{10})}\\
&\quad + \sum_{i=2,4,7}\|\phi_i^1 - \phi_i^2\|_{L^2(0,T;H^{-\frac{1}{2}}(\Gamma_i))} + \sum_{i=3,5,6}\|\phi_i^1 - \phi_i^2\|_{L^2(0,T;\mathbf{H}^{-\frac{1}{2}}(\Gamma_i))}\Big),
\end{aligned} \tag{6.142}$$

where U_j, $j = 1, 2$, are the functions in Assumption 6.1 with h_i^j instead of h_i.

Proof By (6.111) and (6.113), we have

$$\langle A_3 w, w\rangle \ge c_1\|w\|_{\mathbf{V}}^2 - c_2\|w\|^2, \quad c_1 > 0,\ c_2 > 0. \tag{6.143}$$

Also, we can easily prove

$$|\langle A_3 w, u\rangle| \le c_3\|w\|_{\mathbf{V}}\|u\|_{\mathbf{V}}. \tag{6.144}$$

By definition of F_3 in (6.20), we know

$$\begin{aligned}
\|F_3\|_{L^2(0,T;\mathbf{V}^*)} \le\, & c\Big(\Big\|\frac{\partial U}{\partial t}\Big\|_{L^2(0,T;(\mathbf{H}^1)^*)} + \|U\|_{L^2(0,T;(\mathbf{H}^1)^*)} + \|f\|_{L^2(0,T;\mathbf{V}^*)}\\
& + \sum_{i=2,4,7}\|\phi_i\|_{L^2(0,T;H^{-\frac{1}{2}}(\Gamma_i))} + \sum_{i=3,5,6}\|\phi_i\|_{L^2(0,T;\mathbf{H}^{-\frac{1}{2}}(\Gamma_i))}\Big).
\end{aligned} \tag{6.145}$$

By the condition $w_0|_{\cup_{i=8-11}\Gamma_i} = 0$, we know $\Phi(w_0) = 0$. Since $\Phi(u) \ge 0$, we have that $\partial\Phi(w_0) = 0$. Thus, taking into account Remark 6.5, by Proposition 6.1 we have existence of a unique solution and estimate (6.141).

If $v_1 = w_1 + U_1$, $v_2 = w_2 + U_2$ are solutions corresponding to the given data, we get

$$\begin{aligned}&\left\langle\frac{\partial w_1}{\partial t}+A_3w_1,u-w_1\right\rangle+\Phi_1(u)-\Phi_1(w_1)\geq\langle F_{31},u-w_1\rangle,\\&\left\langle\frac{\partial w_2}{\partial t}+A_3w_2,u-w_2\right\rangle+\Phi_2(u)-\Phi_2(w_2)\geq\langle F_{32},u-w_2\rangle,\end{aligned}\tag{6.146}$$

where $\Phi_j(u)$, F_{3j}, $j=1,2$, are one corresponding to $U_j, g_{\tau j}, g_{nj}, g_{+nj}, g_{-nj}, f_j, h_i^j, \phi_i^j$. Putting $u=w_2$ and $u=w_1$, respectively, in the first and second one of (6.146), and adding those, we have

$$\begin{aligned}\left\langle\frac{\partial(w_1-w_2)}{\partial t}+A_3(w_1-w_2),w_2-w_1\right\rangle+\Phi_1(w_2)-\Phi_1(w_1)+\Phi_2(w_1)-\Phi_2(w_2)\\\geq\langle F_{31}-F_{32},w_2-w_1\rangle.\end{aligned}\tag{6.147}$$

From (6.143) and (6.147) we have

$$\begin{aligned}\frac{\partial\|w_1-w_2\|^2}{\partial t}+2c_1\|w_1-w_2\|_{\mathbf{V}}^2\leq\Big(&\|F_{31}-F_{32}\|_{\mathbf{V}^*}^2+\|w_2-w_1\|_{\mathbf{V}}^2\\&+2|\Phi_1(w_2)-\Phi_1(w_1)+\Phi_2(w_1)-\Phi_2(w_2)|\Big)+2c_2\|w_1-w_2\|^2.\end{aligned}\tag{6.148}$$

Since $w_1, w_2\in K(\Omega)$,

$$\begin{aligned}\Phi_1(w_2)-\Phi_1(w_1)=&\int_{\Gamma_8}g_{\tau1}(|w_{2\tau}|-|w_{1\tau}|)\,ds+\int_{\Gamma_9}g_{n1}(|w_{2n}|-|w_{1n}|)\,ds\\&+\int_{\Gamma_{10}}g_{+n1}(w_{2n}-w_{1n})\,ds-\int_{\Gamma_{11}}g_{-n1}(w_{2n}-w_{1n})\,ds,\\\Phi_2(w_2)-\Phi_2(w_1)=&\int_{\Gamma_8}g_{\tau2}(|w_{2\tau}|-|w_{1\tau}|)\,ds+\int_{\Gamma_9}g_{n2}(|w_{2n}|-|w_{1n}|)\,ds\\&+\int_{\Gamma_{10}}g_{+n2}(w_{2n}-w_{1n})\,ds-\int_{\Gamma_{11}}g_{-n2}(w_{2n}-w_{1n})\,ds.\end{aligned}\tag{6.149}$$

Subtracting two formulas in (6.149) yields

$$|\Phi_1(w_2)-\Phi_1(w_1)+\Phi_2(w_1)-\Phi_2(w_2)|\leq G_{12}+\frac{c_1}{2}\|w_2-w_1\|_{\mathbf{V}}^2,\tag{6.150}$$

where

$$\begin{aligned}G_{12}=\frac{2}{c_1}\Big(&\|g_{\tau1}-g_{\tau2}\|_{\mathbf{L}_\tau^2(\Gamma_8)}^2+\|g_{n1}-g_{n2}\|_{L^2(\Gamma_9)}^2\\&+\|g_{+n1}-g_{+n2}\|_{L^2(\Gamma_{10})}^2+\|g_{-n1}-g_{-n2}\|_{L^2(\Gamma_{11})}^2\Big).\end{aligned}$$

By (6.148) and (6.150) we have

$$\|w_1(t) - w_2(t)\|^2 + \frac{c_1}{2}\int_0^t \|w_1 - w_2\|_{\mathbf{V}}^2 \le$$
$$\|w_{10} - w_{20}\|^2 + \int_0^t \|F_{31}(s) - F_{32}(s)\|_{\mathbf{V}^*}^2\, ds + G_{12}t + 2c_2\int_0^t \|w_1(s) - w_2(s)\|^2\, ds. \tag{6.151}$$

By Gronwall's inequality, we have from (6.151)

$$\|w_1(t) - w_2(t)\|^2 \le \Big(\|w_{10} - w_{20}\|^2 + \int_0^T \|F_{31}(s) - F_{32}(s)\|_{\mathbf{V}^*}^2\, ds + G_{12}T\Big)e^{2c_2T},$$
$$\|w_1(t) - w_2(t)\|_{L^2(0,T;\mathbf{V})}^2 \le K\Big(\|w_{10} - w_{20}\|^2 + \int_0^T \|F_{31}(s) - F_{32}(s)\|_{\mathbf{V}^*}^2\, ds + G_{12}\Big). \tag{6.152}$$

Also $\|F_{31} - F_{32}\|_{\mathbf{V}^*}$ is estimated as (6.145), and so from (6.152) we get estimate (6.142). □

6.4 Bibliographical Remarks

The content of Chap. 6 is taken from [4]. One mistake in [4] is corrected here.

In [5–7] the non-steady Stokes equations with the homogeneous Dirichlet boundary condition and leak boundary condition were studied.

In [8] the existence of a weak solution and a local-in-time strong solution to the non-steady Navier-Stokes problem was studied when boundary consists of a portion with homogeneous Dirichlet boundary condition and another portion with the leak condition based on the total stress. In [2] a variational inequality for the Navier-Stokes problem with homogeneous Dirichlet boundary condition and one-sided leak condition based on the total stress was considered. Also, in [9] a variational inequality for the Navier-Stokes problems with homogeneous Dirichlet boundary condition and one-sided boundary conditions based on total pressure was studied.

When boundary consists of separated portions with homogeneous Dirichlet boundary condition and with nonlinear slip condition or leak condition, in [10] existence of a unique strong solution to the non-steady Navier-Stokes problem was studied. In the case of nonlinear slip condition, existence of a unique strong solution to 2-D problem and a unique local-in-time strong solution to 3-D problem was proved. In the case of leak condition, existence of a unique local-in-time strong solution to 3-D problem was proved.

The semi-discrete finite element approximation to the time-dependent 2-D Navier-Stokes equations with mixture of homogeneous Dirichlet boundary condition and Tresca slip boundary conditions was discussed in [11].

For other kinds of non-steady fluid equations with mixture of friction slip boundary conditions and Dirichlet condition, we refer to [12–15].

References

1. R. Temam, *Navier-Stokes Equations* (North-Holland, Amsterdam, 1985)
2. J. Naumann, On evolution inequalities of Navier-Stokes type in three dimensions. Ann. Math. Pure Appl. **124**(4), 107–125 (1980)
3. V. Barbu, *Nonlinear Differential Equations of Monotone Types in Banach Spaces* (Springer, Berlin, 2010)
4. T. Kim, F. Huang, The non-steady Navier-Stokes systems with mixed boundary conditions including friction conditions. Methods Appl. Anal. **25**, 13–50 (2018)
5. H. Fujita, Non-stationary Stokes flows under leak boundary conditions of friction type. J. Comput. Math. **19**, 1–8 (2001)
6. H. Fujita, Variational inequalities and nonlinear semi-groups applied to certain nonlinear problems for the Stokes equation. RIMS Kokyuroku **1234**, 70–85 (2001)
7. H. Fujita, A coherent analysis of Stokes flows under boundary conditions of friction type. J. Comput. Appl. Math. **149**, 57–69 (2002)
8. R. An, Y. Li, K. Li, Solvability of Navier-Stokes equations with leak boundary conditions. Acta Math. Appl. Sin. Engl. Ser. **25**, 225–234 (2009)
9. D.S. Konovalova, Subdifferential boundary value problems for the nonstationary Navier-Stokes and Stokes equations. Differ. Equ. **36**, 875–885 (2000)
10. T. Kashiwabara, On a strong solution of the non-stationary Navier-Stokes equations under slip or leak boundary conditions of friction. J. Differ. Equ. **254**, 756–778 (2013)
11. Y. Li, R. An, Semi-discrete stabilized finite element methods for Navier-Stokes equations with nonlinear slip boundary conditions based on regularization procedure. Numer. Math. **117**, 1–36 (2011)
12. M. Boukrouche, I. Boussetouan, L. Paoli, Non-isothermal Navier-Stokes system with mixed boundary conditions and friction law: uniqueness and regularity properties. Nonlinear Anal. **102**, 168–185 (2014)
13. M. Boukrouche, I. Boussentoua, L. Paoli, Existence for non-isothermal fluid flows with Tresca's friction and Cattaneo's heat law. J. Math. Anal. Appl. **427**, 499–514 (2015)
14. M. Boukrouche, G. Łukaszewicz, On global in time dynamics of a planar Bingham flow subject to a subdifferential boundary conditions. Discret. Contin. Dyn. Syst. **34**, 3969–3983 (2014)
15. J.K. Djoko, P.A. Razafimandimby, Analysis of the Brinkman-Forchheimer equations with slip boundary conditions. Appl. Anal. **93**, 1477–1494 (2014)

Chapter 7
The Steady Boussinesq System

In this chapter we are concerned with the steady Boussinesq system with mixed boundary conditions. The boundary conditions for fluid may include Tresca slip, leak condition, one-sided leak conditions, velocity, pressure, vorticity, stress together and the conditions for temperature may include Dirichlet, Neumann and Robin conditions together. We will get variational formulations consisting of a variational inequality for velocity and a variational equation for temperature, which are equivalent to the original PDE problems for smooth solutions.

Then, we will study the existence of solutions to the variational problems. For the problem with boundary conditions involving the static pressure and stress, it is proved that if the data of problem are small enough, then there exists a unique solution. For the problem with boundary conditions involving the total pressure and total stress, the existence of a solution is proved as well without smallness of the data.

7.1 Problems and Variational Formulations

Let us consider the steady Boussinesq system

$$\begin{cases} -2\nabla\cdot\big(\mu(\theta)\mathcal{E}(v)\big)+(v\cdot\nabla)v+\nabla p=(1-\alpha_0\theta)f, \\ \operatorname{div} v=0, \\ -\nabla\cdot\big(\kappa(\theta)\nabla\theta\big)+v\cdot\nabla(\gamma(\theta)\theta)=g \end{cases} \tag{7.1}$$

T. Kim and D. Cao, *Equations of Motion for Incompressible Viscous Fluids*, Advances in Mathematical Fluid Mechanics,
https://doi.org/10.1007/978-3-030-78659-5_7

under mixed boundary conditions. Here v, p and θ are, respectively, velocity, pressure and temperature, and α_0—parameter for *buoyancy* effect, f—body force, g—heat source, $\kappa(\theta)$—thermal conductivity, $\gamma(\theta)$—specific heat of the fluid. The viscosity, thermal conductivity and specific heat of fluid depend on the temperature.

Let Ω be a bounded domain of $\mathbb{R}^l$, $l = 2, 3$. $\partial\Omega \in C^{0,1}$,

$$\partial\Omega = \cup_{i=1}^{11} \overline{\Gamma}_i = \overline{\Gamma}_D \cup \overline{\Gamma}_R$$

and $\Gamma_D \cap \Gamma_R = \emptyset$, $\Gamma_i \cap \Gamma_j = \emptyset$ for $i \neq j$, $\Gamma_i = \bigcup_j \Gamma_{ij}$, where Γ_{ij} are connected open subsets of $\partial\Omega$, $\Gamma_{ij} \in C^2$ for $i = 2, 3, 7$ and $\Gamma_{ij} \in C^1$ for others. (See Remark 6.1.)

For temperature we are concerned with the boundary conditions

$$\begin{aligned}
&(1) \quad \theta|_{\Gamma_D} = 0,\\
&(2) \quad \left(\kappa(\theta)\frac{\partial\theta}{\partial n} + \beta(x)\theta\right)\Big|_{\Gamma_R} = g_R(x), \quad \beta(x),\ g_R(x) - \text{given functions on } \Gamma_R.
\end{aligned} \tag{7.2}$$

According to boundary conditions for fluid, the Problems I and II are distinguished. Problem I is the one with the following boundary conditions (the case of static pressure)

$$\begin{aligned}
&(1) \quad v|_{\Gamma_1} = 0,\\
&(2) \quad v_\tau|_{\Gamma_2} = 0, \ -p|_{\Gamma_2} = \phi_2,\\
&(3) \quad v_n|_{\Gamma_3} = 0, \ \operatorname{rot} v \times n|_{\Gamma_3} = \phi_3/\mu(\theta),\\
&(4) \quad v_\tau|_{\Gamma_4} = 0, \ (-p + 2\mu(\theta)\varepsilon_{nn}(v))|_{\Gamma_4} = \phi_4,\\
&(5) \quad v_n|_{\Gamma_5} = 0, \ 2(\mu(\theta)\varepsilon_{n\tau}(v) + \alpha v_\tau)|_{\Gamma_5} = \phi_5, \ \ \alpha : \text{a matrix},\\
&(6) \quad (-pn + 2\mu(\theta)\varepsilon_n(v))|_{\Gamma_6} = \phi_6,\\
&(7) \quad v_\tau|_{\Gamma_7} = 0, \ (-p + \mu(\theta)\frac{\partial v}{\partial n} \cdot n)|_{\Gamma_7} = \phi_7,\\
&(8) \quad v_n = 0, \ |\sigma_\tau(\theta, v)| \le g_\tau, \ \sigma_\tau(\theta, v) \cdot v_\tau + g_\tau|v_\tau| = 0 \ \text{ on } \Gamma_8,\\
&(9) \quad v_\tau = 0, \ |\sigma_n(\theta, v, p)| \le g_n, \ \sigma_n(\theta, v, p)v_n + g_n|v_n| = 0 \ \text{ on } \Gamma_9,\\
&(10) \quad v_\tau = 0, \ v_n \ge 0, \ \sigma_n(\theta, v, p) + g_{+n} \ge 0, \ (\sigma_n(\theta, v, p) + g_{+n})v_n = 0 \ \text{ on } \Gamma_{10},\\
&(11) \quad v_\tau = 0, \ v_n \le 0, \ \sigma_n(\theta, v, p) - g_{-n} \le 0, \ (\sigma_n(\theta, v, p) - g_{-n})v_n = 0 \ \text{ on } \Gamma_{11},
\end{aligned} \tag{7.3}$$

and Problem II is the one with the following boundary conditions (the case of total pressure)

$$
\begin{aligned}
&(1)\ \ v|_{\Gamma_1}=0,\\
&(2)\ \ v_\tau|_{\Gamma_2}=0,\ \ -(p+1/2|v|^2)|_{\Gamma_2}=\phi_2,\\
&(3)\ \ v_n|_{\Gamma_3}=0,\ \ \operatorname{rot} v\times n|_{\Gamma_3}=\phi_3/\mu(\theta),\\
&(4)\ \ v_\tau|_{\Gamma_4}=0,\ \ (-p-1/2|v|^2+2\mu(\theta)\varepsilon_{nn}(v))|_{\Gamma_4}=\phi_4,\\
&(5)\ \ v_n|_{\Gamma_5}=0,\ \ 2(\mu(\theta)\varepsilon_{n\tau}(v)+\alpha v_\tau)|_{\Gamma_5}=\phi_5,\ \ \alpha:\text{a matrix},\\
&(6)\ \ (-pn-1/2|v|^2n+2\mu(\theta)\varepsilon_n(v))|_{\Gamma_6}=\phi_6,\\
&(7)\ \ v_\tau|_{\Gamma_7}=0,\ (-p-1/2|v|^2+\mu(\theta)\frac{\partial v}{\partial n}\cdot n)|_{\Gamma_7}=\phi_7,\\
&(8)\ \ v_n=0,\ \ |\sigma_\tau^t(\theta,v)|\le g_\tau,\ \ \sigma_\tau^t(\theta,v)\cdot v_\tau+g_\tau|v_\tau|=0\ \ \text{on } \Gamma_8,\\
&(9)\ \ v_\tau=0,\ \ |\sigma_n^t(\theta,v,p)|\le g_n,\ \ \sigma_n^t(\theta,v,p)v_n+g_n|v_n|=0\ \ \text{on } \Gamma_9,\\
&(10)\ \ v_\tau=0,\ v_n\ge 0,\ \sigma_n^t(\theta,v,p)+g_{+n}\ge 0,\ (\sigma_n^t(\theta,v,p)+g_{+n})v_n=0\ \ \text{on } \Gamma_{10},\\
&(11)\ \ v_\tau=0,\ v_n\le 0,\ \sigma_n^t(\theta,v,p)-g_{-n}\le 0,\ (\sigma_n^t(\theta,v,p)-g_{-n})v_n=0\ \ \text{on } \Gamma_{11}.
\end{aligned}
\tag{7.4}
$$

Difference between (7.3) and (5.2), or (7.4) and (5.3), is that μ in (7.3) and (7.4) depends on θ.

We use the following assumption.

Assumption 7.1 Assume the followings.

(1) Γ_{2j}, Γ_{3j} and Γ_{7j} are convex. $\Gamma_1\neq\emptyset$, $\Gamma_D\neq\emptyset$ and

$$\Gamma_R\subset\left(\cup_{i=1,3,5,8}\Gamma_i\right). \tag{7.5}$$

(2) (3) of Assumption 5.1 holds.

(3) For the functions of (7.1) $f\in\mathbf{L}^3(\Omega)$, $g\in L^{6/5}(\Omega)$ and

$$
\begin{aligned}
&\mu\in C(\mathbb{R}),\ \ 0<\mu_0\le\mu(\xi)\le\mu_1<\infty\quad\forall\xi\in\mathbb{R};\\
&\kappa\in C(\mathbb{R}),\ \ 0<\kappa_0\le\kappa(\xi)\le\kappa_1<\infty\quad\forall\xi\in\mathbb{R};\\
&\gamma\in C(\mathbb{R}),\ \ |\gamma(\xi)|\le\gamma_0\quad\forall\xi\in\mathbb{R}.
\end{aligned}
\tag{7.6}
$$

(4) For the functions of (7.2), (7.3) and (7.4),

$$
\begin{aligned}
&g_R\in L^{4/3}(\Gamma_R);\\
&\beta_0\ge\beta(x)\ge 0,\ \ \beta_0-\text{a constant},\ \ \beta(x)-\text{measurable};\\
&\phi_i\in H^{-\frac12}(\Gamma_i), i=2,4,7,\ \ \phi_i\in\mathbf{H}^{-\frac12}(\Gamma_i), i=3,5,6;\\
&\text{the matrix }\alpha\text{ is positive},\ \ \alpha_{ij}\in L^\infty(\Gamma_5).
\end{aligned}
\tag{7.7}
$$

Remark 7.1 For the first part of (2) in Assumption 7.1 we refer to Remark 6.3. In this chapter this assumption is also necessary to guarantee equivalence between Problem I-VE and Problem I-VI, and between Problem II-VE and Problem II-VI.

Assuming that $f\in\mathbf{L}^2(\Omega)$, $g\in L^2(\Omega)$, $g_R\in L^2(\Gamma_R)$, $\phi_i\in L^2(\Gamma_i)$, $i=2,4,7$, and $\phi_i\in\mathbf{L}^2(\Gamma_i)$, $i=3,5,6$, formally we introduce the following

Definition 7.1 A function $(v, p, \theta) \in \mathbf{H}^2(\Omega) \times H^1(\Omega) \times H^2(\Omega)$ is called a solution to Problem I (or Problem II) if the first two equations and the last equation of (7.1), respectively, hold in $\mathbf{L}^2(\Omega)$ and $L^2(\Omega)$, and each of (7.2) and (7.3) (or (7.4)), respectively, holds in $L^2(\Gamma_D)$, $L^2(\Gamma_R)$, $L^2(\Gamma_i)$ and $\mathbf{L}^2(\Gamma_i)$.

Let $\mathbf{V}$, $K(\Omega)$ be the same as in (5.5) and

$$W^{1,2}_{\Gamma_D}(\Omega) = \{y \in W^{1,2}(\Omega) : y|_{\Gamma_D} = 0\}.$$

Since $\Gamma_1 \neq \emptyset$ and $\Gamma_D \neq \emptyset$, by Korn's and Poincaré's inequalities we use

$$(v, u)_{\mathbf{V}} = (\mathcal{E}(v), \mathcal{E}(u)), \quad (y, z)_{W^{1,2}_{\Gamma_D}(\Omega)} = (\nabla y, \nabla z). \tag{7.8}$$

As in Chap. 5, we will get variational formulations for Problems I, II.

Integrating by parts, taking into account (2.79), and applying Theorems 3.1 and 3.2 on Γ_{ij} $(i = 2, 3, 7)$, for $v \in \mathbf{H}^2(\Omega) \cap \mathbf{V}$, $\theta \in H^2(\Omega)$ and $u \in \mathbf{V}$ we have

$$\begin{aligned}
-2\langle\nabla \cdot (\mu(\theta)\mathcal{E}(v)), u\rangle &= 2(\mu(\theta)\mathcal{E}(v), \mathcal{E}(u)) - 2(\mu(\theta)\mathcal{E}(v)n, u)_{\cup_{i=2}^{11}\Gamma_i}\\
&= 2(\mu(\theta)\mathcal{E}(v), \mathcal{E}(u)) + 2(\mu(\theta)k(x)v, u)_{\Gamma_2} - (\mu(\theta)\mathrm{rot}\, v \times n, u)_{\Gamma_3}\\
&\quad + 2(\mu(\theta)S\tilde{v}, \tilde{u})_{\Gamma_3} - 2(\mu(\theta)\varepsilon_{nn}(v), u_n)_{\Gamma_4} - 2(\mu(\theta)\varepsilon_{n\tau}(v), u)_{\Gamma_5}\\
&\quad - 2(\mu(\theta)\varepsilon_n(v), u)_{\Gamma_6} - \Big(\mu(\theta)\frac{\partial v}{\partial n}, u\Big)_{\Gamma_7} + (\mu(\theta)k(x)v, u)_{\Gamma_7} - 2(\mu(\theta)\varepsilon_{n\tau}(v), u)_{\Gamma_8}\\
&\quad - 2(\mu(\theta)\varepsilon_{nn}(v), u_n)_{\Gamma_9} - 2(\mu(\theta)\varepsilon_{nn}(v), u_n)_{\Gamma_{10}} - 2(\mu(\theta)\varepsilon_{nn}(v), u_n)_{\Gamma_{11}}.
\end{aligned} \tag{7.9}$$

For $p \in H^1(\Omega)$ and $u \in \mathbf{V}$ we have

$$(\nabla p, u) = (p, u_n)_{\cup_{i=2}^{11}\Gamma_i} = (p, u_n)_{\Gamma_2} + (p, u_n)_{\Gamma_4\cup\Gamma_7\cup\Gamma_9\cup\Gamma_{10}\cup\Gamma_{11}} + (pn, u)_{\Gamma_6}, \tag{7.10}$$

where $u_n |_{\Gamma_3\cup\Gamma_5\cup\Gamma_8}= 0$ was used. For $\theta \in H^2(\Omega)$ and $\varphi \in W^{1,2}_{\Gamma_D}(\Omega)$, by (7.2) we have

$$\langle -\nabla \cdot (\kappa(\theta)\nabla\theta), \varphi\rangle = (\kappa(\theta)\nabla\theta, \nabla\varphi) - \Big(\kappa(\theta)\frac{\partial\theta}{\partial n}, \varphi\Big)_{\Gamma_R} = (\kappa(\theta)\nabla\theta, \nabla\varphi) + (\beta\theta - g_R, \varphi)_{\Gamma_R}. \tag{7.11}$$

By (7.5), $v_n = 0$ on Γ_R, and so for $v \in \mathbf{V}$, $\theta \in H^1(\Omega)$ and $\varphi \in W^{1,2}_{\Gamma_D}(\Omega)$ we have

$$\langle v \cdot \nabla(\gamma(\theta)\theta), \varphi\rangle = (v_n\gamma(\theta)\theta, \varphi)_{\Gamma_R} - (\gamma(\theta)\theta v, \nabla\varphi) = -(\gamma(\theta)\theta v, \nabla\varphi). \tag{7.12}$$

7.1.1 Variational Formulation: The Case of Static Pressure

By (7.9)–(7.12), we can see that solutions in the sense of Definition 7.1 (v, p, θ) of the problem (7.1), (7.2), (7.3) satisfy the following (cf. Remark 3.9).

$$\begin{cases} 2(\mu(\theta)\mathcal{E}(v), \mathcal{E}(u)) + \langle (v\cdot\nabla)v, u\rangle + 2(\mu(\theta)k(x)v, u)_{\Gamma_2} \\ \qquad + 2(\mu(\theta)S\tilde{v}, \tilde{u})_{\Gamma_3} + 2(\alpha(x)v, u)_{\Gamma_5} + (\mu(\theta)k(x)v, u)_{\Gamma_7} \\ \qquad - 2(\mu(\theta)\varepsilon_{n\tau}(v), u)_{\Gamma_8} + (p - 2\mu(\theta)\varepsilon_{nn}(v), u_n)_{\Gamma_9\cup\Gamma_{10}\cup\Gamma_{11}} \\ \quad = \langle (1-\alpha_0\theta)f, u\rangle + \sum_{i=2,4,7} \langle \phi_i, u_n\rangle_{\Gamma_i} + \sum_{i=3,5,6} \langle \phi_i, u\rangle_{\Gamma_i} \quad \forall u \in \mathbf{V}, \\ (\kappa(\theta)\nabla\theta, \nabla\varphi) - (\gamma(\theta)\theta v, \nabla\varphi) + (\beta\theta, \varphi)_{\Gamma_R} = \langle g_R, \varphi\rangle_{\Gamma_R} + \langle g, \varphi\rangle \quad \forall \varphi \in W^{1,2}_{\Gamma_D}(\Omega), \\ |\sigma_\tau(\theta, v)| \le g_\tau, \ \ \sigma_\tau(\theta, v)\cdot v_\tau + g_\tau |v_\tau| = 0 \ \text{ on } \Gamma_8, \\ |\sigma_n(\theta, v, p)| \le g_n, \ \ \sigma_n(\theta, v, p)v_n + g_n|v_n| = 0 \ \text{ on } \Gamma_9, \\ \sigma_n(\theta, v, p) + g_{+n} \ge 0, \ \ (\sigma_n(\theta, v, p) + g_{+n})v_n = 0 \ \text{ on } \Gamma_{10}, \\ \sigma_n(\theta, v, p) - g_{-n} \le 0, \ \ (\sigma_n(\theta, v, p) - g_{-n})v_n = 0 \ \text{ on } \Gamma_{11}. \end{cases} \tag{7.13}$$

Define $a_0(\theta; \cdot, \cdot)$, $a_1(\cdot, \cdot, \cdot)$ and $f_1 \in \mathbf{V}^*$ by

$$\begin{aligned} & a_0(\theta; w, u) = 2(\mu(\theta)\mathcal{E}(w), \mathcal{E}(u)) + 2(\mu(\theta)k(x)w, u)_{\Gamma_2} + 2(\mu(\theta)S\tilde{w}, \tilde{u})_{\Gamma_3} \\ & \qquad + 2(\alpha(x)w, u)_{\Gamma_5} + (\mu(\theta)k(x)w, u)_{\Gamma_7} \quad \forall w, u \in \mathbf{V}, \theta \in W^{1,2}(\Omega), \\ & a_1(v, w, u) = \langle (v\cdot\nabla)w, u\rangle \quad \forall v, w, u \in \mathbf{V}, \\ & \langle f_1, u\rangle = \sum_{i=2,4,7} \langle \phi_i, u_n\rangle_{\Gamma_i} + \sum_{i=3,5,6} \langle \phi_i, u\rangle_{\Gamma_i} \quad \forall u \in \mathbf{V}. \end{aligned} \tag{7.14}$$

Define $b_0(\tilde{\theta}; \cdot, \cdot)$ and $g_1 \in (W^{1,2}_{\Gamma_D}(\Omega))^*$ by

$$\begin{aligned} & b_0(\tilde{\theta}; \theta, \varphi) = (\kappa(\tilde{\theta})\nabla\theta, \nabla\varphi) + (\beta(x)\theta, \varphi)_{\Gamma_R} \quad \forall\tilde{\theta}, \theta \in W^{1,2}(\Omega), \forall\varphi \in W^{1,2}_{\Gamma_D}(\Omega), \\ & \langle g_1, \varphi\rangle = \langle g_R, \varphi\rangle_{\Gamma_R} + \langle g, \varphi\rangle \quad \forall\varphi \in W^{1,2}_{\Gamma_D}(\Omega). \end{aligned} \tag{7.15}$$

Then, taking into account $\sigma_\tau(\theta, v) = 2\mu(\theta)\varepsilon_{n\tau}(v)$, $\sigma_n(\theta, v, p) = -p + 2\mu(\theta)\varepsilon_{nn}(v)$ and (7.13), we introduce the following variational formulation for the problem (7.1)–(7.3).

Problem I-VE. Find $(v, \theta, \sigma_\tau, \sigma_n, \sigma_{+n}, \sigma_{-n}) \in K(\Omega) \times W^{1,2}_{\Gamma_D}(\Omega) \times \mathbf{L}^2_\tau(\Gamma_8) \times L^2(\Gamma_9) \times H^{-1/2}(\Gamma_{10}) \times H^{-1/2}(\Gamma_{11})$ such that

$$
\begin{cases}
a_0(\theta; v, u) + a_1(v, v, u) - (\sigma_\tau, u_\tau)_{\Gamma_8} - (\sigma_n, u_n)_{\Gamma_9} \\
\qquad - \langle \sigma_{+n}, u_n \rangle_{\Gamma_{10}} - \langle \sigma_{-n}, u_n \rangle_{\Gamma_{11}} - \langle f - \alpha_0 \theta f, u \rangle = \langle f_1, u \rangle \quad \forall u \in \mathbf{V}, \\
b_0(\theta; \theta, \varphi) - \langle \gamma(\theta)\theta v, \nabla\varphi \rangle = \langle g_1, \varphi \rangle \quad \forall \varphi \in W^{1,2}_{\Gamma_D}(\Omega), \\
|\sigma_\tau| \le g_\tau, \ \sigma_\tau \cdot v_\tau + g_\tau |v_\tau| = 0 \quad \text{on } \Gamma_8, \\
|\sigma_n| \le g_n, \ \sigma_n v_n + g_n |v_n| = 0 \quad \text{on } \Gamma_9, \\
\sigma_{+n} + g_{+n} \ge 0, \ \langle \sigma_{+n} + g_{+n}, v_n \rangle_{\Gamma_{10}} = 0 \quad \text{on } \Gamma_{10}, \\
\sigma_{-n} - g_{-n} \le 0, \ \langle \sigma_{-n} - g_{-n}, v_n \rangle_{\Gamma_{11}} = 0 \quad \text{on } \Gamma_{11},
\end{cases}
\tag{7.16}
$$

where $\mathbf{L}^2_\tau(\Gamma_8)$ is the subspace of $\mathbf{L}^2(\Gamma_8)$ consisting of functions such that $(u, n)_{\mathbf{L}^2(\Gamma_8)} = 0$.

Remark 7.2 Under (4) of Assumption 7.1 the duality products $\langle f_1, u \rangle$ of (7.14) is meaningful (see Remark 5.1).

Theorem 7.1 *Let Assumption 7.1 hold. If (v, p, θ) is a solution in the sense of Definition 7.1 of the problem (7.1)–(7.3), then $(v, \theta, \sigma_\tau|_{\Gamma_8}, \sigma_n|_{\Gamma_9}, \sigma_n|_{\Gamma_{10}}, \sigma_n|_{\Gamma_{11}})$ is a solution to Problem I-VE.*

Conversely, if $f \in \mathbf{L}^2(\Omega)$ and Problem I-VE has a smooth solution $(v, \theta, \sigma_\tau, \sigma_n, \sigma_+, \sigma_-)$ such that $v \in \mathbf{H}^2(\Omega)$, $\theta \in H^2(\Omega)$, $\sigma_+ \in L^2(\Gamma_{10})$ and $\sigma_- \in L^2(\Gamma_{11})$, then there exists $p \in H^1(\Omega)$ such that (v, p, θ) is a solution to the problem (7.1)–(7.3). Moreover, if at least one of the sets $\Gamma_i, i = 2, 4, 6, 7$, is nonempty, then p is unique.

Proof From the problem (7.1)–(7.3) we deduced Problem I-VE, and thus it is suffices to prove conversion from Problem I-VE to the problem (7.1)–(7.3). By Theorem 5.1 there exists a p such that (v, p) satisfies (7.1) and the boundary condition (7.3), and p is unique under the additional condition above. In a routine way (see Sect. 1, Ch. 2 of [1]) we can prove that θ satisfies (7.1) and the boundary condition (7.2). □

We will find another variational formulation consisting of a variational inequality and a variational equation, which is equivalent to Problem I-VE.

Let $(v, \theta, \sigma_\tau, \sigma_n, \sigma_{+n}, \sigma_{-n})$ be a solution of Problem I-VE. Subtracting the first formula of (7.16) with $u = v$ from the first formula of (7.16), we get

$$
\begin{aligned}
a_0(\theta; v, u - v) + a_1(v, v, u - v) - (\sigma_\tau, u_\tau - v_\tau)_{\Gamma_8} - (\sigma_n, u_n - v_n)_{\Gamma_9} - \langle \sigma_{+n}, u_n - v_n \rangle_{\Gamma_{10}} \\
- \langle \sigma_{-n}, u_n - v_n \rangle_{\Gamma_{11}} - \langle f - \alpha_0 \theta f, u - v \rangle = \langle f_1, u - v \rangle \quad \forall u \in \mathbf{V}.
\end{aligned}
\tag{7.17}
$$

Let $\Phi : \mathbf{V} \to \overline{\mathbb{R}}$ be the functional defined by (5.22).Then, Φ is proper, convex, weakly lower semi-continuous and nonnegative. By Theorem 5.2, under Assumption 7.1 for fixed θ the problem

$$
\begin{cases}
a_0(\theta; v, u) + a_1(v, v, u) - (\sigma_\tau, u_\tau)_{\Gamma_8} - (\sigma_n, u_n)_{\Gamma_9} \\
\quad - \langle \sigma_{+n}, u_n \rangle_{\Gamma_{10}} - \langle \sigma_{-n}, u_n \rangle_{\Gamma_{11}} - \langle f - \alpha_0 \theta f, u \rangle = \langle f_1, u \rangle, \\
|\sigma_\tau| \le g_\tau, \ \sigma_\tau \cdot v_\tau + g_\tau |v_\tau| = 0 \quad \text{on } \Gamma_8, \\
|\sigma_n| \le g_n, \ \sigma_n v_n + g_n |v_n| = 0 \quad \text{on } \Gamma_9, \\
\sigma_{+n} + g_{+n} \ge 0, \ \langle \sigma_{+n} + g_{+n}, v_n \rangle_{\Gamma_{10}} = 0 \quad \text{on } \Gamma_{10}, \\
\sigma_{-n} - g_{-n} \le 0, \ \langle \sigma_{-n} - g_{-n}, v_n \rangle_{\Gamma_{11}} = 0 \quad \text{on } \Gamma_{11}
\end{cases}
\tag{7.18}
$$

is equivalent to the following variational inequality.

Find $v \in \mathbf{V}$ such that

$$
\begin{aligned}
a_0(\theta; v, u - v) + a_1(v, v, u - v) + \Phi(u) - \Phi(v) - \langle f - \alpha_0 \theta f, u - v \rangle& \\
\ge \langle f_1, u - v \rangle \quad \forall u \in \mathbf{V}.&
\end{aligned}
\tag{7.19}
$$

Therefore, we have the following variational formulation equivalent to Problem I-VE which consists of a variational inequality for velocity and a variational equation for temperature.

Problem I-VI. Find $(v, \theta) \in \mathbf{V} \times W^{1,2}_{\Gamma_D}(\Omega)$ such that

$$
\begin{cases}
a_0(\theta; v, u - v) + a_1(v, v, u - v) + \Phi(u) - \Phi(v) - \langle f - \alpha_0 \theta f, u - v \rangle \\
\qquad \ge \langle f_1, u - v \rangle \quad \forall u \in \mathbf{V}, \\
b_0(\theta; \theta, \varphi) - \langle \gamma(\theta) \theta v, \nabla \varphi \rangle = \langle g_1, \varphi \rangle \quad \forall \varphi \in W^{1,2}_{\Gamma_D}(\Omega).
\end{cases}
\tag{7.20}
$$

7.1.2 Variational Formulation: The Case of Total Pressure

Taking $(v \cdot \nabla) v = \operatorname{rot} v \times v + \frac{1}{2} \operatorname{grad} |v|^2$ into account, by (7.9)–(7.12) we can see that solutions in the sense of Definition 7.1 (v, p, θ) of problem (7.1), (7.2), (7.4) satisfy the following.

$$
\begin{cases}
2(\mu(\theta) \mathcal{E}(v), \mathcal{E}(u)) + \langle \operatorname{rot} v \times v, u \rangle + 2(\mu(\theta) k(x) v, u)_{\Gamma_2} \\
\qquad + 2(\mu(\theta) S \tilde{v}, \tilde{u})_{\Gamma_3} + 2(\alpha(x) v, u)_{\Gamma_5} + (\mu(\theta) k(x) v, u)_{\Gamma_7} \\
\qquad - 2(\mu(\theta) \varepsilon_{n\tau}(v), u)_{\Gamma_8} + (p + \frac{1}{2} |v|^2 - 2\mu(\theta) \varepsilon_{nn}(v), u_n)_{\Gamma_9 \cup \Gamma_{10} \cup \Gamma_{11}} \\
\quad = \langle (1 - \alpha_0 \theta) f, u \rangle + \sum_{i=2,4,7} \langle \phi_i, u_n \rangle_{\Gamma_i} + \sum_{i=3,5,6} \langle \phi_i, u \rangle_{\Gamma_i} \quad \forall u \in \mathbf{V}, \\
(\kappa(\theta) \nabla \theta, \nabla \varphi) - (\gamma(\theta) \theta v, \nabla \varphi) + (\beta \theta, \varphi)_{\Gamma_R} = \langle g_R, \varphi \rangle_{\Gamma_R} + \langle g, \varphi \rangle \quad \forall \varphi \in W^{1,2}_{\Gamma_D}(\Omega), \\
|\sigma^t_\tau(\theta, v)| \le g_\tau, \ \sigma^t_\tau(\theta, v) \cdot v_\tau + g_\tau |v_\tau| = 0 \quad \text{on } \Gamma_8, \\
|\sigma^t_n(\theta, v, p)| \le g_n, \ \sigma^t_n(\theta, v, p) v_n + g_n |v_n| = 0 \quad \text{on } \Gamma_9, \\
\sigma^t_n(\theta, v, p) + g_{+n} \ge 0, \ (\sigma^t_n(\theta, v, p) + g_{+n}) v_n = 0 \quad \text{on } \Gamma_{10}, \\
\sigma^t_n(\theta, v, p) - g_{-n} \le 0, \ (\sigma^t_n(\theta, v, p) - g_{-n}) v_n = 0 \quad \text{on } \Gamma_{11}.
\end{cases}
\tag{7.21}
$$

Define $a_2(\cdot,\cdot,\cdot)$ by

$$a_2(v,u,w) = \langle \operatorname{rot} v \times u, w\rangle \quad \forall v,u,w \in \mathbf{V}. \tag{7.22}$$

Then, taking into account

$$\sigma_\tau^t(\theta,v) = 2\mu(\theta)\varepsilon_{n\tau}(v), \quad \sigma_n^t(\theta,v,p) = -(p+\frac{1}{2}|v|^2) + 2\mu(\theta)\varepsilon_{nn}(v)$$

and (7.21), we introduce the following variational formulation for problem (7.1), (7.2) (7.4).

Problem II-VE. Find $(v,\theta,\sigma_\tau^t,\sigma_n^t,\sigma_{+n}^t,\sigma_{-n}^t) \in K(\Omega) \times W^{1,2}_{\Gamma_D}(\Omega) \times \mathbf{L}^2_\tau(\Gamma_8) \times L^2(\Gamma_9) \times H^{-1/2}(\Gamma_{10}) \times H^{-1/2}(\Gamma_{11})$ such that

$$\begin{cases} a_0(\theta;v,u) + a_2(v,v,u) - (\sigma_\tau^t,u_\tau)_{\Gamma_8} - (\sigma_n^t,u_n)_{\Gamma_9} \\ \quad -\left\langle\sigma_{+n}^t,u_n\right\rangle_{\Gamma_{10}} - \left\langle\sigma_{-n}^t,u_n\right\rangle_{\Gamma_{11}} - \langle f-\alpha_0\theta f,u\rangle = \langle f_1,u\rangle \quad \forall u\in\mathbf{V}, \\ b_0(\theta;\theta,\varphi) - \langle\gamma(\theta)\theta v,\nabla\varphi\rangle = \langle g_1,\varphi\rangle \quad \forall\varphi\in W^{1,2}_{\Gamma_D}(\Omega), \\ |\sigma_\tau^t| \le g_\tau,\ \sigma_\tau^t\cdot v_\tau + g_\tau|v_\tau| = 0 \quad \text{on } \Gamma_8, \\ |\sigma_n^t| \le g_n,\ \sigma_n^t v_n + g_n|v_n| = 0 \quad \text{on } \Gamma_9, \\ \sigma_{+n}^t + g_{+n} \ge 0,\ \left\langle\sigma_{+n}^t + g_{+n}, v_n\right\rangle_{\Gamma_{10}} = 0 \quad \text{on } \Gamma_{10}, \\ \sigma_{-n}^t - g_{-n} \le 0,\ \left\langle\sigma_{-n}^t - g_{-n}, v_n\right\rangle_{\Gamma_{11}} = 0 \quad \text{on } \Gamma_{11}. \end{cases} \tag{7.23}$$

Relying on Theorem 5.3, in the same way as in Theorem 7.1 we have

Theorem 7.2 *Let Assumption 7.1 hold. If* (v,p,θ) *is a solution in the sense of Definition 7.1 of the problem* (7.1), (7.2), (7.4), *then* $(v,\theta,\sigma_\tau^t|_{\Gamma_8},\sigma_n^t|_{\Gamma_9},\sigma_n^t|_{\Gamma_{10}},\sigma_n^t|_{\Gamma_{11}})$ *is a solution to Problem I-VE.*

Conversely, if $f\in\mathbf{L}^2(\Omega)$ *and Problem I-VE has a smooth solution* $(v,\theta,\sigma_\tau^t,\sigma_n^t,\sigma_+^t,\sigma_-^t)$ *such that* $v\in\boldsymbol{H}^2(\Omega)$, $\theta\in H^2(\Omega)$, $\sigma_+^t\in L^2(\Gamma_{10})$ *and* $\sigma_-^t\in L^2(\Gamma_{11})$, *then there exists* $p\in H^1(\Omega)$ *such that* (v,p,θ) *is a solution to the problem* (7.1), (7.1), (7.4). *Moreover, if at least one of the sets* $\Gamma_i, i=2,4,6,7$, *is nonempty, then* p *is unique.*

Then, using Theorem 5.4, in the same way as used for Problem I-VI we get Problem II-VI equivalent to Problem II-VE which consists of a variational inequality for velocity and a variational equation for temperature.

Problem II-VI. Find $(v,\theta)\in\mathbf{V}\times W^{1,2}_{\Gamma_D}(\Omega)$ such that

$$\begin{cases} a_0(\theta;v,u-v) + a_2(v,v,u-v) + \Phi(u) - \Phi(v) - \langle f-\alpha_0\theta f,u-v\rangle \\ \qquad\qquad \ge \langle f_1,u-v\rangle \quad \forall u\in\mathbf{V}, \\ b_0(\theta;\theta,\varphi) - \langle\gamma(\theta)\theta v,\nabla\varphi\rangle = \langle g_1,\varphi\rangle \quad \forall\varphi\in W^{1,2}_{\Gamma_D}(\Omega). \end{cases} \tag{7.24}$$

where $a_2(\cdot,\cdot,\cdot)$ is the one in (7.22).

In view of the results above, we will study Problem I-VI and Problem II-VI in the following sections.

7.2 Existence and Uniqueness of Solutions: The Case of Static Pressure

The aim of this section is to prove the following

Theorem 7.3 *Under Assumption 7.1 assume that $f, \phi_i, i = 2, \cdots, 7, g, g_R$ are small enough (depending on α_0) in the spaces in (3), (4) of Assumption 7.1 (see (7.51)). Then, there exists a solution (v, θ) to Problem I-VI such that*

$$\begin{aligned} &\|v\|_{\mathbf{V}} \leq \frac{\mu_0}{K}, \\ &\|\theta\|_{W^{1,2}(\Omega)} \leq c\Big(\|g_R\|_{L^{4/3}(\Gamma_R)} + \|g\|_{L^{6/5}(\Omega)}\Big), \end{aligned} \tag{7.25}$$

where K is the one in (7.26) below.

If $\mu(\theta), \kappa(\theta)$ and $\gamma(\theta)$ are independent of θ and $\|f\|_{\mathbf{L}^3}$ is small enough, then the solution satisfying $\|v\|_{\mathbf{V}} \leq c$, $\|\theta\|_{W^{1,2}_{\Gamma_D}(\Omega)} \leq c$ for a constant c small enough is unique.

7.2.1 Existence of a Solution to an Auxiliary Problem

Since

$$|a_1(v, v, u)| = \big|\big((v\cdot\nabla)v, u\big)\big| \leq K\|v\|_{\mathbf{V}}^2\|u\|_{\mathbf{V}} \quad \forall v, u \in \mathbf{V}, \tag{7.26}$$

define $\overline{a}_1(v) \in \mathbf{V}^*$ by

$$\langle \overline{a}_1(v), u\rangle = a_1(v, v, u) \quad \forall v, u \in \mathbf{V}.$$

Define $\gamma_\varepsilon(t)$ by

$$\gamma_\varepsilon(t) := \frac{\gamma(t)t}{(1+\varepsilon|\gamma(t)|)(1+\varepsilon|t|)} \quad t \in \mathbb{R},\ \varepsilon > 0.$$

Then,

$$|\gamma_\varepsilon(t)| \leq \frac{1}{\varepsilon^2}, \quad |\gamma_\varepsilon(t)| \leq |\gamma(t)||t| \leq \gamma_0|t|, \quad \gamma_\varepsilon(t) \to \gamma(t)t \quad \text{as } \varepsilon \to 0. \tag{7.27}$$

For every $\varepsilon > 0$, let Φ_ε be the Moreau-Yosida approximation of Φ and $\nabla\Phi_\varepsilon$ be Fréchet derivative of Φ_ε.

We first consider an auxiliary problem involving two parameters δ, ζ concerned with the norm of velocity (which is useful when there is fluid flux across a portion of boundary), one parameter λ concerned with the norm of temperature (which is useful to deal with buoyancy effect) and a parameter ε for approximation.

Problem I-VIA. Let $\delta > 0$, $\zeta > 0$, $\lambda > 0$ and $\varepsilon > 0$. Find $(v, \theta) \in \mathbf{V} \times W^{1,2}_{\Gamma_D}(\Omega)$ such that

$$\begin{cases} a_0(\theta; v, u) + \dfrac{\delta}{\max\{\delta, \|\overline{a}_1(v)\|_{\mathbf{V}^*}\}} a_1(v, v, u) + \big\langle \nabla\Phi_\varepsilon(v), u\big\rangle \\ \qquad\qquad - \Big\langle\Big(1 - \dfrac{\lambda}{\max\{\lambda, \|\theta\|\}}\alpha_0\theta\Big) f, u\Big\rangle = \langle f_1, u\rangle \quad \forall u \in \mathbf{V}, \\ b_0(\theta; \theta, \varphi) - \dfrac{\zeta}{\max\{\zeta, \|v\|_{\mathbf{V}}\}} \langle \gamma_\varepsilon(\theta) v, \nabla\varphi\rangle = \langle g_1, \varphi\rangle \quad \forall \varphi \in W^{1,2}_{\Gamma_D}(\Omega). \end{cases} \tag{7.28}$$

Theorem 7.4 *There exists a solution $(v_\varepsilon, \theta_\varepsilon) \in \mathbf{V} \times W^{1,2}_{\Gamma_D}(\Omega)$ to Problem I-VIA.*

Proof Let $\mathscr{H} = \mathbf{V} \times W^{1,2}_{\Gamma_D}(\Omega)$. Define an operator $\mathscr{A} : \mathscr{H} \to \mathscr{H}^*$ by

$$\begin{aligned} \big\langle \mathscr{A}(v, \theta), (u, \phi)\big\rangle =& a_0(\theta; v, u) + \frac{\delta}{\max\{\delta, \|\overline{a}_1(v)\|_{\mathbf{V}^*}\}} a_1(v, v, u) + \big\langle \nabla\Phi_\varepsilon(v), u\big\rangle \\ & - \Big\langle\Big(1 - \frac{\alpha_0\lambda}{\max\{\lambda, \|\theta\|\}}\theta\Big) f, u\Big\rangle + b_0(\theta; \theta, \phi) \\ & - \frac{\zeta}{\max\{\zeta, \|v\|_{\mathbf{V}}\}} \langle \gamma_\varepsilon(\theta) v, \nabla\phi\rangle \quad \forall (v, \theta), (u, \phi) \in \mathscr{H}. \end{aligned} \tag{7.29}$$

Let us first check that operator $\mathscr{A}$ is well-defined. By the definition of Φ_ε, $\Phi_\varepsilon(0_{\mathbf{V}}) = 0$ and $\nabla\Phi_\varepsilon(0_{\mathbf{V}}) = 0$. Since $\nabla\Phi_\varepsilon$ is Lipschitz continuous with the constant ε^{-1} (see Remark 1.16),

$$\big|\big\langle \nabla\Phi_\varepsilon(v), u\big\rangle\big| = \big|\big\langle \nabla\Phi_\varepsilon(v) - \nabla\Phi_\varepsilon(0_{\mathbf{V}}), u\big\rangle\big| \le \varepsilon^{-1}\|v\|_{\mathbf{V}}\|u\|_{\mathbf{V}}. \tag{7.30}$$

By (7.26) we have

$$\Big|\frac{\delta}{\max\{\delta, \|\overline{a}_1(v)\|_{\mathbf{V}^*}\}} a_1(v, v, u)\Big| \le \delta\|u\|_{\mathbf{V}}. \tag{7.31}$$

Also, by Hölder's inequality we have

$$\Big|\Big\langle \frac{\alpha_0\lambda}{\max\{\lambda, \|\theta\|\}}\theta f, u\Big\rangle\Big| \le c\alpha_0\lambda\|f\|_{\mathbf{L}^3}\|u\|_{\mathbf{V}}. \tag{7.32}$$

By (7.27) we have

$$\left|\frac{\zeta}{\max\{\zeta,\, \|v\|_{\mathbf{V}}\}} \langle \gamma_\varepsilon(\theta) v,\, \nabla\phi\rangle\right| \le \frac{c\zeta}{\varepsilon^2}\|\phi\|_{W^{1,2}}. \tag{7.33}$$

Estimation of other terms of (7.29) is easy, and so operator $\mathscr{A}$ is well defined.

Then, existence of a solution to Problem I-VIA is equivalent to existence of a solution to

$$\mathscr{A}(v,\theta) = \mathscr{F}, \quad \mathscr{F} = \begin{pmatrix} f_1 \\ g_1 \end{pmatrix}.$$

We will use Theorem 1.45 to prove the existence of a solution to the equation above. To this end, we need to show that $\mathscr{A}$ satisfies the requirement in Theorem 1.45.

(i) Let us prove that $\mathscr{A}$ is coercive, i.e.

$$\frac{1}{\|(v,\theta)\|_{\mathscr{H}}} \big\langle \mathscr{A}(v,\theta),\, (v,\theta)\big\rangle \to \infty \quad \text{as } \|(v,\theta)\|_{\mathscr{H}} \to \infty.$$

Since $\Gamma_{2j}, \Gamma_{3j}, \Gamma_{7j}$ are convex (see Lemma 3.1) and the matrix α is positive, we have from (7.14)

$$a_0(\theta;\, v, v) \ge 2\mu_0 \|v\|_{\mathbf{V}}^2. \tag{7.34}$$

Since the operator $\nabla\Phi_\varepsilon$ is monotone and $\nabla\Phi_\varepsilon(0_{\mathbf{V}}) = 0$,

$$\big\langle \nabla\Phi_\varepsilon(v),\, v\big\rangle = \big\langle \nabla\Phi_\varepsilon(v) - \nabla\Phi_\varepsilon(0_{\mathbf{V}}),\, v - 0_{\mathbf{V}}\big\rangle \ge 0. \tag{7.35}$$

Taking into account (7.31)–(7.35), we have from (7.29)

$$\begin{aligned}
\big\langle \mathscr{A}(v,\theta), (v,\theta)\big\rangle &= a_0(\theta;\, v, v) + \frac{\delta}{\max\{\delta,\, \|\overline{a}_1(v)\|_{\mathbf{V}^*}\}} a_1(v, v, v) + \big\langle \nabla\Phi_\varepsilon(v),\, v\big\rangle \\
&\quad - \Big\langle\Big(1 - \frac{\alpha_0\lambda}{\max\{\lambda,\, \|\theta\|\}}\theta\Big) f,\, v\Big\rangle + b_0(\theta;\, \theta, \theta) - \frac{\zeta}{\max\{\zeta,\, \|v\|_{\mathbf{V}}\}} \langle \gamma_\varepsilon(\theta) v,\, \nabla\theta\rangle \\
&\ge \min\{2\mu_0, \kappa_0\}\big(\|v\|_{\mathbf{V}}^2 + \|\theta\|^2_{W^{1,2}_{\Gamma_D}}\big) - \delta\|v\| - \|f\|_{\mathbf{V}^*}\|v\|_{\mathbf{V}} - c\lambda\|f\|_{\mathbf{L}^3}\|v\|_{\mathbf{V}} - \frac{c\zeta}{\varepsilon^2}\|\theta\|_{W^{1,2}_{\Gamma_D}} \\
&\quad \forall (v,\theta) \in \mathscr{H},
\end{aligned} \tag{7.36}$$

which implies coercive property of $\mathscr{A}$.

(ii) Taking into account (7.30)–(7.33), we have from (7.29)

$$\begin{aligned}
\|\mathscr{A}(v,\theta)\|_{\mathscr{H}^*} &= \sup_{\|(u,\phi)\|_{\mathscr{H}} = 1} \big\langle \mathscr{A}(v,\theta),\, (u,\phi)\big\rangle \\
&\le c\Big(\|v\|_{\mathbf{V}} + \delta + \frac{1}{\varepsilon}\|v\|_{\mathbf{V}} + \|f\|_{\mathbf{L}^3} + \lambda\|f\|_{\mathbf{L}^3} + \|\theta\|_{W^{1,2}_{\Gamma_D}} + \frac{\zeta}{\varepsilon^2}\Big) \\
&\quad \forall (v,\theta) \in \mathscr{H},
\end{aligned} \tag{7.37}$$

which shows that $\mathscr{A}$ maps bounded sets of $\mathscr{H}$ into bounded sets of $\mathscr{H}^*$.

(iii) Let $\{(v_k, \eta_k)\}$ be a sequence such that

$$(v_k, \eta_k) \rightharpoonup (v, \eta) \quad \text{in } \mathscr{H},$$
$$\limsup_{k\to\infty} \langle \mathscr{A}(v_k, \eta_k), (v_k, \eta_k) - (v, \eta)\rangle \leq 0.$$

By taking a subsequence and denoting with the same subindex if necessary, we may assume

$$\begin{aligned} &v_k \to v \quad \text{in } \mathbf{L}^s(\Omega)(1 \leq s < 6) \text{ and a.e. in } \Omega, \\ &\eta_k \to \eta \quad \text{in } L^s(\Omega)(1 \leq s < 6) \text{ and a.e. in } \Omega \qquad \text{as } k \to \infty. \end{aligned} \tag{7.38}$$

Since

$$a_0(\eta_k; v_k - v, v_k - v) = a_0(\eta_k; v_k, v_k - v) - a_0(\eta_k; v, v_k - v),$$
$$b_0(\eta_k; \eta_k - \eta, \eta_k - \eta) = b_0(\eta_k; \eta_k, \eta_k - \eta) - b_0(\eta_k; \eta, \eta_k - \eta),$$

by (7.29) we have

$$\begin{aligned} \min\{\mu_0, \kappa_0\}\big(\|v_k - v\|_{\mathbf{V}}^2 + \|\eta_k - \eta\|_{W^{1,2}_{\Gamma_D}}\big) &\leq \langle \mathscr{A}(v_k, \eta_k), (v_k, \eta_k) - (v, \eta)\rangle \\ &- a_0(\eta_k; v, v_k - v) - b_0(\eta_k; \eta, \eta_k - \eta) \\ &- \frac{\delta}{\max\{\delta, \|\overline{a}_1(v_k)\|_{\mathbf{V}^*}\}} a_1(v_k, v_k, v_k - v) - \big\langle \nabla\Phi_\varepsilon(v_k), v_k - v\big\rangle \\ &+ \Big\langle\Big(1 - \frac{\alpha_0 \lambda}{\max\{\lambda, \|\eta_k\|\}}\eta_k\Big) f, v_k - v\Big\rangle + \frac{\zeta}{\max\{\zeta, \|v_k\|_{\mathbf{V}}\}} \langle \gamma_\varepsilon(\eta_k) v_k, \nabla(\eta_k - \eta)\rangle. \end{aligned} \tag{7.39}$$

By Corollary 1.1 we have that

$$\begin{aligned} a_0(\eta_k; v, v_k - v) &= 2(\mu(\eta_k)\mathcal{E}(v), \mathcal{E}(v_k - v)) + 2(\mu(\eta_k)k(x)v, v_k - v)_{\Gamma_2} \\ &\quad + 2(\mu(\eta_k)S\tilde{v}, \tilde{v}_k - \tilde{v})_{\Gamma_3} + 2(\alpha(x)v, v_k - v)_{\Gamma_5} + (\mu(\eta_k)k(x)v, v_k - v)_{\Gamma_7} \to 0, \\ b_0(\eta_k; \eta, \eta_k - \eta) &= \big(\kappa(\eta_k)\nabla(\eta), \nabla(\eta_k - \eta)\big) + \big(\beta(x)\eta, (\eta_k - \eta)\big)_{\Gamma_R} \to 0 \end{aligned} \tag{7.40}$$

as $k \to \infty$. Also,

$$\Big|\frac{\delta}{\max\{\delta, \|\overline{a}_1(v_k)\|_{\mathbf{V}^*}\}} a_1(v_k, v_k, v_k - v)\Big| \leq \|v_k\|_{\mathbf{L}^4}\|\nabla v_k\|_{\mathbf{L}^2}\|v_k - v\|_{\mathbf{L}^4} \to 0 \quad \text{as } k \to \infty. \tag{7.41}$$

Since $\nabla\Phi_\varepsilon$ is monotone,

$$\begin{aligned} -\big\langle \nabla\Phi_\varepsilon(v_k), v_k - v\big\rangle &= -\big\langle \nabla\Phi_\varepsilon(v_k) - \nabla\Phi_\varepsilon(v), v_k - v\big\rangle - \big\langle \nabla\Phi_\varepsilon(v), v_k - v\big\rangle \\ &\leq -\big\langle \nabla\Phi_\varepsilon(v), v_k - v\big\rangle \to 0 \quad \text{as } k \to \infty. \end{aligned} \tag{7.42}$$

By (7.38), the followings hold.

$$
\begin{aligned}
&\left|\left\langle\frac{\alpha_0\lambda}{\max\{\lambda,\|\eta_k\|\}}\eta_k f, v_k - v\right\rangle\right| \le c\|\eta_k\|_{\mathbf{L}^3}\|f\|_{\mathbf{L}^3}\|v_k - v\|_{\mathbf{L}^3} \to 0,\\
&\left|\frac{\zeta}{\max\{\zeta,\|v_k\|_{\mathbf{V}}\}}\langle\gamma_\varepsilon(\eta_k)v_k, \nabla(\eta_k-\eta)\rangle\right| \le |\langle\big(\gamma_\varepsilon(\eta_k)-\gamma_\varepsilon(\eta)\big)v_k, \nabla(\eta_k-\eta)\rangle|\\
&\qquad\qquad + |\langle\gamma_\varepsilon(\eta)(v_k - v), \nabla(\eta_k-\eta)\rangle| + |\langle\gamma_\varepsilon(\eta)v, \nabla(\eta_k-\eta)\rangle|\\
&\qquad \le c\|\gamma_\varepsilon(\eta_k)-\gamma_\varepsilon(\eta)\|_{L^3}\|v_k\|_{\mathbf{V}}\|\eta_k-\eta\|_{W^{1,2}_{\Gamma_D}}\\
&\qquad\qquad + c\|\gamma_\varepsilon(\eta)\|_{L^6}\|(v_k - v)\|_{\mathbf{L}^3}\|\eta_k-\eta\|_{W^{1,2}_{\Gamma_D}}\\
&\qquad\qquad + \langle\gamma_\varepsilon(\eta)v, \nabla(\eta_k-\eta)\rangle \to 0
\end{aligned}
\tag{7.43}
$$

as $k\to\infty$, where the fact that by Lemma 1.3 $\gamma_\varepsilon(\eta_k)\to\gamma_\varepsilon(\eta)$ in $L^3(\Omega)$ as $k\to\infty$ was used. It is easy to prove convergence of other terms on the right hand side of (7.39). Thus, by (7.39)–(7.43) we have

$$
\begin{aligned}
&\limsup_{k\to\infty}\min\{\mu_0,\kappa_0\}|\big(\|v_k - v\|^2_{\mathbf{V}} + \|\eta_k-\eta\|_{W^{1,2}_{\Gamma_D}}\big)\\
&\qquad\qquad \le \limsup_{k\to\infty}\langle\mathscr{A}(v_k,\eta_k),(v_k,\eta_k)-(v,\eta)\rangle \le 0,
\end{aligned}
$$

which implies

$$
\begin{aligned}
&(v_k,\eta_k)\to(v,\eta) \quad \text{in } \mathscr{H} \ \text{ as } k\to\infty,\\
&v_k\to v,\ \ \eta_k\to\eta \quad \text{a.e. in } \Omega \ \text{ as } k\to\infty.
\end{aligned}
\tag{7.44}
$$

By definition of $\mathscr{A}$,

$$
\begin{aligned}
\langle\mathscr{A}(v_k,\eta_k),&(v_k,\eta_k)-(u,\phi)\rangle = a_0(\eta_k; v_k, v_k - u)\\
&+\frac{\delta}{\max\{\delta,\|\overline{a}_1(v_k)\|_{\mathbf{V}^*}\}}a_1(v_k,v_k,v_k-u) + \big\langle\nabla\Phi_\varepsilon(v_k), v_k - u\big\rangle\\
&-\Big\langle\Big(1-\frac{\alpha_0\lambda}{\max\{\lambda,\|\eta_k\|\}}\eta_k\Big)f, v_k - u\Big\rangle + b_0(\eta_k;\eta_k,\eta_k-\phi)\\
&-\langle\frac{\zeta}{\max\{\zeta,\|v_k\|_{\mathbf{V}}\}}\gamma(\eta_k)v_k, \nabla(\eta_k-\phi)\rangle \quad \forall(u,\phi)\in\mathscr{H}.
\end{aligned}
\tag{7.45}
$$

Taking into account (7.44), by Corollary 1.2 we have

$$
\begin{aligned}
&\liminf_{k\to\infty} a_0(\eta_k; v_k, v_k) \ge a_0(\eta; v, v),\\
&\liminf_{k\to\infty} b_0(\eta_k;\eta_k,\eta_k) \ge b_0(\eta;\eta,\eta).
\end{aligned}
\tag{7.46}
$$

The sequence $\{\frac{\alpha_0\lambda}{\max\{\lambda,\|\eta_k\|\}}\eta_k\}$ converges a.e. in Ω to $\frac{\alpha_0\lambda}{\max\{\lambda,\|\eta\|\}}\eta$ and $\left\|\frac{\alpha_0\lambda}{\max\{\lambda,\|\eta_k\|\}}\eta_k\right\|_{L^2}\le$ $\alpha_0\lambda$, and so this sequence weakly converges to $\frac{\alpha_0\lambda}{\max\{\lambda,\|\eta\|\}}\eta$ in L^2 (see Lemma 1.1). Since $f\in\mathbf{L}^3$ and $(v-u)\in\mathbf{L}^6$, we have $f\cdot(v-u)\in L^2$. Thus, taking into account the first formula of (7.43), we have

$$\begin{aligned}\lim_{k\to\infty}\Big\langle\frac{\alpha_0\lambda}{\max\{\lambda,\|\eta_k\|\}}\eta_k f, v_k-u\Big\rangle\\&=\lim_{k\to\infty}\Big\langle\frac{\alpha_0\lambda}{\max\{\lambda,\|\eta_k\|\}}\eta_k f, v_k-v\Big\rangle+\lim_{k\to\infty}\Big\langle\frac{\alpha_0\lambda}{\max\{\lambda,\|\eta_k\|\}}\eta_k f, v-u\Big\rangle\\&=\Big\langle\frac{\alpha_0\lambda}{\max\{\lambda,\|\eta\|\}}\eta f, v-u\Big\rangle.\end{aligned}\tag{7.47}$$

Using the second formula in (7.43), in the same way we get

$$\lim_{k\to\infty}\Big\langle\frac{\zeta}{\max\{\zeta,\|v_k\|_{\mathbf{V}}\}}\gamma_\varepsilon(\eta_k)v_k,\nabla(\eta_k-\phi)\Big\rangle=\Big\langle\frac{\zeta}{\max\{\zeta,\|v\|_{\mathbf{V}}\}}\gamma_\varepsilon(\eta)v,\nabla(\eta-\phi)\Big\rangle.\tag{7.48}$$

Similarly, we have

$$\lim_{k\to\infty}\frac{\delta}{\max\{\delta,\|\overline{a}_1(v_k)\|_{\mathbf{V}^*}\}}a_1(v_k,v_k,v_k-u)=\frac{\delta}{\max\{\delta,\|\overline{a}_1(v)\|_{\mathbf{V}^*}\}}a_1(v,v,v-u).\tag{7.49}$$

Since $\nabla\Phi_\varepsilon$ is monotone and

$$\begin{aligned}\langle\nabla\Phi_\varepsilon(v_k),v_k-u\rangle&=\langle\nabla\Phi_\varepsilon(v_k),v_k-v\rangle+\langle\nabla\Phi_\varepsilon(v_k),v-u\rangle\\&\geq\langle\nabla\Phi_\varepsilon(v_k)-\nabla\Phi_\varepsilon(v),v_k-v\rangle+\langle\nabla\Phi_\varepsilon(v),v_k-v\rangle+\langle\nabla\Phi_\varepsilon(v_k),v-u\rangle,\end{aligned}$$

we have

$$\lim_{k\to\infty}\langle\nabla\Phi_\varepsilon(v_k),v_k-u\rangle\geq\langle\nabla\Phi_\varepsilon(v),v-u\rangle.\tag{7.50}$$

Thus, by (7.46)–(7.50) we have existence of a subsequence $\{(v_k,\eta_k)\}$ such that

$$\liminf_{k\to\infty}\langle\mathscr{A}(v_k,\eta_k),(v_k,\eta_k)-(u,\phi)\rangle\geq\langle\mathscr{A}(v,\eta),(v,\eta)-(u,\phi)\rangle.$$

Therefore, by virtue of Theorem 1.45 we come to the conclusion. □

7.2.2 *Existence and Estimates of Solutions to the Approximate Problem*

In this section the following theorem is proved.

Theorem 7.5 *If*

$$\|g_1\|_{(W^{1,2}_{\Gamma_D})^*}\|f\|_{\mathbf{L}^3}+\|f+f_1\|_{\mathbf{V}^*}\leq\frac{\mu_0^2}{Kc_{\alpha_0}},\tag{7.51}$$

where K is the one in (7.26) and c_{α_0} is the one in (7.63) below, then there exists a solution $(v_\varepsilon,\theta_\varepsilon)\in\mathbf{V}\times W^{1,2}_{\Gamma_D}(\Omega)$ to the following problem

$$\begin{cases} a_0(\theta_\varepsilon; v_\varepsilon, u) + a_1(v_\varepsilon, v_\varepsilon, u) + \big\langle \nabla \Phi_\varepsilon(v_\varepsilon), u\big\rangle - \big\langle (1-\alpha_0\theta_\varepsilon) f, u\big\rangle = \langle f_1, u\rangle \quad \forall u \in \mathbf{V}, \\ b_0(\theta_\varepsilon; \theta_\varepsilon, \varphi) - \langle \gamma_\varepsilon(\theta_\varepsilon) v_\varepsilon, \nabla\varphi\rangle = \langle g_1, \varphi\rangle \qquad \forall \varphi \in W^{1,2}_{\Gamma_D}(\Omega), \end{cases} \tag{7.52}$$

and the solution satisfies:

$$\begin{aligned} &\|v_\varepsilon\|_V \le \frac{\mu_0}{K}, \\ &\|\theta_\varepsilon\|_{W^{1,2}_{\Gamma_D}(\Omega)} \le c\|g_1\|_{(W^{1,2}_{\Gamma_D})^*}. \end{aligned} \tag{7.53}$$

Proof Let $(v_\varepsilon, \theta_\varepsilon)$ be solutions to (7.28). Putting $\varphi = \theta_\varepsilon$ in the second formula of (7.28), we have

$$(\kappa(\theta_\varepsilon)\nabla\theta_\varepsilon, \nabla\theta_\varepsilon) + (\beta(x)\theta_\varepsilon, \theta_\varepsilon)_{\Gamma_R} - \frac{\zeta}{\max\{\zeta, \|v\|_{\mathbf{V}}\}}\langle \gamma_\varepsilon(\theta_\varepsilon) v_\varepsilon, \nabla\theta_\varepsilon\rangle = \langle g_1, \theta_\varepsilon\rangle. \tag{7.54}$$

Let us first prove

$$\langle \gamma_\varepsilon(\theta_\varepsilon) v_\varepsilon, \nabla\theta_\varepsilon\rangle = 0. \tag{7.55}$$

To this end, define

$$\Psi(t) := \int_0^t \gamma_\varepsilon(s)\, ds, \quad t \in \mathbb{R}.$$

Then, $\Psi \in C^1(\mathbb{R})$ and

$$\begin{aligned} &\nabla\Psi(\theta) = \gamma_\varepsilon(\theta)\nabla\theta, \quad \Psi(\theta) \in W^{1,2}(\Omega) \quad \forall \theta \in W^{1,2}(\Omega), \\ &\Psi(\theta)|_{\Gamma_D} = 0 \quad \forall \theta \in W^{1,2}_{\Gamma_D}(\Omega). \end{aligned} \tag{7.56}$$

Taking into account $v_\varepsilon \cdot n|_{\Gamma_R} = 0$, by (7.56) we have

$$\langle \gamma_\varepsilon(\theta) v_\varepsilon, \nabla\theta_\varepsilon\rangle = \int_\Omega \gamma_\varepsilon(\theta_\varepsilon) v_\varepsilon \cdot \nabla\theta_\varepsilon\, dx = \int_\Omega v_\varepsilon \cdot \nabla\Psi(\theta_\varepsilon) = 0,$$

which means (7.55).

Next, we estimate other terms in (7.54).

$$\begin{aligned} &\big|\langle g_1, \theta_\varepsilon\rangle\big| \le \frac{\kappa_0}{4}\|\theta_\varepsilon\|^2_{H^1} + c\|g_1\|^2_{(W^{1,2}_{\Gamma_D})^*}, \\ &(\beta(x)\theta_\varepsilon, \theta_\varepsilon)_{\Gamma_R} \ge 0. \end{aligned} \tag{7.57}$$

By (7.54), (7.54) and (7.57), we have

$$\|\theta_\varepsilon\|^2_{W^{1,2}_{\Gamma_D}} \le \frac{2c}{\kappa_0}\|g_1\|^2_{(W^{1,2}_{\Gamma_D})^*}, \tag{7.58}$$

which implies

$$\|\theta_\varepsilon\| \le c_1 \|g_1\|_{(W^{1,2}_{\Gamma_D})^*}. \tag{7.59}$$

Putting

$$\lambda = c_1 \|g_1\|_{(W^{1,2}_{\Gamma_D})^*}$$

and taking into account (7.59), we have

$$\frac{\lambda}{\max\{\lambda, \|\theta_\varepsilon\|\}} = 1, \tag{7.60}$$

$$\left|\left\langle \frac{\alpha_0 \lambda}{\max\{\lambda, \|\theta_\varepsilon\|\}} \theta_\varepsilon f, u \right\rangle\right| \le c_1 \alpha_0 \|g_1\|_{(W^{1,2}_{\Gamma_D})^*} \|f\|_{\mathbf{L}^3} \|u\|_{\mathbf{V}}. \tag{7.61}$$

Putting $u = v_\varepsilon$ in the first equation of (7.28), we have

$$\begin{aligned} a_0(\theta_\varepsilon; v_\varepsilon, v_\varepsilon) &+ \frac{\delta}{\max\{\delta, \|\overline{a}_1(v_\varepsilon)\|_{\mathbf{V}^*}\}} a_1(v_\varepsilon, v_\varepsilon, v_\varepsilon) + \big\langle \nabla \Phi_\varepsilon(v_\varepsilon), v_\varepsilon \big\rangle \\ &- \left\langle \left(1 - \frac{\alpha_0 \lambda}{\max\{\lambda, \|\theta_\varepsilon\|\}} \theta_\varepsilon\right) f, v_\varepsilon \right\rangle = \langle f_1, v_\varepsilon \rangle. \end{aligned} \tag{7.62}$$

Thus taking into account (7.26), (7.34), (7.35) and (7.61), we have from (7.62)

$$\begin{aligned} 2\mu_0 \|v_\varepsilon\|^2_{\mathbf{V}} &\le a_0(\theta_\varepsilon; v_\varepsilon, v_\varepsilon) \\ &\le \frac{\delta}{\max\{\delta, \|\overline{a}_1(v_\varepsilon)\|_{\mathbf{V}^*}\}} |a_1(v_\varepsilon, v_\varepsilon, v_\varepsilon)| + |\alpha_0 \langle \theta_\varepsilon f, v_\varepsilon \rangle| + |\langle f + f_1, v_\varepsilon \rangle| \\ &\le K \|v_\varepsilon\|^3_{\mathbf{V}} + c_{\alpha_0} \big(\|g_1\|_{(W^{1,2}_{\Gamma_D})^*} \|f\|_{\mathbf{L}^3} + \|f + f_1\|_{\mathbf{V}^*} \big) \|v_\varepsilon\|_{\mathbf{V}}. \end{aligned} \tag{7.63}$$

Note that the estimate above is independent of δ, but c_{α_0} depends on α_0. This implies

$$0 \le K \|v_\varepsilon\|^2_{\mathbf{V}} - 2\mu_0 \|v_\varepsilon\|_{\mathbf{V}} + c_{\alpha_0} \big(\|g_1\|_{(W^{1,2}_{\Gamma_D})^*} \|f\|_{\mathbf{L}^3} + \|f + f_1\|_{\mathbf{V}^*} \big).$$

Let us consider a quadratic polynomial for $x > 0$ related to the inequality above

$$Kx^2 - 2\mu_0 x + a.$$

If $0 \le Ka \le \mu_0^2$, then there exists a nonnegative minimum root $x_1 (\le \frac{\mu_0}{K})$ and a maximum root x_2. Thus, we can see that if

$$\|g_1\|_{(W^{1,2}_{\Gamma_D})^*} \|f\|_{\mathbf{L}^3} + \|f + f_1\|_{\mathbf{V}^*} \le \frac{\mu_0^2}{K c_{\alpha_0}}, \tag{7.64}$$

then

$$\|v_\varepsilon\|_{\mathbf{V}} \leq \frac{\mu_0}{K} \quad \text{or} \quad \|v_\varepsilon\|_{\mathbf{V}} \geq x_2. \tag{7.65}$$

On the other hand, we have from (7.63) another estimation under consideration of δ

$$2\mu_0\|v_\varepsilon\|_{\mathbf{V}}^2 \leq a_0(\theta_\varepsilon; v_\varepsilon, v_\varepsilon) \leq \delta\|v_\varepsilon\|_{\mathbf{V}} + c_{\alpha_0}\big(\|g_1\|_{(W^{1,2}_{\Gamma_D})^*}\|f\|_{\mathbf{L}^3} + \|f + f_1\|_{\mathbf{V}^*}\big)\|v_\varepsilon\|_{\mathbf{V}},$$

which implies

$$\|v_\varepsilon\|_{\mathbf{V}} \leq \frac{1}{2\mu_0}\Big(\delta + c_{\alpha_0}\big(\|g_1\|_{(W^{1,2}_{\Gamma_D})^*}\|f\|_{\mathbf{L}^3} + \|f + f_1\|_{\mathbf{V}^*}\big)\Big). \tag{7.66}$$

In view of (7.65), let us take $\delta = K\left(\frac{\mu_0}{K}\right)^2 = \frac{\mu_0^2}{K}$.

Thus, in view of (7.51), we have from (7.66)

$$\|v_\varepsilon\|_{\mathbf{V}} \leq \frac{\delta}{2\mu_0} + \frac{1}{2\mu_0}\frac{\mu_0^2}{K} = \frac{\mu_0}{K}. \tag{7.67}$$

By (7.67) under the condition (7.51) we have that $\|\bar{a}_1(v_\varepsilon)\|_{\mathbf{V}^*} \leq K\|v_\varepsilon\|_{\mathbf{V}}^2 \leq \frac{\mu_0^2}{K}$ (see (7.26)), and so we get

$$\frac{\delta}{\max\{\delta, \|\bar{a}_1(v_\varepsilon)\|_{\mathbf{V}^*}\}} = 1. \tag{7.68}$$

Taking $\zeta = \frac{\mu_0}{K}$, by (7.67) we get

$$\frac{\zeta}{\max\{\zeta, \|v_\varepsilon\|_{\mathbf{V}}\}} = 1. \tag{7.69}$$

By (7.60), (7.68) and (7.69), we see that under condition (7.64) $(v_\varepsilon, \theta_\varepsilon)$ satisfies (7.52). By virtue of (7.67) and (7.58) we get (7.53). □

7.2.3 Existence and Uniqueness of a Solution

First, by passing to the limit of solutions in Theorem 7.5, we will prove existence of a solution to Problem I-VI. Owing to (7.53) we can extract subsequences, which are denoted as before, such that

$$\begin{aligned}
&v_\varepsilon \rightharpoonup v \quad \text{in } \mathbf{V},\\
&v_\varepsilon \to v \quad \text{in } \mathbf{L}^q, 1 \leq q < 6,\\
&\theta_\varepsilon \rightharpoonup \theta \quad \text{in } H^1(\Omega),\\
&\theta_\varepsilon \to \theta \quad \text{in } L^q(\Omega), 1 \leq q < 6,
\end{aligned} \tag{7.70}$$

as $\varepsilon \to 0$.

Subtracting the first formula of (7.52) with $u = v_\varepsilon$ from the first formula of (7.52), we have

$$\begin{aligned} a_0(\theta_\varepsilon; v_\varepsilon, u - v_\varepsilon) + a_1(v_\varepsilon, v_\varepsilon, u - v_\varepsilon) + \big\langle \nabla \Phi_\varepsilon(v_\varepsilon), u - v_\varepsilon \big\rangle \\ - \big\langle (1 - \alpha_0 \theta_\varepsilon) f, u - v_\varepsilon \big\rangle = \langle f_1, u - v_\varepsilon \rangle \quad \forall u \in \mathbf{V}. \end{aligned} \tag{7.71}$$

By Corollaries 1.1 and 1.2, we have

$$\begin{aligned} &a_0(\theta_\varepsilon; v_\varepsilon, u) \to a_0(\theta; v, u) \quad \text{as } \varepsilon \to 0, \\ &\liminf_{\varepsilon \to 0} a_0(\theta_\varepsilon; v_\varepsilon, v_\varepsilon) \geq a_0(\theta; v, v), \end{aligned}$$

which imply that

$$\limsup_{\varepsilon \to 0} a_0(\theta_\varepsilon; v_\varepsilon, u - v_\varepsilon) \leq a_0(\theta; v, u - v). \tag{7.72}$$

It is easy to prove

$$a_1(v_\varepsilon, v_\varepsilon, u - v_\varepsilon) \to a_1(v, v, u - v) \quad \text{as } \varepsilon \to 0. \tag{7.73}$$

Since Φ_ε is convex, continuous and Fréchet differentiable, we have

$$\Phi_\varepsilon(u) - \Phi_\varepsilon(v_\varepsilon) \geq \langle \nabla \Phi_\varepsilon(v_\varepsilon), u - v_\varepsilon \rangle \quad \forall u \in \mathbf{V}, \tag{7.74}$$

which, together with (1.37), implies

$$\Phi_\varepsilon(u) - \Phi(J_\varepsilon v_\varepsilon) \geq \langle \nabla \Phi_\varepsilon(v_\varepsilon), u - v_\varepsilon \rangle \quad \forall u \in \mathbf{V}. \tag{7.75}$$

Since $\Phi(0_{\mathbf{V}}) = 0$, by (1.37) we get $\Phi_\varepsilon(0_{\mathbf{V}}) = 0$, and so we have from (7.74)

$$\Phi_\varepsilon(v_\varepsilon) \leq \langle \nabla \Phi_\varepsilon(v_\varepsilon), v_\varepsilon \rangle. \tag{7.76}$$

On the other hand, putting $u = v_\varepsilon$ in the first formula of (7.52), we have

$$a_0(\theta_\varepsilon; v_\varepsilon, v_\varepsilon) + a_1(v_\varepsilon, v_\varepsilon, v_\varepsilon) + \big\langle \nabla \Phi_\varepsilon(v_\varepsilon), v_\varepsilon \big\rangle = \big\langle (1 - \alpha_0 \theta_\varepsilon) f, v_\varepsilon \big\rangle + \langle f_1, v_\varepsilon \rangle. \tag{7.77}$$

From (7.76) and (7.77) we have

$$a_0(\theta_\varepsilon; v_\varepsilon, v_\varepsilon) + a_1(v_\varepsilon, v_\varepsilon, v_\varepsilon) + \Phi_\varepsilon(v_\varepsilon) \leq \big\langle (1 - \alpha_0 \theta_\varepsilon) f, v_\varepsilon \big\rangle + \langle f_1, v_\varepsilon \rangle,$$

from which we get

$$|\Phi_\varepsilon(v_\varepsilon)| \le c\big(\|g_1\|_{(W^{1,2}_{\Gamma_D})^*}\|f\|_{\mathbf{L}^3} + \|f\|_{\mathbf{L}^3} + \|f_1\|_{\mathbf{V}^*}\big)\|v_\varepsilon\|_{\mathbf{V}} + |a_1(v_\varepsilon, v_\varepsilon, v_\varepsilon)|. \tag{7.78}$$

By virtue of (1.36), (7.26), (7.66) and (7.78), we have

$$\|v_\varepsilon - J_\varepsilon v_\varepsilon\|^2_{\mathbf{V}} \le \Big[c\big(\|g_1\|_{(W^{1,2}_{\Gamma_D})^*}\|f\|_{\mathbf{L}^3} + \|f\|_{\mathbf{L}^3} + \|f_1\|_{\mathbf{V}^*}\big)\|v_\varepsilon\|_{\mathbf{V}}\big)\frac{\mu_0}{K} + \frac{\mu_0^3}{K^2}\Big]2\varepsilon,$$

which shows that since $v_\varepsilon \rightharpoonup v$ in $\mathbf{V}$,

$$J_\varepsilon v_\varepsilon \rightharpoonup v \text{ in } \mathbf{V} \quad \text{as } \varepsilon \to 0.$$

Then, by virtue of lower weak semi-continuity of $\Phi(v)$,

$$\liminf_{\varepsilon\to 0} \Phi(J_\varepsilon v_\varepsilon) \ge \Phi(v). \tag{7.79}$$

By (1.37) we have

$$\Phi_\varepsilon(u) \to \Phi(u) \quad \text{as } \varepsilon \to 0. \tag{7.80}$$

Taking into account (7.79) and (7.80), we have from (7.75)

$$\Phi(u) - \Phi(v) \ge \limsup_{\varepsilon\to 0}\langle \nabla\Phi_\varepsilon(v_\varepsilon), u - v_\varepsilon\rangle \quad \forall u \in \mathbf{V}. \tag{7.81}$$

Using

$$\begin{aligned}|\big(\theta_\varepsilon f, v_\varepsilon\big) - \big(\theta f, v\big)| &\le |\big(\theta_\varepsilon f, v_\varepsilon\big) - \big(\theta f, v_\varepsilon\big)| + |\big(\theta f, v_\varepsilon\big) - \big(\theta f, v\big)|\\ &\le \|\theta_\varepsilon - \theta\|_{L^3}\|f\|_{\mathbf{L}^2}\|v_\varepsilon\|_{\mathbf{L}^6} + \|\theta\|_{L^6}\|f\|_{\mathbf{L}^2}\|v_\varepsilon - v\|_{\mathbf{L}^3},\end{aligned} \tag{7.82}$$

we can prove

$$\big((1-\alpha_0\theta_\varepsilon)f, u - v_\varepsilon\big) \to \big((1-\alpha_0\theta)f, u - v\big) \quad \text{as } \varepsilon \to 0. \tag{7.83}$$

It is easy to prove

$$\langle f_1, u - v_\varepsilon\rangle \to \langle f_1, u - v\rangle \quad \text{as } \varepsilon \to 0. \tag{7.84}$$

By virtue of (7.72), (7.73), (7.81), (7.83) and (7.84), from (7.71) we get

$$a_0(\theta; v, u - v) + a_1(v, v, u - v) + \Phi(u) - \Phi(v) - \big((1-\alpha_0\theta)f, u - v\big) \ge \langle f_1, u - v\rangle \quad \forall u \in \mathbf{V},$$

which is the first formula in (7.20).

We will get the second equation in (7.20). By Corollary 1.1, we have

$$b_0(\theta_\varepsilon; \theta_\varepsilon, \varphi) \to b_0(\theta; \theta, \varphi) \quad \forall \varphi \in W^{1,2}_{\Gamma_D}(\Omega) \quad \text{as } \varepsilon \to 0. \tag{7.85}$$

Let us prove

$$\langle \gamma_\varepsilon(\theta_\varepsilon)v_\varepsilon, \nabla\varphi\rangle \to \langle \gamma(\theta)\theta v, \nabla\varphi\rangle \quad \forall\varphi \in W^{1,2}_{\Gamma_D}(\Omega) \quad \text{as } \varepsilon \to 0. \tag{7.86}$$

By Hölder's inequality we have

$$\begin{aligned} &|\langle \gamma_\varepsilon(\theta_\varepsilon)v_\varepsilon, \nabla\varphi\rangle - \langle \gamma(\theta)\theta v, \nabla\varphi\rangle| \\ &\quad\le |\langle \gamma_\varepsilon(\theta_\varepsilon)v_\varepsilon, \nabla\varphi\rangle - \langle \gamma(\theta)\theta v_\varepsilon, \nabla\varphi\rangle| + |\langle \gamma(\theta)\theta v_\varepsilon, \nabla\varphi\rangle - \langle \gamma(\theta)\theta v, \nabla\varphi\rangle| \\ &\quad\le \|\gamma_\varepsilon(\theta_\varepsilon) - \gamma(\theta)\theta\|_{L^3}\|v_\varepsilon\|_{\mathbf{L}^6}\|\nabla\varphi\|_{\mathbf{L}^2} + \|\gamma(\theta)\theta\|_{L^4}\|v_\varepsilon - v\|_{\mathbf{L}^4}\|\nabla\varphi\|_{\mathbf{L}^2}. \end{aligned} \tag{7.87}$$

By the definition of $\gamma_\varepsilon(t)$ we have

$$\begin{aligned} \|\gamma_\varepsilon(\theta_\varepsilon) - \gamma(\theta)\theta\|_{L^3} &\le \left\| \frac{\gamma(\theta_\varepsilon)\theta_\varepsilon}{(1+\varepsilon|\gamma(\theta_\varepsilon)|)(1+\varepsilon|\theta_\varepsilon|)} - \gamma(\theta)\theta \right\|_{L^3} \\ &\le \|\gamma(\theta_\varepsilon)\theta_\varepsilon - \gamma(\theta)\theta\|_{L^3} + \varepsilon \left\|\gamma(\theta)\theta\big(|\gamma(\theta_\varepsilon)| + |\theta_\varepsilon| + \varepsilon|\gamma(\theta_\varepsilon)||\theta_\varepsilon|\big)\right\|_{L^3}. \end{aligned} \tag{7.88}$$

By virtue of Lemma 1.3 we know that $\gamma(\theta_\varepsilon)$ converges to $\gamma(\theta)$ in space $L^p(\Omega)$ ($\forall p,\ 1 < p < \infty,$) as ε goes to zero. Thus, from (7.87) and (7.88) we get (7.86).

By virtue of (7.85) and (7.86), from the second formula in (7.52) we get the second formula in (7.20).

Estimates (7.25) follow from (7.53).

Next, let us prove uniqueness of solutions. Suppose that there are two solutions (v_1, θ_1) and (v_2, θ_2). Since μ is independent of θ by the condition, denoting $a_0(\cdot; v, u)$ by $a_0(v, u)$ yields

$$\begin{aligned} a_0(v_1, v_2 - v_1) + a_1(v_1, v_1, v_2 - v_1) + \Phi(v_2) - \Phi(v_1) - \langle f - \alpha_0\theta_1 f, v_2 - v_1\rangle \\ \ge \langle f_1, v_2 - v_1\rangle, \\ a_0(v_2, v_1 - v_2) + a_1(v_2, v_2, v_1 - v_2) + \Phi(v_1) - \Phi(v_2) - \langle f - \alpha_0\theta_2 f, v_1 - v_2\rangle \\ \ge \langle f_1, v_1 - v_2\rangle, \end{aligned}$$

which imply

$$\begin{aligned} a_0\big(v_1 - v_2, v_1 - v_2\big) \le |\alpha_0||(\theta_1 - \theta_2)f, v_1 - v_2)| \\ + |a_1(v_1, v_1, v_1 - v_2) - a_1(v_2, v_2, v_1 - v_2)|. \end{aligned} \tag{7.89}$$

By virtue of (7.89), we have

$$\begin{aligned} &2\mu\|v_1 - v_2\|^2_{\mathbf{V}} \\ &\quad\le \frac{\mu}{2}\|v_1 - v_2\|^2_{\mathbf{V}} + c\|f\|^2_{\mathbf{L}^3}\|\theta_1 - \theta_2\|^2 \\ &\qquad + |a_1(v_1 - v_2, v_1, v_1 - v_2) + a_1(v_2, v_1 - v_2, v_1 - v_2)| \\ &\quad\le \frac{\mu}{2}\|v_1 - v_2\|^2_{\mathbf{V}} + c\|f\|^2_{\mathbf{L}^3}\|\theta_1 - \theta_2\|^2 + c_1(\|v_1\|_{\mathbf{V}} + \|v_2\|_{\mathbf{V}})\|v_1 - v_2\|^2_{\mathbf{V}} \end{aligned}$$

and so

$$\frac{3\mu}{2}\|v_1 - v_2\|_{\mathbf{V}}^2 \leq c\|f\|_{\mathbf{L}^3}^2\|\theta_1 - \theta_2\|^2 + c_1(\|v_1\|_{\mathbf{V}} + \|v_2\|_{\mathbf{V}})\|v_1 - v_2\|_{\mathbf{V}}^2. \tag{7.90}$$

Since $\kappa(\theta)$, $\gamma(\theta)$ are independent of θ, put $\kappa(\theta) = \kappa$, $\gamma(\theta) = c_v$. Then, from

$$\begin{aligned}
&(\kappa\nabla\theta_1, \nabla\varphi) + (\beta(x)\theta_1, \varphi)_{\Gamma_R} - c_v\langle v_1\theta_1, \nabla\varphi_1\rangle = \langle g_1, \varphi\rangle,\\
&(\kappa\nabla\theta_2, \nabla\varphi) + (\beta(x)\theta_2, \varphi)_{\Gamma_R} - c_v\langle v_2\theta_2, \nabla\varphi\rangle = \langle g_1, \varphi\rangle
\end{aligned}$$

we have

$$\begin{aligned}
&\kappa(\nabla\theta_1 - \nabla\theta_2, \nabla\theta_1 - \nabla\theta_2) + (\beta(x)(\theta_1 - \theta_2), \theta_1 - \theta_2)_{\Gamma_R}\\
&\quad - c_v\langle v_1(\theta_1 - \theta_2), \nabla(\theta_1 - \theta_2)\rangle - c_v\langle (v_1 - v_2)\theta_2, \nabla(\theta_1 - \theta_2)\rangle = 0.
\end{aligned} \tag{7.91}$$

Using $c_v\langle v_1(\theta_1 - \theta_2), \nabla(\theta_1 - \theta_2)\rangle = 0$ (see (7.12) with $\gamma(\theta)$=const), we obtain from (7.91)

$$\kappa\|\nabla\theta_1 - \nabla\theta_2\|^2 \leq \frac{c_v c}{\kappa}\|v_1 - v_2\|_{\mathbf{V}}^2\|\theta_2\|_{W_{\Gamma_D}^{1,2}}^2 + \frac{\kappa}{2}\|\nabla\theta_1 - \nabla\theta_2\|^2,$$

and so

$$\frac{\kappa}{2}\|\theta_1 - \theta_2\|_{W^{1,2}}^2 \leq \frac{c_v c}{\kappa}\|v_1 - v_2\|_{\mathbf{V}}^2\|\theta_2\|_{W_{\Gamma_D}^{1,2}}^2. \tag{7.92}$$

Therefore adding (7.90) and (7.92), we get

$$\begin{aligned}
&\min\{\frac{3\mu}{2}, \frac{\kappa}{2}\}(\|v_1 - v_2\|_{\mathbf{V}}^2 + \|\theta_1 - \theta_2\|_{W^{1,2}}^2)\\
&\leq c\|f\|_{\mathbf{L}^3}^2\|\theta_1 - \theta_2\|^2 + c_1(\|v_1\|_{\mathbf{V}} + \|v_2\|_{\mathbf{V}})\|v_1 - v_2\|_{\mathbf{V}}^2 + \frac{c_v c}{\kappa}\|v_1 - v_2\|_{\mathbf{V}}^2\|\theta_2\|_{W_{\Gamma_D}^{1,2}}^2.
\end{aligned} \tag{7.93}$$

Thus, if $\|v_i\|_{\mathbf{V}}$, $\|\theta_2\|_{W_{\Gamma_D}^{1,2}}$ and $\|f\|_{\mathbf{L}^3}$ are small, then we have from the above that $v_1 = v_2$ and $\theta_1 = \theta_2$. □

7.3 Existence of a Solution: The Case of Total Pressure

Theorem 7.6 *Under Assumption 7.1 there exists a solution (v, θ) to Problem II-VI such that*

$$\begin{aligned}
&\|v\|_{\mathbf{V}} \leq c\big(\|f\|_{\mathbf{L}^3} + \sum_{i=2-7}\|\phi_i\|_{\Gamma_i} + \|g_R\|_{L^{4/3}(\Gamma_R)} + \|g\|_{L^{6/5}(\Omega)}\big),\\
&\|\theta\|_{W^{1,2}(\Omega)} \leq c\Big(\|g_R\|_{L^{4/3}(\Gamma_R)} + \|g\|_{L^{6/5}(\Omega)}\Big).
\end{aligned} \tag{7.94}$$

First, we look for solutions to an auxiliary problem:

Problem II-VIA. Let $\zeta > 0, \lambda > 0$ and $\varepsilon > 0$. Find $(v, \theta) \in \mathbf{V} \times W^{1,2}(\Omega)$ such that $\eta = \theta - \theta_D \in W^{1,2}_{\Gamma_D}(\Omega)$ and

$$\begin{cases} a_0(\theta; v, u) + a_2(v, v, u) + \big\langle \nabla \Phi_\varepsilon(v), u \big\rangle - \Big\langle \Big(1 - \dfrac{\alpha_0 \lambda}{\max\{\lambda, \|\theta\|\}}\theta\Big) f, u \Big\rangle \\ \qquad\qquad\qquad\qquad\qquad = \langle f_1, u \rangle \quad \forall u \in \mathbf{V}, \\ b_0(\theta; \theta, \varphi) - \dfrac{\zeta}{\max\{\zeta, \|v_\varepsilon\|_{\mathbf{V}}\}} \langle \gamma_\varepsilon(\theta) v, \nabla \varphi \rangle = \langle g_1, \varphi \rangle \quad \forall \varphi \in W^{1,2}_{\Gamma_D}(\Omega). \end{cases} \tag{7.95}$$

Theorem 7.7 *There exists a solution $(v_\varepsilon, \theta_\varepsilon) \in \mathbf{V} \times W^{1,2}(\Omega)$ to Problem II-VIA.*

Proof Let $\mathscr{H} = \mathbf{V} \times W^{1,2}_{\Gamma_D}(\Omega)$. Define an operator $\mathscr{A} : \mathscr{H} \to \mathscr{H}^*$ by

$$\begin{aligned} \big\langle \mathscr{A}(v, \eta), (u, \phi) \big\rangle =& a_0(\eta; v, u) + a_2(v, v, u) + \big\langle \nabla \Phi_\varepsilon(v), u \big\rangle \\ & - \Big\langle \Big(1 - \frac{\alpha_0 \lambda}{\max\{\lambda, \|\theta\|\}}\theta\Big) f, u \Big\rangle + b_0(\theta; \theta, \varphi) \\ & - \frac{\zeta}{\max\{\zeta, \|v\|_{\mathbf{V}}\}} \langle \gamma_\varepsilon(\eta + \theta_D) v, \nabla \varphi \rangle \qquad \forall (v, \eta), (u, \phi) \in \mathscr{H}. \end{aligned}$$

Using

$$\begin{aligned} & a_2(v, v, v) = 0, \\ & |a_2(v, v, u)| \le K \|v\|^2_{\mathbf{V}} \|u\|_{\mathbf{V}}, \\ & |a_2(v_\varepsilon, v_\varepsilon, v_\varepsilon - u)| \le c \|\nabla v_\varepsilon\|_{\mathbf{L}^2} \|v_\varepsilon\|_{\mathbf{L}^4} \|v_\varepsilon - u\|_{\mathbf{L}^4}, \end{aligned}$$

respectively, in (7.36), (7.37) and (7.41), we can verify that the proof of Theorem 7.4 for Problem I-VIA is valid for Problem II-VIA. Thus, we come to the asserted conclusion. □

Theorem 7.8 *There exists a solution $(v_\varepsilon, \theta_\varepsilon) \in \mathbf{V} \times W^{1,2}(\Omega)$ to the following problem*

$$\begin{cases} a_0(\theta_\varepsilon; v_\varepsilon, u) + a_2(v_\varepsilon, v_\varepsilon, u) + \big\langle \nabla \Phi_\varepsilon(v_\varepsilon), u \big\rangle - \big\langle (1 - \alpha_0 \theta_\varepsilon) f, u \big\rangle = \langle f_1, u \rangle \quad \forall u \in \mathbf{V}, \\ b_0(\theta_\varepsilon; \theta_\varepsilon, \varphi) - \langle \gamma_\varepsilon(\theta_\varepsilon) v_\varepsilon, \nabla \varphi \rangle = \langle g_1, \varphi \rangle \quad \forall \varphi \in W^{1,2}_{\Gamma_D}(\Omega), \end{cases} \tag{7.96}$$

and the solution satisfies:

$$\begin{aligned} & \|v_\varepsilon\|_{\mathbf{V}} \le \frac{c}{2\mu_0} \big(\|g_1\|_{(W^{1,2}_{\Gamma_D})^*} \|f\|_{\mathbf{L}^3} + \|f\|_{\mathbf{V}^*} + \|f_1\|_{\mathbf{V}^*} \big), \\ & \|\theta_\varepsilon\|_{W^{1,2}_{\Gamma_D}(\Omega)} \le c \|g_1\|_{(W^{1,2}_{\Gamma_D})^*}. \end{aligned} \tag{7.97}$$

Proof Let $(v_\varepsilon, \theta_\varepsilon)$ be solutions to (7.95). In the same way as in (7.54)–(7.58) we have

$$\|\theta_\varepsilon\|^2_{W^{1,2}_{\Gamma_D}} \le \frac{2c}{\kappa_0}\|g_1\|^2_{(W^{1,2}_{\Gamma_D})^*}, \tag{7.98}$$

which implies

$$\|\theta_\varepsilon(x)\| \le c_1\|g_1\|_{(W^{1,2}_{\Gamma_D})^*}. \tag{7.99}$$

Putting $\lambda = c_1\|g_1\|_{(W^{1,2}_{\Gamma_D})^*}$, we have from the first equation of (7.95)

$$a_0(\theta_\varepsilon; v_\varepsilon, u) + a_2(v_\varepsilon, v_\varepsilon, u) + \big\langle \nabla\Phi_\varepsilon(v_\varepsilon), u\big\rangle - \Big\langle \big(1-\alpha_0\theta_\varepsilon\big) f, v_\varepsilon\Big\rangle = \langle f_1, u\rangle \quad \forall u \in \mathbf{V}, \tag{7.100}$$

which is the first formula of (7.96).

Putting $u = v_\varepsilon$ in (7.100), we have

$$a_0(\theta_\varepsilon; v_\varepsilon, v_\varepsilon) + a_2(v_\varepsilon, v_\varepsilon, v_\varepsilon) + \big\langle \nabla\Phi_\varepsilon(v_\varepsilon), v_\varepsilon\big\rangle - \big\langle \big(1-\alpha_0\theta_\varepsilon\big) f, v_\varepsilon\big\rangle = \langle f_1, v_\varepsilon\rangle. \tag{7.101}$$

Taking into account $a_2(v_\varepsilon, v_\varepsilon, v_\varepsilon) = 0$, (7.34), (7.35) and (7.99), we have from (7.101)

$$2\mu_0\|v_\varepsilon\|^2_{\mathbf{V}} \le a_0(\theta; v, v) \le c\big(\alpha_0 c_1\|g_1\|_{(W^{1,2}_{\Gamma_D})^*}\|f\|_{\mathbf{L}^3} + \|f\|_{\mathbf{V}^*} + \|f_1\|_{\mathbf{V}^*}\big)\|v_\varepsilon\|_{\mathbf{V}},$$

which implies

$$\|v_\varepsilon\|_{\mathbf{V}} \le \frac{c}{2\mu_0}\big(\alpha_0 c_1\|g_1\|_{(W^{1,2}_{\Gamma_D})^*}\|f\|_{\mathbf{L}^3} + \|f\|_{\mathbf{V}^*} + \|f_1\|_{\mathbf{V}^*}\big). \tag{7.102}$$

Taking the right hand side of (7.102) as ζ in (7.95), we get the second equation of (7.96). By (7.98) and (7.102), we get (7.97). □

Now repeating the arguments in Sect. 7.2.3 with the solutions of Theorem 7.8, we complete proof of Theorem 7.6. □

Remark 7.3 The Eq. (9.1) of Chap. 9 is more general than (7.1), and from the results of Chap. 9 some results for (7.1) with the boundary conditions (7.2), (7.3) or (7.2), (7.4) can be obtained. However the result for the case of total pressure demands that the parameter for buoyancy effect α_0 is small enough in accordance with the data of problem (see (9.138), and the result for the case of static pressure demands that the data of problem satisfy two smallness conditions together (see (9.50) and (9.97)).

7.4 Bibliographical Remarks

The content of Chap. 7 is taken from [2].

Several papers are concerned with (7.1). In [3, 4] under homogeneous Dirichlet boundary condition for velocity and mixture of non-homogeneous Dirichlet and Neumann conditions for temperature existence of a solution to (7.1) was studied. In [5] under non-homogeneous Dirichlet boundary condition for velocity and mixture of non-homogeneous Dirichlet and Neumann conditions for temperature, where smoothness of boundary data is weaker than [3, 4], existence of a solution to (7.1) was obtained.

In [6] under mixture of non-homogeneous Dirichlet, total pressure and vorticity boundary conditions for fluid and mixture of non-homogeneous Dirichlet, Neumann and Robin conditions for temperature the existence of a solution was established. In [7] variational inequalities for Navier-Stokes type operators were studied, which can describe (7.1) with one-sided flow boundary conditions for fluid and heat on a portion of boundary. In [8] under homogeneous Dirichlet boundary condition for velocity and mixture of non-homogeneous Dirichlet and homogeneous Neumann conditions for temperature the existence and uniqueness and smoothness of a weak solution were discussed. In [9] when the boundary consists of several connected components, boundary value problem of (7.1) with non-homogeneous Dirichlet boundary condition was studied. In [10] when the boundary consists of several connected components, Dirichlet problem of (7.1) under a weaker condition than [9] was investigated.

In [11] Dirichlet problem of (7.1) for arbitrarily large and very weak boundary data was studied.

References

1. H. Gajewski, K. Gröger, K. Zacharias, *Nichtlineare Operatorgleichungen und Operatordifferentialgleichungen* (Academic-Verlag Berlin, 1974) (Russian 1978)
2. T. Kim, The steady Boussinesq system with mixed boundary conditions including conditions of friction type. Applications of Mathematics, to appear
3. H. Morimoto, On the existence of weak solutions of equations of natural convection. J. Fac. Sci. Univ. Tokyo, Sect. IA **36**, 87–102 (1989)
4. H. Morimoto, On the existence and uniqueness of the stationary solution to the equation of natural convection. Tokyo J. Math. **14**, 217–226 (1991)
5. E.J. Villamizar-Roa, M.A. Rodríguez-Bellido, M.A. Rojas-Medar, The Boussinesq system with mixed nonsmooth boundary data. C. R. Acad. Sci. Paris, Ser. I **343**, 191–196 (2006)
6. G.V. Alekseev, A.B. Smishliaev, Solvability of the boundary value problems for Boussinesq equations with inhomogeneous boundary conditions. J. Math. Fluid Mech. **3**(1), 18–39 (2001)
7. A.Yu. Chebotarev, Variational inequalities for Navier-Stokes type operators and one-sided problems for equations of viscous heat-conducting fluids. Math. Notes **70**, 264–274 (2001)
8. D.A. Kovtunov, Solvability of the stationary heat convection problem for a high-viscosity fluid. Diff. Equat. **45**(1), 73–85 (2009)
9. H. Morimoto, Heat convection equation with nonhomogeneous boundary conditions. Funkcialai Ekvacioj **53**, 213–229 (2010)
10. P. Acevedo, C. Amrouche, C. Conca, Boussinesq system with non-homogeneous boundary conditions. Appl. Math. Lett. **53**, 39–44 (2016)
11. H. Kim, The existence and uniqueness of very weak solutions of the stationary Boussinesq system. Nonlinear Anal. **75**, 317–330 (2012)

Chapter 8
The Non-steady Boussinesq System

In this chapter we are concerned with the non-steady Boussinesq problem corresponding to the steady problem in Chap. 7. The formulations consist of a non-steady variational inequality for velocity and a non-steady variational equation for temperature.

For the problem with boundary conditions involving the static pressure and stress, it is proved that if the data of problem are small enough and compatibility conditions at the initial time for velocity and temperature are satisfied, then there exists a unique solution on the given interval. For the problem with boundary conditions involving the total pressure and total stress, the existence of a solution is proved without restriction on the data of problem.

8.1 Problems and Assumptions

The non-steady Boussinesq system with initial condition for heat convection is as follows.

$$\begin{cases} \dfrac{\partial v}{\partial t} - 2\nabla \cdot \big(\mu(\theta)\mathcal{E}(v)\big) + (v \cdot \nabla)v + \nabla p = (1 - \alpha_0\theta) f, \\ \operatorname{div} v = 0, \\ \dfrac{\partial \theta}{\partial t} - \nabla \cdot \big(\kappa(\theta)\nabla\theta\big) + v \cdot \nabla\theta = g, \\ v(0) = v_0, \ \ \theta(0) = \theta_0. \end{cases} \tag{8.1}$$

Let Ω be as in Chap. 7 with $\Gamma_{ij} \in C^{2,1}$ for $i = 2, 3, 7$ instead of $\Gamma_{ij} \in C^2$. (See Remark 6.1) Let $Q = \Omega \times (0, T)$, $0 < T < \infty$.

T. Kim and D. Cao, *Equations of Motion for Incompressible Viscous Fluids*, Advances in Mathematical Fluid Mechanics,
https://doi.org/10.1007/978-3-030-78659-5_8

The boundary conditions for temperature are as follows:

$$\begin{aligned}
&(1)\quad \theta|_{\Gamma_D}=0,\\
&(2)\quad \Big(\kappa(\theta)\frac{\partial\theta}{\partial n}+\beta(x)\theta\Big)\Big|_{\Gamma_R}=g_R(t,x),\quad \beta(x),\ g_R(t,x)-\text{given functions.}
\end{aligned}\tag{8.2}$$

Problems I and II are distinguished according to boundary conditions for fluids. Problem I is the one with boundary condition (7.3) assumed that μ and κ are independent of θ (the case of static pressure) and Problem II is the one with boundary condition (7.4) (the case of total pressure).

Let $\mathbf{V}$, $K(\Omega)$ be the same as in (5.5) and

$$\begin{aligned}
&H:\text{completion in } \mathbf{L}^2(\Omega) \text{ of } \mathbf{V},\\
&H_K:\text{closure in } \mathbf{L}^2(\Omega) \text{ of } K(\Omega),\\
&K(Q)=\{u\in L^2(0,T;\mathbf{V}):u'\in L^1(0,T;\mathbf{V}^*);\ u_n|_{\Gamma_{10}}\geq 0,\ u_n|_{\Gamma_{11}}\leq 0\},\\
&W^{1,2}_{\Gamma_D}(\Omega)=\{y\in W^{1,2}(\Omega):y|_{\Gamma_D}=0\}.
\end{aligned}$$

We use the inner products in (7.8) for $\mathbf{V}$ and $W^{1,2}_{\Gamma_D}(\Omega)$.

We will use the following two kinds of assumptions according to different boundary conditions. The assumption for Problem I is stronger than the one for Problem II, and consequently smoothness of solution with respect to t for Problem I is stronger than the one in Problem II.

Assumption 8.1 *(for the case of static pressure)* We assume the followings.

(1) $\Gamma_1\neq\emptyset$, $\Gamma_D\neq\emptyset$ and

$$\Gamma_R\subset\left(\cup_{i=1,3,5,8}\Gamma_i\right).\tag{8.3}$$

(2) (3) of Assumption 5.1 holds.

(3) For the functions of (8.1)

$$\begin{aligned}
&f\in W^{1,\infty}(0,T;\mathbf{L}^3(\Omega)),\\
&g\in W^{1,2}\big(0,T;(W^{1,2}_{\Gamma_D})^*\big),\\
&\mu \text{ and } \kappa \text{ are independent of } \theta.
\end{aligned}\tag{8.4}$$

(4) For the functions of (8.2) and (7.3),

$$\begin{aligned}
&g_R\in W^{1,2}(0,T;L^{4/3}(\Gamma_R)),\\
&\beta_1\geq\beta(x)\geq 0,\ \ \beta_1-\text{a constant},\ \ \beta(x)-\text{measurable},\\
&\phi_i\in W^{1,\infty}(0,T;H^{-1/2}(\Gamma_i)),\ i=2,4,7,\\
&\phi_i\in W^{1,\infty}(0,T;\mathbf{H}^{-1/2}(\Gamma_i)),\ i=3,5,6,\\
&\alpha_{ij}\in L^\infty(\Gamma_5).
\end{aligned}\tag{8.5}$$

Assumption 8.2 (*for the case of total pressure*) Let (1) and (2) of Assumption 8.1 hold and suppose the followings.

(3') For the functions of (8.1),

$$\begin{aligned} &f \in L^\infty(0,T;\mathbf{L}^3(\Omega)),\\ &g \in L^2\big(0,T;(W^{1,2}_{\Gamma_D})^*\big),\\ &\mu \in C(\mathbb{R}),\ 0<\mu_0 \le \mu(\xi) \le \mu_1 < \infty \quad \forall \xi \in \mathbb{R},\\ &\kappa \in C(\mathbb{R}),\ 0<\kappa_0 \le \kappa(\xi) \le \kappa_1 < \infty \quad \forall \xi \in \mathbb{R}. \end{aligned} \tag{8.6}$$

(4') For the functions of (8.2) and (7.4),

$$\begin{aligned} &g_R \in L^2(0,T;L^{4/3}(\Gamma_R));\\ &\beta_1 \ge \beta(x) \ge 0,\ \beta_1 - \text{a constant},\ \beta(x) - \text{measurable};\\ &\phi_i \in L^2(0,T;H^{-\frac{1}{2}}(\Gamma_i)), i=2,4,7,\\ &\phi_i \in L^2(0,T;\mathbf{H}^{-\frac{1}{2}}(\Gamma_i)), i=3,5,6;\\ &\alpha_{ij} \in L^\infty(\Gamma_5). \end{aligned} \tag{8.7}$$

Remark 8.1 For (2) of Assumption 8.1 we refer to Remark 6.3. In this chapter this assumption is also necessary to guarantee equivalence between Problem I-VE and Problem I-VI (in Remark 8.4), and between Problem II-VE and Problem II-VI.

8.2 Variational Formulations for Problems

In this section we first give variational formulations for the problems above. Taking into account (8.3) and $v_n|_{\Gamma_3\cup\Gamma_5\cup\Gamma_8}=0$, for $v \in \mathbf{V}$, $\theta \in W^{1,2}(\Omega)$ and $\varphi \in W^{1,2}_{\Gamma_D}(\Omega)$, we have

$$(v\cdot\nabla\theta,\varphi) = (v_n\theta,\varphi)_{\Gamma_R} - (\theta v,\nabla\varphi) = -(\theta v,\nabla\varphi). \tag{8.8}$$

8.2.1 Variational Formulations: The Case of Static Pressure

By (7.9)–(7.11) and (8.8), as in Sect. 7.1 we can see that smooth solutions (v,p,θ) of problem (8.1), (8.2), (7.3) satisfy the following system.

$$\begin{cases} (\frac{\partial v}{\partial t}, u) + 2(\mu\mathcal{E}(v), \mathcal{E}(u)) + \langle (v\cdot\nabla)v, u\rangle + 2(\mu k(x)v, u)_{\Gamma_2} + 2(\mu S\tilde{v}, \tilde{u})_{\Gamma_3} \\ \quad + 2(\alpha(x)v, u)_{\Gamma_5} + (\mu k(x)v, u)_{\Gamma_7} - 2(\mu\varepsilon_{n\tau}(v), u)_{\Gamma_8} + (p - 2\mu\varepsilon_{nn}(v), u_n)_{\Gamma_9\cup\Gamma_{10}\cup\Gamma_{11}} \\ \quad = \langle (1-\alpha_0\theta)f, u\rangle + \sum_{i=2,4,7}\langle \phi_i, u_n\rangle_{\Gamma_i} + \sum_{i=3,5,6}\langle \phi_i, u\rangle_{\Gamma_i} \quad \forall u \in \mathbf{V}, \\ (\frac{\partial \theta}{\partial t}, \varphi) + (\kappa\nabla\theta, \nabla\varphi) - (\theta v, \nabla\varphi) + (\beta\theta, \varphi)_{\Gamma_R} = \langle g_R, \varphi\rangle_{\Gamma_R} + \langle g, \varphi\rangle \quad \forall\varphi \in W^{1,2}_{\Gamma_D}(\Omega), \\ |\sigma_\tau(\theta, v)| \le g_\tau, \;\; \sigma_\tau(\theta, v)\cdot v_\tau + g_\tau|v_\tau| = 0 \;\; \text{on } \Gamma_8, \\ |\sigma_n(v, p)| \le g_n, \;\; \sigma_n(v, p)v_n + g_n|v_n| = 0 \;\; \text{on } \Gamma_9, \\ \sigma_n(v, p) + g_{+n} \ge 0, \;\; (\sigma_n(v, p) + g_{+n})v_n = 0 \;\; \text{on } \Gamma_{10}, \\ \sigma_n(v, p) - g_{-n} \le 0, \;\; (\sigma_n(v, p) - g_{-n})v_n = 0 \;\; \text{on } \Gamma_{11}. \end{cases} \tag{8.9}$$

Define $a_{01}(\cdot,\cdot), a_{11}(\cdot,\cdot,\cdot)$ and $f_1(t) \in \mathbf{V}^*$ by

$$\begin{aligned} &a_{01}(w, u) = 2(\mu\mathcal{E}(w), \mathcal{E}(u)) + 2(\mu k(x)w, u)_{\Gamma_2} + 2(\mu S\tilde{w}, \tilde{u})_{\Gamma_3} \\ &\qquad + 2(\alpha(x)w, u)_{\Gamma_5} + (\mu k(x)w, u)_{\Gamma_7} \quad \forall w, u \in \mathbf{V}, \\ &a_{11}(v, u, w) = \langle (v\cdot\nabla)u, w\rangle \quad \forall v, u, w \in \mathbf{V}, \\ &\langle f_1(t), u\rangle = \sum_{i=2,4,7}\langle \phi_i(t), u_n\rangle_{\Gamma_i} + \sum_{i=3,5,6}\langle \phi_i(t), u\rangle_{\Gamma_i} \quad \forall u \in \mathbf{V}. \end{aligned} \tag{8.10}$$

Define $b_1(\cdot,\cdot)$ and $g_1(t) \in (W^{1,2}_{\Gamma_D}(\Omega))^*$ by

$$\begin{aligned} &b_1(\theta, \varphi) = (\kappa\nabla\theta, \nabla\varphi) + (\beta(x)\theta, \varphi)_{\Gamma_R} \quad \forall\theta \in W^{1,2}(\Omega), \varphi \in W^{1,2}_{\Gamma_D}(\Omega), \\ &\langle g_1(t), \varphi\rangle = \langle g_R(t), \varphi\rangle_{\Gamma_R} + \langle g(t), \varphi\rangle \quad \forall\varphi \in W^{1,2}_{\Gamma_D}(\Omega). \end{aligned} \tag{8.11}$$

Remark 8.2 Under (4) of Assumption 8.1 the duality product $\langle f_1, u\rangle$ of (8.10) is meaningful (see Remark 5.1). By (8.4) and (8.5),

$$f_1 \in W^{1,\infty}(0, T; \mathbf{V}^*), \quad g_1 \in W^{1,2}(0, T; (W^{1,2}_{\Gamma_D})^*). \tag{8.12}$$

Then, taking into account

$$\sigma_\tau(\theta, v) = 2\mu\varepsilon_{n\tau}(v), \quad \sigma_n(\theta, v, p) = -p + 2\mu\varepsilon_{nn}(v)$$

and (8.9), we introduce the following variational formulation for problem (8.1), (8.2), (7.3).

Problem I-VE. Find $v \in K(Q)$, $\theta \in L^2(0, T; W^{1,2}_{\Gamma_D}(\Omega))$ and $(\sigma_\tau, \sigma_n, \sigma_{+n}, \sigma_{-n}) \in \mathbf{L}^2_\tau(\Gamma_8) \times L^2(\Gamma_9) \times H^{-\frac{1}{2}}(\Gamma_{10}) \times H^{-\frac{1}{2}}(\Gamma_{11})$ for a.e. $t \in (0, T)$ such that $v(0) = v_0$, $\theta(0) = \theta_0$ and

$$
\begin{cases}
\langle \dfrac{\partial v}{\partial t}, u\rangle + a_{01}(v,u) + a_{11}(v,v,u) - (\sigma_\tau, u_\tau)_{\Gamma_8} - (\sigma_n, u_n)_{\Gamma_9} \\
\qquad - \langle \sigma_{+n}, u_n\rangle_{\Gamma_{10}} - \langle \sigma_{-n}, u_n\rangle_{\Gamma_{11}} - \langle f - \alpha_0\theta f, u\rangle = \langle f_1, u\rangle \quad \forall u \in L^2(0,T;\mathbf{V}), \\
\langle \dfrac{\partial \theta}{\partial t}, \varphi\rangle + b_1(\theta,\varphi) - \langle \theta v, \nabla\varphi\rangle = \langle g_1, \varphi\rangle \quad \forall \varphi \in L^2(0,T; W^{1,2}_{\Gamma_D}(\Omega)), \\
|\sigma_\tau| \le g_\tau, \ \sigma_\tau \cdot v_\tau + g_\tau |v_\tau| = 0 \quad \text{on } \Gamma_8, \\
|\sigma_n| \le g_n, \ \sigma_n v_n + g_n |v_n| = 0 \quad \text{on } \Gamma_9, \\
\sigma_{+n} + g_{+n} \ge 0, \ \langle \sigma_{+n} + g_{+n}, v_n\rangle_{\Gamma_{10}} = 0 \quad \text{on } \Gamma_{10}, \\
\sigma_{-n} - g_{-n} \le 0, \ \langle \sigma_{-n} - g_{-n}, v_n\rangle_{\Gamma_{11}} = 0 \quad \text{on } \Gamma_{11},
\end{cases}
\tag{8.13}
$$

where $\mathbf{L}^2_\tau(\Gamma_8)$ is the subspace of $\mathbf{L}^2(\Gamma_8)$ consisting of functions such that $(u, n)_{\mathbf{L}^2(\Gamma_8)} = 0$.

Remark 8.3 We showed that smooth solutions to the problem (8.1), (8.2), (7.3) are solutions to Problem I-VE. Assume that a solution to Problem I-VE is smooth enough $\big(v \in L^2(0,T;\mathbf{H}^2(\Omega)),\ v' \in L^1(0,T;\mathbf{L}^2(\Omega)),\ \theta \in L^2(0,T;H^2(\Omega)),\ \theta' \in L^1(0,T;L^2(\Omega)),\ \sigma_\tau \in L^2(0,T;\mathbf{L}^2(\Gamma_8)),\ \sigma_n \in L^2(0,T;L^2(\Gamma_9)),\ \sigma_+ \in L^2(0,T; L^2(\Gamma_{10}))$ and $\sigma_- \in L^2(0,T;L^2(\Gamma_{11}))\big)$ and $f \in L^2(0,T;\mathbf{L}^2(\Omega))$. Then, for a.e. $t \in (0,T)$ there exists $p(t) \in H^1(\Omega)$ (see Theorem 5.3) such that $p \in L^1(0,T;H^1(\Omega))$ and (v, p, θ) satisfies the first two equations of (8.1) and the boundary condition (7.3) in suitable spaces. By a standard way, it is proved that (v, θ) satisfies the third equation of (8.1), the boundary condition (8.2) and the initial conditions.

In this sense, Problem I-VE is equivalent to the problem (8.1), (8.2), (7.3).

We will find another variational formulation consisting of a variational inequality and a variational equation, which is equivalent to Problem I-VE if the solution is smooth enough (see Remark 8.4).

For fixed θ, let us consider the problem

$$
\begin{cases}
\langle \dfrac{\partial v}{\partial t}, u\rangle + a_{01}(v,u) + a_{11}(v,v,u) - (\sigma_\tau, u_\tau)_{\Gamma_8} - (\sigma_n, u_n)_{\Gamma_9} - \langle \sigma_{+n}, u_n\rangle_{\Gamma_{10}} \\
\qquad - \langle \sigma_{-n}, u_n\rangle_{\Gamma_{11}} - \langle (f - \alpha_0\theta f, u\rangle = \langle f_1, u\rangle, \quad \forall u \in L^2(0,T;\mathbf{V}), \\
|\sigma_\tau| \le g_\tau, \ \sigma_\tau \cdot v_\tau + g_\tau |v_\tau| = 0 \quad \text{on } \Gamma_8, \\
|\sigma_n| \le g_n, \ \sigma_n v_n + g_n |v_n| = 0 \quad \text{on } \Gamma_9, \\
\sigma_{+n} + g_{+n} \ge 0, \ \langle \sigma_{+n} + g_{+n}, v_n\rangle_{\Gamma_{10}} = 0 \quad \text{on } \Gamma_{10}, \\
\sigma_{-n} - g_{-n} \le 0, \ \langle \sigma_{-n} - g_{-n}, v_n\rangle_{\Gamma_{11}} = 0 \quad \text{on } \Gamma_{11}.
\end{cases}
\tag{8.14}
$$

Subtracting the first formula of (8.14) with $u = v$ from the first one of (8.14), we get

$$
\begin{aligned}
&\langle \frac{\partial v}{\partial t}, u - v\rangle + a_{01}(v, u-v) + a_{11}(v,v,u-v) - (\sigma_\tau, u_\tau - v_\tau)_{\Gamma_8} - (\sigma_n, u_n - v_n)_{\Gamma_9} \\
&\qquad - \langle \sigma_{+n}, u_n - v_n\rangle_{\Gamma_{10}} - \langle \sigma_{-n}, u_n - v_n\rangle_{\Gamma_{11}} - \langle f - \alpha_0\theta f, u - v\rangle \\
&\quad = \langle f_1, u - v\rangle \quad \forall u \in \mathbf{V}.
\end{aligned}
\tag{8.15}
$$

Let $\Phi : \mathbf{V} \to \overline{\mathbb{R}}$ be the functional defined by (5.22). Then, the functional Φ is proper, convex, lower weak semi-continuous and nonnegative.

Define a functional $\Psi(u)$ by

$$\Psi(u) = \begin{cases} \int_0^T \Phi(u(t))\, dt & \text{if } \Phi(u(t)) \in L^1(0,T), \\ +\infty & \text{otherwise.} \end{cases} \tag{8.16}$$

In the same way as Problem I of Sect. 5.1, from (8.15) we get

$$\begin{aligned} \left\langle \frac{\partial v}{\partial t}, u - v \right\rangle + a_{01}(v, u - v) + a_{11}(v, v, u - v) + \Phi(u) - \Phi(v) \\ \geq \langle (1 - \alpha_0 \theta) f, u - v \rangle + \langle f_1, u - v \rangle. \end{aligned} \tag{8.17}$$

Define operators $A_1 : \mathbf{V} \to \mathbf{V}^*$, $B_1 : \mathbf{V} \times \mathbf{V} \to \mathbf{V}^*$ and $C_1 : W^{1,2}_{\Gamma_D}(\Omega) \to (W^{1,2}_{\Gamma_D}(\Omega))^*$, respectively, by

$$\begin{aligned} &\langle A_1 v, u \rangle = a_{01}(v, u) \quad \forall v, u \in \mathbf{V}, \\ &\langle B_1(v, u), w \rangle = a_{11}(v, u, w) \quad \forall v, u, w \in \mathbf{V}, \\ &\langle C_1 \theta, \varphi \rangle = b_1(\theta, \varphi) \quad \forall \theta, \varphi \in W^{1,2}_{\Gamma_D}(\Omega). \end{aligned} \tag{8.18}$$

If v is a solutions to (8.14), then we can see that the solution satisfies (see (6.18))

$$\begin{aligned} \int_0^T \langle v'(t) + A_1 v(t) + B_1(v(t), v(t)) - (1 - \alpha_0)\theta(t) f(t) - f_1(t), u(t) - v(t) \rangle\, dt \\ + \Psi(u) - \Psi(v) \geq 0 \qquad \forall u \in L^4(0, T; \mathbf{V}). \end{aligned} \tag{8.19}$$

Therefore, we have the following variational formulation of the problem (8.1), (8.2), (7.3) which consists of a variational inequality for velocity and a variational equation for temperature.

Problem I-VI. Find $(v, \theta) \in \big(L^\infty(0, T; H) \cap L^2(0, T; \mathbf{V})\big) \times \big(L^\infty(0, T; L^2(\Omega)) \cap L^2(0, T; W^{1,2}_{\Gamma_D}(\Omega))\big)$ such that

$$\begin{cases} \int_0^T \langle v' + A_1 v(t) + B_1(v(t), v(t)) - (1 - \alpha_0 \theta) f - f_1, u(t) - v(t) \rangle\, dt \\ \qquad + \Psi(u) - \Psi(v) \geq 0 \quad \forall u \in L^4(0, T; \mathbf{V}), \\ \int_0^T \Big[\Big\langle \frac{\partial \theta}{\partial t}, \varphi \Big\rangle + \langle C_1 \theta(t), \varphi \rangle - \langle \theta v, \nabla \varphi \rangle - \langle g_1, \varphi \rangle \Big]\, dt = 0 \quad \forall \varphi \in L^2(0, T; W^{1,2}_{\Gamma_D}), \\ v(0) = v_0, \ \theta(0) = \theta_0. \end{cases} \tag{8.20}$$

Remark 8.4 If the solution to Problem I-VI is smooth as much as $v \in L^2(0,T;\mathbf{V})$, $v' \in L^2(0,T;\mathbf{V}^*)$, then the first one of (8.20) is equivalent to

$$\langle v'(t) + A_1 v(t) + B_1(v(t), v(t)) - (1-\alpha_0\theta)f - f_1, u - v(t)\rangle + \Phi(u) - \Phi(v(t)) \geq 0$$
$$\text{for a.e. } t \in [0,T], \quad \forall u \in K(\Omega) \qquad (8.21)$$

(Remark, p. 114 of [1]). In (8.17) putting $\langle F_1, u - v\rangle = \langle -\frac{\partial v}{\partial t} - (1-\alpha_0\theta)f - f_1, u - v\rangle$, by Theorem 5.4 we can see the existence of $(\sigma_\tau, \sigma_n, \sigma_{+n}, \sigma_{-n}) \in \mathbf{L}^2_\tau(\Gamma_8) \times L^2(\Gamma_9) \times H^{-1/2}(\Gamma_{10}) \times H^{-1/2}(\Gamma_{11})$ for a.e. $t \in (0,T)$ such that $(v, \theta, \sigma_\tau, \sigma_n, \sigma_{+n}, \sigma_{-n})$ is a solution to Problem I-VE.

8.2.2 Variational Formulations: The Case of Total Pressure

Since $(v\cdot\nabla)v = \operatorname{rot} v \times v + \frac{1}{2}\operatorname{grad}|v|^2$, by (7.9)–(7.11) and (8.8), we can see that smooth solutions (v, p, θ) of problem (8.1), (8.2), (7.4) satisfy the following.

$$\begin{cases} \Big(\dfrac{\partial v}{\partial t}, u\Big) + 2(\mu(\theta)\mathcal{E}(v), \mathcal{E}(u)) + \langle \operatorname{rot} v \times v, u\rangle + 2(\mu(\theta)k(x)v, u)_{\Gamma_2} \\ \qquad + 2(\mu(\theta)S\tilde{v}, \tilde{u})_{\Gamma_3} + 2(\alpha(x)v, u)_{\Gamma_5} + (\mu(\theta)k(x)v, u)_{\Gamma_7} \\ \qquad - 2(\mu(\theta)\varepsilon_{n\tau}(v), u)_{\Gamma_8} + (p + \dfrac{1}{2}|v|^2 - 2\mu(\theta)\varepsilon_{nn}(v), u_n)_{\Gamma_9\cup\Gamma_{10}\cup\Gamma_{11}} \\ \quad = \langle (1-\alpha_0\theta)f, u\rangle + \displaystyle\sum_{i=2,4,7}\langle \phi_i, u_n\rangle_{\Gamma_i} + \sum_{i=3,5,6}\langle \phi_i, u\rangle_{\Gamma_i} \quad \forall u \in \mathbf{V}, \\ \Big(\dfrac{\partial \theta}{\partial t}, \varphi\Big) + (\kappa(\theta)\nabla\theta, \nabla\varphi) - (\theta v, \nabla\varphi) + (\beta\theta, \varphi)_{\Gamma_R} = \langle g_R, \varphi\rangle_{\Gamma_R} + \langle g, \varphi\rangle \quad \forall \varphi \in W^{1,2}_{\Gamma_D}(\Omega), \\ |\sigma^t_\tau(\theta, v)| \leq g_\tau, \;\; \sigma^t_\tau(\theta, v)\cdot v_\tau + g_\tau|v_\tau| = 0 \;\text{ on } \Gamma_8, \\ |\sigma^t_n(\theta, v, p)| \leq g_n, \;\; \sigma^t_n(\theta, v, p)v_n + g_n|v_n| = 0 \;\text{ on } \Gamma_9, \\ \sigma^t_n(\theta, v, p) + g_{+n} \geq 0, \;\; (\sigma^t_n(\theta, v, p) + g_{+n})v_n = 0 \;\text{ on } \Gamma_{10}, \\ \sigma^t_n(\theta, v, p) - g_{-n} \leq 0, \;\; (\sigma^t_n(\theta, v, p) - g_{-n})v_n = 0 \;\text{ on } \Gamma_{11}. \end{cases} \tag{8.22}$$

Define $a_{02}(\theta;\cdot,\cdot)$, $a_{12}(\cdot,\cdot,\cdot)$ and $f_2 \in V^*$ by

$$\begin{aligned} a_{02}(\tilde\theta; w, u) &= 2(\mu(\tilde\theta)\mathcal{E}(w), \mathcal{E}(u)) + 2(\mu(\tilde\theta)k(x)w, u)_{\Gamma_2} + 2(\mu(\tilde\theta)S\tilde{w}, \tilde{u})_{\Gamma_3} \\ &\quad + 2(\alpha(x)w, u)_{\Gamma_5} + (\mu(\tilde\theta)k(x)w, u)_{\Gamma_7} \quad \forall w, u \in \mathbf{V}, \; \forall \tilde\theta \in W^{1,2}_{\Gamma_D}(\Omega), \\ a_{12}(v, u, w) &= \langle \operatorname{rot} v \times u, w\rangle \quad \forall v, u, w \in \mathbf{V}, \\ \langle f_2, u\rangle &= \sum_{i=2,4,7}\langle \phi_i, u_n\rangle_{\Gamma_i} + \sum_{i=3,5,6}\langle \phi_i, u\rangle_{\Gamma_i} \quad \forall u \in \mathbf{V}. \end{aligned} \tag{8.23}$$

Define $b_2(\theta;\cdot,\cdot)$ and $g_2 \in (W^{1,2}_{\Gamma_D}(\Omega))^*$ by

$$
\begin{aligned}
&b_2(\tilde{\theta};\theta,\varphi)=(\kappa(\tilde{\theta})\nabla\theta,\nabla\varphi)+(\beta(x)\theta,\varphi)_{\Gamma_R}\quad \forall\tilde{\theta},\theta,\varphi\in W^{1,2}_{\Gamma_D}(\Omega),\\
&\langle g_2,\varphi\rangle=\langle g_R,\varphi\rangle_{\Gamma_R}+\langle g,\varphi\rangle\quad \forall\varphi\in W^{1,2}_{\Gamma_D}(\Omega).
\end{aligned}
\tag{8.24}
$$

By (8.6) and (8.7),

$$
f_2\in L^2(0,T;\mathbf{V}^*),\quad g_2\in L^2(0,T;(W^{1,2}_{\Gamma_D})^*).\tag{8.25}
$$

Then, taking into account

$$
\sigma^t_\tau(\theta,v)=2\mu(\theta)\varepsilon_{n\tau}(v),\quad \sigma^t_n(\theta,v,p)=-(p+\frac{1}{2}|v|^2)+2\mu(\theta)\varepsilon_{nn}(v)
$$

and (8.22), we are led to the following variational formulation for problem (8.1), (8.2) (7.4).

Problem II-VE. Find $v\in K(Q)$, $\theta\in L^\infty(0,T;L^2(\Omega))\cap L^2(0,T;W^{1,2}_{\Gamma_D}(\Omega))$ and $(\sigma^t_\tau,\sigma^t_n,\sigma^t_{+n},\sigma^t_{-n})\in \mathbf{L}^2_\tau(\Gamma_8)\times L^2(\Gamma_9)\times H^{-\frac{1}{2}}(\Gamma_{10})\times H^{-\frac{1}{2}}(\Gamma_{11})$ for a.e. $t\in(0,T)$ such that $v(0)=v_0$, $\theta(0)=\theta_0$ and

$$
\begin{cases}
\left\langle\dfrac{\partial v}{\partial t},u\right\rangle+a_{02}(\theta;v,u)+a_{12}(v,v,u)-(\sigma^t_\tau,u_\tau)_{\Gamma_8}-(\sigma^t_n,u_n)_{\Gamma_9}\\
\qquad-\left\langle\sigma^t_{+n},u_n\right\rangle_{\Gamma_{10}}-\left\langle\sigma^t_{-n},u_n\right\rangle_{\Gamma_{11}}-\langle f-\alpha_0\theta f,u\rangle=\langle f_2,u\rangle\quad\forall u\in\mathbf{V},\\
\left\langle\dfrac{\partial\theta}{\partial t},\varphi\right\rangle+b_2(\theta;\theta,\varphi)-\langle\theta v,\nabla\varphi\rangle=\langle g_2,\varphi\rangle\quad\forall\varphi\in W^{1,2}_{\Gamma_D}(\Omega),\\
|\sigma^t_\tau|\le g_\tau,\ \sigma^t_\tau\cdot v_\tau+g_\tau|v_\tau|=0\quad\text{on }\Gamma_8,\\
|\sigma^t_n|\le g_n,\ \sigma^t_n v_n+g_n|v_n|=0\quad\text{on }\Gamma_9,\\
\sigma^t_{+n}+g_{+n}\ge 0,\ \left\langle\sigma^t_{+n}+g_{+n},v_n\right\rangle_{\Gamma_{10}}=0\quad\text{on }\Gamma_{10},\\
\sigma^t_{-n}-g_{-n}\le 0,\ \left\langle\sigma^t_{-n}-g_{-n},v_n\right\rangle_{\Gamma_{11}}=0\quad\text{on }\Gamma_{11},
\end{cases}
\tag{8.26}
$$

where $\mathbf{L}^2_\tau(\Gamma_8)$ is the subspace of $\mathbf{L}^2(\Gamma_8)$ consisting of functions such that $(u,n)_{\mathbf{L}^2(\Gamma_8)}=0$.

Remark 8.5 Similarly to Remark 8.3, Problem II-VE is equivalent to the problem (8.1), (8.2), (7.4).

Define operators $A_2(\tilde{\theta}):\mathbf{V}\to\mathbf{V}^*$ and $B_2:\mathbf{V}\times\mathbf{V}\to\mathbf{V}^*$, respectively, by

$$
\begin{aligned}
&\langle A_2(\tilde{\theta})v,u\rangle=a_{02}(\tilde{\theta};v,u)\quad\forall v,u\in\mathbf{V},\tilde{\theta}\in W^{1,2}_{\Gamma_D}(\Omega),\\
&\langle B_2(v,u),w\rangle=a_{12}(v,u,v)\quad\forall v,u,w\in\mathbf{V}.
\end{aligned}
\tag{8.27}
$$

Let functional Ψ be defined by (8.16). Then, in the same way as in Problem I-VI of Sect. 6.1 we find a variational inequality for velocity. Then we have another variational formulation consisting of a variational inequality for velocity and a variational equation for temperature, which is equivalent to Problem II-VE if the solution is smooth enough.

Problem II-VI. Find $(v,\theta)\in\big(L^\infty(0,T;H)\cap L^2(0,T;\mathbf{V})\big)\times\big(L^\infty(0,T;L^2(\Omega))\cap L^2(0,T;W^{1,2}_{\Gamma_D}(\Omega))\big)$ such that

$$\begin{cases}\displaystyle\int_0^T\langle u'+A_2(\theta)v(t)+B_2(v(t),v(t))-(1-\alpha_0\theta)f-f_2,u(t)-v(t)\rangle\,dt\\ \displaystyle\quad+\Psi(u)-\Psi(v)\ge-\frac{1}{2}\|v_0-u(0)\|^2\quad\forall u\in L^4(0,T;\mathbf{V})\text{ with }u'\in L^2(0,T;\mathbf{V}^*),\\ \displaystyle\int_0^T\Big[\Big\langle-\theta,\frac{\partial\varphi}{\partial t}\Big\rangle+b_2(\theta;\theta,\varphi)-\langle\theta v,\nabla\varphi\rangle-\langle g_2,\varphi\rangle\Big]\,dt\\ \quad=\langle\theta_0(x),\varphi(x,0)\rangle\quad\forall\varphi\in C^1([0,T];W^{1,2}_{\Gamma_D}(\Omega))\text{ with }\varphi(\cdot,T)=0.\end{cases}\tag{8.28}$$

8.3 Existence and Uniqueness of Solutions: The Case of Static Pressure

The aim of this section is to prove the following

Theorem 8.1 (the case of static pressure) *Let Assumption 8.1 be satisfied. Suppose that*

(1) The norms of f, $\phi_i, i=2,\cdots,6$, g, g_R in the spaces that they belong to are small enough (depending on α_0);

(2) $v_0\in\mathbf{V}$, $\Phi(v_0)=0$;

(3) $\theta_0\in W^{1,2}_{\Gamma_D}(\Omega)$;

(4) $\big(A_1v_0+B_1(v_0,v_0)-f_1(0)\big)\in H$ (compatibility condition at the initial time for velocity);

(5) $\|v_0\|_V$, $\|A_1v_0+B_1(v_0,v_0)-(1-\alpha_0\theta_0)f(0)-f_1(0)\|$ are small enough;

(6) $\big(C_1\theta_0+v_0\cdot\nabla\theta_0-g_1(0)\big)\in L^2(\Omega)$ (compatibility condition at the initial time for temperature);

(7) $\|\theta_0\|_{W^{1,2}_{\Gamma_D}(\Omega)}$ and $\|C_1\theta_0+v_0\cdot\nabla\theta_0-g_1(0)\|_{L^2(\Omega)}$ are small enough.

Then there exists a solution (v,θ) to (8.20) such that

$$\begin{aligned}&v\in C([0,T];\mathbf{V}),\quad v'\in L^2(0,T;\mathbf{V})\cap L^\infty(0,T;H),\\&\theta\in C([0,T];W^{1,2}_{\Gamma_D}),\quad\theta'\in L^2(0,T;W^{1,2}_{\Gamma_D}(\Omega))\cap L^\infty(0,T;L^2(\Omega)).\end{aligned}\tag{8.29}$$

Furthermore, the solution satisfying $\|v\|_V\le c$, $\|\theta\|_{W^{1,2}_{\Gamma_D}}\le c$ for a constant $c>0$ small enough is unique.

Theorem 8.1 will be proved in the next two subsections.

8.3.1 Existence and Estimation of Solutions to an Approximate Problem

We first consider an approximate problem for (8.20).

For every $\varepsilon > 0$, let Φ_ε be the Moreau-Yosida approximation of Φ and $\nabla\Phi_\varepsilon$ be Fréchet derivative of Φ_ε (see Sect. 1.6).

By the fact that $\Gamma_{2j}, \Gamma_{3j}, \Gamma_{7j}$ are in $C^{2,1}(\Gamma_{ij})$ and 4) of Assumption 8.1, there exists a constant M such that

$$\|S(x)\|_\infty,\ \|k(x)\|_\infty,\ \|\alpha\|_{L^\infty(\Gamma_5)} \le M.$$

Thus, there exists c_* such that

$$2\left|\mu(k(x)z, z)_{\Gamma_2} + 2\mu(S\tilde{z}, \tilde{z})_{\Gamma_3} + (\alpha(x)z, z)_{\Gamma_5} + \mu(k(x)z, z)_{\Gamma_7}\right| \le \frac{\mu}{4}\|z\|_{\mathbf{V}}^2 + c_*\|z\|^2 \quad \forall z \in \mathbf{V} \tag{8.30}$$

(see Theorem 1.27). Thus, we have

$$\begin{aligned} &\langle A_1 u, u\rangle \ge \frac{7\mu}{4}\|u\|_{\mathbf{V}}^2 - c_*\|u\|^2 \quad \forall u \in \mathbf{V},\\ &|\langle A_1 u, v\rangle| \le c_1\|u\|_{\mathbf{V}}\|v\|_{\mathbf{V}} \quad \exists c_1 > 0,\ \forall u, v \in \mathbf{V} \end{aligned} \tag{8.31}$$

and

$$|\langle B_1(v, u), w\rangle| \le c_2\|v\|_{\mathbf{V}}\|u\|_{\mathbf{V}}\|w\|_{\mathbf{V}}, \tag{8.32}$$

where the operators A_1, B_1 are the ones in (8.18).

Let $\{u_j, j = 1, 2, \cdots\}$ and $\{\varphi_j, j = 1, 2, \cdots\}$ be, respectively, bases of the space $\mathbf{V}$ and $W^{1,2}_{\Gamma_D}(\Omega)$. Without loss of generality, we assume that $u_1 = v_0$, $\varphi_1 = \theta_0$ as in [1]. We are to find a solution $v_m = \sum_{j=1}^m g_{jm}(t)u_j$, $\theta_m = \sum_{j=1}^m r_{jm}(t)\varphi_j$ to problem

$$\begin{cases} \left(\dfrac{\partial v_m}{\partial t}, u_j\right) + 2\big(\mu\mathcal{E}(v_m), \mathcal{E}(u_j)\big) + ((v_m\cdot\nabla)v_m, u_j) + 2(\mu k(x)v_m, u_j)_{\Gamma_2}\\ \qquad + 2(\mu S\tilde{v}_m, \tilde{u}_j)_{\Gamma_3} + 2(\alpha(x)v_m, u_j)_{\Gamma_5} + (\mu k(x)v_m, u_j)_{\Gamma_7} + \big\langle\nabla\Phi_\varepsilon(v_m(t)), u_j\big\rangle\\ \quad = \langle(1-\alpha_0\theta_m)f, u_j\rangle + \langle f_1, u_j\rangle,\\ \left(\dfrac{\partial\theta_m}{\partial t}, \varphi_j\right) + (\kappa\nabla\theta_m, \nabla\varphi_j) + (\beta(x)\theta_m, \varphi_j)_{\Gamma_R} - \langle v_m\theta_m, \nabla\varphi_j\rangle = \langle g_1, \varphi_j\rangle,\\ v_m(0) = v_0, \quad \theta_m(0) = \theta_0. \end{cases} \tag{8.33}$$

which gives us an ordinary differential system for $g_{jm}(t)$, $r_{jm}(t)$, $j = 1, \cdots, m$. The solutions to (8.33) depend on ε, but for convenience of notation hereafter we use subindex m instead of subindex m, ε. For a t_m there exist absolute continuous functions $g_{jm}(t)$ and $r_{jm}(t)$ on $[0, t_m)$. Since $f \in W^{1,\infty}(0, T; \mathbf{L}^3(\Omega))$,

$f_1 \in W^{1,\infty}(0,T;\mathbf{V}^*)$, $g_1 \in W^{1,2}(0,T;(W^{1,2}_{\Gamma_D})^*)$ and $\nabla\Phi_\varepsilon$ is Lipschitz continuous, $g'_{jm}(t)$ and $r'_{jm}(t)$ are in fact absolute continuous. If $\|v_m(t)\|$, $\|\theta_m(t)\|$ are bounded and $v_m(t), \theta_m(t)$ are integrable, then $g_{jm}(t)$, $r_{jm}(t)$ are prolonged over t_m. Under smallness of the data of problem and compatibility condition of the data at the initial instant we will find estimates for $\|v_m(t)\|$ and $\|\theta_m(t)\|$ below, by which we obtain (8.82) and see that $t_m = T$.

Multiplying the first and second equation of (8.33), respectively, by $g_{jm}(t), \varphi_{jm}(t)$ and adding for $i = 1, \cdots, m$, we get

$$\begin{cases} \left\langle \dfrac{\partial v_m}{\partial t}, v_m \right\rangle + 2\big(\mu\mathcal{E}(v_m), \mathcal{E}(v_m)\big) + \langle (v_m \cdot \nabla) v_m, v_m\rangle + 2(\mu k(x) v_m, v_m)_{\Gamma_2} \\ \qquad + 2(\mu S\tilde{v}_m, \tilde{v}_m)_{\Gamma_3} + 2(\alpha(x)v_m, v_m)_{\Gamma_5} + (\mu k(x) v_m, v_m)_{\Gamma_7} + \langle \nabla\Phi_\varepsilon(v_m(t)), v_m\rangle \\ \quad = \langle (1-\alpha_0\theta_m) f, v_m\rangle + \langle f_1, v_m\rangle, \\ \left\langle \dfrac{\partial \theta_m}{\partial t}, \theta_m \right\rangle + (\kappa\nabla\theta_m, \nabla\theta_m) + (\beta(x)\theta_m, \theta_m)_{\Gamma_R} - \langle v_m\theta_m, \nabla\theta_m\rangle = \langle g_1, \theta_m\rangle, \\ v_m(0) = v_0, \quad \theta_m(0) = \theta_0. \end{cases} \tag{8.34}$$

We will find a priori estimate for

$$I(t) := \|v_m(t)\|^2 + \|v'_m(t)\|^2 + \|\theta_m(t)\|^2 + \|\theta'_m(t)\|^2.$$

Since Φ_ε is convex, continuous and Fréchet differentiable, we have

$$\Phi_\varepsilon(u) - \Phi_\varepsilon(v_m(t)) \ge \langle \nabla\Phi_\varepsilon(v_m(t)), u - v_m(t)\rangle \quad \forall u \in \mathbf{V}. \tag{8.35}$$

On the other hand, by condition (2) of the theorem and (1.37), we have that $\Phi_\varepsilon(0_\mathbf{V}) = 0$. Thus

$$0 \le \Phi_\varepsilon(v_m(t)) \le \langle \nabla\Phi_\varepsilon(v_m(t)), v_m(t)\rangle. \tag{8.36}$$

Also,

$$2|\langle -\alpha_0\theta_m(t) f, v_m(t)\rangle| \le c|\alpha_0|\,\|\theta_m(t)\|^2_{W^{1,2}_{\Gamma_D}}\|f\|^2_{\mathbf{L}^3} + \frac{\mu}{4}\|v_m(t)\|^2_{\mathbf{V}}. \tag{8.37}$$

By virtue of (8.31), (8.32), (8.36) and (8.37), we have from the first equation of (8.34)

$$\begin{aligned} &\frac{d}{dt}\|v_m(t)\|^2 + \frac{7\mu}{2}\|v_m(t)\|^2_{\mathbf{V}} - 2c_2\|v_m(t)\|^3_{\mathbf{V}} + 2\Phi_\varepsilon(v_m(t)) \\ &\qquad \le c|\alpha_0|\,\|\theta_m(t)\|^2_{W^{1,2}_{\Gamma_D}}\|f\|^2_{\mathbf{L}^3} + c\|f\|^2_{\mathbf{L}^3} + c\|f_1\|^2_{\mathbf{V}^*} + \frac{\mu}{2}\|v_m(t)\|^2_{\mathbf{V}} + 2c_*\|v_m(t)\|^2, \end{aligned}$$

where c_* and c_2 are, respectively, the one in (8.30) and (8.32), and so

$$\frac{d}{dt}\|v_m(t)\|^2 + \big(3\mu - 2c_2\|v_m(t)\|_{\mathbf{V}}\big)\|v_m(t)\|_{\mathbf{V}}^2 + 2\Phi_\varepsilon(v_m(t))$$
$$\leq c|\alpha_0|\|\theta_m(t)\|^2_{W^{1,2}_{\Gamma_D}}\|f\|^2_{\mathbf{L}^3} + c\|f\|^2_{\mathbf{L}^3} + c\|f_1\|^2_{\mathbf{V}^*} + 2c_*\|v_m(t)\|^2. \tag{8.38}$$

Here and the rest of this section constants independent of the data of problem are written by c with the exceptions of c_*, c_2.

Setting $t = 0$ in the first equation of (8.33), multiplying the resulting equation by $g'_{jm}(0)$ and adding for $j = 1, \cdots, m$, we get

$$\|v'_m(0)\|^2 + \langle A_1 v_m(0), v'_m(0)\rangle + \langle B_1(v_m(0), v_m(0)), v'_m(0)\rangle + \langle\nabla\Phi_\varepsilon(v_0), v'_m(0)\rangle$$
$$= \langle(1-\alpha_0\theta_0)f(0), v'_m(0)\rangle + \langle f_1(0), v'_m(0)\rangle. \tag{8.39}$$

By condition (2) of theorem, for any $u \in \mathbf{V}$ we have $\Phi(u) \geq \Phi(v_0) = 0$, which by (1.37) implies $\Phi_\varepsilon(v_0) = 0$ and $\nabla\Phi_\varepsilon(v_0) = 0$. Then, we have from (8.39)

$$\|v'_m(0)\| \leq \|A_1 v_0 + B_1(v_0, v_0) - (1-\alpha_0\theta_0)f(0) - f_1(0)\|, \tag{8.40}$$

which is valid by the compatibility condition at the initial time for velocity (condition (4)) and the conditions for θ_0, f. On the other hand, taking into account (8.31), (8.32) and (8.36), we have from the first equation of (8.34)

$$\frac{7\mu}{2}\|v_m(t)\|^2_{\mathbf{V}} \leq 2c_2\|v_m\|^3_{\mathbf{V}} + 2\langle(1-\alpha_0\theta_m(t))f(t), v_m(t)\rangle$$
$$+ 2\langle f_1(t), v_m(t)\rangle + 2c_*\|v_m\|^2 - 2(v_m(t), v'_m(t)),$$

and so

$$3\mu\|v_m(t)\|_{\mathbf{V}} \leq 2c_2\|v_m(t)\|^2_{\mathbf{V}} + c\|f(t)\|_{\mathbf{L}^3} + c|\alpha_0|\|\theta_m(t)\|\|f(t)\|_{\mathbf{L}^3}$$
$$+ c\|f_1(t)\|_{\mathbf{V}^*} + (2\delta\|v'_m(t)\| + 2c_*\delta\|v_m(t)\|), \tag{8.41}$$

where δ is such that $\|\cdot\| \leq \delta\|\cdot\|_{\mathbf{V}}$.

Since $\langle v_m\theta_m, \nabla\theta_m\rangle = 0$ by (8.8), we get from the second equation of (8.34)

$$\frac{d}{dt}\|\theta_m(t)\|^2 + 2\kappa\|\theta_m(t)\|^2_{W^{1,2}_{\Gamma_D}} + 2(\beta(x)\theta_m, \theta_m)_{\Gamma_R} = 2\langle g_1, \theta_m(t)\rangle. \tag{8.42}$$

We have from (8.42)

$$\frac{d}{dt}\|\theta_m(t)\|^2 + \kappa\|\theta_m(t)\|^2_{W^{1,2}_{\Gamma_D}} + (\beta(x)\theta_m, \theta_m)_{\Gamma_R} \leq \frac{1}{\kappa}\|g_1\|^2_{(W^{1,2}_{\Gamma_D})^*} \tag{8.43}$$

and

$$\|\theta_m(t)\|^2 + \int_0^t\int_\Omega \kappa|\nabla\theta_m(s)|^2\,dxds \leq \|\theta_0\|^2 + \frac{1}{\kappa}\int_0^t \|g_1(s)\|^2_{(W^{1,2}_{\Gamma_D})^*}\,ds. \tag{8.44}$$

Setting $t=0$ in the second equation of (8.33), multiplying the resulting equation by $r'_{jm}(0)$ and adding for $j=1,\cdots,m$, we get

$$\|\theta'_m(0)\|^2 + b_1(\theta_0,\theta'_m(0)) + (v_0\cdot\nabla\theta_0,\theta'_m(0)) = \langle g_1(0),\theta'_m(0)\rangle, \tag{8.45}$$

where $-(v_0\theta_0,\nabla\theta'_m(0)) = (v_0\cdot\nabla\theta_0,\theta'_m(0))$ was used. We have from (8.45)

$$\|\theta'_m(0)\| \le \|C_1\theta_0 + v_0\cdot\nabla\theta_0 - g_1(0)\|, \tag{8.46}$$

which is valid by the compatibility condition at the initial time for temperature (condition (6)).

On the other hand, taking into account $\langle v_m\theta_m,\nabla\theta_m\rangle = 0$, we have from the second equation of (8.34)

$$\kappa\|\theta_m(t)\|^2_{W^{1,2}_{\Gamma_D}} \le \|g_1\|_{(W^{1,2}_{\Gamma_D})^*}\|\theta_m(t)\|_{W^{1,2}_{\Gamma_D}} + \delta_1\|\theta_m(t)\|_{W^{1,2}_{\Gamma_D}}\|\theta'_m(t)\|,$$

where δ_1 is such that $\|\cdot\|\le\delta_1\|\cdot\|_{W^{1,2}_{\Gamma_D}}$, and so

$$\|\theta_m(t)\|_{W^{1,2}_{\Gamma_D}} \le \frac{1}{\kappa}\big(\|g_1\|_{(W^{1,2}_{\Gamma_D})^*} + \delta_1\|\theta'_m(t)\|\big). \tag{8.47}$$

substituting (8.44) into (8.41), we have

$$\begin{aligned}3\mu\|v_m(t)\|_{\mathbf{V}} \le{}& c_2\|v_m(t)\|^2_{\mathbf{V}} + c\|f\|_{L^\infty(0,T;\mathbf{L}^3)}\\ &+ c|\alpha_0|\left(\|\theta_0\|^2 + \frac{1}{\kappa}\int_0^T \|g_1\|^2_{(W^{1,2}_{\Gamma_D})^*}\,ds\right)^{1/2}\|f\|_{L^\infty(0,T;\mathbf{L}^3)}\\ &+ \|f_1\|_{L^\infty(0,T;\mathbf{V}^*)} + \max\{2\delta,2c_*\delta\}(\|v'_m(t)\| + \|v_m(t)\|).\end{aligned} \tag{8.48}$$

Differentiating the first equality of (8.33) with respect to t yields

$$\begin{aligned}&\big\langle v''_m(t),v_j\big\rangle + \big\langle A_1v'_m(t),v_j\big\rangle + \big\langle (B_1(v_m(t),v_m(t))',v_j\big\rangle + \big\langle(\nabla\Phi_\varepsilon(v_m))',v_j\big\rangle\\ &\quad= \big\langle -\alpha_0\theta'_m(t)f,v_j\big\rangle - \big\langle(1-\alpha_0\theta_m(t))f',v_j\big\rangle + \big\langle f'_1,v_j\big\rangle.\end{aligned} \tag{8.49}$$

Multiplying (8.49) by $g'_{jm}(t)$ and summing for j yield

$$\begin{aligned}&\langle v''_m(t),v'_m(t)\rangle + \langle A_1v'_m(t),v'_m(t)\rangle + \langle (B_1v_m(t),v_m(t))',v'_m(t)\rangle + \langle(\nabla\Phi_\varepsilon(v_m))',v'_m(t)\rangle\\ &\quad= \langle -\alpha_0\theta'_m(t)f,v'_m(t)\rangle - \langle(1-\alpha_0\theta_m(t))f',v'_m(t)\rangle + \langle f'_1,v'_m(t)\rangle.\end{aligned} \tag{8.50}$$

Calculating $(B((v_m(t),v_m(t)))'$, we have

$$\begin{aligned}|\langle(B_1v_m(t),v_m(t))',v'_m(t)\rangle| &= |\langle B_1(v'_m,v_m),v'_m\rangle + \langle B_1(v_m,v'_m),v'_m\rangle|\\ &\le 2c_2\|v_m\|_{\mathbf{V}}\|v'_m\|^2_{\mathbf{V}},\end{aligned} \tag{8.51}$$

where c_2 is the one in (8.32). Also, by Hölder's and Young's inequalities we have

$$
\begin{aligned}
&2|\langle \alpha_0\theta_m'(t)f, v_m'(t)\rangle| \leq c|\alpha_0|\|\theta_m'(t)\|^2_{W^{1,2}_{\Gamma_D}}\|f\|^2_{\mathbf{L}^3}+\frac{\mu}{8}\|v_m'(t)\|^2_{\mathbf{V}},\\
&2|\langle \alpha_0\theta_m(t)f', v_m'(t)\rangle| \leq c|\alpha_0|\|\theta_m(t)\|^2_{W^{1,2}_{\Gamma_D}}\|f'\|^2_{\mathbf{L}^3}+\frac{\mu}{8}\|v_m'(t)\|^2_{\mathbf{V}},\\
&2|\langle f'(t), v_m'(t)\rangle| \leq c\|f'(t)\|^2_{\mathbf{L}^3}+\frac{\mu}{8}\|v_m'(t)\|^2_{\mathbf{V}},\\
&2|\langle f_1'(t), v_m'(t)\rangle| \leq c\|f_1'(t)\|^2_{\mathbf{V}^*}+\frac{\mu}{8}\|v_m'(t)\|^2_{\mathbf{V}}.
\end{aligned}
\tag{8.52}
$$

Taking into account (8.31), (8.51), (8.52) and the fact that $\langle(\nabla\Phi_\varepsilon(v_m))', v_m'\rangle \geq 0$ (see (1.38)), we have from (8.50)

$$
\begin{aligned}
&\frac{d}{dt}\|v_m'(t)\|^2+(3\mu-4c_2\|v_m(t)\|_{\mathbf{V}})\|v_m'(t)\|^2_{\mathbf{V}}+\frac{\mu}{2}\|v_m'(t)\|^2_{\mathbf{V}}\\
&\qquad\leq c\alpha_0\|\theta_m'(t)\|^2_{W^{1,2}_{\Gamma_D}}\|f\|^2_{\mathbf{L}^3}+c|\alpha_0|\|\theta_m(t)\|^2_{W^{1,2}_{\Gamma_D}}\|f'\|^2_{\mathbf{L}^3}\\
&\qquad\quad+c\|f'\|^2_{\mathbf{L}^3}+c\|f_1'\|^2_{\mathbf{V}^*}+2c_*\|v_m'(t)\|^2+\frac{\mu}{2}\|v_m'(t)\|^2_{\mathbf{V}},
\end{aligned}
$$

that is,

$$
\begin{aligned}
&\frac{d}{dt}\|v_m'(t)\|^2+(3\mu-4c_2\|v_m(t)\|_{\mathbf{V}})\|v_m'(t)\|^2_{\mathbf{V}}\\
&\qquad\leq c|\alpha_0|\|\theta_m'(t)\|^2_{W^{1,2}_{\Gamma_D}}\|f\|^2_{\mathbf{L}^3}+c|\alpha_0|\|\theta_m(t)\|^2_{W^{1,2}_{\Gamma_D}}\|f'\|^2_{\mathbf{L}^3}\\
&\qquad\quad+c\|f'\|^2_{\mathbf{L}^3}+c\|f_1'\|^2_{\mathbf{V}^*}+2c_*\|v_m'(t)\|^2.
\end{aligned}
\tag{8.53}
$$

Differentiating the second equality of (8.33) with respect to t, we have

$$
\langle\theta_m''(t),\varphi_j\rangle+\langle C_1\theta_m'(t),\varphi_j\rangle-\langle v_m'\theta,\nabla\varphi_j\rangle-\langle v_m\theta',\nabla\varphi_j\rangle=\langle g_1',\varphi_j\rangle. \tag{8.54}
$$

Multiplying (8.54) by $r_{jm}'(t)$ and adding for j, we have

$$
\langle\theta_m''(t),\theta_m'(t)\rangle+\langle C_1\theta_m'(t),\theta_m'(t)\rangle-\langle v_m'\theta_m,\nabla\theta_m'(t)\rangle-\langle v_m\theta',\nabla\theta_m'(t)\rangle=\langle g_1',\theta_m'(t)\rangle. \tag{8.55}
$$

On the other hand, we get

$$
2|\langle v_m'\theta,\nabla\theta_m'(t)\rangle| \leq \frac{c}{\kappa}\|v_m'\|^2_{\mathbf{V}}\|\theta\|^2_{W^{1,2}_{\Gamma_D}}+\kappa\|\theta_m'(t)\|^2_{W^{1,2}_{\Gamma_D}}. \tag{8.56}
$$

Taking into account $\langle v_m\theta',\nabla\theta_m'(t)\rangle=0$ (see (8.8)) and (8.56), we have from (8.55)

$$
\begin{aligned}
&\frac{d}{dt}\|\theta_m'(t)\|^2 + 2\kappa\|\theta_m'(t)\|^2_{W^{1,2}_{\Gamma_D}} \\
&\qquad \le \frac{c}{\kappa}\|v_m'\|^2_{\mathbf{V}}\|\theta\|^2_{W^{1,2}_{\Gamma_D}} + \kappa\|\theta_m'(t)\|^2_{W^{1,2}_{\Gamma_D}} + c\|g_1'\|^2_{(W^{1,2}_{\Gamma_D})^*} + \frac{\kappa}{4}\|\theta_m'(t)\|^2_{W^{1,2}_{\Gamma_D}}.
\end{aligned} \tag{8.57}
$$

Rewriting (8.57) yields

$$
\frac{d}{dt}\|\theta_m'(t)\|^2 + \frac{3\kappa}{4}\|\theta_m'(t)\|^2_{W^{1,2}_{\Gamma_D}} \le \frac{c}{\kappa}\|v_m'\|^2_{\mathbf{V}}\|\theta\|^2_{W^{1,2}_{\Gamma_D}} + c\|g_1'\|^2_{(W^{1,2}_{\Gamma_D})^*}. \tag{8.58}
$$

Adding (8.38), (8.53), (8.43) and (8.58), we have

$$
\begin{aligned}
&\frac{d}{dt}I(t) + (2\mu - 4c_2\|v_m(t)\|_{\mathbf{V}})\left(\|v_m(t)\|^2_{\mathbf{V}} + \|v_m'(t)\|^2_{\mathbf{V}}\right) + \left(\mu - \frac{c}{\kappa}\|\theta_m\|^2_{W^{1,2}_{\Gamma_D}}\right)\|v_m'(t)\|^2_{\mathbf{V}} \\
&\qquad + (\kappa - c|\alpha_0|\|f\|^2_{\mathbf{L}^3} - c|\alpha_0|\|f'\|^2_{\mathbf{L}^3})\|\theta_m\|^2_{W^{1,2}_{\Gamma_D}} + \left(\frac{3\kappa}{4} - c|\alpha_0|\|f\|^2_{\mathbf{L}^3}\right)\|\theta_m'\|^2_{W^{1,2}_{\Gamma_D}} \\
&\qquad \le c(\|f\|^2_{\mathbf{L}^3} + \|f'\|^2_{\mathbf{L}^3}) + c(\|f_1\|^2_{\mathbf{V}^*} + \|f_1'\|^2_{\mathbf{V}^*}) + c(\|g_1\|^2_{(W^{1,2}_{\Gamma_D})^*} + \|g_1'\|^2_{(W^{1,2}_{\Gamma_D})^*}) \\
&\qquad + 2c_*(\|v_m(t)\|^2 + \|v_m'(t)\|^2).
\end{aligned} \tag{8.59}
$$

Integrating (8.59) yields

$$
\begin{aligned}
&I(t) + \int_0^t \Big[(2\mu - 4c_2\|v_m(s)\|_{\mathbf{V}})\left(\|v_m(t)\|^2_{\mathbf{V}} + \|v_m'(s)\|^2_{\mathbf{V}}\right) + \left(\mu - \frac{c}{\kappa}\|\theta_m\|^2_{W^{1,2}_{\Gamma_D}}\right)\|v_m'(s)\|^2_{\mathbf{V}} \\
&\qquad + (\kappa - c|\alpha_0|\|f\|^2_{\mathbf{L}^3} - c|\alpha_0|\|f'\|^2_{\mathbf{L}^3})\|\theta_m\|^2_{W^{1,2}_{\Gamma_D}} + \left(\frac{3\kappa}{4} - c|\alpha_0|\|f\|^2_{\mathbf{L}^3}\right)\|\theta_m'\|^2_{W^{1,2}_{\Gamma_D}}\Big]\,ds \\
&\qquad \le I(0) + F(t) + 2c_*\int_0^t (\|v_m(s)\|^2 + \|v_m'(s)\|^2)\,ds,
\end{aligned} \tag{8.60}
$$

where

$$
F(t) := ct(\|f\|^2_{W^{1,\infty}(0,T;\mathbf{L}^3)} + \|f_1\|^2_{W^{1,\infty}(0,T;\mathbf{V}^*)}) + c\|g_1\|^2_{W^{1,2}(0,t;(W^{1,2}_{\Gamma_D})^*)}. \tag{8.61}
$$

By (8.40) and (8.46) we have

$$
\begin{aligned}
I(0) \le &\|v_0\|^2 + \|A_1 v_0 + B_1(v_0, v_0) - (1 - \alpha_0\theta_0)f(0) - f_1(0)\|^2 \\
&+ \|\theta_0\|^2 + \|C_1\theta_0 + v_0\cdot\nabla\theta_0 - g_1(0)\|^2.
\end{aligned} \tag{8.62}
$$

By the condition of theorem, we can assume $\|f\|_{W^{1,\infty}(0,T;\mathbf{L}^3)}$ to be so small that

$$
\begin{aligned}
&\kappa - c|\alpha_0|\|f(t)\|^2_{\mathbf{L}^3} - c|\alpha_0|\|f'(t)\|^2_{\mathbf{L}^3} \ge 0 \quad \text{at a.e. } t \in [0, T], \\
&\frac{3\kappa}{4} - c|\alpha_0|\|f(t)\|^2_{\mathbf{L}^3} \ge 0 \quad \text{at a.e. } t \in [0, T].
\end{aligned} \tag{8.63}
$$

If

$$\|v_m(0)\|_{\mathbf{V}} = \|v_0\|_{\mathbf{V}} < \frac{\mu}{2c_2} \tag{8.64}$$

and

$$\|\theta_m(0)\|_{W^{1,2}_{\Gamma_D}} = \|\theta_0\|_{W^{1,2}_{\Gamma_D}} < \frac{\mu\kappa}{c} \tag{8.65}$$

are valid, then there exists a t_m such that on $[0, t_m]$

$$\begin{aligned} &2\mu - 4c_2\|v_m(t)\|_{\mathbf{V}} \geq 0, \\ &\mu - \frac{c}{\kappa}\|\theta_m(t)\|_{W^{1,2}_{\Gamma_D}} \geq 0. \end{aligned} \tag{8.66}$$

Therefore, taking into account (8.62), by Gronwall's inequality we have

$$\begin{aligned} I(t) \leq &\Big(\|v_0\|^2 + \|A_1 v_0 + B_1(v_0, v_0) - (1 - \alpha_0\theta_0) f(0) - f_1(0)\|^2 \\ &+ \|\theta_0\|^2 + \|C_1\theta_0 + v_0 \cdot \nabla\theta_0 - g_1(0)\|^2 + F(T)\Big) e^{2c_* t} \end{aligned} \tag{8.67}$$

on the all intervals of t satisfying (8.66).

Using this estimate, we will obtain a quadratic inequality satisfied by $\|v_m(t)\|_{\mathbf{V}}$. Put

$$\begin{aligned} \beta := &\Big(\|v_0\|^2 + \|A_1 v_0 + B_1(v_0, v_0) - (1 - \alpha_0\theta_0) f(0) - f_1(0)\|^2 \\ &+ \|\theta_0\|^2 + \|C_1\theta_0 + v_0 \cdot \nabla\theta_0 - g_1(0)\|^2 + F(T)\Big) e^{2c_* T}. \end{aligned} \tag{8.68}$$

Obviously, β depends only on the data of problem.

Then, when f satisfies (8.63), we can get from (8.67)

$$\begin{aligned} &\left(\|v_m(t)\| + \|v_m'(t)\|\right) \leq \sqrt{2}\left(\|v_m(t)\|^2 + \|v_m'(t)\|^2\right)^{\frac{1}{2}} \leq \sqrt{2\beta}, \\ &\|\theta_m'(t)\| \leq \sqrt{\beta} \end{aligned} \tag{8.69}$$

on $[0, t_m]$ where (8.66) holds. Let the data of problem be so small that

$$\frac{c}{\kappa^2}(\|g_1\|_{W^{1,2}(0,T;(W^{1,2}_{\Gamma_D})^*)} + \delta_1\sqrt{\beta}) \leq \frac{\mu}{2}. \tag{8.70}$$

By (8.47) and (8.69), for the small data of problem we have on $[0, t_m]$

$$\frac{c}{\kappa}\|\theta_m(t)\|_{W^{1,2}_{\Gamma_D}} \leq \frac{c}{\kappa^2}(\|g_1\|_{W^{1,2}(0,T;(W^{1,2}_{\Gamma_D}))^*} + \delta_1\sqrt{\beta}) \leq \frac{\mu}{2},$$

which implies

$$\mu - \frac{c}{\kappa}\|\theta_m(t)\|_{W^{1,2}_{\Gamma_D}} \geq \frac{\mu}{2} \quad \forall t \in [0, t_m]. \tag{8.71}$$

Therefore, for such small data of problem that (8.70) is valid, if

$$2\mu - 4c_2\|v_m(t)\|_{\mathbf{V}} \geq 0 \quad \forall t \in [0, t_m + \gamma], \ \gamma > 0, \ t_m + \gamma \leq T,$$

then owing to (8.71), step by step, we have

$$\mu - \frac{c}{\kappa}\|\theta_m(t)\|_{W^{1,2}_{\Gamma_D}} \geq \frac{\mu}{2} \quad \forall t \in [0, t_m + \gamma]. \tag{8.72}$$

From above we see that for the small data of problem satisfying (8.63)–(8.65) and (8.70),

$$\mu - \frac{c}{\kappa}\|\theta_m(t)\|_{W^{1,2}_{\Gamma_D}} \geq \frac{\mu}{2} \tag{8.73}$$

is valid on the interval where the first inequality of (8.66) holds.

Put

$$\begin{aligned} \gamma :=& \|f(t)\|_{L^\infty(0,T;\mathbf{L}^3)} + c|\alpha_0|\Big(\|\theta_0\|^2 + \frac{1}{\kappa}\int_0^T \|g_1\|^2_{(W^{1,2}_{\Gamma_D})^*}\, ds\Big)^{1/2}\|f(t)\|_{L^\infty(0,T;\mathbf{L}^3)} \\ &+ \|f_1(t)\|_{L^\infty(0,T;V^*)} + \max\{2\delta, 2c_*\delta\}\sqrt{2\beta}. \end{aligned} \tag{8.74}$$

By (8.44), (8.69) and (8.41), for the small data satisfying (8.63)–(8.65) and (8.70) we have a quadratic inequality for $\|v_m(t)\|_{\mathbf{V}}$, which is the one we want,

$$0 \leq \gamma - 3\mu\|v_m(t)\|_{\mathbf{V}} + 2c_2\|v_m(t)\|^2_{\mathbf{V}} \tag{8.75}$$

on the intervals where the first inequality of (8.66) is satisfied.

By the conditions of theorem, we can assume that the data of problem are so small that (8.63)–(8.65) and (8.70) are valid and γ satisfies the following inequality

$$9\mu^2 - 8c_2\gamma > 4\mu^2. \tag{8.76}$$

Now, let us prove that if

$$\|v_0\|_{\mathbf{V}} \leq \frac{3\mu - \sqrt{9\mu^2 - 8c_2\gamma}}{4c_2} \quad \Big(< \frac{\mu}{4c_2}\Big), \tag{8.77}$$

then for any m

$$2\mu - 4c_2\|v_m(t)\|_{\mathbf{V}} \geq \mu \quad \forall t \in [0, T]. \tag{8.78}$$

Since $2\mu - 4c_2\|v_0\|_{\mathbf{V}} > \mu$, on a interval $[0, t_m]$

$$2\mu - 4c_2\|v_m(t)\|_{\mathbf{V}} \geq \mu.$$

Let us prove that if the first inequality of (8.66) is valid on an interval $[0, \bar{t}_m]$, then more strong estimate

$$2\mu - 4c_2\|v_m(t)\|_{\mathbf{V}} \geq \mu \quad \forall t \in [0, \bar{t}_m] \tag{8.79}$$

is valid. To this end, putting $y = \|v_m(t)\|_{\mathbf{V}}$ in (8.75) (which is valid on the interval where the first inequality of (8.66) holds when (8.63)–(8.65) and(8.70) are valid), we get

$$0 \leq \gamma - 3\mu y + 2c_2 y^2 \quad \text{on } [0, \bar{t}_m]. \tag{8.80}$$

By virtue of (8.76), there exist two real roots of polynomial $\gamma - 3\mu y + c_2 y^2$

$$y_1 = \frac{3\mu - \sqrt{9\mu^2 - 8c_2\gamma}}{4c_2} \quad \text{and} \quad y_2 = \frac{3\mu + \sqrt{9\mu^2 - 8c_2\gamma}}{4c_2},$$

and on the intervals $[0, y_1]$ and $[y_2, +\infty)$ (8.80) holds. Thus, by continuity of $\|v_m(t)\|_{\mathbf{V}}$ with respect to t we have from $\|v(0)\|_{\mathbf{V}} \in [0, y_1]$ that $\|v_m(t)\|_{\mathbf{V}} \in [0, y_1]$ $\forall t \in [0, \bar{t}_m]$, that is,

$$\|v_m(t)\|_{\mathbf{V}} \leq \frac{3\mu - \sqrt{9\mu^2 - 8c_2\gamma}}{4c_2} < \frac{\mu}{4c_2} \quad \forall t \in [0, \bar{t}_m].$$

Thus,

$$2\mu - 4c_2\|v_m(t)\|_{\mathbf{V}} > \mu \quad \forall t \in [0, \bar{t}_m], \tag{8.81}$$

which shows (8.79). Thus, step by step we see that the first inequality of (8.66) is valid on $[0, T]$ and so is (8.78).

If (8.77) is valid, then so is (8.64). Therefore, for the small data satisfying (8.63), (8.65), (8.70), (8.76) and (8.77), we have also (8.73) on $[0, T]$. By (8.78) and (8.73), we have

$$\begin{aligned} &\|v_m(t)\|_{\mathbf{V}} \leq \frac{\mu}{4c_2} \quad \forall t \in [0, T], \ \forall m, \ \forall \varepsilon > 0, \\ &\frac{c}{\kappa}\|\theta_m(t)\|_{W^{1,2}_{\Gamma_D}} \leq \frac{\mu}{2} \quad \forall t \in [0, T], \ \forall m, \ \forall \varepsilon > 0. \end{aligned} \tag{8.82}$$

Note that β, γ depend on α_0, and the restriction of data depends on α_0.

Then, by (8.67) and (8.60), we have

$$\begin{aligned} &\|v'_m(t)\| \leq \text{const} \quad \forall t \in [0, T], \ \forall m, \ \forall \varepsilon > 0, \\ &\|v'_m\|_{L^2(0,T;\mathbf{V})} \leq \text{const} \quad \forall m, \ \forall \varepsilon > 0, \\ &\|\theta'_m(t)\| \leq \text{const} \quad \forall t \in [0, T], \ \forall m, \ \forall \varepsilon > 0, \\ &\|\theta'_m\|_{L^2(0,T;W^{1,2}_{\Gamma_D})} \leq \text{const} \quad \forall m, \ \forall \varepsilon > 0. \end{aligned} \tag{8.83}$$

By (8.82),

$$\int_0^T \Phi_\varepsilon(v_m(t))\,dt \le \text{const} \quad \forall m,\ \forall \varepsilon > 0, \tag{8.84}$$

and so by (1.36) and (1.37),

$$\int_0^T \|v_m(t) - J_\varepsilon(v_m(t))\|_{\mathbf{V}}^2\,dt \le c\varepsilon \quad \forall m,\ \forall \varepsilon > 0. \tag{8.85}$$

8.3.2 Existence and Uniqueness of a Solution

In this subsection we complete proof of Theorem 8.1. Let us first prove existence of a solution. Due to (8.82) and (8.83), we can extract a subsequence from $\{(v_m, \theta_m)\}$ obtained in preceding subsection, which is still denoted with the subindex as before, such that

$$\begin{aligned} v_m &\to v \quad \text{in } C([0,T];\mathbf{V}),\\ v'_m &\rightharpoonup v' \quad \text{in } L^2(0,T;\mathbf{V}),\\ v'_m &\overset{*}{\rightharpoonup} v' \quad \text{in } L^\infty(0,T;H),\\ \theta_m &\to \theta \quad \text{in } C([0,T];W^{1,2}_{\Gamma_D}),\\ \theta'_m &\rightharpoonup \theta' \quad \text{in } L^2(0,T;W^{1,2}_{\Gamma_D}),\\ \theta'_m &\overset{*}{\rightharpoonup} \theta' \quad \text{in } L^\infty(0,T;L^2(\Omega)) \end{aligned} \tag{8.86}$$

when $m \to \infty$ and $\varepsilon \to 0$ (note that for convenience in the preceding subsection we used subindex m instead of m, ε).

Put $u = \sum_{j=1}^M k_j(t)u_j$, where $k_j(t) \in C^1[0,T]$, M is a positive integer. Multiply the first equation of (8.33) by $k_j(t)$ and sum for $j = 1, \cdots, M$. Then, multiply the first equation of (8.33) by $g_{jm}(t)$ and sum for $j = 1, \cdots, m$. Subtracting the resulting equations yields

$$\begin{aligned} \langle v'_m(t) + A_1 v_m(t) + B_1(v_m(t), v_m(t)) + \nabla\Phi_\varepsilon(v_m), u(t) - v_m(t)\rangle \\ = \langle (1-\alpha_0\theta) f + f_1, u(t) - v_m(t)\rangle. \end{aligned} \tag{8.87}$$

Since Φ_ε is convex, continuous and Fréchet differentiable, we have

$$\Phi_\varepsilon(u(t)) - \Phi_\varepsilon(v_m(t)) \ge \langle \nabla\Phi_\varepsilon(v_m(t)), u - v_m(t)\rangle. \tag{8.88}$$

Taking into account (8.88), we have from (8.87)

$$\int_0^T \langle v_m'(t) + A_1 v_m(t) + B_1(v_m(t), v_m(t)) - (1-\alpha_0\theta_m)f - f_1, u(t) - v_m(t)\rangle\, dt$$
$$+ \int_0^T \big(\Phi_\varepsilon(u(t)) - \Phi_\varepsilon(v_m(t))\big)\, dt \geq 0. \tag{8.89}$$

Since $\Phi(u) \geq \Phi_\varepsilon(u)$ and $\Phi(J_\varepsilon w_m(t)) \leq \Phi_\varepsilon(w_m(t))$ (see (1.37)), we have from (8.89)

$$\int_0^T \langle v_m'(t) + A_1 v_m(t) + B_1(v_m(t), v_m(t)) - (1-\alpha_0\theta_m)f - f_1, u(t) - v_m(t)\rangle\, dt$$
$$+ \Psi(u(t)) - \int_0^T \Phi(J_\varepsilon v_m(t))\, dt \geq 0. \tag{8.90}$$

By (8.85), we have that $J_\varepsilon v_m \to v$ in $L^2(0,T;\mathbf{V})$ as $m \to \infty$ and $\varepsilon \to 0$, and so by lower semi-continuity of Φ we have

$$\lim_{m\to\infty,\varepsilon\to 0} \int_0^T \Phi(J_\varepsilon v_m(t))\, dt \geq \int_0^T \Phi(v(t))\, dt. \tag{8.91}$$

By standard way we can prove that

$$\int_0^T \langle B_1(v_m(t), v_m(t)), v(t)\rangle\, dt \to \int_0^T \langle B_1(v(t), v(t)), v(t)\rangle\, dt \tag{8.92}$$

as $m \to \infty$ and $\varepsilon \to 0$.

Since $\lim_{m_k\to\infty,\varepsilon\to 0}\langle A_1 v_m(t), v_m(t)\rangle = \langle A_1 v(t), v(t)\rangle$, by (8.91) and (8.92), we have from (8.90)

$$\int_0^T \langle v'(t) + A_1 v(t) + B_1(v(t), v(t)) - (1-\alpha_0\theta)f - f_1, u(t) - v(t)\rangle\, dt$$
$$+ \Psi(u) - \Psi(v) \geq 0. \tag{8.93}$$

Since $v_m(0) = v_0$, by (8.86) it is obvious that $v(0) = v_0$.

$B_1(v(t), v(t)) \in L^\infty(0,T;\mathbf{V}^*)$ and the set $\{u = \sum_{j=1}^M k_j(t)u_j;\ k_j(t) \in C^1[0,T],\ M : \text{positive integer}\}$ is dense in $L^4(0,T;\mathbf{V})$, and so (8.93) is valid for all $u \in L^4(0,T;\mathbf{V})$.

By (8.86), from the second equation of (8.33) we can get

$$\int_0^T \langle \frac{\partial\theta_m}{\partial t}, \varphi\rangle + (\kappa\nabla\theta_m, \nabla\varphi) + (\beta(x)\theta_m, \varphi)_{\Gamma_R} - \langle v_m\theta_m, \nabla\varphi\rangle\, dt = \int_0^T \langle g_1, \varphi\rangle\, dt,$$
$$\varphi \in L^2(0,T; W^{1,2}_{\Gamma_D}). \tag{8.94}$$

It is easy to see that

$$(\kappa \nabla \theta_m, \nabla \varphi) \to (\kappa \nabla \theta, \nabla \varphi) \quad \text{for a.e. } t \in [0, T].$$

Also,

$$\begin{aligned}
&\int_0^T \left|\langle v_m \theta_m, \nabla \varphi\rangle - \langle v\theta, \nabla \varphi\rangle\right| dt \\
&\leq \int_0^T \|v_m - v\|_{\mathbf{L}^6} \|\theta_m\|_{L^3} \|\nabla \varphi\|_{\mathbf{L}^2}\, dt + \int_0^T \langle v(\theta_m - \theta), \nabla \varphi\rangle\, dt \\
&\leq \|v_m - v\|_{L^\infty(0,T;\mathbf{V})} \|\theta_m\|_{L^2(0;T;L^3)} \|\varphi\|_{L^2(0,T;W^{1,2}_{\Gamma_D})} + \int_0^T |\langle v(\theta_m - \theta), \nabla \varphi\rangle|\, dt.
\end{aligned} \tag{8.95}$$

Since $v\nabla \varphi \in L^2(0, T; L^{6/5}(\Omega))$, by (8.86) we have that as $m \to \infty$ and $\varepsilon \to 0$, $\int_0^T |\langle v(\theta_m - \theta), \nabla \varphi\rangle|\, dt \to 0$. Thus,

$$\int_0^T \langle v_m \theta_m, \nabla \varphi\rangle\, dt \to \int_0^T \langle v\theta, \nabla \varphi\rangle\, dt.$$

It is easy to prove that

$$\int_0^T (\beta(x)\theta_m, \varphi)_{\Gamma_R}\, dt \to \int_0^T (\beta(x)\theta, \varphi)_{\Gamma_R}\, dt.$$

Therefore, we have from (8.94)

$$\int_0^T \Big(\langle \frac{\partial \theta}{\partial t}, \varphi\rangle + (\kappa \nabla \theta, \nabla \varphi) + (\beta(x)\theta, \varphi)_{\Gamma_R} - \langle v\theta, \nabla \varphi\rangle\Big)\, dt = \int_0^T \langle g_1, \varphi\rangle\, dt. \tag{8.96}$$

Since $\theta_m(0) = \theta_0$, by (8.86) it is obvious that $\theta(0) = \theta_0$.

Therefore, we have proved the existence of a solution.

Next, let us prove uniqueness of a solution. Let (v_1, θ_1), (v_2, θ_2) be two solutions to Problem I-VI satisfying inequality (8.82) instead of approximate solutions. Then, taking (8.29) into account (Remark 8.4), we have from (8.21)

$$\begin{aligned}
&\langle v_1'(t) + A_1 v_1(t) + B_1(v_1(t), v_1(t)) - (1 - \alpha_0 \theta_1) f - f_1, v_2(t) - v_1(t)\rangle \\
&\qquad\qquad + \Phi(v_2(t)) - \Phi(v_1(t)) \geq 0, \\
&\langle v_2'(t) + A_1 v_2(t) + B_1(v_2(t), v_2(t)) - (1 - \alpha_0 \theta_2) f - f_1, v_1(t) - v_2(t)\rangle \\
&\qquad\qquad + \Phi(v_1(t)) - \Phi(v_2(t)) \geq 0,
\end{aligned}$$

which imply

$$
\begin{aligned}
&\langle v_1'(t) - v_2'(t), v_1(t) - v_2(t)\rangle + \langle A_1(v_1(t) - v_2(t)), v_1(t) - v_2(t)\rangle \\
&\le |\alpha_0||(\theta_1 - \theta_2) f, v_1(t) - v_2(t))| + |\langle B_1(v_1(t), v_1(t)) - B_1(v_2(t), v_2(t)), v_1(t) - v_2(t)\rangle|.
\end{aligned} \tag{8.97}
$$

By virtue of (8.31) and (8.32), we have

$$
\begin{aligned}
\frac{d}{dt}&(\|(v_1(t) - v_2(t)\|^2) + \frac{7\mu}{2}\|v_1(t) - v_2(t)\|_{\mathbf{V}}^2 \\
&\le 2c_*\|v_1(t) - v_2(t)\|^2 + \frac{\mu}{2}\|v_1(t) - v_2(t)\|_{\mathbf{V}}^2 + c\|f\|_{\mathbf{L}^3}^2\|\theta_1 - \theta_2\|^2 \\
&\quad + 2|\langle B(v_1(t) - v_2(t), v_1(t)), v_1(t) - v_2(t)\rangle| \\
&\quad + 2|\langle B(v_2(t), v_1(t) - v_2(t)), v_1(t) - v_2(t)\rangle| \\
&\le 2c_*\|v_1(t) - v_2(t)\|^2 + \frac{\mu}{2}\|v_1(t) - v_2(t)\|_{\mathbf{V}}^2 + c\|f\|_{\mathbf{L}^3}^2\|\theta_1 - \theta_2\|^2 \\
&\quad + 2c_2(\|v_1(t)\|_{\mathbf{V}} + \|v_2(t)\|_{\mathbf{V}})\|v_1(t) - v_2(t)\|_{\mathbf{V}}^2,
\end{aligned}
$$

where c_2 is the one in (8.32). By (8.82),

$$
2c_2(\|v_1(t)\|_{\mathbf{V}} + \|v_2(t)\|_{\mathbf{V}}) \le \mu,
$$

and so we have

$$
\frac{d\|v_1(t) - v_2(t)\|^2}{dt} + 2\mu\|v_1(t) - v_2(t)\|_{\mathbf{V}}^2 \le 2c_*\|v_1(t) - v_2(t)\|^2 + c\|f\|_{\mathbf{L}^3}^2\|\theta_1 - \theta_2\|^2. \tag{8.98}
$$

Also, from

$$
\begin{aligned}
&\left\langle\frac{\partial\theta_1}{\partial t}, \varphi\right\rangle + (\kappa\nabla\theta_1, \nabla\varphi) + (\beta(x)\theta_1, \varphi)_{\Gamma_R} - \langle v_1\theta_1, \nabla\varphi_1\rangle = \langle g_1, \varphi\rangle, \\
&\left\langle\frac{\partial\theta_2}{\partial t}, \varphi\right\rangle + (\kappa\nabla\theta_2, \nabla\varphi) + (\beta(x)\theta_2, \varphi)_{\Gamma_R} - \langle v_2\theta_2, \nabla\varphi\rangle = \langle g_1, \varphi\rangle
\end{aligned}
$$

we have

$$
\begin{aligned}
\left\langle\frac{\partial\theta_1 - \theta_2}{\partial t}, \theta_1 - \theta_2\right\rangle &+ \kappa(\nabla\theta_1 - \nabla\theta_2, \nabla\theta_1 - \nabla\theta_2) + (\beta(x)(\theta_1 - \theta_2), \theta_1 - \theta_2)_{\Gamma_R} \\
&- \langle v_1(\theta_1 - \theta_2), \nabla(\theta_1 - \theta_2)\rangle - \langle(v_1 - v_2)\theta_2, \nabla(\theta_1 - \theta_2)\rangle = 0.
\end{aligned} \tag{8.99}
$$

Taking into account $\langle v_1(\theta_1 - \theta_2), \nabla(\theta_1 - \theta_2)\rangle = 0$ (see (8.8)), by (8.56) we have

$$
\frac{d}{dt}\|\theta_1 - \theta_2\|^2 + 2\kappa\|\nabla\theta_1 - \nabla\theta_2\|^2 \le \frac{c}{\kappa}\|v_1(t) - v_2(t)\|_{\mathbf{V}}^2\|\theta_2(t)\|_{W_{\Gamma_D}^{1,2}}^2 + \kappa\|\nabla\theta_1 - \nabla\theta_2\|^2,
$$

and so

$$\frac{d}{dt}(\|\theta_1(t)-\theta_2(t)\|^2) \le \frac{c}{\kappa}\|v_1(t)-v_2(t)\|_{\mathbf{V}}^2 \|\theta_2(t)\|_{W^{1,2}_{\Gamma_D}}^2. \tag{8.100}$$

By (8.82),

$$\frac{c}{\kappa}\|\theta_2(t)\|_{W^{1,2}_{\Gamma_D}}^2 \le \frac{\mu}{2}.$$

Therefore adding (8.98) and (8.100), we get

$$\begin{aligned}\frac{d}{dt}&(\|v_1(t)-v_2(t)\|^2+\|\theta_1(t)-\theta_2(t)\|^2)\\ &\le \big(2c_*+c\|f\|_{\mathbf{L}^3}^2\big)\big(\|v_1(t)-v_2(t)\|^2+\|\theta_1(t)-\theta_2(t)\|^2\big).\end{aligned} \tag{8.101}$$

We have from (8.101)

$$\begin{aligned}&\|v_1(t)-v_2(t)\|^2+\|\theta_1(t)-\theta_2(t)\|^2\\ &\le \int_0^t \big(2c_*+2c\|f\|_{\mathbf{L}^3}^2\big)\big(\|v_1(s)-v_2(s)\|^2+\|\theta_1(s)-\theta_2(s)\|^2\big)\,ds,\end{aligned} \tag{8.102}$$

which implies $v_1(t)=v_2(t)$, $\theta_1(t)=\theta_2(t)$ for all $t\in[0,T]$, and the proof is complete. □

8.4 Existence of a Solution: The Case of Total Pressure

Theorem 8.2 (the case of total pressure) *Let Assumption 8.2 be satisfied, $v_0 \in H_K$ and $\theta_0 \in L^2(\Omega)$. Then there exists a solution $(v,\theta)\in\big(L^\infty(0,T;H)\cap L^2(0,T;\mathbf{V})\big)\times\big(L^\infty(0,T;L^2(\Omega))\cap L^2(0,T;W^{1,2}_{\Gamma_D}(\Omega))\big)$ to* (8.28).

As in the preceding section, we will show Theorem 8.2 in the next two subsections.

8.4.1 Existence of a Solution to an Approximate Problem

We first consider an approximate problem for (8.28). For every $0<\varepsilon<1$, let a functional Φ_ε be Moreau-Yosida approximation of Φ.

Let $\{u_j, j=1,2,\cdots\}$ and $\{\varphi_j, j=1,2,\cdots\}$ be, respectively, bases of the space $\mathbf{V}$ and $W^{1,2}_{\Gamma_D}(\Omega)$. Without loss of generality, we assume that $u_1=v_0$, $\varphi_1=\theta_0$. We find a solution $v_m=\sum_{j=1}^m g_{jm}(t)u_j$, $\theta_m=\sum_{j=1}^m r_{jm}(t)\varphi_j$ to the following problem

$$\begin{cases} \left\langle \dfrac{\partial v_m}{\partial t}, u_j \right\rangle + 2\big\langle \mu(\theta_m)\mathcal{E}(v_m), \mathcal{E}(u_j)\big\rangle + \langle \mathrm{rot} v_m \times v_m, u_j\rangle + 2(\mu(\theta_m)k(x)v_m, u_j)_{\Gamma_2} \\ \quad + 2(\mu(\theta_m)S\tilde{v}_m, \tilde{u}_j)_{\Gamma_3} + 2(\alpha(x)v_m, u_j)_{\Gamma_5} + (\mu(\theta_m)k(x)v_m, u_j)_{\Gamma_7} + \langle \nabla\Phi_\varepsilon(v_m(t)), u_j\rangle \\ \qquad = \langle (1-\alpha_0\theta_m)f, u_j\rangle + \langle f_2, u_j\rangle, \\ \left\langle \dfrac{\partial \theta_m}{\partial t}, \varphi_j \right\rangle + (\kappa(\theta_m)\nabla\theta_m, \nabla\varphi_j) + (\beta(x)\theta_m, \varphi_j)_{\Gamma_R} - \langle v_m\theta_m, \nabla\varphi_j\rangle = \langle g_2, \varphi_j\rangle, \\ v_m(0) = v_0, \quad \theta_m(0) = \theta_0. \end{cases} \tag{8.103}$$

which gives us a system for $g_{jm}(t)$, $r_{jm}(t)$, $j = 1, \cdots, m$. The solutions to (8.103) depend on ε, but for convenience of notation here and in what follows of this section we use subindex m. For a t_m there exist absolute continuous functions $g_{jm}(t)$ and $r_{jm}(t)$ on $[0, t_m)$. If $\|v_m(t)\|$, $\|\theta_m(t)\|$ are bounded and $v_m(t)$, $\theta_m(t)$ are integrable, then $g_{jm}(t)$, $r_{jm}(t)$ are prolonged over t_m. We will find estimates (8.118) below, by which we see that $t_m = T$.

Multiplying the first and second equations of (8.103), respectively, by $g_{jm}(t)$ and $\varphi_{jm}(t)$ and adding for $i = 1, \cdots, m$, we get

$$\begin{cases} \left\langle \dfrac{\partial v_m}{\partial t}, v_m \right\rangle + 2\big\langle \mu(\theta_m)\mathcal{E}(v_m), \mathcal{E}(v_m)\big\rangle + \langle \mathrm{rot}\, v_m \times v_m, v_m\rangle + 2(\mu(\theta_m)k(x)v_m, v_m)_{\Gamma_2} \\ \qquad + 2(\mu(\theta_m)S\tilde{v}_m, \tilde{v}_m)_{\Gamma_3} + 2(\alpha(x,t)v_m, v_m)_{\Gamma_5} + (\mu(\theta_m)k(x)v_m, v_m)_{\Gamma_7} \\ \qquad + \langle \nabla\Phi_\varepsilon(v_m(t)), v_m\rangle + \langle \alpha_0\theta_m f, v_m\rangle = \langle f, v_m\rangle + \langle f_2, v_m\rangle, \\ \left\langle \dfrac{\partial \theta_m}{\partial t}, \theta_m \right\rangle + (\kappa(\theta_m)\nabla\theta_m, \nabla\theta_m) + (\beta(x)\theta_m, \theta_m)_{\Gamma_R} - \langle v_m\theta_m, \nabla\theta_m\rangle = \langle g_2, \theta_m\rangle, \\ v_m(0) = v_0, \quad \theta_m(0) = \theta_0. \end{cases} \tag{8.104}$$

Let us estimate terms on the left hand side above. It is easy to see

$$2\Big\langle \mu(\theta_m)\mathcal{E}(v_m), \mathcal{E}(v_m)\Big\rangle dt \geq 2\mu_0\|v_m\|_{\mathbf{V}}^2. \tag{8.105}$$

By the fact that $\Gamma_{2j}, \Gamma_{3j}, \Gamma_{7j}$ are in $C^{2,1}(\Gamma_{ij})$ and Assumption 8.2, there exists a constant M such that

$$\|S(x)\|_\infty, \ \|k(x)\|_\infty, \ \|\alpha\|_{\mathbf{L}^\infty(\Gamma_5)} \leq M. \tag{8.106}$$

Thus,

$$\begin{aligned} &\Big|\big(2(\mu(\theta_m)k(x)v, v)_{\Gamma_2} + 2(\mu(\theta_m)S\tilde{v}_m, \tilde{v}_m)_{\Gamma_3} + 2(\alpha(x)v_m, v_m)_{\Gamma_5} + (\mu(\theta_m)k(x)v_m, v_m)_{\Gamma_7}\big)\Big| \\ &\qquad \leq \frac{\mu_0}{2}\|v_m\|_{\mathbf{V}}^2 + k_{11}\|v_m\|^2 \end{aligned} \tag{8.107}$$

(see Theorem 1.27). Obviously,

$$\langle \mathrm{rot}\, v_m \times v_m, v_m\rangle = 0. \tag{8.108}$$

Since Φ_ε is convex, continuous and Fréchet differentiable, we have

$$\Phi_\varepsilon(y) - \Phi_\varepsilon(x) \geq \langle \nabla\Phi_\varepsilon(x), y - x\rangle \quad \forall x, y \in \mathbf{V}. \tag{8.109}$$

Thus,

$$\langle \nabla\Phi_\varepsilon(v_m(t)), 0_{\mathbf{V}} - v_m(t)\rangle \leq \Phi_\varepsilon(0_{\mathbf{V}}) - \Phi_\varepsilon(v_m(t)) \leq -\Phi_\varepsilon(v_m(t)),$$

and

$$\langle \nabla\Phi_\varepsilon(v_m(t)), v_m(t)\rangle \geq \Phi_\varepsilon(v_m(t)). \tag{8.110}$$

Also, by the Hölder inequality we have

$$\left|\langle \alpha_0\theta_m f, v_m\rangle\, dt\right| \leq k_{12}\|v_m(t)\|^2 + \frac{\kappa_0}{4}\|\theta_m(t)\|^2_{W^{1,2}_{\Gamma_D}}, \tag{8.111}$$

where $k_{12} = c|\alpha_0|\|f\|^2_{L^\infty(0,T;\mathbf{L}^3)}$.

$$|\langle f, w_m\rangle + \langle f_2, w_m\rangle| \leq \frac{\mu_0}{2}\|w_m\|^2_{\mathbf{V}} + c(\|f\|^2_{\mathbf{L}^3} + \|f_2\|^2_{\mathbf{V}^*}). \tag{8.112}$$

Also, we have

$$\begin{aligned}&\big\langle \kappa(\theta_m)\nabla\theta_m, \nabla\theta_m)\big\rangle \geq \kappa_0\|\theta_m\|^2_{W^{1,2}_{\Gamma_D}},\\ &(\beta(x)\theta_m, \theta_m)_{\Gamma_R} \geq 0.\end{aligned} \tag{8.113}$$

By (8.8) we have

$$\langle w_m\theta_m, \nabla\theta_m\rangle = 0. \tag{8.114}$$

Also,

$$|\langle g_2, \theta_m\rangle| \leq \frac{\kappa_0}{4}\|\theta_m(t)\|^2_{W^{1,2}_{\Gamma_D}} + c\|g_2\|^2_{(W^{1,2}_{\Gamma_D})^*}. \tag{8.115}$$

Taking

$$k_1 = k_{11} + k_{12} \tag{8.116}$$

and using (8.105)–(8.115), we have from (8.104)

$$\begin{aligned}\frac{d}{dt}\|v_m(t)\|^2 + &\frac{d}{dt}\|\theta_m(t)\|^2 + 2\mu_0\|v_m\|^2_{\mathbf{V}} + \kappa_0\|\theta_m\|^2_{W^{1,2}_{\Gamma_D}} + \Phi_\varepsilon(v_m(t))\\ &\leq c\Big(\|f(t)\|^2_{\mathbf{L}^3} + \|f_2(t)\|^2_{\mathbf{V}^*} + \|g_2(t)\|^2_{(W^{1,2}_{\Gamma_D})^*}\Big) + 2k_1\|v_m(t)\|^2.\end{aligned} \tag{8.117}$$

Applying Gronwall's inequality, we have from (8.117)

$$
\begin{aligned}
&\|v_m(t)\|^2 + \|\theta_m(t)\|^2 \\
&\quad \le \Big[\|v_0\|^2 + \|\theta_0\|^2 + \int_0^t \Big(\|f(s)\|^2_{\mathbf{L}^3} + \|f_2(s)\|^2_{\mathbf{V}^*} + \|g_2(s)\|^2_{(W^{1,2}_{\Gamma_D})^*}\Big)\,ds\Big]e^{2k_1 t},\\
&\|v_m\|^2_{L^2(0,T;\mathbf{V})} + \|\theta_m\|^2_{L^2(0,t;W^{1,2}_{\Gamma_D})}\\
&\quad \le c\Big(\|v_0\|^2 + \|\theta_0\|^2 + \int_0^T \Big(\|f(t)\|^2_{\mathbf{L}^3} + \|f_2(t)\|^2_{\mathbf{V}^*} + \|g_2(t)\|^2_{(W^{1,2}_{\Gamma_D})^*}\Big)\,dt\Big),\\
&\int_0^T \Phi_\varepsilon(v_m(t))\,dt \le c\Big(\|v_0\|^2 + \|\theta_0\|^2 + \int_0^T \Big(\|f(t)\|^2_{\mathbf{L}^3} + \|f_2(t)\|^2_{\mathbf{V}^*} + \|g_2(t)\|^2_{(W^{1,2}_{\Gamma_D})^*}\Big)\,dt\Big).
\end{aligned}
\tag{8.118}
$$

Note that c in (8.118) depends on T and f (via k_{12}), but independent of m and ε.

Since Φ is nonnegative, by (1.36) and the third inequality of (8.118), we have

$$
\int_0^T \|v_m(t) - J_\varepsilon(v_m(t))\|^2_{\mathbf{V}}\,dt \le c\varepsilon \tag{8.119}
$$

with c independent of ε.

Multiplying the first equation of (8.103) by $g_{jm}(t) - g_{jm}(s)$, summing for j and taking into account (8.109), we have

$$
\begin{aligned}
&\frac{1}{2}\frac{d\|v_m(t) - v_m(s)\|^2}{dt}\\
&\quad + \langle A_2(\theta_m)v_m(t) + B_2(v_m(t), v_m(t)) + \alpha_0\theta_m(t)f(t) - f(t) - f_2(t), v_m(t) - v_m(s)\rangle\\
&= \langle \nabla\Phi_\varepsilon(v_m(t)), v_m(s) - v_m(t)\rangle \le \Phi_\varepsilon(v_m(s)) - \Phi_\varepsilon(v_m(t)) \le \Phi_\varepsilon(v_m(s)),
\end{aligned}
\tag{8.120}
$$

where the operators $A_2(\theta_m)$, B_2 are the ones in (8.27).

By (8.107) and (8.106), we have

$$
\begin{aligned}
&\langle A_2(\theta_m)v_m(t), v_m(t)\rangle \ge \frac{3\mu_0}{2}\|v_m(t)\|^2_{\mathbf{V}} - k_{11}\|v_m(t)\|^2,\\
&|\langle A_2(\theta_m)v_m(t), v_m(s)\rangle| \le c\|v_m(t)\|_{\mathbf{V}}\|v_m(s)\|_{\mathbf{V}},
\end{aligned}
\tag{8.121}
$$

where k_{11} is the one in (8.107). Taking into account (8.121) and the fact that $\langle B_2(v_m(t),$
$v_m(t)), v_m(t)\rangle = 0$, we have from (8.120)

$$
\begin{aligned}
\frac{1}{2}\frac{d\|v_m(t) - v_m(s)\|^2}{dt} \le &\Phi_\varepsilon(v_m(s)) + \langle A_2(\theta_m)v_m(t), v_m(s)\rangle + \langle B_2(v_m(t), v_m(t)), v_m(s)\rangle\\
&+ \langle -\alpha_0\theta_m(t)f(t) + f(t) + f_2(t), v_m(t) - v_m(s)\rangle + k_{11}\|v_m(t)\|^2.
\end{aligned}
\tag{8.122}
$$

Let us integrate every term of (8.122) first with respect to t from s to $s+h$ and then with respect to s from 0 to T, where $v_m(t) = 0$ when $t \in (T, T+h)$. It is clear that

$$\int_0^T \int_s^{s+h} \frac{d\|v_m(t) - v_m(s)\|^2}{dt}\, dtds = \int_0^T \|v_m(s+h) - v_m(s)\|^2\, ds. \tag{8.123}$$

By the third inequality of (8.118),

$$\int_0^T \int_s^{s+h} \Phi_\varepsilon(v_m(s))\, dtds \le h \int_0^T \Phi_\varepsilon(v_m(s))\, ds \le c_1 h. \tag{8.124}$$

By (8.118) and (8.122), we have

$$\begin{aligned}\Big|\int_0^T \int_s^{s+h} \langle A_2(\theta_m)v_m(t), v_m(s)\rangle\, dtds\Big| &\le c \int_0^T \|v_m(s)\|_{\mathbf{V}} \int_s^{s+h} \|v_m(t)\|_{\mathbf{V}}\, dt\, ds \\ &\le c \int_0^T \|v_m(s)\|_{\mathbf{V}} (\sqrt{h}\|v_m\|_{L^2(0,T;\mathbf{V})})\, ds \le c_2\sqrt{h}.\end{aligned} \tag{8.125}$$

Since $\|w\|_{\mathbf{L}^3} \le K\|w\|_{\mathbf{L}^2}^{\frac{1}{2}} \|w\|_{\mathbf{L}^6}^{\frac{1}{2}}$ (see Theorem 1.12), we have

$$\begin{aligned}|\langle B_2(v, w), z\rangle| = |\langle \mathrm{rot}\, v \times w, z\rangle| &\le K\|\mathrm{rot}\, v\|_{\mathbf{L}^2} \|w\|_{\mathbf{L}^3} \|z\|_{\mathbf{L}^6} \\ &\le K\|v\|_{\mathbf{V}} \|w\|^{\frac{1}{2}} \|w\|_{\mathbf{V}}^{\frac{1}{2}} \|z\|_{\mathbf{V}},\end{aligned} \tag{8.126}$$

and so by (8.118) we have

$$\begin{aligned}&\Big|\int_0^T \int_s^{s+h} \langle B_2(v_m(t), v_m(t))v_m(s)\rangle\, dtds\Big| \\ &\quad\le K \int_0^T \int_s^{s+h} \|v_m(t)\|_{\mathbf{V}}^{\frac{3}{2}} \|v_m(t)\|^{\frac{1}{2}} \|v_m(s)\|_{\mathbf{V}}\, dtds \\ &\quad\le K \int_0^T \|v_m(s)\|_{\mathbf{V}} \Big(\int_s^{s+h} \|v_m(t)\|_{\mathbf{V}}^2\, dt\Big)^{\frac{3}{4}} \Big(\int_s^{s+h} \|v_m(t)\|^2\, dt\Big)^{\frac{1}{4}} ds \\ &\quad\le c_3 h^{\frac{1}{4}}.\end{aligned} \tag{8.127}$$

Also, by (8.118) we have

$$\begin{aligned}\Big|\int_0^T \int_s^{s+h} \langle (f+f_2)(t), v_m(t)\rangle\, dtds\Big| &\le \int_0^T |\langle (f+f_2)(t), v_m(t)\rangle| \Big(\int_{t-h}^t ds\Big) dt \le c_4 h, \\ \Big|\int_0^T \int_s^{s+h} \langle (f+f_2)(t), -v_m(s)\rangle\, dtds\Big| &\le K \int_0^T \|v_m(s)\|_{\mathbf{V}} \int_s^{s+h} \|(f+f_2)(t)\|_{\mathbf{V}^*}\, dtds \\ &\le c_5\sqrt{h}.\end{aligned} \tag{8.128}$$

In the same way, we get

$$
\begin{aligned}
&\left|\int_0^T \int_s^{s+h} \langle \alpha_0 \theta_m f, v_m(t) - v_m(s)\rangle \, dt ds\right| \le c_6 h + c_7 \sqrt{h},\\
&\left|\int_0^T \int_s^{s+h} k_{11} \|v_m(t)\|^2 \, dt ds\right| \le c_8 h.
\end{aligned}
\tag{8.129}
$$

Note that constants $c_i, i = 1, \cdots, 8$, are independent of m and ε. By virtue of (8.123)–(8.129), uniformly with respect to m, ε

$$
\int_0^T \|v_m(s+h) - v_m(s)\|^2 \, ds \le O(h^{\frac{1}{4}}),
\tag{8.130}
$$

and the set $\{v_m\}$ is relatively compact in $L^2(0, T; \mathbf{W}^{\frac{9}{10},2}(\Omega))$ (see Theorem 1.38).

Also, we have

$$
\begin{aligned}
&|(\kappa(\theta_m)\nabla\theta_m, \nabla\varphi)| \le \kappa_1 \|\nabla\theta_m\|_{\mathbf{L}^2} \|\varphi\|_{W^{1,2}_{\Gamma_D}(\Omega)},\\
&|(\beta(x)\theta_m, \varphi)_{\Gamma_R}| \le c\|\theta_m\|_{W^{1,2}_{\Gamma_D}(\Omega)} \|\varphi\|_{W^{1,2}_{\Gamma_D}(\Omega)},\\
&|\langle v_m\theta_m, \nabla\varphi\rangle| \le c\|v_m\|_{\mathbf{V}} \|\theta_m\|_{L^3} \|\varphi\|_{W^{1,2}_{\Gamma_D}(\Omega)}.
\end{aligned}
\tag{8.131}
$$

By (8.131), we have from the second equation of (8.103)

$$
\begin{aligned}
\left|\left\langle \frac{\partial\theta_m}{\partial t}, \varphi\right\rangle\right| \le c\Big(\|\theta_m\|_{W^{1,2}_{\Gamma_D}(\Omega)} + \|v_m\|^2_{\mathbf{V}} + \|\theta_m\|^2_{W^{1,2}_{\Gamma_D}(\Omega)} + \|g_2\|_{(W^{1,2}_{\Gamma_D})^*}\Big)\|\varphi\|_{W^{1,2}_{\Gamma_D}}\\
\forall\varphi \in W^{1,2}_{\Gamma_D}(\Omega).
\end{aligned}
$$

Hence, by (8.118) we see that

$$
\begin{aligned}
&\theta'_m \in L^1\big(0, T; (W^{1,\tau}_{\Gamma_D})^*\big),\\
&\|\theta'_m\|_{L^1\left(0,T;(W^{1,\tau}_{\Gamma_D})^*\right)} \le c,
\end{aligned}
\tag{8.132}
$$

where c is independent of m and ε. Thus, the set $\{\theta_\varepsilon\}$ is relatively compact in $L^2(0, T; W^{\frac{9}{10},2}(\Omega))$ (see Theorem 1.39).

8.4.2 Existence of a Solution

Having obtained solutions to (8.103), which are actually a sequence of approximating solutions to (8.28), we can extract subsequences, which are still denoted as before, such that as $m \to \infty$, $\varepsilon \to 0$,

$$
\begin{aligned}
v_m &\rightharpoonup v \quad \text{in } L^2(0,T;\mathbf{V}),\\
v_m &\overset{*}{\rightharpoonup} v \quad \text{in } L^\infty(0,T;H),\\
v_m &\to v \quad \text{in } L^2(0,T;\mathbf{W}^{\frac{9}{10},2}(\Omega)),\\
\theta_m &\rightharpoonup \theta \quad \text{in } L^2(0,T;W^{1,2}_D(\Omega)),\\
\theta_m &\overset{*}{\rightharpoonup} \theta \quad \text{in } L^\infty(0,T;L^2(\Omega)),\\
\theta_m &\to \theta \quad \text{in } L^2(0,T;W^{\frac{9}{10},2}(\Omega)).
\end{aligned}
\tag{8.133}
$$

On the other hand, putting $u=\sum_{j=1}^M k_j(t)u_j$, where $k_j(t)\in C^1[0,T]$ and M is a positive integer, let us multiply the first equation of (8.103) by $k_j(t)$ and sum for $j=1,\cdots,M$. Then, let us multiply the first equation of (8.103) by $g_{jm}(t)$ and sum for $j=1,\cdots,m$. Substituting the resulting equations yields

$$
\begin{aligned}
&\int_0^T \Big\langle \frac{\partial v_m}{\partial t} + A_2(\theta_m)v_m(t) + B_2(v_m(t),v_m(t)) + \nabla\Phi_\varepsilon(v_m(t)), u(t)-v_m(t)\Big\rangle\, dt\\
&\qquad = \int_0^T \langle -\alpha_0\theta_m f + f + f_2, u(t)-v_m(t)\rangle\, dt.
\end{aligned}
\tag{8.134}
$$

Since

$$
\begin{aligned}
&\int_0^T \langle v_m'(t), u(t)-v_m(t)\rangle\, dt\\
&\quad = \int_0^T \langle u'(t), u(t)-v_m(t)\rangle\, dt - \frac{1}{2}\|v_m(T)-u(T)\|^2 + \frac{1}{2}\|v_m(0)-u(0)\|^2,
\end{aligned}
$$

taking into account (8.109), we have from (8.134)

$$
\begin{aligned}
&\int_0^T \langle u'(t) + A_2(\theta_m)v_m(t) + B_2(v_m(t),v_m(t)), u(t)-v_m(t)\rangle\, dt\\
&\quad - \int_0^T \langle -\alpha_0\theta_m f + f + f_2, u(t)-v_m(t)\rangle\, dt + \int_0^T \big(\Phi_\varepsilon(u(t)) - \Phi_\varepsilon(v_m(t))\big)\, dt\\
&\ge -\frac{1}{2}\|v_m(0)-u(0)\|^2.
\end{aligned}
\tag{8.135}
$$

Since $\Phi_\varepsilon(u)\le\Phi(u)$ and $\Phi(J_\varepsilon v_m(t))\le\Phi_\varepsilon(v_m(t))$ (see (1.37)), we have from (8.135)

$$\int_0^T \langle u'(t) + A_2(\theta_m)v_m(t) + B_2(v_m(t), v_m(t)), u(t) - v_m(t)\rangle\, dt$$
$$- \int_0^T \langle -\alpha_0\theta_m f + f + f_2, u(t) - v_m(t)\rangle\, dt + \Psi(u) - \int_0^T \Phi(J_\varepsilon v_m(t)))\, dt$$
$$\geq -\frac{1}{2}\|v_m(0) - u(0)\|^2. \tag{8.136}$$

By (8.133) and Corollary 1.1, we get

$$\int_0^T \langle A_2(\theta_m)v_m(t), u(t)\rangle\, dt \equiv \int_0^T a_{02}(\theta_m(t); v_m(t), u(t))\, dt \to \int_0^T \langle A_2(\theta)v(t), u(t)\rangle\, dt. \tag{8.137}$$

Owing to (8.133),

$$v_m \to v \text{ in } L^2(0, T; \mathbf{L}^2(\partial\Omega)).$$

Thus taking a subsequence if necessary, by Lemma 1.3 we deduce

$$2(\mu(\theta_m)k(x)v_m, v_m)_{\Gamma_2} + 2(\mu(\theta_m)S\tilde{v}_m, \tilde{v}_m)_{\Gamma_3} + 2(\alpha(x)v_m, v_m)_{\Gamma_5} + (\mu(\theta_m)k(x)v_m, v_m)_{\Gamma_7}$$
$$\to 2(\mu(\theta)k(x)v, v)_{\Gamma_2} + 2(\mu(\theta)S\tilde{v}, \tilde{v})_{\Gamma_3} + 2(\alpha(x)v, v)_{\Gamma_5} + (\mu(\theta)k(x)v, v)_{\Gamma_7}.$$

Therefore, taking into account

$$\liminf 2(\mu(\theta_m)\mathcal{E}(v_m), \mathcal{E}(v_m)) \geq 2(\mu(\theta)\mathcal{E}(v), \mathcal{E}(v))$$

(see Corollary 1.2), we have

$$\liminf \langle A_2(\theta_m)v_m(t), v_m(t)\rangle \geq \langle A_2(\theta)v(t), v(t)\rangle. \tag{8.138}$$

From (8.137) and (8.138), it follows

$$\liminf_{\substack{m\to\infty \\ \varepsilon\to 0}} \int_0^T \langle A_2(\theta_m)v_m(t), u(t) - v_m(t)\rangle\, dt \leq \int_0^T \langle A_2(\theta)v(t), u(t) - v(t)\rangle\, dt. \tag{8.139}$$

By (8.119) and (8.133), we have that $J_\varepsilon(v_m) \rightharpoonup v$ in $L^2(0, T; \mathbf{V})$ as $m \to \infty$, $\varepsilon \to 0$. Since the functional $\Phi : \mathbf{V} \to \overline{\mathbb{R}}$ is weakly lower semi-continuous, we have

$$\liminf_{\substack{m\to\infty \\ \varepsilon\to 0}} \int_0^T \Phi(J_\varepsilon v_m(t))\, dt \geq \int_0^T \Phi(v(t))\, dt \equiv \Psi(v). \tag{8.140}$$

It is not difficult to prove that

$$\int_0^T \langle B_2(v_m(t), v_m(t)), u(t)\rangle\, dt \to \int_0^T \langle B_2(v(t), v(t)), u(t)\rangle\, dt \tag{8.141}$$

as $m \to \infty$, $\varepsilon \to 0$ (see (6.46)–(6.49)).

Let us prove that

$$\int_0^T \langle \alpha_0\theta_m f, u(t) - v_m(t)\rangle\, dt \to \int_0^T \langle \alpha_0\theta f, u(t) - v(t)\rangle\, dt \to 0 \quad \text{as } m \to \infty, \varepsilon \to 0. \tag{8.142}$$

By (8.133), we have

$$\int_0^T \langle (\alpha_0\theta_m - \alpha_0\theta) f, u(t)\rangle\, dt \to 0 \tag{8.143}$$

and

$$\begin{aligned}&\int_0^T \big(\langle \alpha_0\theta_m f, v_m(t)\rangle - \langle \alpha_0\theta f, v(t)\rangle\big)\, dt \\ &= \int_0^T \langle \alpha_0(\theta_m - \theta) f, v_m(t)\rangle\, dt + \int_0^T \langle \alpha_0\theta f, v_m(t) - v(t)\rangle\, dt \to 0.\end{aligned} \tag{8.144}$$

By (8.143) and (8.144) we get (8.142). It is obvious that $\langle B_2(v(t), v(t)), v(t)\rangle = 0$.

Therefore, by (8.139)–(8.141) and (8.142), we have from (8.136)

$$\begin{aligned}\int_0^T \langle u'(t) + A_2(\theta)v(t) + B_2(v(t), v(t)) - (1 - \alpha_0\theta) f - f_2, u(t) - v(t)\rangle\, dt \\ + \Psi(u) - \Psi(v) \geq -\frac{1}{2}\|v_0 - u(0)\|^2.\end{aligned} \tag{8.145}$$

Since $\|B_2(v(t), v(t))\|_{\mathbf{V}^*} \leq K\|v(t)\|_{\mathbf{V}}^{\frac{3}{2}}\|v(t)\|^{\frac{1}{2}}$ (see (8.126)) and $v \in L^\infty(0, T; \mathbf{L}^2(\Omega))$, we see that $B_2(v, v) \in L^{\frac{4}{3}}(0, T; \mathbf{V}^*)$. Therefore, by density of the set $\{u = \sum_{j=1}^M k_j(t)u_j,\ k_j(t) \in C^1[0, T], M{=}1, 2, \cdots\}$ in $\{L^4(0, T; \mathbf{V}) : u' {\in} L^2(0, T; \mathbf{V}^*)\}$ (see Remark 1.8), the inequality (8.145) is valid for all $u \in \{L^4(0, T; \mathbf{V}) : u' \in L^2(0, T; \mathbf{V}^*)\}$.

Thus, the first formula of (8.28) holds.

Putting $\varphi(t) = \sum_{j=1}^M k_j(t)\varphi_j$, where $k_j(t) \in C^1[0, T]$, $k_j(T) = 0$, multiplying the second equation of (8.103) by $k_j(t)$ and summing up for $j = 1, \cdots, M$, we obtain

$$\begin{aligned}\big(\theta_m(t), \varphi(t)\big) - \int_0^t \Big(\theta_m, \frac{\partial\varphi}{\partial t}\Big)\, ds + \int_0^t (\kappa(\theta_m\nabla\theta_m, \nabla\varphi)\, ds + \int_0^t (\beta(x)\theta_m, \varphi)_{\Gamma_R}\, ds \\ - \int_0^t \langle v_m\theta_m, \nabla\varphi\rangle\, ds = \big(\theta_m(0), \varphi(0)\big) + \int_0^t \big(g_2, \varphi\big)\, ds \quad \forall t \in [0, T].\end{aligned} \tag{8.146}$$

By Corollary 1.1 we have

$$\int_0^t (\kappa(\theta_m)\nabla\theta_m, \nabla\varphi)\, ds \to \int_0^t (\kappa(\theta)\nabla\theta, \nabla\varphi)\, ds \quad \text{as } m \to \infty,\ \varepsilon \to 0. \tag{8.147}$$

Since $W^{\frac{9}{10},2}(\Omega) \subset L^4(\Omega)$ (see Theorem 1.20), we have

$$v_m \to v \text{ in } L^2(0, T; \mathbf{L}^4(\Omega)), \qquad \theta_m \to \theta \text{ in } L^2(0, T; L^4(\Omega)). \tag{8.148}$$

By (8.148), we have

$$\begin{aligned}
&\int_0^t |\langle v_m\theta_m, \nabla\varphi\rangle - \langle v\theta, \nabla\varphi\rangle|\, ds \\
&\le \int_0^t |\langle (v_m - v)\theta_m, \nabla\varphi\rangle|\, ds + \int_0^t |\langle v(\theta_m - \theta), \nabla\varphi\rangle|\, ds \\
&\le \left(\|v_m - v\|_{L^2(0,T;\mathbf{L}^4)}\|\theta_m\|_{L^2(0,T;L^4)} + \|v\|_{L^2(0,T;\mathbf{L}^4)}\|\theta_m - \theta\|_{L^2(0,T;L^4)}\|\right)\|\nabla\varphi\|_{L^\infty(0,T;\mathbf{L}^2)} \to 0,
\end{aligned}$$

which implies

$$\int_0^t \langle v_m\theta_m, \nabla\varphi\rangle\, ds \to \int_0^t \langle v\theta, \nabla\varphi\rangle\, ds \quad \text{as } m \to \infty,\ \varepsilon \to 0. \tag{8.149}$$

Therefore taking into account (8.147) and (8.149), we have from (8.146)

$$\begin{aligned}
-\int_0^T \left\langle\theta, \frac{\partial\varphi}{\partial t}\right\rangle dt + \int_0^T (\kappa(\theta)\nabla\theta, \nabla\varphi)\, dt + \int_0^T (\beta(x)\theta, \varphi)_{\Gamma_R}\, dt \\
- \int_0^T \langle v\theta, \nabla\varphi\rangle\, dt = \langle\theta(0), \varphi(0)\rangle + \int_0^t \langle g_2, \varphi\rangle\, ds.
\end{aligned} \tag{8.150}$$

Since the set $\{\varphi(t) = \sum_{j=1}^M k_j(t)\varphi_j : k_j(t) \in C^1[0, T],\ k_j(T) = 0,\ M = 1, 2, \cdots\}$ is dense in $\{\varphi \in C^1([0, T]; W^{1,2}_{\Gamma_D}) : \varphi(T) = 0\}$ (see Remark 1.8), we deduce from (8.150) the second equation of (8.28). □

Remark 8.6 The Eq. (10.1) of Chap. 10 is more general than (8.1), and from the result of Chap. 10 a result for (8.1) with the boundary conditions (8.2), (7.4) can be got, however the result demands that the parameter for buoyancy effect α_0 is small enough in accordance with the data of problem and the solution includes "defect measure" as in [2].

8.5 Bibliographical Remarks

The content of Chap. 8 is taken from [3].

In [4, 5] existence and uniqueness (for 2-D) of a solution to the problem were shown under homogeneous Dirichlet boundary condition for velocity and mixture of non-homogeneous Dirichlet and Neumann boundary conditions for temperature.

In [6] for the problem with non-homogeneous Dirichlet boundary conditions for velocity and temperature the existence of time periodic solution was shown (see [7]). In [8–11] problem (8.1) on the time dependent domain was studied under non-homogeneous Dirichlet boundary condition for velocity and temperature. In [12, 13] the problem on exterior domains with homogeneous Dirichlet boundary condition for velocity and non-homogeneous Dirichlet boundary condition for temperature was discussed. In [14] problem (8.1) was studied under mixture of non-homogeneous Dirichlet and stress boundary conditions for fluid and mixture of non-homogeneous Dirichlet, Neumann and Robin boundary conditions for temperature. They proved existence of a unique local-in-time solution under a compatibility condition at initial time (see (27) and (31) of [14]). In [15] problem (8.1) in cylindrical pipe with inflow and outflow was studied under slip boundary conditions for velocity and the Neumann conditions for temperature. There it was proved the existence of a solution on the given interval when norms of derivatives in the direction along the cylinder of the initial velocity, initial temperature and the external force are small enough. In [16] the existence of a solution to problem (8.1) on the time dependent domain was obtained under mixture of Dirichlet condition of velocity, total pressure and vorticity boundary conditions for fluid and mixture of Dirichlet, Neumann and Robin boundary conditions for temperature.

In [17] on 3-D channel under mixture of homogeneous Dirichlet boundary condition and "do nothing" condition for fluid and mixture of homogeneous Dirichlet and Neumann boundary conditions for temperature, local existence and global uniqueness of strong solutions were studied.

References

1. J. Naumann, On evolution inequalities of Navier-Stokes type in three dimensions. Ann. Math. Pure Appl. **124**(4), 107–125 (1980)
2. J. Naumann, J. Wolf, Existence of weak solutions to the equations of natural convection with dissipative heating. *Advances in Mathematical Fluid Mechanics*, pp. 367–384 (Springer, 2010)
3. T. Kim, The non-steady Boussinesq system with mixed boundary conditions including conditions of friction type. Inter. J. Diff. Equat. **2020** Article ID 6096531 (2020)
4. H. Morimoto, On non-stationary Boussinesq equations. Proc. Japan Acad. **67** Ser. A, 159–161 (1991)
5. H. Morimoto, Non-stationary Boussinesq equations. J. Fac. Sci. Univ. Tokyo Sect. IA Math. **39**, 61–75 (1992)
6. H. Morimoto, Heat convection equation with nonhomogeneous boundary conditions. Funkcialai Ekvacioj **53**, 213–229 (2010)
7. H. Morimoto, Survey on time periodic problem for fluid flow under inhomogeneous boundary condition. Discrete Cont. Dyn. Syst. Ser. S **5**, 631–639 (2012)
8. H. Inoue, M. Ôtani, Strong solutions of initial boundary value problems for heat convection equations in noncylindrical domains. Nonlinear Anal. **24**, 1061–1090 (1995)
9. H. Inoue, M. Ôtani, Periodic problems for heat convection equations in noncylindrical domains. Funkcialaj Ekvacioj **40**, 19–39 (1997)
10. K. Oeda, On the initial value problem for the heat convection equation of Boussinesq approximation in a time-dependent domain. Proc. Japan Acad. **64** Ser. A, 143–146 (1988)

11. K. Oeda, Remarks on the periodic solution of the heat convection equation in a perturbed annulus domain. Proc. Japan Acad. **73** Set. A, 21–25 (1997)
12. K. Oeda, Periodic solutions of the heat convection equations in exterior domains. Proc. Japan Acad. **73** Set. A, 49–54 (1997)
13. K. Oeda, N. Matsuda, Initial value problems for the heat convection equations in exterior domains. Tokyo J. Math. **21**, 359–375 (1998)
14. Z. Skalák, P. Kučera, An existence theorem for the Boussinesq equations with non-Dirichlet boundary conditions. Appl. Math. **45**(2), 81–98 (2000)
15. P. Kacprzyk, Long-time existence of solutions to the Navier-Stokes equations with inflow-outflow and heat convection. Math. Meth. Appl. Sci. **35**, 1000–1013 (2012)
16. T. Kim, D. Cao, Existence of solutions to the heat convection equations in a time-dependent domain with mixed boundary conditions. J. Math. Sci. Univ. Tokyo **22**, 531–568 (2015)
17. M. Beneš, The "do nothing" problem for Boussinesq fluids. Appl. Math. Lett. **31**, 25–28 (2014)

Chapter 9
The Steady Equations for Heat-Conducting Fluids

In this chapter we are concerned with the equation for steady flow of heat-conducting incompressible Newtonian fluids with dissipative heating under mixed boundary conditions. The boundary conditions for fluid may include Tresca slip, leak condition, one-sided leak conditions, velocity, pressure, vorticity, stress together and the conditions for temperature may include Dirichlet, Neumann and Robin conditions together. On the basis of results of Sect. 3.1, we get variational formulations consisting of a variational inequality for velocity and a variational equation for temperature, which are equivalent to the original PDE problems for smooth solutions.

Then, we study the existence of solutions to the variational problems. To this end, we first study the existence of solutions to auxiliary problems including one parameter for approximation and two or three parameters concerned with the norms of velocity and temperature. Then we determine the parameters concerned with the norms of velocity and temperature in accordance with the data of problems, and we get the existence of solutions by passing to limits as the parameter for approximation goes to zero.

For the problem of the case of static pressure it is proved that if the body force and boundary data for fluid are small enough and buoyancy effect and energy dissipation effect due to expansion are small enough, then there exists a solution. However, for the problem of the case of total pressure it is proved that if buoyancy effect and energy dissipation effect due to expansion are small enough, then there exists a solution.

9.1 Problems and Assumptions

The system studied in this chapter is as follows:

$$\begin{cases} -2\nabla\cdot\big(\mu(\theta)\mathcal{E}(v)\big)+(v\cdot\nabla)v+\nabla p=(1-\alpha_0\theta)f, \\ \operatorname{div} v=0, \\ -\nabla\cdot(\kappa(\theta)\nabla\theta)+v\cdot\nabla(\gamma(\theta)\theta)-\alpha_2\mu(\theta)\mathcal{E}(v):\mathcal{E}(v)=\alpha_1\theta f\cdot v+g. \end{cases} \tag{9.1}$$

T. Kim and D. Cao, *Equations of Motion for Incompressible Viscous Fluids*, Advances in Mathematical Fluid Mechanics,
https://doi.org/10.1007/978-3-030-78659-5_9

Here α_1 is parameter for dissipation of energy due to expansion, α_2 is a nonnegative real number and others are the same as in Chap. 7. The term $\mu(\theta)\mathcal{E}(v) : \mathcal{E}(v)$ represents the *dissipation of energy* due to viscosity (the Joule effect).

Let Ω be as in Chap. 7. (See Remark 6.1.)

For temperature we consider the following boundary conditions

$$\begin{aligned}
&(1)\quad \theta|_{\Gamma_D} = \theta_D|_{\Gamma_D}, \quad \theta_D - \text{a given function on } \Omega,\\
&(2)\quad \Big(\kappa(\theta)\frac{\partial\theta}{\partial n} + \beta(x)\theta\Big)\Big|_{\Gamma_R} = g_R(x), \quad \beta(x),\ g_R(x) - \text{given functions on } \Gamma_R.
\end{aligned} \tag{9.2}$$

As before, Problems I and II are distinguished according to boundary conditions for fluid. Problem I is the one with the boundary conditions (7.3) (the case of static pressure) and Problem II is the one with boundary conditions (7.4) (the case of total pressure).

We use the following assumption.

Assumption 9.1 We assume the followings.

(1) Γ_{2j}, Γ_{3j} and Γ_{7j} are convex. $\Gamma_1 \neq \emptyset$, $\Gamma_D \neq \emptyset$ and

$$\Gamma_R \subset \left(\cup_{i=1,3,5,8}\Gamma_i\right). \tag{9.3}$$

(2) (3) of Assumption 5.1 holds.

(3) For the functions of (9.1) $f \in \mathbf{L}^t(\Omega), t > 3,\ \ g \in L^{6/5}(\Omega)$ and

$$\begin{aligned}
&\mu \in C(\mathbb{R}),\ 0 < \mu_0 \le \mu(\xi) \le \mu_1 < \infty \quad \forall \xi \in \mathbb{R};\\
&\kappa \in C(\mathbb{R}),\ 0 < \kappa_0 \le \kappa(\xi) \le \kappa_1 < \infty \quad \forall \xi \in \mathbb{R};\\
&\gamma \in C(\mathbb{R}),\ |\gamma(\xi)| \le \gamma_0 \quad \forall \xi \in \mathbb{R}.
\end{aligned} \tag{9.4}$$

(4) For the functions of (9.2), (7.3) and (7.4),

$$\begin{aligned}
&\theta_D \in W^{1,2}(\Omega),\ \nabla\theta_D \in L^\infty(\Omega),\ \theta_D \ge 0, \quad g_R \in L^{4/3}(\Gamma_R);\\
&\beta_0 \ge \beta(x) \ge 0,\ \beta_0 - \text{a constant},\ \beta(x) - \text{measurable};\\
&\phi_i \in H^{-\frac{1}{2}}(\Gamma_i), i = 2, 4, 7,\ \phi_i \in \mathbf{H}^{-\frac{1}{2}}(\Gamma_i), i = 3, 5, 6;\\
&\text{the matrix } \alpha \text{ is positive},\ \alpha_{jk} \in L^\infty(\Gamma_5).
\end{aligned} \tag{9.5}$$

Remark 9.1 For (2) of Assumption 9.1 we refer to Remark 7.1.

9.2 Variational Formulations for Problems

In this section for every problem above we first obtain a variational formulation which consists of six formulas with six unknown functions. Then, for every problem we get another variational formulation equivalent to the variational formulation obtained, which consists of one variational inequality for velocity and a variational equation for temperature.

Let $\mathbf{V}$, $K(\Omega)$ be the same as in (5.5) and

$$W^{1,p}_{\Gamma_D}(\Omega) = \{y \in W^{1,p}(\Omega) : y|_{\Gamma_D} = 0\}.$$

We will use inner products in (7.8) for $\mathbf{V}$ and $W^{1,2}_{\Gamma_D}(\Omega)$.

9.2.1 Variational Formulation: The Case of Static Pressure

By (7.9)–(7.12), we can see that smooth solutions (v, p, θ) of problem (9.1), (9.2), (7.3) satisfy the following (cf. Remark 3.9).

$$\left\{\begin{aligned}
&2(\mu(\theta)\mathcal{E}(v), \mathcal{E}(u)) + \langle (v\cdot\nabla)v, u\rangle + 2(\mu(\theta)k(x)v, u)_{\Gamma_2}\\
&\qquad + 2(\mu(\theta)S\tilde{v}, \tilde{u})_{\Gamma_3} + 2(\alpha(x)v, u)_{\Gamma_5} + (\mu(\theta)k(x)v, u)_{\Gamma_7}\\
&\qquad - 2(\mu(\theta)\varepsilon_{n\tau}(v), u)_{\Gamma_8} + (p - 2\mu(\theta)\varepsilon_{nn}(v), u_n)_{\Gamma_9\cup\Gamma_{10}\cup\Gamma_{11}}\\
&\quad = \langle (1-\alpha_0\theta)f, u\rangle + \sum_{i=2,4,7}\langle \phi_i, u_n\rangle_{\Gamma_i} + \sum_{i=3,5,6}\langle \phi_i, u\rangle_{\Gamma_i} \quad \forall u \in \mathbf{V},\\
&(\kappa(\theta)\nabla\theta, \nabla\varphi) - (\gamma(\theta)\theta v, \nabla\varphi) - (\alpha_2\mu(\theta)|\mathcal{E}(v)|^2, \varphi) + (\beta\theta, \varphi)_{\Gamma_R} - (\alpha_1\theta f\cdot v, \varphi)\\
&\quad = \langle g_R, \varphi\rangle_{\Gamma_R} + \langle g, \varphi\rangle \qquad \forall\varphi \in W^{1,\infty}_{\Gamma_D}(\Omega),\\
&|\sigma_\tau(\theta, v)| \le g_\tau,\ \ \sigma_\tau(\theta, v)\cdot v_\tau + g_\tau|v_\tau| = 0 \ \text{ on } \Gamma_8,\\
&|\sigma_n(\theta, v, p)| \le g_n,\ \ \sigma_n(\theta, v, p)v_n + g_n|v_n| = 0 \ \text{ on } \Gamma_9,\\
&\sigma_n(\theta, v, p) + g_{+n} \ge 0,\ \ (\sigma_n(\theta, v, p) + g_{+n})v_n = 0 \ \text{ on } \Gamma_{10},\\
&\sigma_n(\theta, v, p) - g_{-n} \le 0,\ \ (\sigma_n(\theta, v, p) - g_{-n})v_n = 0 \ \text{ on } \Gamma_{11},\\
&\theta|_{\Gamma_D} = \theta_D|_{\Gamma_D} \ \text{ on } \Gamma_D.
\end{aligned}\right. \tag{9.6}$$

Define $a_0(\theta; \cdot, \cdot)$, $a_1(\cdot, \cdot, \cdot)$ and $f_1 \in \mathbf{V}^*$ by

$$\begin{aligned}
&a_0(\theta; w, u) = 2(\mu(\theta)\mathcal{E}(w), \mathcal{E}(u)) + 2(\mu(\theta)k(x)w, u)_{\Gamma_2} + 2(\mu(\theta)S\tilde{w}, \tilde{u})_{\Gamma_3}\\
&\qquad + 2(\alpha(x)w, u)_{\Gamma_5} + (\mu(\theta)k(x)w, u)_{\Gamma_7} \quad \forall w, u \in \mathbf{V}, \forall\theta \in W^{1,2}(\Omega),\\
&a_1(v, w, u) = \langle (v\cdot\nabla)w, u\rangle \quad \forall v, w, u \in \mathbf{V},\\
&\langle f_1, u\rangle = \sum_{i=2,4,7}\langle \phi_i, u_n\rangle_{\Gamma_i} + \sum_{i=3,5,6}\langle \phi_i, u\rangle_{\Gamma_i} \quad \forall u \in \mathbf{V}.
\end{aligned} \tag{9.7}$$

Define $b_0(\tilde{\theta}; \cdot, \cdot)$ and $f_2 \in (W^{1,2}_{\Gamma_D}(\Omega))^*$ by

$$
\begin{aligned}
&b_0(\tilde{\theta}; \theta, \varphi) = (\kappa(\tilde{\theta})\nabla\theta, \nabla\varphi) + (\beta(x)\theta, \varphi)_{\Gamma_R} \quad \forall\tilde{\theta}, \theta \in W^{1,2}(\Omega), \varphi \in W^{1,2}_{\Gamma_D}(\Omega),\\
&\langle f_2, \varphi\rangle = \langle g_R, \varphi\rangle_{\Gamma_R} + \langle g, \varphi\rangle \quad \forall\varphi \in W^{1,2}_{\Gamma_D}(\Omega).
\end{aligned}
\tag{9.8}
$$

Then, taking into account

$$
\sigma_\tau(\theta, v) = 2\mu(\theta)\varepsilon_{n\tau}(v), \quad \sigma_n(\theta, v, p) = -p + 2\mu(\theta)\varepsilon_{nn}(v)
$$

and (9.6), we introduce the following variational formulation for problem (9.1), (9.2), (7.3).

Problem I-VE. Find $(v, \theta, \sigma_\tau, \sigma_n, \sigma_{+n}, \sigma_{-n}) \in K(\Omega) \times \left(\bigcap_{1\le r<\frac{3}{2}} W^{1,r}(\Omega)\right) \times \mathbf{L}^2_\tau(\Gamma_8) \times L^2(\Gamma_9) \times H^{-1/2}(\Gamma_{10}) \times H^{-1/2}(\Gamma_{11})$ such that $\theta|_{\Gamma_D} = \theta_D|_{\Gamma_D}$ and

$$
\begin{cases}
a_0(\theta; v, u) + a_1(v, v, u) - (\sigma_\tau, u_\tau)_{\Gamma_8} - (\sigma_n, u_n)_{\Gamma_9}\\
\qquad - \langle\sigma_{+n}, u_n\rangle_{\Gamma_{10}} - \langle\sigma_{-n}, u_n\rangle_{\Gamma_{11}} - \langle f - \alpha_0\theta f, u\rangle = \langle f_1, u\rangle \quad \forall u \in \mathbf{V},\\
b_0(\theta; \theta, \varphi) - (\gamma(\theta)\theta v, \nabla\varphi) - \langle\alpha_2\mu(\theta)|\mathcal{E}(v)|^2, \varphi\rangle - \langle\alpha_1\theta f \cdot v, \varphi\rangle = \langle f_2, \varphi\rangle\\
\qquad\qquad\qquad\qquad \forall\varphi \in W^{1,\infty}_{\Gamma_D}(\Omega),\\
|\sigma_\tau| \le g_\tau, \ \sigma_\tau \cdot v_\tau + g_\tau|v_\tau| = 0 \quad \text{on } \Gamma_8,\\
|\sigma_n| \le g_n, \ \sigma_n v_n + g_n|v_n| = 0 \quad \text{on } \Gamma_9,\\
\sigma_{+n} + g_{+n} \ge 0, \ \langle\sigma_{+n} + g_{+n}, v_n\rangle_{\Gamma_{10}} = 0 \quad \text{on } \Gamma_{10},\\
\sigma_{-n} - g_{-n} \le 0, \ \langle\sigma_{-n} - g_{-n}, v_n\rangle_{\Gamma_{11}} = 0 \quad \text{on } \Gamma_{11},
\end{cases}
\tag{9.9}
$$

where $\mathbf{L}^2_\tau(\Gamma_8)$ is the subspace of $\mathbf{L}^2(\Gamma_8)$ consisting of functions such that $(u, n)_{\mathbf{L}^2(\Gamma_8)} = 0$.

Remark 9.2 As in Definition 7.1, except that the third equality of (9.1) holds in $L^1(\Omega)$, define a solution to the problem (9.1), (9.2), (7.3). Then, as in Theorem 7.1, Problem I-VE is equivalent to the problem (9.1), (9.2), (7.3), which shows that Problem I-VE is well-defined.

We will find another variational formulation consisting of a variational inequality and a variational equation, which is equivalent to Problem I-VE.

Let $(v, \theta, \sigma_\tau, \sigma_n, \sigma_{+n}, \sigma_{-n})$ be a solution of Problem I-VE. Subtracting the first formula of (9.9) with $u = v$ from the first formula of (9.9), we get

$$
\begin{aligned}
&a_0(\theta; v, u - v) + a_1(v, v, u - v) - (\sigma_\tau, u_\tau - v_\tau)_{\Gamma_8} - (\sigma_n, u_n - v_n)_{\Gamma_9} - \langle\sigma_{+n}, u_n - v_n\rangle_{\Gamma_{10}}\\
&\qquad - \langle\sigma_{-n}, u_n - v_n\rangle_{\Gamma_{11}} - \langle f - \alpha_0\theta f, u - v\rangle = \langle f_1, u - v\rangle \quad \forall u \in \mathbf{V}.
\end{aligned}
\tag{9.10}
$$

Let $\Phi : \mathbf{V} \to \overline{\mathbb{R}}$ be the functional defined by (5.22). Then, Φ is proper, convex lower weak semi-continuous and nonnegative.

By Theorem 5.2, under Assumption 9.1 for a fixed θ, the problem

$$\begin{cases} a_0(\theta; v, u) + a_1(v, v, u) - (\sigma_\tau, u_\tau)_{\Gamma_8} - (\sigma_n, u_n)_{\Gamma_9} \\ \qquad - \langle \sigma_{+n}, u_n \rangle_{\Gamma_{10}} - \langle \sigma_{-n}, u_n \rangle_{\Gamma_{11}} - \langle f - \alpha_0 \theta f, u \rangle = \langle f_1, u \rangle, \\ |\sigma_\tau| \le g_\tau, \ \sigma_\tau \cdot v_\tau + g_\tau |v_\tau| = 0 \quad \text{on } \Gamma_8, \\ |\sigma_n| \le g_n, \ \sigma_n v_n + g_n |v_n| = 0 \quad \text{on } \Gamma_9, \\ \sigma_{+n} + g_{+n} \ge 0, \ \langle \sigma_{+n} + g_{+n}, v_n \rangle_{\Gamma_{10}} = 0 \quad \text{on } \Gamma_{10}, \\ \sigma_{-n} - g_{-n} \le 0, \ \langle \sigma_{-n} - g_{-n}, v_n \rangle_{\Gamma_{11}} = 0 \quad \text{on } \Gamma_{11} \end{cases} \tag{9.11}$$

is equivalent to the following variational inequality.

Find $v \in \mathbf{V}$ such that

$$\begin{aligned} a_0(\theta; v, u - v) + a_1(v, v, u - v) + \Phi(u) - \Phi(v) - \langle f - \alpha_0 \theta f, u - v \rangle \\ \ge \langle f_1, u - v \rangle \quad \forall u \in \mathbf{V}. \end{aligned} \tag{9.12}$$

Therefore, we have the following variational formulation equivalent to Problem I-VE, which consists of a variational inequality for velocity and a variational equation for temperature.

Problem I-VI. Find $(v, \theta) \in \mathbf{V} \times \left(\bigcap_{1 \le r < \frac{3}{2}} W^{1,r}(\Omega) \right)$ such that $\theta|_{\Gamma_D} = \theta_D|_{\Gamma_D}$ and

$$\begin{cases} a_0(\theta; v, u - v) + a_1(v, v, u - v) + \Phi(u) - \Phi(v) - \langle f - \alpha_0 \theta f, u - v \rangle \ge \langle f_1, u - v \rangle \\ \qquad \qquad \forall u \in \mathbf{V}, \\ b_0(\theta; \theta, \varphi) - \langle \gamma(\theta) \theta v, \nabla \varphi \rangle - \langle \alpha_2 \mu(\theta) |\mathcal{E}(u)|^2, \varphi \rangle - \langle \alpha_1 \theta f \cdot v, \varphi \rangle = \langle f_2, \varphi \rangle \\ \qquad \qquad \forall \varphi \in W^{1,\infty}_{\Gamma_D}(\Omega). \end{cases} \tag{9.13}$$

9.2.2 *Variational Formulation: The Case of Total Pressure*

Since $(v \cdot \nabla) v = \operatorname{rot} v \times v + \frac{1}{2} \operatorname{grad} |v|^2$, by (7.9)–(7.12) we can verify that smooth solutions (v, p, θ) of problem (9.1), (9.2), (7.4) satisfy the following.

$$\begin{cases} 2(\mu(\theta)\mathcal{E}(v), \mathcal{E}(u)) + \langle \operatorname{rot} v \times v, u\rangle + 2(\mu(\theta)k(x)v, u)_{\Gamma_2} \\ \qquad + 2(\mu(\theta)S\tilde{v}, \tilde{u})_{\Gamma_3} + 2(\alpha(x)v, u)_{\Gamma_5} + (\mu(\theta)k(x)v, u)_{\Gamma_7} \\ \qquad - 2(\mu(\theta)\varepsilon_{n\tau}(v), u)_{\Gamma_8} + (p + \frac{1}{2}|v|^2 - 2\mu(\theta)\varepsilon_{nn}(v), u_n)_{\Gamma_9\cup\Gamma_{10}\cup\Gamma_{11}} \\ \quad = \langle(1-\alpha_0\theta)f, u\rangle + \sum_{i=2,4,7} \langle\phi_i, u_n\rangle_{\Gamma_i} + \sum_{i=3,5,6} \langle\phi_i, u\rangle_{\Gamma_i} \quad \forall u \in \mathbf{V}, \\ (\kappa(\theta)\nabla\theta, \nabla\varphi) - (\gamma(\theta)\theta v, \nabla\varphi) - (\alpha_2\mu(\theta)|\mathcal{E}(v)|^2, \varphi) + (\beta\theta, \varphi)_{\Gamma_R} - (\alpha_1\theta f\cdot v, \varphi) \\ \quad = \langle g_R, \varphi\rangle_{\Gamma_R} + \langle g, \varphi\rangle \quad \forall\varphi \in W^{1,\infty}_{\Gamma_D}(\Omega), \\ |\sigma^t_\tau(\theta, v)| \le g_\tau, \ \sigma^t_\tau(\theta, v)\cdot v_\tau + g_\tau|v_\tau| = 0 \quad \text{on } \Gamma_8, \\ |\sigma^t_n(\theta, v, p)| \le g_n, \ \sigma^t_n(\theta, v, p)v_n + g_n|v_n| = 0 \quad \text{on } \Gamma_9, \\ \sigma^t_n(\theta, v, p) + g_{+n} \ge 0, \ (\sigma^t_n(\theta, v, p) + g_{+n})v_n = 0 \quad \text{on } \Gamma_{10}, \\ \sigma^t_n(\theta, v, p) - g_{-n} \le 0, \ (\sigma^t_n(\theta, v, p) - g_{-n})v_n = 0 \quad \text{on } \Gamma_{11}, \\ \theta|_{\Gamma_D} = \theta_D|_{\Gamma_D}. \end{cases} \tag{9.14}$$

Let us define $a_2(\cdot,\cdot,\cdot)$ by

$$a_2(v, u, w) = \langle \operatorname{rot} v \times u, w\rangle \quad \forall v, u, w \in \mathbf{V}. \tag{9.15}$$

Then, taking into account (9.14) and

$$\sigma^t_\tau(\theta, v) = 2\mu(\theta)\varepsilon_{n\tau}(v), \quad \sigma^t_n(\theta, v, p) = -(p + \frac{1}{2}|v|^2) + 2\mu(\theta)\varepsilon_{nn}(v),$$

we get the following variational formulation for problem (9.1), (9.2), (7.4).

Problem II-VE. Find $(v, \theta, \sigma^t_\tau, \sigma^t_n, \sigma^t_{+n}, \sigma^t_{-n}) \in K(\Omega) \times \left(\bigcap_{1\le r<\frac{3}{2}} W^{1,r}(\Omega)\right) \times \mathbf{L}^2_\tau(\Gamma_8) \times L^2(\Gamma_9) \times H^{-1/2}(\Gamma_{10}) \times H^{-1/2}(\Gamma_{11})$ such that $\theta|_{\Gamma_D} = \theta_D|_{\Gamma_D}$ and

$$\begin{cases} a_0(\theta; v, u) + a_2(v, v, u) - (\sigma^t_\tau, u_\tau)_{\Gamma_8} - (\sigma^t_n, u_n)_{\Gamma_9} - \left\langle\sigma^t_{+n}, u_n\right\rangle_{\Gamma_{10}} \\ \qquad - \left\langle\sigma^t_{-n}, u_n\right\rangle_{\Gamma_{11}} - \langle f - \alpha_0\theta f, u\rangle = \langle f_1, u\rangle \quad \forall u \in \mathbf{V}, \\ b_0(\theta; \theta, \varphi) - \langle\gamma(\theta)\theta v, \nabla\varphi\rangle - \langle\alpha_2\mu(\theta)|\mathcal{E}(v)|^2, \varphi\rangle - \langle\alpha_1\theta f\cdot v, \varphi\rangle \\ \qquad\qquad = \langle f_2, \varphi\rangle \quad \forall\varphi \in W^{1,\infty}_{\Gamma_D}(\Omega), \\ |\sigma^t_\tau| \le g_\tau, \ \sigma^t_\tau\cdot v_\tau + g_\tau|v_\tau| = 0 \quad \text{on } \Gamma_8, \\ |\sigma^t_n| \le g_n, \ \sigma^t_n v_n + g_n|v_n| = 0 \quad \text{on } \Gamma_9, \\ \sigma^t_{+n} + g_{+n} \ge 0, \ \left\langle\sigma^t_{+n} + g_{+n}, v_n\right\rangle_{\Gamma_{10}} = 0 \quad \text{on } \Gamma_{10}, \\ \sigma^t_{-n} - g_{-n} \le 0, \ \left\langle\sigma^t_{-n} - g_{-n}, v_n\right\rangle_{\Gamma_{11}} = 0 \quad \text{on } \Gamma_{11}. \end{cases} \tag{9.16}$$

Remark 9.3 Similarly to Remark 9.2, Problem II-VE is equivalent to problem (9.1), (9.2), (7.4).

Then, using Theorem 5.4, in the same way as Problem I-VI we get Problem II-VI equivalent to Problem II-VE which consists of a variational inequality for velocity and a variational equation for temperature.

Problem II-VI. Find $(v, \theta) \in \mathbf{V} \times \left(\bigcap_{1\le r<\frac{3}{2}} W^{1,r}(\Omega)\right)$ such that $\theta|_{\Gamma_D} = \theta_D|_{\Gamma_D}$ and

$$\begin{cases} a_0(\theta; v, u-v) + a_2(v, v, u-v) + \Phi(u) - \Phi(v) - \langle f - \alpha_0\theta f, u-v\rangle \\ \qquad\qquad \ge \langle f_1, u-v\rangle \quad \forall u \in \mathbf{V}, \\ b_0(\theta; \theta, \varphi) - \langle \gamma(\theta)\theta v, \nabla\varphi\rangle - \langle \alpha_2\mu(\theta)|\mathcal{E}(v)|^2, \varphi\rangle - \langle \alpha_1\theta f \cdot v, \varphi\rangle \\ \qquad\qquad = \langle f_2, \varphi\rangle \quad \forall \varphi \in W^{1,\infty}_{\Gamma_D}(\Omega), \end{cases} \tag{9.17}$$

where $a_2(\cdot,\cdot,\cdot)$ is the one defined in (9.15) and Φ is defined by (5.22).

9.3 Existence and Uniqueness of Solutions: The Case of Static Pressure

In this section we prove the main result for the case of static pressure.

Theorem 9.1 *Let Assumption 9.1 hold and assume that*

(1) $f, \phi_i, i = 2, \cdots, 7$, are small enough in the spaces in (3), (4) of Assumption 9.1 (see (9.50)),

(2) $\max\{|\alpha_0|, |\alpha_1|\}$ is small enough in accordance with f, ϕ_i $(i = 2, \cdots, 7)$, g, g_R, θ_D (see (9.97)).

Then, there exists a solution (v, θ) to Problem I-VI such that

$$\|v\|_{\mathbf{V}} \le \frac{\mu_0}{K},$$
$$\|\theta^-\|_{W^{1,2}(\Omega)} \le c\Big(\frac{\mu_0}{K}\|f\|_{L^t} + \|g_R\|_{L^{4/3}(\Gamma_R)} + \|g\|_{L^{6/5}(\Omega)}\Big),$$
$$\int_\Omega |\nabla\theta|^r\,dx \le cK_\sigma^{r/(1-\sigma)}\big(1 + \|\theta_D\|^2_{W^{1,2}}\big) \quad \forall r\ (1 < r < 3/2),$$

where $\theta^-(x) = \min\{\theta(x), 0\}$, K is the constant in (9.18) below, K_σ is the one in (9.90) and $\sigma = \frac{3-2r}{3-r}$.

9.3.1 Existence of a Solution to an Auxiliary Problem

Since

$$|a_1(v, v, u)| = |((v\cdot\nabla)v, u)| \le K\|v\|^2_{\mathbf{V}}\|u\|_{\mathbf{V}} \quad \forall v, u \in \mathbf{V}, \tag{9.18}$$

define $\overline{a}_1(v) \in \mathbf{V}^*$ by $\langle \overline{a}_1(v), u\rangle = a_1(v, v, u) \quad \forall v, u \in \mathbf{V}$.
Define $\gamma_\varepsilon(t)$ by

$$\gamma_\varepsilon(t) := \frac{\gamma(t)t}{(1+\varepsilon|\gamma(t)|)(1+\varepsilon|t|)} \quad t \in \mathbb{R},\ \varepsilon > 0.$$

Then,

$$|\gamma_\varepsilon(t)| \le \frac{1}{\varepsilon^2}, \quad |\gamma_\varepsilon(t)| \le |\gamma(t)||t| \le \gamma_0|t|, \quad \gamma_\varepsilon(t) \to \gamma(t)t \text{ as } \varepsilon \to 0. \tag{9.19}$$

For every $\varepsilon > 0$, let Φ_ε be the Moreau-Yosida approximation of Φ and $\nabla\Phi_\varepsilon$ be Fréchet derivative of Φ_ε.

Similarly to Sect. 7.2.1, we first consider an auxiliary problem involving two parameters δ, ζ concerned with the norm of velocity (which is useful when there is fluid flux across a portion of boundary), one parameter λ concerned with the norm of temperature (which is useful to deal with buoyancy effect and energy dissipation effect due to expansion) and a parameter ε for approximation.

Problem I-VIA. Let $\delta > 0, \zeta > 0, \lambda > 0, \varepsilon > 0$ and $q \in (\frac{12}{5}, 6)$. Find $(v, \theta) \in \mathbf{V} \times W^{1,2}(\Omega)$ such that $\eta = \theta - \theta_D \in W^{1,2}_{\Gamma_D}(\Omega)$ and

$$\begin{cases} a_0(\theta; v, u) + \dfrac{\delta}{\max\{\delta, \|\overline{a}_1(v)\|_{\mathbf{V}^*}\}} a_1(v, v, u) + \big\langle \nabla\Phi_\varepsilon(v), u\big\rangle \\ \qquad -\Big\langle\Big(1 - \dfrac{\alpha_0\lambda}{\max\{\lambda, \|\theta\|^2_{L^q}\}}\theta\Big) f_\varepsilon, u\Big\rangle = \langle f_1, u\rangle \quad \forall u \in \mathbf{V}, \\ b_0(\theta; \theta, \varphi) - \dfrac{\zeta}{\max\{\zeta, \|v\|_{\mathbf{V}}\}}\langle \gamma_\varepsilon(\theta)v, \nabla\varphi\rangle - \Big\langle \alpha_2\mu(\theta)\dfrac{|\mathcal{E}(v)|^2}{1+\varepsilon|\mathcal{E}(v)|^2}, \varphi\Big\rangle \\ \qquad -\Big\langle \dfrac{\alpha_1\lambda}{\max\{\lambda, \|\theta\|^2_{L^q}\}}\theta f_\varepsilon \cdot v, \varphi\Big\rangle = \langle f_2, \varphi\rangle \quad \forall \varphi \in W^{1,2}_{\Gamma_D}(\Omega), \end{cases} \tag{9.20}$$

where $f_\varepsilon \in \mathbf{L}^\infty(\Omega)$ is such that $\|f - f_\varepsilon\|_{\mathbf{L}^t} \le \varepsilon$.

Theorem 9.2 *There exists a solution $(v_\varepsilon, \theta_\varepsilon) \in V \times W^{1,2}(\Omega)$ to Problem I-VIA.*

Proof Let r be such that

$$\frac{2}{q} + \frac{1}{r} = 1. \tag{9.21}$$

Since $q \in (\frac{12}{5}, 6)$, we see that $\frac{3}{2} < r < 6$. Let $\mathscr{H} = \mathbf{V} \times W^{1,2}_{\Gamma_D}(\Omega)$. Define an operator $\mathscr{A} : \mathscr{H} \to \mathscr{H}^*$ by

$$
\begin{aligned}
\big\langle \mathscr{A}(v,\eta),(u,\phi)\big\rangle &= a_0(\eta+\theta_D; v,u) + \frac{\delta}{\max\{\delta, \|\overline{a}_1(v)\|_{\mathbf{V}^*}\}} a_1(v,v,u) + \big\langle \nabla\Phi_\varepsilon(v), u\big\rangle \\
&- \Big\langle \Big(1 - \frac{\alpha_0\lambda}{\max\{\lambda, \|\eta+\theta_D\|^2_{L^q}\}}(\eta+\theta_D)\Big) f_\varepsilon, u\Big\rangle + b_0(\eta+\theta_D; \eta+\theta_D, \phi) \\
&- \frac{\zeta}{\max\{\zeta, \|v\|_{\mathbf{V}}\}} \langle \gamma_\varepsilon(\eta+\theta_D)v, \nabla\phi\rangle - \Big\langle \alpha_2\mu(\eta+\theta_D)\frac{|\mathcal{E}(v)|^2}{1+\varepsilon|\mathcal{E}(v)|^2}, \phi\Big\rangle \\
&- \Big\langle \frac{\alpha_1\lambda}{\max\{\lambda, \|\eta+\theta_D\|^2_{L^q}\}}(\eta+\theta_D) f_\varepsilon\cdot v, \phi\Big\rangle \quad \forall (v,\eta),(u,\phi)\in\mathscr{H}.
\end{aligned}
\tag{9.22}
$$

Let us show that the operator $\mathscr{A}$ is well-defined. By definition of Φ_ε it follows that $\Phi_\varepsilon(0_{\mathbf{V}}) = 0$ and $\nabla\Phi_\varepsilon(0_{\mathbf{V}}) = 0$. Since $\nabla\Phi_\varepsilon$ is Lipschitz continuous with the constant ε^{-1} (see Remark 1.16),

$$
\big|\big\langle \nabla\Phi_\varepsilon(v), u\big\rangle\big| = \big|\big\langle \nabla\Phi_\varepsilon(v) - \nabla\Phi_\varepsilon(0_{\mathbf{V}}), u\big\rangle\big| \le \varepsilon^{-1}\|v\|_{\mathbf{V}}\|u\|_{\mathbf{V}}. \tag{9.23}
$$

By (9.18) we have

$$
\Big|\frac{\delta}{\max\{\delta, \|\overline{a}_1(v)\|_{\mathbf{V}^*}\}} a_1(v,v,u)\Big| \le \delta\|u\|_{\mathbf{V}}. \tag{9.24}
$$

On the other hand, it is easily checked that by (9.19)

$$
\Big|\frac{\zeta}{\max\{\zeta, \|v\|_{\mathbf{V}}\}} \langle \gamma_\varepsilon(\eta+\theta_D)v, \nabla\phi\rangle\Big| \le \frac{c\zeta}{\varepsilon^2}\|\phi\|_{W^{1,2}}. \tag{9.25}
$$

The trivial inequality $ab \le \max\{a^2, b^2\}$ implies that

$$
\frac{\sqrt{\lambda}}{\max\{\lambda, \|\eta+\theta_D\|^2_{L^q}\}}\|\eta+\theta_D\|_{L^q} \le 1. \tag{9.26}
$$

Taking into account (9.21) and (9.26), we have

$$
\begin{aligned}
&\Big|\Big\langle \frac{\alpha_0\lambda}{\max\{\lambda, \|\eta+\theta_D\|^2_{L^q}\}}(\eta+\theta_D) f_\varepsilon, u\Big\rangle\Big| \le \frac{\alpha_0\lambda}{\max\{\lambda, \|\eta+\theta_D\|^2_{L^q}\}}\|\eta+\theta_D\|_{L^q}\|f_\varepsilon\|_{\mathbf{L}^q}\|u\|_{\mathbf{L}^r} \\
&\qquad\qquad \le c\alpha_0\sqrt{\lambda}\|f_\varepsilon\|_{\mathbf{L}^q}\|u\|_{\mathbf{L}^r}, \\
&\Big|\Big\langle \frac{\alpha_1\lambda}{\max\{\lambda, \|\eta+\theta_D\|^2_{L^q}\}}(\eta+\theta_D) f_\varepsilon\cdot v, \phi\Big\rangle\Big| \\
&\le \frac{\alpha_1\lambda}{\max\{\lambda, \|\eta+\theta_D\|^2_{L^q}\}}\|\eta+\theta_D\|_{L^q}\|f_\varepsilon\|_{\mathbf{L}^\infty}\|v\|_{\mathbf{L}^r}\|\phi\|_{L^q} \le c\alpha_1\sqrt{\lambda}\|f_\varepsilon\|_{\mathbf{L}^\infty}\|v\|_{\mathbf{L}^r}\|\phi\|_{L^q}.
\end{aligned}
\tag{9.27}
$$

Estimation of other terms is easy, and so operator $\mathscr{A}$ is well defined.

Then, the existence of a solution to Problem I-VIA is equivalent to the one of a solution to

$$\mathscr{A}(v,\eta) = \mathscr{F}, \quad \mathscr{F} = \begin{pmatrix} f_1 \\ f_2 \end{pmatrix}.$$

By applying Theorem 1.45, we will prove the existence of a solution to the equation above. To this end, we verify that $\mathscr{A}$ satisfies assumptions of Theorem 1.45.

(i) Let us first prove that $\mathscr{A}$ is coercive, i.e.,

$$\frac{1}{\|(v,\eta)\|_{\mathscr{H}}}\big\langle \mathscr{A}(v,\eta),(v,\eta)\big\rangle \to \infty \quad \text{as } \|(v,\eta)\|_{\mathscr{H}} \to \infty.$$

Since $\Gamma_{2j}, \Gamma_{3j}, \Gamma_{7j}$ are convex and the matrix α is positive, it follows from (9.7) (see Lemma 3.1) that

$$a_0(\eta+\theta_D; v, v) \ge 2\mu_0\|v\|^2_{\mathbf{V}}. \tag{9.28}$$

Taking into account (9.24), (9.25), (9.28) and the first formula of (9.27), we have

$$\begin{aligned}
&\big\langle \mathscr{A}(v,\eta),(v,\eta)\big\rangle = a_0(\eta+\theta_D; v, v) + \frac{\delta}{\max\{\delta, \|\overline{a}_1(v)\|_{\mathbf{V}^*}\}} a_1(v,v,v)\\
&+\big\langle \nabla\Phi_\varepsilon(v), v\big\rangle - \Big\langle\Big(1-\frac{\alpha_0\lambda}{\max\{\lambda, \|\eta+\theta_D\|^2_{L^q}\}}(\eta+\theta_D)\Big)f_\varepsilon, v\Big\rangle\\
&+b_0(\eta+\theta_D;\eta,\eta) + b_0(\eta+\theta_D;\theta_D,\eta) - \frac{\zeta}{\max\{\zeta, \|v\|_{\mathbf{V}}\}}\langle \gamma_\varepsilon(\eta+\theta_D)v, \nabla\eta\rangle\\
&-\Big\langle \alpha_2\mu(\eta+\theta_D)\frac{|\mathcal{E}(v)|^2}{1+\varepsilon|\mathcal{E}(v)|^2}, \eta\Big\rangle - \Big\langle \frac{\alpha_1\lambda}{\max\{\lambda, \|\eta+\theta_D\|^2_{L^q}\}}(\eta+\theta_D)f_\varepsilon\cdot v, \eta\Big\rangle\\
&\ge \frac{3}{4}\min\{2\mu_0, \kappa_0\}(\|v\|^2_{\mathbf{V}} + \|\eta\|^2_{W^{1,2}_{\Gamma_D}}) - \delta\|v\| - c(\|f_\varepsilon\|^2_{\mathbf{L}^q} + \|\theta_D\|^2_{L^2} + \frac{1}{\varepsilon^2}) + \langle \nabla\Phi_\varepsilon(v), v\rangle\\
&-\frac{c\zeta}{\varepsilon^2}\|\eta\|_{W^{1,2}_{\Gamma_D}} - \Big\langle \frac{\alpha_1\lambda}{\max\{\lambda, \|\eta+\theta_D\|^2_{L^q}\}}(\eta+\theta_D)f_\varepsilon\cdot v, \eta\Big\rangle \quad \forall (v,\eta)\in\mathscr{H}.
\end{aligned} \tag{9.29}$$

Since the operator $\nabla\Phi_\varepsilon$ is monotone and $\nabla\Phi_\varepsilon(0_{\mathbf{V}}) = 0$, we have

$$\big\langle \nabla\Phi_\varepsilon(v), v\big\rangle = \big\langle \nabla\Phi_\varepsilon(v) - \nabla\Phi_\varepsilon(0_{\mathbf{V}}), v - 0_{\mathbf{V}}\big\rangle \ge 0. \tag{9.30}$$

Since $\langle(\eta+\theta_D)f\cdot v, \eta\rangle = \langle(\eta+\theta_D)f\cdot v, (\eta+\theta_D)\rangle - \langle(\eta+\theta_D)f\cdot v, \theta_D\rangle$, by (9.21) and (9.26) we have

$$\begin{aligned}
&\Big|\Big\langle \frac{\alpha_1\lambda}{\max\{\lambda, \|\eta+\theta_D\|^2_{L^q}\}}(\eta+\theta_D)f_\varepsilon\cdot v, \eta\Big\rangle\Big|\\
&\le c\frac{\lambda\|\eta+\theta_D\|^2_{L^q}}{\max\{\lambda, \|\eta+\theta_D\|^2_{L^q}\}}\|f_\varepsilon\|_{\mathbf{L}^\infty}\|v\|_{\mathbf{L}^r} + c\frac{\lambda\|\eta+\theta_D\|_{L^q}}{\max\{\lambda, \|\eta+\theta_D\|^2_{L^q}\}}\|f_\varepsilon\|_{\mathbf{L}^\infty}\|v\|_{\mathbf{L}^r}\|\theta_D\|_{L^q}\\
&\le c\lambda\|f_\varepsilon\|_{\mathbf{L}^\infty}\|v\|_{\mathbf{L}^r} + c\sqrt{\lambda}\|f_\varepsilon\|_{\mathbf{L}^\infty}\|v\|_{\mathbf{L}^r}\|\theta_D\|_{L^q}.
\end{aligned} \tag{9.31}$$

By virtue of (9.29)–(9.31), we conclude that

$$\langle \mathscr{A}(v,\eta),(v,\eta)\rangle \geq \frac{1}{4}\min\{2\mu_0,\kappa_0\}\Big(\|v\|^2_{\mathbf{V}}+\|\eta\|^2_{W^{1,2}_{\Gamma_D}}\Big)-\delta\|v\|-\frac{c\zeta}{\varepsilon^2}\|\eta\|_{W^{1,2}_{\Gamma_D}}$$
$$-c\lambda\|f_\varepsilon\|_{\mathbf{L}^\infty}\|v\|_{\mathbf{V}}-c\sqrt{\lambda}\|f_\varepsilon\|_{\mathbf{L}^\infty}\|v\|_{\mathbf{V}}\|\theta_D\|_{L^q}-c\Big(\|f_\varepsilon\|^2_{\mathbf{L}^t}+\|\theta_D\|^2_{L^2}+\frac{1}{\varepsilon^2}\Big)$$
$$\forall (v,\eta)\in\mathscr{H},$$

which implies coercivity of $\mathscr{A}$.

(ii) Taking into account (9.23)–(9.27), we have from (9.22)

$$\|\mathscr{A}(v,\eta)\|_{\mathscr{H}^*}=\sup_{\|(u,\phi)\|_{\mathscr{H}}=1}\langle \mathscr{A}(v,\eta),(u,\phi)\rangle$$
$$\leq c\Big(\|v\|_{\mathbf{V}}+\delta+\frac{c\zeta}{\varepsilon^2}+\frac{1}{\varepsilon}\|v\|_{\mathbf{V}}+\|f_\varepsilon\|_{\mathbf{L}^2}+\sqrt{\lambda}\|f_\varepsilon\|_{\mathbf{L}^q}+\|\eta+\theta_D\|_{W^{1,2}_{\Gamma_D}} \tag{9.32}$$
$$+\|v\|_{\mathbf{L}^2}+\frac{1}{\varepsilon}+\sqrt{\lambda}\|f_\varepsilon\|_{\mathbf{L}^\infty}\|v\|_{\mathbf{V}}\Big)\quad \forall (v,\eta),(u,\phi)\in\mathscr{H},$$

which shows that $\mathscr{A}$ maps bounded sets of $\mathscr{H}$ into bounded sets of $\mathscr{H}^*$.

(iii) Let $\{(v_k,\eta_k)\}$ be a sequence such that

$$(v_k,\eta_k)\rightharpoonup (v,\eta)\ \text{in}\ \mathscr{H},$$
$$\limsup_{k\to\infty}\langle \mathscr{A}(v_k,\eta_k),(v_k,\eta_k)-(v,\eta)\rangle\leq 0.$$

By taking a subsequence and denoting with same subindex if necessary, we may assume

$$\begin{aligned} &v_k\to v\quad \text{in } \mathbf{L}^s(\Omega)\ (1\leq s<6) \text{ and a.e. in } \Omega,\\ &\eta_k\to \eta\quad \text{in } L^s(\Omega)\ (1\leq s<6) \text{ and a.e. in } \Omega \qquad \text{as } k\to\infty. \end{aligned} \tag{9.33}$$

Since

$$a_0(\eta_k+\theta_D;v_k-v,v_k-v)=a_0(\eta_k+\theta_D;v_k,v_k-v)-a_0(\eta_k+\theta_D;v,v_k-v),$$
$$b_0(\eta_k+\theta_D;\eta_k-\eta,\eta_k-\eta)=b_0(\eta_k+\theta_D;\eta_k+\theta_D,\eta_k-\eta)$$
$$-b_0(\eta_k+\theta_D;\eta+\theta_D,\eta_k-\eta),$$

by (9.22) we have

$$
\begin{aligned}
\min\{\mu_0, \kappa_0\}\big(\|v_k - v\|_{\mathbf{V}}^2 + \|\eta_k - \eta\|_{W^{1,2}_{\Gamma_D}}\big) &\le \langle \mathscr{A}(v_k, \eta_k), (v_k, \eta_k) - (v, \eta)\rangle \\
&- a_0(\eta_k + \theta_D; v, v_k - v) - b_0(\eta_k + \theta_D; \eta + \theta_D, \eta_k - \eta) \\
&- \frac{\delta}{\max\{\delta, \|\overline{a}_1(v_k)\|_{\mathbf{V}^*}\}} a_1(v_k, v_k, v_k - v) - \big\langle \nabla \Phi_\varepsilon(v_k), v_k - v\big\rangle \\
&+ \Big\langle \Big(1 - \frac{\alpha_0 \lambda}{\max\{\lambda, \|\eta_k + \theta_D\|_{L^q}^2\}}(\eta_k + \theta_D)\Big) f_\varepsilon, v_k - v\Big\rangle \\
&+ \frac{\zeta}{\max\{\zeta, \|v_k\|_{\mathbf{V}}\}} \langle \gamma_\varepsilon(\eta_k + \theta_D) v_k, \nabla(\eta_k - \eta)\rangle \\
&+ \Big\langle \alpha_2 \mu(\theta) \frac{|\mathcal{E}(v_k)|^2}{1 + \varepsilon |\mathcal{E}(v_k)|^2}, \eta_k - \eta\Big\rangle \\
&+ \Big\langle \frac{\alpha_1 \lambda}{\max\{\lambda, \|\eta_k + \theta_D\|_{L^q}^2\}} (\eta_k + \theta_D) f_\varepsilon \cdot v_k, \eta_k - \eta\Big\rangle.
\end{aligned}
\tag{9.34}
$$

Now let us estimate all terms except the first one on the right hand side of (9.34). By Corollary 1.1 it follows that as $k \to \infty$

$$
\begin{aligned}
a_0(\eta_k + \theta_D; v, v_k - v) &= 2(\mu(\eta_k + \theta_D)\mathcal{E}(v), \mathcal{E}(v_k - v)) \\
&\quad + 2(\mu(\eta_k + \theta_D)k(x)v, v_k - v)_{\Gamma_2} + 2(\mu(\eta_k + \theta_D)S\tilde{v}, \tilde{v}_k - \tilde{v})_{\Gamma_3} \\
&\quad + 2(\alpha(x)v, v_k - v)_{\Gamma_5} + (\mu(\eta_k + \theta_D)k(x)v, v_k - v)_{\Gamma_7} \to 0, \\
b_0(\eta_k + \theta_D; \eta + \theta_D, \eta_k - \eta) &= \big(\kappa(\eta_k + \theta_D)\nabla(\eta + \theta_D), \nabla(\eta_k - \eta)\big) \\
&\quad + \big(\beta(x)(\eta + \theta_D), (\eta_k - \eta)\big)_{\Gamma_R} \to 0.
\end{aligned}
\tag{9.35}
$$

Also, as $k \to \infty$

$$
\begin{aligned}
&\Big|\frac{\delta}{\max\{\delta, \|\overline{a}_1(v_k)\|_{\mathbf{V}^*}\}} a_1(v_k, v_k, v_k - v)\Big| \\
&\qquad \le \frac{\delta}{\max\{\delta, \|\overline{a}_1(v_k)\|_{\mathbf{V}^*}\}} \|v_k\|_{\mathbf{L}^4} \|\nabla v_k\|_{\mathbf{L}^2} \|v_k - v\|_{\mathbf{L}^4} \to 0.
\end{aligned}
\tag{9.36}
$$

Since $\nabla \Phi_\varepsilon$ is monotone, as $k \to \infty$

$$
\begin{aligned}
-\big\langle \nabla \Phi_\varepsilon(v_k), v_k - v\big\rangle &= -\big\langle \nabla \Phi_\varepsilon(v_k) - \nabla \Phi_\varepsilon(v), v_k - v\big\rangle - \big\langle \nabla \Phi_\varepsilon(v), v_k - v\big\rangle \\
&\le -\big\langle \nabla \Phi_\varepsilon(v), v_k - v\big\rangle \to 0.
\end{aligned}
\tag{9.37}
$$

By (9.26) and (9.33), the followings hold.

$$
\begin{aligned}
&\Big|\Big\langle \frac{\alpha_0\lambda}{\max\{\lambda, \|\eta_k+\theta_D\|^2_{L^q}\}}(\eta_k+\theta_D)f_\varepsilon, v_k - v\Big\rangle\Big| \le c\sqrt{\lambda}\|f_\varepsilon\|_{\mathbf{L}^\infty}\|v_k-v\|_{\mathbf{L}^3} \to 0,\\
&\Big|\Big\langle \frac{\alpha_1\lambda}{\max\{\lambda, \|\eta_k+\theta_D\|^2_{L^q}\}}(\eta_k+\theta_D)f_\varepsilon\cdot v_k, \eta_k - \eta\Big\rangle\Big| \le c\sqrt{\lambda}\|f_\varepsilon\|_{\mathbf{L}^\infty}\|v_k\|_{\mathbf{L}^6}\|\eta_k-\eta\|_{L^r} \to 0,\\
&\Big|\frac{\zeta}{\max\{\zeta, \|v_k\|_{\mathbf{V}}\}}\langle \gamma_\varepsilon(\eta_k+\theta_D)v_k, \nabla(\eta_k-\eta)\rangle\Big|\\
&\qquad\le |\langle(\gamma_\varepsilon(\eta_k+\theta_D)-\gamma_\varepsilon(\eta+\theta_D))v_k, \nabla(\eta_k-\eta)\rangle|\\
&\qquad\quad + |\langle\gamma_\varepsilon(\eta+\theta_D)(v_k-v), \nabla(\eta_k-\eta)\rangle| + |\langle\gamma_\varepsilon(\eta+\theta_D)v, \nabla(\eta_k-\eta)\rangle|\\
&\qquad\le c\|\gamma_\varepsilon(\eta_k+\theta_D)-\gamma_\varepsilon(\eta+\theta_D)\|_{L^3}\|v_k\|_{\mathbf{V}}\|\eta_k-\eta\|_{W^{1,2}_{\Gamma_D}}\\
&\qquad\quad + c\|\gamma_\varepsilon(\eta+\theta_D)\|_{L^6}\|(v_k-v)\|_{\mathbf{L}^3}\|\eta_k-\eta\|_{W^{1,2}_{\Gamma_D}}\\
&\qquad\quad + \langle\gamma_\varepsilon(\eta+\theta_D)v, \nabla(\eta_k-\eta)\rangle \to 0
\end{aligned}
\tag{9.38}
$$

as $k \to \infty$, where the fact that by Lemma 1.3 $\gamma_\varepsilon(\eta_k+\theta_D) \to \gamma_\varepsilon(\eta+\theta_D)$ in $L^3(\Omega)$ as $k \to \infty$ was used. It is easy to prove convergence to zero of other terms in the right hand side of (9.34). Thus, by (9.33)–(9.38) we conclude that

$$
\begin{aligned}
\limsup_{k\to\infty} \min\{\mu_0, \kappa_0\}|\big(\|v_k-v\|^2_{\mathbf{V}} + \|\eta_k-\eta\|_{W^{1,2}_{\Gamma_D}}\big)&\\
\le \limsup_{k\to\infty}\langle \mathscr{A}(v_k,\eta_k), (v_k,\eta_k)-(v,\eta)\rangle \le 0,&
\end{aligned}
$$

which implies

$$
\begin{aligned}
&(v_k,\eta_k) \to (v,\eta) \quad \text{in } \mathscr{H} \text{ as } k\to\infty,\\
&v_k \to v, \quad \eta_k \to \eta \quad \text{a.e. in } \Omega \text{ as } k\to\infty.
\end{aligned}
\tag{9.39}
$$

By the definition of $\mathscr{A}$, for $(u,\phi) \in \mathscr{H}$

$$
\begin{aligned}
&\langle \mathscr{A}(v_k,\eta_k), (v_k,\eta_k)-(u,\phi)\rangle\\
&= a_0(\eta_k+\theta_D; v_k, v_k-u) + \frac{\delta}{\max\{\delta, \|\overline{a}_1(v_k)\|_{\mathbf{V}^*}\}}a_1(v_k, v_k, v_k-u)\\
&+ \big\langle\nabla\Phi_\varepsilon(v_k), v_k-u\big\rangle - \Big\langle\Big(1-\frac{\alpha_0\lambda}{\max\{\lambda, \|\eta_k+\theta_D\|^2_{L^q}\}}(\eta_k+\theta_D)\Big)f, v_k-u\Big\rangle\\
&+ b_0(\eta_k+\theta_D; \eta_k+\theta_D, \eta_k-\phi) - \Big\langle\frac{\zeta}{\max\{\zeta, \|v_k\|_{\mathbf{V}}\}}\gamma(\eta_k+\theta_D)v_k, \nabla(\eta_k-\phi)\Big\rangle\\
&- \Big\langle\alpha_2\mu(\theta_k)\frac{|\mathcal{E}(v_k)|^2}{1+\varepsilon|\mathcal{E}(v_k)|^2}, \eta_k-\phi\Big\rangle - \Big\langle\frac{\alpha_1\lambda}{\max\{\lambda, \|\eta_k+\theta_D\|^2_{L^q}\}}(\eta_k+\theta_D)f_\varepsilon\cdot v_k, \eta_k-\phi\Big\rangle.
\end{aligned}
\tag{9.40}
$$

Taking into account (9.39), by Corollary 1.2 we have

$$
\begin{aligned}
&\liminf_{k\to\infty} a_0(\eta_k+\theta_D; v_k, v_k) \ge a_0(\eta+\theta_D; v, v),\\
&\liminf_{k\to\infty} b_0(\eta_k+\theta_D; \eta_k, \eta_k) \ge b_0(\eta+\theta_D; \eta, \eta).
\end{aligned}
\tag{9.41}
$$

The sequence $\{\frac{\alpha_0\lambda}{\max\{\lambda,\|\eta_k+\theta_D\|^2_{L^q}\}}(\eta_k+\theta_D)\}$ converges a.e. in Ω to $\frac{\alpha_0\lambda}{\max\{\lambda,\|\eta+\theta_D\|^2_{L^q}\}}(\eta+\theta_D)$ and $\left\|\frac{\alpha_0\lambda}{\max\{\lambda,\|\eta_k+\theta_D\|^2_{L^q}\}}(\eta_k+\theta_D)\right\|_{L^q}\le\alpha_0\sqrt{\lambda}$, and so this sequence weakly converges in L^q (see Lemma 1.2). Then by virtue of this fact and the first formula of (9.38), we have

$$\begin{aligned}\lim_{k\to\infty}&\Big\langle\frac{\alpha_0\lambda}{\max\{\lambda,\,\|\eta_k+\theta_D\|^2_{L^q}\}}(\eta_k+\theta_D)f_\varepsilon,\,v_k-u\Big\rangle\\&=\lim_{k\to\infty}\Big\langle\frac{\alpha_0\lambda}{\max\{\lambda,\,\|\eta_k+\theta_D\|^2_{L^q}\}}(\eta_k+\theta_D)f_\varepsilon,\,v_k-v\Big\rangle\\&\quad+\lim_{k\to\infty}\Big\langle\frac{\alpha_0\lambda}{\max\{\lambda,\,\|\eta_k+\theta_D\|^2_{L^q}\}}(\eta_k+\theta_D)f_\varepsilon,\,v-u\Big\rangle\\&=\Big\langle\frac{\alpha_0\lambda}{\max\{\lambda,\,\|\eta+\theta_D\|^2_{L^q}\}}(\eta+\theta_D)f_\varepsilon,\,v-u\Big\rangle,\end{aligned}\tag{9.42}$$

where the fact that owing to (9.21) $f_\varepsilon\cdot(v-u)\in L^3\subset L^{q'},\ \frac{1}{q}+\frac{1}{q'}=1$, was used. Using the second formula in (9.38), in the same way we get

$$\begin{aligned}\lim_{k\to\infty}&\Big\langle\frac{\alpha_1\lambda}{\max\{\lambda,\,\|\eta_k+\theta_D\|^2_{L^q}\}}(\eta_k+\theta_D)f_\varepsilon\cdot v_k,\,\eta_k-\phi\Big\rangle\\&=\Big\langle\frac{\alpha_1\lambda}{\max\{\lambda,\,\|\eta+\theta_D\|^2_{L^q}\}}(\eta+\theta_D)f_\varepsilon\cdot v,\,\eta-\phi\Big\rangle.\end{aligned}\tag{9.43}$$

It is easy to prove convergence of other terms on the right hand side of (9.40). Thus, by (9.41)–(9.43) we have existence of a subsequence $\{(v_k,\eta_k)\}$ such that

$$\liminf_{k\to\infty}\langle\mathscr{A}(v_k,\eta_k),(v_k,\eta_k)-(u,\phi)\rangle\ge\langle\mathscr{A}(v,\eta),(v,\eta)-(u,\phi)\rangle.$$

Therefore, we can apply Theorem 1.45 to obtain the conclusion. □

9.3.2 A Priori Estimates of Solutions to the Auxiliary Problem

Let us choose q_0 such that

$$\frac{1}{q_0}+\frac{1}{t}+\frac{1}{3}\le 1,\quad \frac{12}{5}<q_0<3.\tag{9.44}$$

Since $t>3$ (see (3) of Assumption 9.1), such a choice is possible. Then, q_0 satisfies the condition for q in Problem I-VIA.

Lemma 9.1 *If $v \in \mathbf{V}$ and $|\alpha_1|\sqrt{\lambda} \le 1$, then for all θ_ε satisfying the second formula of (9.20) and the condition $\theta_\varepsilon - \theta_D \in W^{1,2}_{\Gamma_D}(\Omega)$ the following estimate holds.*

$$\|\theta_\varepsilon^-\|_{W^{1,2}_{\Gamma_D}(\Omega)} \le c\Big(\|f_\varepsilon\|_{\mathbf{L}^t}\|v\|_{\mathbf{V}} + \|g_R\|_{L^{4/3}(\Gamma_R)} + \|g\|_{L^{6/5}(\Omega)}\Big), \tag{9.45}$$

where constant c is independent of ε.

Proof Since $\theta_\varepsilon|_{\Gamma_D} = \theta_D|_{\Gamma_D} \ge 0$, a function $\varphi(x) = \theta_\varepsilon^-(x) \equiv \min\{\theta_\varepsilon(x), 0\}$ is admissible in the second formula of (9.20), and we have

$$\begin{aligned}&(\kappa(\theta_\varepsilon)\nabla\theta_\varepsilon, \nabla\theta_\varepsilon^-) + (\beta(x)\theta_\varepsilon, \theta_\varepsilon^-)_{\Gamma_R} - \frac{\zeta}{\max\{\zeta, \|v\|_{\mathbf{V}}\}}\langle\gamma_\varepsilon(\theta_\varepsilon)v, \nabla\theta_\varepsilon^-\rangle\\ &\quad - \Big\langle\alpha_2\mu(\theta)\frac{|\mathcal{E}(v)|^2}{1+\varepsilon|\mathcal{E}(v)|^2}, \theta_\varepsilon^-\Big\rangle - \Big\langle\frac{\alpha_1\lambda}{\max\{\lambda, \|\theta_\varepsilon\|^2_{L^{q_0}}\}}\theta_\varepsilon f_\varepsilon\cdot v, \theta_\varepsilon^-\Big\rangle = \langle f_2, \theta_\varepsilon^-\rangle.\end{aligned} \tag{9.46}$$

Let us prove

$$\langle\gamma_\varepsilon(\theta_\varepsilon)v, \nabla\theta_\varepsilon^-\rangle = 0. \tag{9.47}$$

To this end, define

$$\Psi(t) := \int_0^t \gamma_\varepsilon(s)\,ds, \quad t \in \mathbb{R}.$$

Then, $\Psi \in C^1(\mathbb{R})$ and

$$\begin{aligned}&\nabla\Psi(\theta) = \gamma_\varepsilon(\theta)\nabla\theta, \quad \Psi(\theta) \in W^{1,2}(\Omega) \quad \forall\theta \in W^{1,2}(\Omega),\\ &\Psi(\theta)|_{\Gamma_D} = 0 \quad \forall\theta \in W^{1,2}_{\Gamma_D}(\Omega).\end{aligned} \tag{9.48}$$

Since $\nabla\theta_\varepsilon^- = 0$ on $\Omega^+ = \{x : \theta(x) \ge 0\}$ and $v\cdot n|_{\Gamma_R} = 0$, by (9.48) we have

$$\langle\gamma_\varepsilon(\theta)v, \nabla\theta_\varepsilon^-\rangle = \int_\Omega \gamma_\varepsilon(\theta_\varepsilon^-)v\cdot\nabla\theta_\varepsilon^-\,dx = \int_\Omega v\cdot\nabla\Psi(\theta_\varepsilon^-) = 0,$$

which means (9.47).

Also,

$$\begin{aligned}&-\Big\langle\alpha_2\mu(\theta)\frac{|\mathcal{E}(v)|^2}{1+\varepsilon|\mathcal{E}(v)|^2}, \theta_\varepsilon^-\Big\rangle \ge 0,\\ &\Big|\Big\langle\frac{\alpha_1\lambda}{\max\{\lambda, \|\theta_\varepsilon\|^2_{L^{q_0}}\}}\theta_\varepsilon f_\varepsilon\cdot v, \theta_\varepsilon^-\Big\rangle\Big| \le c\|f_\varepsilon\|_{\mathbf{L}^t}\|v\|_{\mathbf{V}}\|\theta_\varepsilon^-\|_{W^{1,2}_{\Gamma_D}}\\ &\qquad\qquad \le \frac{\kappa_0}{4}\|\theta_\varepsilon^-\|^2_{W^{1,2}_{\Gamma_D}} + c\|f_\varepsilon\|^2_{\mathbf{L}^t}\|v\|^2_{\mathbf{V}},\\ &|\langle f_2, \theta_\varepsilon^-\rangle| \le \frac{\kappa_0}{4}\|\theta_\varepsilon^-\|^2_{H^1} + c\|f_2\|^2_{(W^{1,2}_{\Gamma_D})^*},\\ &(\beta(x)\theta_\varepsilon, \theta_\varepsilon^-)_{\Gamma_R} \ge 0,\end{aligned} \tag{9.49}$$

where to get the second inequality (9.26) and (9.44) were used. By (9.46)–(9.49) we have

$$\|\theta_\varepsilon^-\|^2_{W^{1,2}_{\Gamma_D}} \le \frac{2c}{\kappa_0}\Big(\|f_\varepsilon\|^2_{\mathbf{L}^t}\|v\|^2_{\mathbf{V}} + \|f_2\|^2_{(W^{1,2}_{\Gamma_D})^*}\Big),$$

which implies (9.45). □

Lemma 9.2 *If* $|\alpha_0|\sqrt{\lambda} \le 1$ *and*

$$(\|f\|_{\mathbf{L}^t} + \|f_1\|_{\mathbf{V}^*}) < \frac{\mu_0^2}{Kc_0}, \tag{9.50}$$

where K *is the constant in* (9.18) *and* c_0 *is the one in* (9.55) *below, then there exist parameters* δ *and* ζ *such that*

$$\frac{\delta}{\max\{\delta,\ \|\overline{a}_1(v_\varepsilon)\|_{\mathbf{V}^*}\}} = 1, \quad \frac{\zeta}{\max\{\zeta,\ \|v_\varepsilon\|_{\mathbf{V}}\}} = 1 \tag{9.51}$$

for all small ε *and solutions of* (9.20) v_ε. *In addition,*

$$\|v_\varepsilon\|_{\mathbf{V}} \le \frac{\mu_0}{K}. \tag{9.52}$$

Proof Since $|\alpha_0|\sqrt{\lambda} \le 1$, we have

$$\Big|\langle\frac{\alpha_0\lambda}{\max\{\lambda,\ \|\theta_\varepsilon\|^2_{L^{q_0}}\}}\theta_\varepsilon f_\varepsilon, u\rangle\Big| \le \frac{c\sqrt{\lambda}}{\max\{\lambda,\ \|\theta_\varepsilon\|^2_{L^{q_0}}\}}\|\theta_\varepsilon\|_{L^{q_0}}\|f_\varepsilon\|_{\mathbf{L}^t}\|u\|_{\mathbf{L}^6} \le c\|f_\varepsilon\|_{\mathbf{L}^t}\|u\|_{\mathbf{V}}, \tag{9.53}$$

where c is independent of α_0. Putting $u = v_\varepsilon$ in the first equation of (9.20) yields

$$\begin{aligned} a_0(\theta_\varepsilon; v_\varepsilon, v_\varepsilon) &+ \frac{\delta}{\max\{\delta,\ \|\overline{a}_1(v_\varepsilon)\|_{\mathbf{V}^*}\}}a_1(v_\varepsilon, v_\varepsilon, v_\varepsilon) + \big\langle\nabla\Phi_\varepsilon(v_\varepsilon), v_\varepsilon\big\rangle \\ &-\Big\langle\Big(1 - \frac{\alpha_0\lambda}{\max\{\lambda,\ \|\theta_\varepsilon\|^2_{L^{q_0}}\}}\theta_\varepsilon\Big) f_\varepsilon, v_\varepsilon\Big\rangle = \langle f_1, v_\varepsilon\rangle. \end{aligned} \tag{9.54}$$

There exists a constant c_0' such that

$$\|f_\varepsilon\|_{\mathbf{L}^{\frac{6}{5}}} + c\|f_\varepsilon\|_{\mathbf{L}^t} \le c_0'\|f_\varepsilon\|_{\mathbf{L}^t},$$

where c is the one in (9.53). Thus taking into account (9.18), (9.28), (9.30) and (9.53), we have from (9.54)

$$\begin{aligned}
2\mu_0\|v_\varepsilon\|_{\mathbf{V}}^2 &\le a_0(\theta_\varepsilon; v_\varepsilon, v_\varepsilon)\\
&\le \frac{\delta}{\max\{\delta, \|\overline{a}_1(v_\varepsilon)\|_{\mathbf{V}^*}\}}|a_1(v_\varepsilon, v_\varepsilon, v_\varepsilon)|\\
&\quad + \left|\left\langle\left(1 - \frac{\alpha_0\lambda}{\max\{\lambda, \|\theta_\varepsilon\|_{L^{q_0}}^2\}}\theta_\varepsilon\right) f_\varepsilon, v_\varepsilon\right\rangle\right| + |\langle f_1, v_\varepsilon\rangle|\\
&\le K\|v_\varepsilon\|_{\mathbf{V}}^3 + \left(c_0'\|f_\varepsilon\|_{\mathbf{L}^t} + \|f_1\|_{\mathbf{V}^*}\right)\|v_\varepsilon\|_{\mathbf{V}}\\
&\le K\|v_\varepsilon\|_{\mathbf{V}}^3 + c_0(\|f_\varepsilon\|_{\mathbf{L}^t} + \|f_1\|_{\mathbf{V}^*})\|v_\varepsilon\|_{\mathbf{V}},
\end{aligned}$$

where

$$c_0 = \max\{c_0', 1\}. \tag{9.55}$$

Note that the estimate above is independent of δ. This implies

$$0 \le K\|v_\varepsilon\|_{\mathbf{V}}^2 - 2\mu_0\|v_\varepsilon\|_{\mathbf{V}} + c_0(\|f_\varepsilon\|_{\mathbf{L}^t} + \|f_1\|_{\mathbf{V}^*}).$$

Let us consider a quadratic polynomial for $x > 0$ concerned with the inequality above

$$Kx^2 - 2\mu_0 x + a.$$

If $0 \le Ka \le \mu_0^2$, then there exists a nonnegative minimum root $x_1(\le \frac{\mu_0}{K})$ and a maximum root x_2. Thus, we can know that if

$$(\|f_\varepsilon\|_{\mathbf{L}^t} + \|f_1\|_{\mathbf{V}^*}) \le \frac{\mu_0^2}{Kc_0},$$

then

$$\|v_\varepsilon\|_{\mathbf{V}} \le \frac{\mu_0}{K} \quad \text{or} \quad \|v_\varepsilon\|_{\mathbf{V}} \ge x_2. \tag{9.56}$$

On the other hand, we have from (9.54) another estimation under consideration of δ

$$2\mu_0\|v_\varepsilon\|_{\mathbf{V}}^2 \le a_0(\theta_\varepsilon; v_\varepsilon, v_\varepsilon) \le \delta\|v_\varepsilon\|_{\mathbf{V}} + c_0(\|f_\varepsilon\|_{\mathbf{L}^t} + \|f_1\|_{\mathbf{V}^*})\|v_\varepsilon\|_{\mathbf{V}},$$

which implies

$$\|v_\varepsilon\|_{\mathbf{V}} \le \frac{1}{2\mu_0}\left(\delta + c_0(\|f_\varepsilon\|_{\mathbf{L}^t} + \|f_1\|_{\mathbf{V}^*})\right). \tag{9.57}$$

In view of (9.56), we can take $\delta = K\left(\frac{\mu_0}{K}\right)^2 = \frac{\mu_0^2}{K}$.

If $(\|f\|_{\mathbf{L}^t} + \|f_1\|_{\mathbf{V}^*}) < \frac{\mu_0^2}{Kc_0}$, then for all ε small enough

$$(\|f_\varepsilon\|_{\mathbf{L}^t} + \|f_1\|_{\mathbf{V}^*}) \le \frac{\mu_0^2}{Kc_0}.$$

Thus, without loss of generality we have from (9.57)

$$\|v_\varepsilon\|_{\mathbf{V}} \leq \frac{\delta}{2\mu_0} + \frac{1}{2\mu_0}\frac{\mu_0^2}{K} = \frac{\mu_0}{K}. \tag{9.58}$$

By (9.58) under the condition (9.50) $\|\overline{a}_1(v_\varepsilon)\|_{\mathbf{V}^*} \leq K\|v_\varepsilon\|_{\mathbf{V}}^2 \leq \frac{\mu_0^2}{K}$ (see (9.18)), and so we get the first one in (9.51). Taking $\zeta = \frac{\mu_0}{K}$, we get the second one in (9.51). □

Lemma 9.3 *If* $\max\{|\alpha_0|, |\alpha_1|\}\sqrt{\lambda} \leq 1$ *and* (9.50) *holds, then under the parameter* ζ *by Lemma 9.2 there exists a* λ_1 *independent of* ε *such that*

$$\|\theta_\varepsilon\|_{L^{q_0}} \leq \sqrt{\lambda_1}. \tag{9.59}$$

Moreover, if $1 < r < \frac{3}{2}$, *then*

$$\int_\Omega |\nabla\theta_\varepsilon|^r\,dx \leq cK_\sigma^{r/(1-\sigma)}\big(1 + \|\theta_D\|_{W^{1,2}}^2\big), \quad \sigma = \frac{3-2r}{3-r}, \tag{9.60}$$

where K_σ *is the constant in* (9.90) *below.*

Proof Using Lemmas 9.1 and 9.2, we will arrive at the conclusion. For simplicity of notation, from now on in this lemma we denote $v_\varepsilon, \theta_\varepsilon$ by v, θ. Set

$$d_0 := \|\theta_D\|_{L^\infty(\Omega)} \geq \|\theta_D\|_{L^\infty(\partial\Omega)}. \tag{9.61}$$

Then, since $(\theta - d_0)^+ \equiv \max\{0, \theta - d_0\} = 0$ on Γ_D, the function

$$\varphi = 1 - \frac{1}{(1 + (\theta - d_0)^+)^\sigma}, \quad \sigma > 0 \tag{9.62}$$

belongs to $W^{1,2}_{\Gamma_D}(\Omega)$ and $0 \leq \varphi(x) \leq 1$ a.e. in Ω.

Taking φ of (9.62) and ζ satisfying (9.51), we have from the second equation of (9.20)

$$\begin{aligned}
&\sigma\int_\Omega \kappa(\theta)\frac{|\nabla(\theta - d_0)^+|^2}{(1 + (\theta - d_0)^+)^{1+\sigma}}\,dx + (\beta\theta, \varphi)_{\Gamma_R} - \sigma\int_\Omega \gamma_\varepsilon(\theta)v\cdot\frac{\nabla(\theta - d_0)^+}{(1 + (\theta - d_0)^+)^{1+\sigma}}\,dx\\
&= \int_\Omega \alpha_2\mu(\theta)\frac{|\mathcal{E}(v)|^2}{1 + \varepsilon|\mathcal{E}(v)|^2}\Big(1 - \frac{1}{(1 + (\theta - d_0)^+)^\sigma}\Big)\,dx\\
&+ \frac{\alpha_1\lambda}{\max\{\lambda, \|\theta\|_{L^{q_0}}^2\}}\int_\Omega \theta f_\varepsilon\cdot v\Big(1 - \frac{1}{(1 + (\theta - d_0)^+)^\sigma}\Big)\,dx + \Big\langle f_2, \Big(1 - \frac{1}{(1 + (\theta - d_0)^+)^\sigma}\Big)\Big\rangle\\
&\equiv I_1 + I_2 + I_3.
\end{aligned} \tag{9.63}$$

Note that $\varphi(x) = 0$ at x such that $\theta(x) \leq d_0$, and $(\beta(x)\theta, \varphi)_{\Gamma_R} \geq 0$. Let us show the third term on the left hand side of (9.63) vanishes. To this end, define

$$\Psi_\sigma(t) := \int_0^t \frac{\gamma_\varepsilon(s+d_0)}{(1+s)^{1+\sigma}}\, ds, \quad t \ge 0.$$

Then, $\Psi_\sigma \in C^1(\mathbb{R})$ and

$$\begin{aligned} &\Psi_\sigma((\theta-d_0)^+)|_{\Gamma_D} = 0,\\ &\nabla\Psi_\sigma((\theta-d_0)^+) = \gamma_\varepsilon(\theta)\frac{\nabla(\theta-d_0)^+}{(1+(\theta-d_0)^+)^{1+\sigma}}, \end{aligned} \tag{9.64}$$

where the fact that if $\theta(x)-d_0 \ge 0$, then $(\theta(x)-d_0)^+ + d_0 = \theta(x)$ was used. Taking into account the fact that $v_n = 0$ on Γ_R (see (9.3)) and the first equality of (9.64), we have

$$\int_\Omega \gamma_\varepsilon(\theta) v \cdot \frac{\nabla(\theta-d_0)^+}{(1+(\theta-d_0)^+)^{1+\sigma}}\, dx = \int_\Omega v\cdot\nabla\Psi_\sigma((\theta-d_0)^+) = 0. \tag{9.65}$$

It is easily seen that

$$|I_1| \le c\|v\|_{\mathbf{V}}^2. \tag{9.66}$$

Since $\max\{|\alpha_0|, |\alpha_1|\}\sqrt{\lambda} \le 1$, by (9.26) we have

$$|I_2| \le \frac{\sqrt{\lambda}}{\max\{\lambda, \|\theta\|^2_{L^{q_0}}\}}\|\theta\|_{q_0}\|f_\varepsilon\|_{\mathbf{L}^t}\|v\|_{\mathbf{L}^3} \le c\|f_\varepsilon\|_{\mathbf{L}^t}\|v\|_{\mathbf{V}}. \tag{9.67}$$

Also,

$$|I_3| \le c\sqrt{\mathrm{mes}\Omega}\|f_2\|_{(W^{1,2}_{\Gamma_D})^*} \le c\big(\|g_R\|_{L^{4/3}(\Gamma_R)} + \|g\|_{L^{6/5}(\Omega)}\big). \tag{9.68}$$

By (9.65)–(9.68), we have from (9.63)

$$\sigma\int_{\{x;\theta\ge d_0\}} \frac{|\nabla\theta|^2}{(1+(\theta-d_0)^+)^{1+\sigma}}\, dx \le c\Big(\|v\|_{\mathbf{V}}^2 + \|f_\varepsilon\|_{\mathbf{L}^t}\|v\|_{\mathbf{V}} + \|g_R\|_{L^{4/3}(\Gamma_R)} + \|g\|_{L^{6/5}(\Omega)}\Big). \tag{9.69}$$

Next, taking $\varphi = \min\{\theta-\theta_D, d_0\}$ admissible in the second formula of (9.20), we have

$$
\begin{aligned}
&\kappa_0 \int_{\{x;\theta-\theta_D<d_0\}} |\nabla\theta|^2 dx \\
&\le \int_{\{x;\theta-\theta_D<d_0\}} \kappa(x)\nabla\theta\cdot\nabla\theta_D\,dx - \int_{\Gamma_R} \beta(x)\theta(x)\min\{\theta(x)-\theta_D(x),d_0\}\,dx \\
&+ \int_\Omega \gamma_\varepsilon(\theta)v\cdot\nabla(\min\{\theta-\theta_D,d_0\})\,dx + \int_\Omega \alpha_2\mu(\theta)\frac{|\mathcal{E}(v)|^2}{1+\varepsilon|\mathcal{E}(v)|^2}\min\{\theta-\theta_D,d_0\}\,dx \\
&+ \frac{\alpha_1\lambda}{\max\{\lambda,\|\theta\|^2_{L^{q_0}}\}}\int_\Omega \theta f_\varepsilon\cdot v\min\{\theta-\theta_D,d_0\}\,dx + \langle f_2,\min\{\theta-\theta_D,d_0\}\rangle \\
&\equiv \overline{I}_1+\overline{I}_2+\overline{I}_3+\overline{I}_4+\overline{I}_5+\overline{I}_6.
\end{aligned}
\tag{9.70}
$$

By Young's inequality,

$$
|\overline{I}_1| \le \frac{\kappa_0}{4}\int_{\{x;\theta-\theta_D<d_0\}} |\nabla\theta|^2\,dx + \frac{\kappa_1^2}{\kappa_0}\int_\Omega |\nabla\theta_D|^2\,dx. \tag{9.71}
$$

Let us estimate $\overline{I}_2$. If $\theta(x)-\theta_D(x)\ge d_0$, then $\varphi(x)=d_0$ and $\theta(x)\ge\theta_D(x)+d_0\ge 0$. Hence,

$$
\int_{\Gamma_R\cap\{x;\theta(x)-\theta_D(x)\ge d_0\}} \beta(x)\theta(x)\min\{\theta(x)-\theta_D(x),d_0\}\,dx \ge 0. \tag{9.72}
$$

On the other hand, if $\theta(x)-\theta_D(x)\le d_0$, then $\varphi(x)=\theta(x)-\theta_D(x)$ and $\theta(x)\le\theta_D(x)+d_0$. Thus

$$
\begin{aligned}
&\int_{\Gamma_R\cap\{x;\theta(x)-\theta_D(x)\le d_0\}} \beta(x)\theta(x)\min\{\theta(x)-\theta_D(x),d_0\}\,dx \\
&= \int_{\Gamma_R\cap\{x;\theta(x)\le\theta_D(x)+d_0\}} \beta(x)\theta(x)\theta(x)\,dx - \int_{\Gamma_R\cap\{x;\theta(x)\le\theta_D(x)+d_0\}} \beta(x)\theta(x)\theta_D(x)\,dx \\
&\ge -\int_{\Gamma_R\cap\{x;0\le\theta(x)\le\theta_D(x)+d_0\}} \beta(x)\theta(x)\theta_D(x)\,dx - \int_{\Gamma_R\cap\{x;\theta(x)\le 0\}} \beta(x)\theta(x)\theta_D(x)\,dx \\
&\ge -\beta_0 2d_0^2\int_{\Gamma_R} dx - \beta_0 d_0\int_{\Gamma_R} |\theta^-(x)|\,dx \\
&\ge -c\big(1+\|\theta^-(x)\|_{W^{1,2}_{\Gamma_D}(\Omega)}\big).
\end{aligned}
\tag{9.73}
$$

By (9.72) and (9.73), we get

$$
\overline{I}_2 \le c\big(1+\|\theta^-(x)\|_{W^{1,2}_{\Gamma_D}(\Omega)}\big). \tag{9.74}
$$

Taking into account (9.19), we have

$$
\begin{aligned}
\overline{I}_3 &= \int_{\{x;\theta-\theta_D\le d_0\}} \gamma_\varepsilon(\theta)v\cdot\nabla\theta\,dx - \int_{\{x;\theta-\theta_D\le d_0\}} \gamma_\varepsilon(\theta)v\cdot\nabla\theta_D\,dx\\
&\le \frac{\kappa_0}{4}\int_{\{x;\theta-\theta_D\le d_0\}} |\nabla\theta|^2\,dx + c\gamma_0^2\int_{\{x;0\le\theta\le d_0+\theta_D\}} |\theta|^2|v|^2\,dx + c\gamma_0^2\int_{\{x;\theta\le 0\}} |\theta|^2|v|^2\,dx\\
&\quad + \gamma_0\int_{\{x;0\le\theta\le d_0+\theta_D\}} |\theta||v\cdot\nabla\theta_D|\,dx + \gamma_0\int_{\{x;\theta\le 0\}} |\theta||v\cdot\nabla\theta_D|\,dx.
\end{aligned}
\tag{9.75}
$$

Let us estimate all terms except the first one on the right hand side of (9.75). Since $|\theta|\le 2d_0$ on $\{x;\,0\le\theta\le d_0+\theta_D\}$,

$$
\int_{\{x;0\le\theta\le d_0+\theta_D\}} \theta^2|v|^2\,dx \le 4d_0^2\int_\Omega |v|^2\,dx \le c\|v\|_{\mathbf{V}}^2. \tag{9.76}
$$

And, by Hölder's inequality we obtain

$$
\int_{\{x;\theta\le 0\}} \theta^2|v|^2\,dx \le \Big(\int_{\{x;\theta\le 0\}} \theta^4\,dx\Big)^{1/2}\Big(\int_{\{x;\theta\le 0\}} |v|^4\,dx\Big)^{1/2} \le c\|\theta^-\|_{W^{1,2}_{\Gamma_D}}^2\|v\|_{\mathbf{V}}^2, \tag{9.77}
$$

$$
\int_{\{x;0\le\theta\le d_0+\theta_D\}} |\theta||v\cdot\nabla\theta_D|\,dx \le 2d_0\|v\|_{\mathbf{L}^2}\|\theta_D\|_{W^{1,2}} \le c\|v\|_{\mathbf{V}}\|\theta_D\|_{W^{1,2}}, \tag{9.78}
$$

$$
\int_{\{x;\theta\le 0\}} |\theta||v\cdot\nabla\theta_D|\,dx \le c\|\theta^-\|_{W^{1,2}_{\Gamma_D}}\|v\|_{\mathbf{V}}\|\theta_D\|_{W^{1,2}}. \tag{9.79}
$$

Combining (9.75)–(9.79) yields

$$
\begin{aligned}
\overline{I}_3 \le& \frac{\kappa_0}{4}\int_{\{x;\theta-\theta_D\le d_0\}} |\nabla\theta|^2\,dx\\
&+ c\Big(\|v\|_{\mathbf{V}}^2 + \|\theta^-\|_{W^{1,2}_{\Gamma_D}}^2\|v\|_{\mathbf{V}}^2 + \|v\|_{\mathbf{V}}\|\theta_D\|_{W^{1,2}} + \|\theta^-\|_{W^{1,2}_{\Gamma_D}}\|v\|_{\mathbf{V}}\|\theta_D\|_{W^{1,2}}\Big).
\end{aligned}
\tag{9.80}
$$

Next, we have

$$
\begin{aligned}
\overline{I}_4 &\le \int_{\{x;\theta-\theta_D\ge d_0\}} \alpha_2\mu(\theta)\frac{|\mathcal{E}(v)|^2}{1+\varepsilon|\mathcal{E}(v)|^2}d_0\,dx + \int_{\{x;0\le\theta\le d_0+\theta_D\}} \alpha_2\mu(\theta)\frac{|\mathcal{E}(v)|^2}{1+\varepsilon|\mathcal{E}(v)|^2}2d_0\,dx\\
&\quad + \int_{\{x;\theta\le 0\}} \alpha_2\mu(\theta)\frac{|\mathcal{E}(v)|^2}{1+\varepsilon|\mathcal{E}(v)|^2}\theta\,dx + \int_{\{x;\theta\le 0\}} \alpha_2\mu(\theta)\frac{|\mathcal{E}(v)|^2}{1+\varepsilon|\mathcal{E}(v)|^2}d_0\,dx\\
&\le \int_\Omega \alpha_2\mu(\theta)|\mathcal{E}(v)|^2 2d_0\,dx \le c\|v\|_{\mathbf{V}}^2.
\end{aligned}
\tag{9.81}
$$

Taking into account (9.44) and applying Hölder's inequality, we have

$$
\begin{aligned}
\overline{I}_5 \leq & \frac{|\alpha_1|\lambda}{\max\{\lambda, \|\theta\|^2_{L^{q_0}}\}} \int_{\{x;\theta-\theta_D\geq d_0\}} |\theta f_\varepsilon \cdot v| d_0 \, dx \\
& + \frac{|\alpha_1|\lambda}{\max\{\lambda, \|\theta\|^2_{L^{q_0}}\}} \int_{\{x;0\leq\theta\leq d_0+\theta_D\}} |\theta f_\varepsilon \cdot v| 2d_0 \, dx \\
& + \frac{\alpha_1\lambda}{\max\{\lambda, \|\theta\|^2_{L^{q_0}}\}} \int_{\{x;\theta\leq 0\}} \theta f_\varepsilon \cdot v\theta \, dx + \frac{|\alpha_1|\lambda}{\max\{\lambda, \|\theta\|^2_{L^{q_0}}\}} \int_{\{x;\theta\leq 0\}} |\theta f_\varepsilon \cdot v| d_0 \, dx \\
\leq & \, 2d_0 \frac{|\alpha_1|\lambda}{\max\{\lambda, \|\theta\|^2_{L^{q_0}}\}} \int_\Omega |\theta f_\varepsilon \cdot v| \, dx + \frac{|\alpha_1|\lambda}{\max\{\lambda, \|\theta\|^2_{L^{q_0}}\}} \int_\Omega |\theta f_\varepsilon \cdot v| |\theta^-| \, dx \\
\leq & \, c\|f_\varepsilon\|_{\mathbf{L}^t} \|v\|_{\mathbf{V}} + c\|f_\varepsilon\|_{\mathbf{L}^t} \|v\|_{\mathbf{V}} \|\theta^-\|_{W^{1,2}} \leq c\|f_\varepsilon\|_{\mathbf{L}^t} \|v\|_{\mathbf{V}} (1 + \|\theta^-\|_{W^{1,2}}).
\end{aligned}
\tag{9.82}
$$

Let us estimate $\overline{I}_6 = (g_R, \varphi)_{\Gamma_R} + (g, \varphi)$, where $\varphi = \min\{\theta(x) - \theta_D(x), d_0\}$. If $\theta(x) - \theta_D(x) \geq d_0$, then $\varphi(x) = d_0$ and

$$
\int_{\Gamma_R \cap \{x;\theta(x)-\theta_D(x)\geq d_0\}} g_R \varphi \, dx \leq d_0 \int_{\Gamma_R \cap \{x;\theta(x)-\theta_D(x)\geq d_0\}} |g_R| \, dx. \tag{9.83}
$$

If $0 \leq \theta(x) \leq d_0 + \theta_D(x)$, then $|\varphi(x)| = |\theta(x) - \theta_D(x)| \leq 2d_0$ and

$$
\int_{\Gamma_R \cap \{x;0\leq\theta(x)\leq d_0+\theta_D(x)\}} g_R \varphi(x) \, dx \leq 2d_0 \int_{\Gamma_R \cap \{x;0\leq\theta(x)\leq d_0+\theta_D(x)\}} |g_R| \, dx. \tag{9.84}
$$

And

$$
\begin{aligned}
\int_{\Gamma_R \cap \{x;\theta(x)\leq 0\}} g_R \varphi(x) \, dx &= \int_{\Gamma_R \cap \{x;\theta(x)\leq 0\}} g_R (\theta^-(x) - d_0) \, dx \\
&\leq d_0 \int_{\Gamma_R \cap \{x;\theta(x)\leq 0\}} |g_R| \, dx + \|g_R\|_{L^{4/3}(\Gamma_R)} \|\theta^-\|_{W^{1,2}_{\Gamma_D}(\Omega)}.
\end{aligned}
\tag{9.85}
$$

Also, we have

$$
\begin{aligned}
\langle g, \varphi \rangle \leq & \int_{\{x;\theta-\theta_D\geq d_0\}} |g| d_0 \, dx + \int_{\{x;0\leq\theta\leq d_0+\theta_D\}} |g| 2d_0 \, dx + \int_{\{x;\theta\leq 0\}} g(\theta^- - d_0) \, dx \\
\leq & \, 2d_0 \|g\|_{L^1(\Omega)} + \|g\|_{L^{6/5}(\Omega)} \|\theta^-\|_{W^{1,2}_{\Gamma_D}(\Omega)} \\
\leq & \, c\|g\|_{L^{6/5}(\Omega)} (1 + \|\theta^-\|_{W^{1,2}_{\Gamma_D}(\Omega)}).
\end{aligned}
\tag{9.86}
$$

By (9.83)–(9.86), we get

$$
\overline{I}_6 \leq c(\|g_R\|_{L^{4/3}(\Gamma_R)} + \|g\|_{L^{6/5}(\Omega)})(1 + \|\theta^-\|_{W^{1,2}_{\Gamma_D}(\Omega)}). \tag{9.87}
$$

Therefore, by virtue of (9.70), (9.71), (9.74), (9.80)–(9.82) and (9.87) we have

$$\int_{\{x;\theta-\theta_D<d_0\}} |\nabla\theta|^2 dx \le c\Big(\|\theta_D\|^2_{W^{1,2}} + \|v\|^2_{\mathbf{V}} + \|\theta^-\|^2_{W^{1,2}_{\Gamma_D}}\|v\|^2_{\mathbf{V}}\Big)$$
$$+ c\Big(1 + \|f_\varepsilon\|_{\mathbf{L}^t}\|v\|_{\mathbf{V}} + \|g_R\|_{L^{4/3}(\Gamma_R)} + \|g\|_{L^{6/5}(\Omega)}\Big)\big(1 + \|\theta^-\|_{W^{1,2}_{\Gamma_D}}\big). \tag{9.88}$$

Combining (9.69), (9.88) and taking into account (9.45), we get

$$\begin{aligned}
\int_\Omega \frac{|\nabla\theta|^2}{(1+(\theta-d_0)^+)^{1+\sigma}}\,dx &\le \int_{\{x;\theta\ge d_0\}} \frac{|\nabla\theta|^2}{(1+(\theta-d_0)^+)^{1+\sigma}}\,dx + \int_{\{x;\theta\le d_0\}} |\nabla\theta|^2\,dx \\
&\le c(1+\frac{1}{\sigma})\Big(1 + \|v\|^2_{\mathbf{V}} + \|f_\varepsilon\|_{\mathbf{L}^t}\|v\|_{\mathbf{V}} + \|g_R\|_{L^{4/3}(\Gamma_R)} + \|g\|_{L^{6/5}(\Omega)}\Big)\big(1+\|\theta^-\|_{W^{1,2}_{\Gamma_D}}\big) \\
&\quad + c\Big(\|\theta_D\|^2_{W^{1,2}} + \|v\|^2_{\mathbf{V}} + \|\theta^-\|^2_{W^{1,2}_{\Gamma_D}}\|v\|^2_{\mathbf{V}}\Big) \\
&\le c\Big[(1+\frac{1}{\sigma})\Big(1 + \|v\|^2_{\mathbf{V}} + \|f_\varepsilon\|_{\mathbf{L}^t}\|v\|_{\mathbf{V}} + \|g_R\|_{L^{4/3}(\Gamma_R)} + \|g\|_{L^{6/5}(\Omega)}\Big)^2 \\
&\quad + \|\theta_D\|^2_{W^{1,2}} + \|v\|^2_{\mathbf{V}} + \Big(\|f_\varepsilon\|_{\mathbf{L}^t}\|v\|_{\mathbf{V}} + \|g_R\|_{L^{4/3}(\Gamma_R)} + \|g\|_{L^{6/5}(\Omega)}\Big)^2\|v\|^2_{\mathbf{V}}\Big].
\end{aligned} \tag{9.89}$$

Since $f_\varepsilon \to f$ in $\mathbf{L}^t(\Omega)$, we may assume that $\|f_\varepsilon\|_{\mathbf{L}^t} \le 1 + \|f\|_{\mathbf{L}^t}$ for all ε. Thus taking into account (9.58), we have from (9.89) that under (9.50)

$$\begin{aligned}
&\int_\Omega \frac{|\nabla\theta|^2}{(1+(\theta-d_0)^+)^{1+\sigma}}\,dx \\
&\le c\Big[(1+\frac{1}{\sigma})\Big(1 + (\frac{\mu_0}{K})^2 + (1+\|f\|_{\mathbf{L}^t})\frac{\mu_0}{K} + \|g_R\|_{L^{4/3}(\Gamma_R)} + \|g\|_{L^{6/5}(\Omega)}\Big)^2 \\
&\quad + \|\theta_D\|^2_{W^{1,2}} + (\frac{\mu_0}{K})^2 + \Big((1+\|f\|_{\mathbf{L}^t})\frac{\mu_0}{K} + \|g_R\|_{L^{4/3}(\Gamma_R)} + \|g\|_{L^{6/5}(\Omega)}\Big)^2(\frac{\mu_0}{K})^2\Big] \\
&\equiv K_\sigma.
\end{aligned} \tag{9.90}$$

To get boundedness of $\theta \in L^{q_0}(\Omega)$ independent of ε, we will use the property $W^{1,r_0}(\Omega) \subset L^{q_0}$, where r_0 is such that $\frac{1}{r_0} - \frac{1}{3} = \frac{1}{q_0}$. Then, $\frac{4}{3} < r_0 < \frac{3}{2}$. To this end, take $\sigma_0 > 0$ such that

$$\frac{r_0(1+\sigma_0)}{2-r_0} = q_0.$$

Then, $\sigma_0 = \frac{3-2r_0}{3-r_0}$, and $0 < \sigma_0 < \frac{1}{5}$. Putting $\sigma = \sigma_0$ in (9.90), by virtue of Hölder's inequality with exponents $\frac{2}{r_0}$, $\frac{2}{2-r_0}$ and the inequality $|a+b|^p \le 2^p(|a|^p + |b|^p)$, $|a| + |b| \le (|a|^{\frac{1}{p}} + |b|^{\frac{1}{p}})^p$, $p \in (1,\infty)$, we have

$$\begin{aligned}
&\int_\Omega |\nabla\theta|^{r_0}\,dx \\
&\le \Big(\int_\Omega \frac{|\nabla\theta|^2}{(1+(\theta-d_0)^+)^{(1+\sigma_0)}}\,dx\Big)^{r_0/2}\Big(\int_\Omega (1+(\theta-d_0)^+)^{(1+\sigma_0)\frac{r_0}{2-r_0}}\,dx\Big)^{(2-r_0)/2} \\
&\le cK_{\sigma_0}^{r_0/2}\Big(1 + \Big(\int_\Omega |\theta|^{3r_0/(3-r_0)}\,dx\Big)^{(2-r_0)/2}\Big),
\end{aligned} \tag{9.91}$$

where $(1+\sigma_0)\frac{r_0}{2-r_0} = 3r_0/(3-r_0)$ was used and K_{σ_0} is the one with $\sigma = \sigma_0$ in K_σ of (9.90). By Sobolev's imbedding theorem and Friedrichs' inequality,

$$\Big(\int_\Omega |\theta|^{3r_0/(3-r_0)}\,dx\Big)^{(2-r_0)/2} \le c\|\theta\|_{W^{1,r_0}}^{(1+\sigma_0)\frac{r_0}{2}} \le c\Big(\int_\Omega |\nabla\theta|^{r_0}\,dx + \|\theta_D\|_{W^{1,r_0}}^{r_0}\Big)^{(1+\sigma_0)/2}$$
$$\le c\bigg(\Big(\int_\Omega |\nabla\theta|^{r_0}\,dx\Big)^{(1+\sigma_0)/2} + \|\theta_D\|_{W^{1,r_0}}^{r_0(1+\sigma_0)/2}\bigg). \tag{9.92}$$

Substituting (9.92) into (9.91) and using Young's inequality and Hölder's inequality with exponents $\frac{2}{1-\sigma_0}, \frac{2}{1+\sigma_0}$ on the right hand side, we have

$$\int_\Omega |\nabla\theta|^{r_0}\,dx \le cK_{\sigma_0}^{r_0/2}\Big(\int_\Omega |\nabla\theta|^{r_0}\,dx\Big)^{(1+\sigma_0)/2} + cK_{\sigma_0}^{r_0/2}\big(1+\|\theta_D\|_{W^{1,r_0}}^{r_0(1+\sigma_0)/2}\big)$$
$$\le \frac{1}{2}\int_\Omega |\nabla\theta|^{r_0}\,dx + cK_{\sigma_0}^{r_0/(1-\sigma_0)} + cK_{\sigma_0}^{r_0/(1-\sigma_0)}\big(1+\|\theta_D\|_{W^{1,r_0}}^{r_0}\big).$$

Thus, we get

$$\int_\Omega |\nabla\theta|^{r_0}\,dx \le cK_{\sigma_0}^{r_0/(1-\sigma_0)}\big(1+\|\theta_D\|_{W^{1,r_0}}^{r_0}\big). \tag{9.93}$$

By virtue of (9.93), Sobolev's imbedding theorem and Friedrichs' inequality, we have

$$\begin{aligned}\|\theta\|_{L^{q_0}} &\le c\|\theta\|_{W^{1,r_0}} \le c\Big(\int_\Omega |\nabla\theta|^{r_0}\,dx + \|\theta_D\|_{W^{1,r_0}}^{r_0}\Big)^{1/r_0}\\ &\le c\,\big(K_{\sigma_0}^{r_0/(1-\sigma_0)}(1+\|\theta_D\|_{W^{1,r_0}}^{r_0}) + \|\theta_D\|_{W^{1,r_0}}^{r_0}\big)^{1/r_0}\\ &\le c\,\big(K_{\sigma_0}^{1/(1-\sigma_0)}(1+\|\theta_D\|_{W^{1,r_0}}) + \|\theta_D\|_{W^{1,r_0}}\big)\\ &\le c\,\big(K_{\sigma_0}^{1/(1-\sigma_0)}(1+\|\theta_D\|_{W^{1,2}}) + \|\theta_D\|_{W^{1,2}}\big) \equiv \overline{K}_{\sigma_0}.\end{aligned} \tag{9.94}$$

$\overline{K}_{\sigma_0}$ is independent of λ, ε and depends on $\|f\|_{\mathbf{L}^t}$, $\|\theta_D\|_{W^{1,2}}$, $\|g_R\|_{L^{4/3}(\Gamma_R)}$, $\|g\|_{L^{6/5}(\Omega)}$ and $\phi_i, i = 2,\cdots,7$, via f_1 in (9.7).

Putting

$$\sqrt{\lambda_1} = \overline{K}_{\sigma_0}, \tag{9.95}$$

we have (9.59).

Now, for all r with $1 < r < \frac{3}{2}$ and all $\varepsilon > 0$ let us get an estimate of $\int_\Omega |\nabla\theta|^r\,dx$ independent of ε. Putting $\sigma = \frac{3-2r}{3-r}$, we have that $0 < \sigma < \frac{1}{2}$. Repeating the argument of (9.91)–(9.93), we have

$$\int_\Omega |\nabla\theta|^r\,dx \le cK_\sigma^{r/(1-\sigma)}\big(1+\|\theta_D\|_{W^{1,r}}^r\big),$$

which implies (9.60). □

When $\lambda = \lambda_1$ and

$$\max\{|\alpha_0|, |\alpha_1|\}\sqrt{\lambda_1} \leq 1,$$

by (9.94) and (9.95) we have

$$\frac{\lambda_1}{\max\{\lambda_1, \|\theta_\varepsilon\|^2_{L^{q_0}}\}} = 1 \quad \forall \varepsilon > 0. \tag{9.96}$$

Therefore, summarizing Lemmas 9.1–9.3, we have

Theorem 9.3 *If*

$$\begin{aligned} &\max\{|\alpha_0|, |\alpha_1|\}\sqrt{\lambda_1} \leq 1, \\ &(\|f\|^2_{\mathbf{L}'} + \|f_1\|^2_{\mathbf{V}^*}) < \frac{\mu_0^2}{Kc_0}, \end{aligned} \tag{9.97}$$

then there exists a solution $(v_\varepsilon, \theta_\varepsilon) \in \mathbf{V} \times W^{1,2}(\Omega)$ to the following problem

$$\begin{cases} a_0(\theta_\varepsilon; v_\varepsilon, u) + a_1(v_\varepsilon, v_\varepsilon, u) + \langle \nabla \Phi_\varepsilon(v_\varepsilon), u\rangle - \langle (1-\alpha_0\theta_\varepsilon) f_\varepsilon, u\rangle = \langle f_1, u\rangle \quad \forall u \in \mathbf{V}, \\ b_0(\theta_\varepsilon; \theta_\varepsilon, \varphi) - \langle \gamma_\varepsilon(\theta_\varepsilon) v_\varepsilon, \nabla\varphi\rangle - \left\langle \alpha_2\mu(\theta_\varepsilon)\dfrac{|\mathcal{E}(v_\varepsilon)|^2}{1+\varepsilon|\mathcal{E}(v_\varepsilon)|^2}, \varphi\right\rangle - \langle \alpha_1\theta_\varepsilon f_\varepsilon \cdot v_\varepsilon, \varphi\rangle \\ \qquad = \langle f_2, \varphi\rangle \quad \forall \varphi \in W^{1,2}_{\Gamma_D}(\Omega), \end{cases} \tag{9.98}$$

$$\theta_\varepsilon - \theta_D \in W^{1,2}_{\Gamma_D}(\Omega), \tag{9.99}$$

and the solution satisfies:

$$\begin{aligned} &\|v_\varepsilon\|_{\mathbf{V}} \leq \frac{\mu_0}{K}, \\ &\|\theta_\varepsilon^-\|_{W^{1,2}_{\Gamma_D}(\Omega)} \leq c\Big(\|f\|_{L'}\|v_\varepsilon\|_V + \|g_R\|_{L^{4/3}(\Gamma_R)} + \|g\|_{L^{6/5}(\Omega)}\Big), \\ &\int_\Omega |\nabla\theta_\varepsilon|^r\,dx \leq cK_\sigma^{r/(1-\sigma)}\big(1 + \|\theta_D\|^2_{W^{1,2}}\big) \quad \forall r, 1 < r < \frac{3}{2}, \end{aligned} \tag{9.100}$$

where $\sigma = \frac{3-2r}{3-r}$.

9.3.3 Passing to Limits

By passing to the limit of solutions in Theorem 9.3, we will prove Theorem 9.1. Owing to (9.100) we can extract subsequences, which are denoted as before, such that as $\varepsilon \to 0$,

$$
\begin{aligned}
&v_\varepsilon \rightharpoonup v \quad \text{in } \mathbf{V}, \\
&v_\varepsilon \to v \quad \text{in } \mathbf{L}^q,\ 1 \le q < 6, \quad \text{and a.e. in } \Omega, \\
&\theta_\varepsilon \rightharpoonup \theta \quad \text{in } W^{1,r}(\Omega) \quad \forall r,\ 1 \le r < \frac{3}{2}, \\
&\theta_\varepsilon \to \theta \quad \text{in } L^s(\Omega) \quad \forall s,\ 1 \le s < 3, \quad \text{and a.e. in } \Omega.
\end{aligned}
\tag{9.101}
$$

By (9.99), we have that $\theta|_{\Gamma_D} = \theta_D|_{\Gamma_D}$.

Subtracting the first formula of (9.98) with $u = v_\varepsilon$ from the first formula of (9.98), we have

$$
\begin{aligned}
a_0(\theta_\varepsilon; v_\varepsilon, u - v_\varepsilon) + a_1(v_\varepsilon, v_\varepsilon, u - v_\varepsilon) + \big\langle \nabla\Phi_\varepsilon(v_\varepsilon), u - v_\varepsilon\big\rangle - \big\langle (1-\alpha_0\theta_\varepsilon) f, u - v_\varepsilon\big\rangle \\
= \langle f_1, u - v_\varepsilon\rangle \qquad \forall u \in \mathbf{V}.
\end{aligned}
\tag{9.102}
$$

By Corollaries 1.1 and 1.2 we have

$$
\begin{aligned}
&a_0(\theta_\varepsilon; v_\varepsilon, u) \to a_0(\theta; v, u) \quad \text{as } \varepsilon \to 0, \\
&\liminf_{\varepsilon\to 0} a_0(\theta_\varepsilon; v_\varepsilon, v_\varepsilon) \ge a_0(\theta; v, v),
\end{aligned}
$$

which imply that

$$
\limsup_{\varepsilon\to 0} a_0(\theta_\varepsilon; v_\varepsilon, u - v_\varepsilon) \le a_0(\theta; v, u - v). \tag{9.103}
$$

It is easy to prove

$$
a_1(v_\varepsilon, v_\varepsilon, u - v_\varepsilon) \to a_1(v, v, u - v) \quad \text{as } \varepsilon \to 0. \tag{9.104}
$$

Since Φ_ε is convex, continuous and Fréchet differentiable, we have

$$
\Phi_\varepsilon(u) - \Phi_\varepsilon(v_\varepsilon) \ge \langle \nabla\Phi_\varepsilon(v_\varepsilon), u - v_\varepsilon\rangle \quad \forall u \in \mathbf{V}, \tag{9.105}
$$

which together with (1.37) implies

$$
\Phi_\varepsilon(u) - \Phi(J_\varepsilon v_\varepsilon) \ge \langle \nabla\Phi_\varepsilon(v_\varepsilon), u - v_\varepsilon\rangle \quad \forall u \in \mathbf{V}. \tag{9.106}
$$

Since $\Phi(0_{\mathbf{V}}) = 0$, by (1.37) we get $\Phi_\varepsilon(0_{\mathbf{V}}) = 0$, and so we have from (9.105)

$$
\Phi_\varepsilon(v_\varepsilon) \le \langle \nabla\Phi_\varepsilon(v_\varepsilon), v_\varepsilon\rangle. \tag{9.107}
$$

On the other hand, putting $u = v_\varepsilon$ in the first formula of (9.98), we have

$$
a_0(\theta_\varepsilon; v_\varepsilon, v_\varepsilon) + a_1(v_\varepsilon, v_\varepsilon, v_\varepsilon) + \big\langle \nabla\Phi_\varepsilon(v_\varepsilon), v_\varepsilon\big\rangle = \big\langle (1-\alpha_0\theta_\varepsilon) f_\varepsilon, v_\varepsilon\big\rangle + \langle f_1, v_\varepsilon\rangle. \tag{9.108}
$$

From (9.107) and (9.108) we have

$$a_0(\theta_\varepsilon; v_\varepsilon, v_\varepsilon) + a_1(v_\varepsilon, v_\varepsilon, v_\varepsilon) + \Phi_\varepsilon(v_\varepsilon) \leq \big\langle (1-\alpha_0\theta_\varepsilon) f_\varepsilon, v_\varepsilon \big\rangle + \langle f_1, v_\varepsilon \rangle,$$

from which we get

$$|\Phi_\varepsilon(v_\varepsilon)| \leq c\big((1+\overline{K}_{\sigma_0})\|f_\varepsilon\|_{\mathbf{L}^t} + \|f_1\|_{\mathbf{V}^*}\big)\|v_\varepsilon\|_{\mathbf{V}} + |a_1(v_\varepsilon, v_\varepsilon, v_\varepsilon)|, \tag{9.109}$$

where (9.94) was used. By virtue of (1.36), (9.18), (9.58) and (9.109), we have

$$\|v_\varepsilon - J_\varepsilon v_\varepsilon\|_{\mathbf{V}}^2 \leq \big[c\big((1+\overline{K}_{\sigma_0})\|f_\varepsilon\|_{\mathbf{L}^t} + \|f_1\|_{\mathbf{V}^*}\big)\frac{\mu_0}{K} + \frac{\mu_0^3}{K^2}\big]2\varepsilon,$$

which shows that

$$J_\varepsilon v_\varepsilon \rightharpoonup v \quad \text{in } \mathbf{V} \quad \text{as } \varepsilon \to 0.$$

Then, by virtue of weak lower semi-continuity of $\Phi(v)$,

$$\liminf_{\varepsilon\to 0} \Phi(J_\varepsilon v_\varepsilon) \geq \Phi(v). \tag{9.110}$$

By (1.37) we have

$$\Phi_\varepsilon(u) \to \Phi(u) \quad \text{as } \varepsilon \to 0. \tag{9.111}$$

Taking into account (9.110) and (9.111), we have from (9.106)

$$\Phi(u) - \Phi(v) \geq \limsup_{\varepsilon\to 0}\langle \nabla\Phi_\varepsilon(v_\varepsilon), u - v_\varepsilon\rangle \quad \forall u \in \mathbf{V}. \tag{9.112}$$

Using

$$\begin{aligned} &|\langle \theta_\varepsilon f_\varepsilon, v_\varepsilon\rangle - \langle \theta f, v\rangle| \\ &\quad \leq |\langle \theta_\varepsilon f_\varepsilon, v_\varepsilon\rangle - \langle \theta f_\varepsilon, v_\varepsilon\rangle| + |\langle \theta f_\varepsilon, v_\varepsilon\rangle - \langle \theta f, v_\varepsilon\rangle| + |\langle \theta f, v_\varepsilon\rangle - \langle \theta f, v\rangle| \\ &\quad \leq \|\theta_\varepsilon - \theta\|_{L^2}\|f_\varepsilon\|_{\mathbf{L}^3}\|v_\varepsilon\|_{\mathbf{L}^6} + \|\theta\|_{L^2}\|f_\varepsilon - f\|_{\mathbf{L}^3}|v_\varepsilon\|_{\mathbf{L}^6} + \|\theta\|_{L^{15/7}}\|f\|_{\mathbf{L}^3}|v_\varepsilon - v\|_{\mathbf{L}^5}, \end{aligned} \tag{9.113}$$

we can prove

$$\big\langle (1-\alpha_0\theta_\varepsilon) f, u - v_\varepsilon\big\rangle \to \big\langle (1-\alpha_0\theta) f, u - v\big\rangle \quad \text{as } \varepsilon \to 0. \tag{9.114}$$

It is easy to prove

$$\langle f_1, u - v_\varepsilon\rangle \to \langle f_1, u - v\rangle \quad \text{as } \varepsilon \to 0. \tag{9.115}$$

By virtue of (9.103), (9.104), (9.112), (9.114) and (9.115), we get from (9.102)

$$a_0(\theta; v, u - v) + a_1(v, v, u - v) + \Phi(u) - \Phi(v) - \big\langle (1-\alpha_0\theta) f, u - v\big\rangle \geq \langle f_1, u - v\rangle \quad \forall u \in \mathbf{V},$$

which is the first formula in (9.13). From the above discussion we know that $v \in K(\Omega)$, i.e. $\Phi(v) < +\infty$. Thus putting $u = v$, we have from (9.112)

$$0 \geq \limsup_{\varepsilon\to 0}\langle \nabla\Phi_\varepsilon(v_\varepsilon), v - v_\varepsilon\rangle. \tag{9.116}$$

Now we will get the second equation of (9.13). By Corollary 1.1, we have

$$b_0(\theta_\varepsilon; \theta_\varepsilon, \varphi) \to b_0(\theta; \theta, \varphi) \quad \forall\varphi \in W^{1,\infty}_{\Gamma_D}(\Omega) \quad \text{as } \varepsilon \to 0. \tag{9.117}$$

Let us prove

$$\langle \gamma_\varepsilon(\theta_\varepsilon)v_\varepsilon, \nabla\phi\rangle \to \langle \gamma(\theta)\theta v, \nabla\varphi\rangle \quad \forall\varphi \in W^{1,\infty}_{\Gamma_D}(\Omega) \quad \text{as } \varepsilon \to 0. \tag{9.118}$$

By Hölder's inequality we have

$$\begin{aligned} &|\langle \gamma_\varepsilon(\theta_\varepsilon)v_\varepsilon, \nabla\varphi\rangle - \langle \gamma(\theta)\theta v, \nabla\varphi\rangle| \\ &\quad\leq |\langle \gamma_\varepsilon(\theta_\varepsilon)v_\varepsilon, \nabla\varphi\rangle - \langle \gamma(\theta)\theta v_\varepsilon, \nabla\varphi\rangle| + |\langle \gamma(\theta)\theta v_\varepsilon, \nabla\varphi\rangle - \langle \gamma(\theta)\theta v, \nabla\varphi\rangle| \\ &\quad\leq \|\gamma_\varepsilon(\theta_\varepsilon) - \gamma(\theta)\theta\|_{L^{6/5}}\|v_\varepsilon\|_{\mathbf{L}^6}\|\nabla\varphi\|_{\mathbf{L}^\infty} + \|\gamma(\theta)\theta\|_{L^4}\|v_\varepsilon - v\|_{\mathbf{L}^4}\|\nabla\varphi\|_{\mathbf{L}^\infty}. \end{aligned} \tag{9.119}$$

By the definition of $\gamma_\varepsilon(t)$ it follows that

$$\begin{aligned} \|\gamma_\varepsilon(\theta_\varepsilon) - \gamma(\theta)\theta\|_{L^{6/5}} &\leq \left\| \frac{\gamma(\theta_\varepsilon)\theta_\varepsilon}{(1+\varepsilon|\gamma(\theta_\varepsilon)|)(1+\varepsilon|\theta_\varepsilon|)} - \gamma(\theta)\theta \right\|_{L^{6/5}} \\ &\leq \|\gamma(\theta_\varepsilon)\theta_\varepsilon - \gamma(\theta)\theta\|_{L^{6/5}} + \varepsilon \left\|\gamma(\theta)\theta\big(|\gamma(\theta_\varepsilon)| + |\theta_\varepsilon| + \varepsilon|\gamma(\theta_\varepsilon)||\theta_\varepsilon|\big)\right\|_{L^{6/5}}. \end{aligned} \tag{9.120}$$

Then, by virtue of Lemma 1.3 we know that $\gamma(\theta_\varepsilon)$ converges to $\gamma(\theta)$ in $L^p(\Omega)$ ($\forall p,\ 1 < p < \infty,$) as ε goes to zero. Thus, from (9.119) and (9.120) we get (9.118).

Let us consider

$$\begin{aligned} \mu_0\|v - v_\varepsilon\|^2_{\mathbf{V}} &\leq a_0(\theta_\varepsilon; v - v_\varepsilon, v - v_\varepsilon) \\ &= a_0(\theta_\varepsilon; v, v - v_\varepsilon) + a_1(v_\varepsilon, v_\varepsilon, v - v_\varepsilon) + \nabla\Phi_\varepsilon(v_\varepsilon)(v - v_\varepsilon) \\ &\quad - \big\langle(1 - \alpha_0\theta_\varepsilon)f, v - v_\varepsilon\big\rangle - \langle f_1, v - v_\varepsilon\rangle, \end{aligned} \tag{9.121}$$

which is obtained from the first formula of (9.98) by putting $u = v - v_\varepsilon$. By virtue of (9.103), (9.104), (9.114) and (9.116), the right hand side of (9.121) converges to 0 as ε goes to 0. Thus, we have

$$v_\varepsilon \to v \quad \text{in } \mathbf{V} \quad \text{as } \varepsilon \to 0. \tag{9.122}$$

On the other hand,

$$\begin{aligned}
&\alpha_2\mu(\theta_\varepsilon)\frac{|\mathcal{E}(v_\varepsilon)|^2}{1+\varepsilon|\mathcal{E}(v_\varepsilon)|^2}-\alpha_2\mu(\theta)|\mathcal{E}(v)|^2\\
&=\left(\alpha_2\mu(\theta_\varepsilon)\frac{|\mathcal{E}(v_\varepsilon)|^2}{1+\varepsilon|\mathcal{E}(v_\varepsilon)|^2}-\alpha_2\mu(\theta_\varepsilon)|\mathcal{E}(v)|^2\right)+\alpha_2\big(\mu(\theta_\varepsilon)-\mu(\theta)\big)|\mathcal{E}(v)|^2\\
&=\alpha_2\mu(\theta_\varepsilon)\frac{|\mathcal{E}(v_\varepsilon)|^2-|\mathcal{E}(v)|^2}{1+\varepsilon|\mathcal{E}(v_\varepsilon)|^2}-\alpha_2\mu(\theta_\varepsilon)\frac{\varepsilon|\mathcal{E}(v_\varepsilon)|^2|\mathcal{E}(v)|^2}{1+\varepsilon|\mathcal{E}(v_\varepsilon)|^2}+\alpha_2\big(\mu(\theta_\varepsilon)-\mu(\theta)\big)|\mathcal{E}(v)|^2\\
&\equiv I_{\varepsilon 1}+I_{\varepsilon 2}+\alpha_2\big(\mu(\theta_\varepsilon)-\mu(\theta)\big)|\mathcal{E}(v)|^2.
\end{aligned}$$

By (9.122) we have

$$\|I_{\varepsilon 1}\|_{L^1(\Omega)}\to 0 \quad \text{as } \varepsilon\to 0.$$

Since $|I_{\varepsilon 2}|\le c|\mathcal{E}(v)|^2$ and (passing to a subsequence if necessary) $I_{\varepsilon 2}\to 0$ a.e. in Ω, by virtue of Lebesgue's theorem (see Theorem 1.9) we have

$$\|I_{\varepsilon 2}\|_{L^1(\Omega)}\to 0 \quad \text{as } \varepsilon\to 0.$$

By Corollary 1.1, we have

$$\mu(\theta_\varepsilon)\varepsilon_{ij}(v)\rightharpoonup\mu(\theta)\varepsilon_{ij}(v) \quad \text{in } L^2(\Omega) \quad \text{as } \varepsilon\to 0,$$

and so

$$\langle\alpha_2\big(\mu(\theta_\varepsilon)-\mu(\theta)\big)|\mathcal{E}(v)|^2,\varphi\rangle\to 0 \quad \forall\varphi\in W^{1,\infty}_{\Gamma_D}(\Omega) \quad \text{as } \varepsilon\to 0.$$

Thus, we have that as $\varepsilon\to 0$

$$\left\langle\alpha_2\mu(\theta_\varepsilon)\frac{|\mathcal{E}(v_\varepsilon)|^2}{1+\varepsilon|\mathcal{E}(v_\varepsilon)|^2}-\alpha_2\mu(\theta)|\mathcal{E}(v)|^2,\varphi\right\rangle\to 0 \quad \forall\varphi\in W^{1,\infty}_{\Gamma_D}(\Omega). \tag{9.123}$$

Taking into account

$$\begin{aligned}
&|\langle\alpha_1\theta_\varepsilon f_\varepsilon\cdot v_\varepsilon,\varphi\rangle-\langle\alpha_1\theta f\cdot v,\varphi\rangle|\\
&\quad\le|\langle\alpha_1\theta_\varepsilon f_\varepsilon\cdot v_\varepsilon,\varphi\rangle-\langle\alpha_1\theta f_\varepsilon\cdot v_\varepsilon,\varphi\rangle|+|\langle\alpha_1\theta f_\varepsilon\cdot v_\varepsilon,\varphi\rangle-\langle\alpha_1\theta f\cdot v_\varepsilon,\varphi\rangle|\\
&\qquad+|\langle\alpha_1\theta f\cdot v_\varepsilon,\varphi\rangle-\langle\alpha_1\theta f\cdot v,\varphi\rangle|\\
&\quad\le c\|\theta_\varepsilon-\theta\|_{L^2}\|f_\varepsilon\|_{\mathbf{L}^3}\|v_\varepsilon\|_{\mathbf{L}^6}\|\varphi\|_{L^\infty}+c\|\theta\|_{L^2}\|f_\varepsilon-f\|_{\mathbf{L}^3}\|v\|_{\mathbf{L}^6}\|\varphi\|_{L^\infty}\\
&\qquad+c\|\theta\|_{L^2}\|f\|_{\mathbf{L}^3}\|v_\varepsilon-v\|_{\mathbf{L}^6}\|\varphi\|_{L^\infty},
\end{aligned}$$

we can prove

$$\langle\alpha_1\theta_\varepsilon f\cdot v_\varepsilon,\varphi\rangle\to\langle\alpha_1\theta f\cdot v,\varphi\rangle \quad \forall\varphi\in W^{1,\infty}_{\Gamma_D}(\Omega) \quad \text{as } \varepsilon\to 0. \tag{9.124}$$

By virtue of (9.117), (9.118), (9.123) and (9.124), from the second formula in (9.98) we get the second formula in (9.13).

The estimates of solutions follow from (9.100) by (9.101). □

9.4 Existence of a Solution: The Case of Total Pressure

In this section we prove the main result for the case of total pressure.

Theorem 9.4 *Let Assumption 9.1 hold and assume that* $\max\{|\alpha_0|, |\alpha_1|\}$ *is small enough in accordance with* $f, \phi_i, i = 2, \cdots, 7, g, g_R, \theta_D$ *(see (9.138)). Then, there exists a solution* (v, θ) *to Problem II-VI such that*

$$\begin{aligned}
&\|v\|_{\mathbf{V}} \le c\big(1 + \|f\|_{\mathbf{L}^{t}} + \|f_1\|_{\mathbf{V}^*}\big),\\
&\|\theta^-\|_{W^{1,2}(\Omega)} \le c\Big(\|f\|_{\mathbf{L}^{t}} + \|f\|^2_{\mathbf{L}^{t}} + \|f_1\|_{\mathbf{V}^*}\|f\|_{\mathbf{L}^{t}} + \|g_R\|_{L^{4/3}(\Gamma_R)} + \|g\|_{L^{6/5}(\Omega)}\Big),\\
&\int_\Omega |\nabla\theta|^r\,dx \le cL_\sigma^{r/(1-\sigma)}\big(1 + \|\theta_D\|^2_{W^{1,2}}\big) \quad \forall r\ (1 < r < 3/2),
\end{aligned} \tag{9.125}$$

where L_σ *is the one in* (9.135) *below and* $\sigma = \frac{3-2r}{3-r}$.

First, we look for solutions to the following auxiliary problem:

Problem II-VIA. Let $\zeta > 0$, $\lambda > 0$, $\varepsilon > 0$ and $q \in (\frac{12}{5}, 6)$. Find $(v, \theta) \in \mathbf{V} \times W^{1,2}(\Omega)$ such that $\theta - \theta_D \in W^{1,2}_{\Gamma_D}(\Omega)$ and

$$\begin{cases}
a_0(\theta; v, u) + a_2(v, v, u) + \big\langle \nabla\Phi_\varepsilon(v), u\big\rangle - \Big\langle\Big(1 - \dfrac{\alpha_0\lambda}{\max\{\lambda, \|\theta\|^2_{L^q}\}}\theta\Big) f_\varepsilon, u\Big\rangle \\
\qquad\qquad = \langle f_1, u\rangle \quad \forall u \in \mathbf{V},\\
b_0(\theta; \theta, \varphi) - \dfrac{\zeta}{\max\{\zeta, \|v_\varepsilon\|_{\mathbf{V}}\}}\langle \gamma_\varepsilon(\theta)v, \nabla\varphi\rangle - \Big\langle \alpha_2\mu(\theta)\dfrac{|\mathcal{E}(v)|^2}{1 + \varepsilon|\mathcal{E}(v)|^2}, \varphi\Big\rangle \\
\qquad - \Big\langle \dfrac{\alpha_1\lambda}{\max\{\lambda, \|\theta\|^2_{L^q}\}}\theta f_\varepsilon \cdot v, \varphi\Big\rangle = \langle f_2, \varphi\rangle \quad \forall\varphi \in W^{1,2}_{\Gamma_D}(\Omega),
\end{cases} \tag{9.126}$$

where $f_\varepsilon \in \mathbf{L}^\infty(\Omega)$ is such that $\|f - f_\varepsilon\|_{\mathbf{L}^{t}} \le \varepsilon$.

Theorem 9.5 *There exists a solution* $(v_\varepsilon, \theta_\varepsilon) \in \boldsymbol{V} \times W^{1,2}(\Omega)$ *to Problem II-VIA.*

Proof Let $\mathscr{H} = \mathbf{V} \times W^{1,2}_{\Gamma_D}(\Omega)$. Define an operator $\mathscr{A} : \mathscr{H} \to \mathscr{H}^*$ by

$$\begin{aligned}
\big\langle \mathscr{A}(v, \eta), (u, \phi)\big\rangle &= a_0(\eta + \theta_D; v, u) + a_2(v, v, u) + \big\langle \nabla\Phi_\varepsilon(v), u\big\rangle \\
&\quad - \Big\langle\Big(1 - \frac{\alpha_0\lambda}{\max\{\lambda, \|\eta + \theta_D\|^2_{L^q}\}}(\eta + \theta_D)\Big) f_\varepsilon, u\Big\rangle + b_0(\eta + \theta_D; \eta + \theta_D, \varphi) \\
&\quad - \frac{\zeta}{\max\{\zeta, \|v_\varepsilon\|_{\mathbf{V}}\}}\langle \gamma_\varepsilon(\eta + \theta_D)v, \nabla\varphi\rangle - \Big\langle \alpha_2\mu(\eta + \theta_D)\frac{|\mathcal{E}(v)|^2}{1 + \varepsilon|\mathcal{E}(v)|^2}, \varphi\Big\rangle \\
&\quad - \Big\langle \frac{\alpha_1\lambda}{\max\{\lambda, \|\eta + \theta_D\|^2_{L^q}\}}(\eta + \theta_D) f_\varepsilon \cdot v, \varphi\Big\rangle \\
&\qquad\qquad\qquad\qquad \forall (v, \eta), (u, \phi) \in \mathscr{H}.
\end{aligned}$$

Note that instead of $\frac{\delta}{\max\{\delta, \|\overline{a}_1(v)\|_{\mathbf{V}^*}\}}a_1(v, v, u)$ in Problem I-VIA, for Problem II-VIA the term $a_2(v, v, u)$ is used.

Using

$$
\begin{aligned}
&a_2(v, v, v) = 0,\\
&|a_2(v, v, u)| \le K\|v\|_{\mathbf{V}}^2\|u\|_{\mathbf{V}},\\
&|a_2(v_\varepsilon, v_\varepsilon, v_\varepsilon - u)| \le c\|\nabla v_\varepsilon\|_{\mathbf{L}^2}\|v_\varepsilon\|_{\mathbf{L}^4}\|v_\varepsilon - u\|_{\mathbf{L}^4},
\end{aligned}
$$

respectively, in (9.29), (9.32) and (9.36), we can see that the proof of Theorem 9.2 for Problem I-VIA is valid for Problem II-VIA. Thus, we come to the asserted conclusion. □

As (9.44) let us choose q_0 such that

$$
\frac{1}{q_0} + \frac{1}{t} + \frac{1}{3} \le 1, \quad \frac{12}{5} < q_0 < 3.
$$

Lemma 9.4 *If $|\alpha_0|\sqrt{\lambda} \le 1$, then there exists a parameter ζ such that*

$$
\frac{\zeta}{\max\{\zeta, \|v_\varepsilon\|_{\mathbf{V}}\}} = 1 \tag{9.127}
$$

for all small ε and solutions $(v_\varepsilon, \theta_\varepsilon)$ of (9.126).

Proof Since $|\alpha_0|\sqrt{\lambda} \le 1$, we have

$$
\Big|\Big\langle \frac{\alpha_0\lambda}{\max\{\lambda, \|\theta_\varepsilon\|_{L^{q_0}}^2\}}\theta_\varepsilon f_\varepsilon, u\Big\rangle\Big| \le \frac{c\sqrt{\lambda}}{\max\{\lambda, \|\theta_\varepsilon\|_{L^{q_0}}^2\}}\|\theta_\varepsilon\|_{L^{q_0}}\|f_\varepsilon\|_{\mathbf{L}^t}\|u\|_{\mathbf{L}^6} \le c\|f_\varepsilon\|_{\mathbf{L}^t}\|u\|_{\mathbf{V}^*}. \tag{9.128}
$$

Putting $u = v_\varepsilon$ in the first equation of (9.126), we have

$$
\begin{aligned}
a_0(\theta_\varepsilon; v_\varepsilon, v_\varepsilon) + a_2(v_\varepsilon, v_\varepsilon, v_\varepsilon) + \big\langle \nabla\Phi_\varepsilon(v_\varepsilon), v_\varepsilon\big\rangle - \Big\langle\Big(1 - \frac{\alpha_0\lambda}{\max\{\lambda, \|\theta_\varepsilon\|_{L^{q_0}}^2\}}\theta_\varepsilon\Big) f_\varepsilon, v_\varepsilon\Big\rangle&\\
= \langle f_1, v_\varepsilon\rangle.&
\end{aligned} \tag{9.129}
$$

Taking into account $a_2(v_\varepsilon, v_\varepsilon, v_\varepsilon) = 0$, (9.30) and (9.128), we have from (9.129)

$$
\begin{aligned}
2\mu_0\|v_\varepsilon\|_{\mathbf{V}}^2 \le a_0(\theta_\varepsilon; v_\varepsilon, v_\varepsilon) &\le \Big\langle\Big(1 - \frac{\alpha_0\lambda}{\max\{\lambda, \|\theta_\varepsilon\|_{L^{q_0}}^2\}}\theta_\varepsilon\Big) f_\varepsilon, v\Big\rangle + \langle f_1, v_\varepsilon\rangle\\
&\le c(\|f_\varepsilon\|_{\mathbf{L}^{\frac{6}{5}}} + \|f_\varepsilon\|_{\mathbf{L}^t} + \|f_1\|_{\mathbf{V}^*})\|v_\varepsilon\|_{\mathbf{V}},
\end{aligned}
$$

which implies

$$
\|v_\varepsilon\|_{\mathbf{V}} \le \frac{c}{2\mu_0}(\|f_\varepsilon\|_{\mathbf{L}^t} + \|f_1\|_{\mathbf{V}^*}).
$$

Since $f_\varepsilon \to f$ in $\mathbf{L}^t(\Omega)$, again we may assume that $\|f_\varepsilon\|_{\mathbf{L}^t} \le 1 + \|f\|_{\mathbf{L}^t}$ for all ε. Therefore,

$$
\|v_\varepsilon\|_{\mathbf{V}} \le \frac{c}{2\mu_0}(1 + \|f\|_{\mathbf{L}^t} + \|f_1\|_{\mathbf{V}^*}). \tag{9.130}
$$

Putting $\zeta = \frac{c}{2\mu_0}(1 + \|f\|_{\mathbf{L}^t} + \|f_1\|_{\mathbf{V}^*})$, we come to the asserted conclusion. □

Lemma 9.5 *If* $\max\{|\alpha_0|, |\alpha_1|\}\sqrt{\lambda} \le 1$, *then under the parameter* ζ *by Lemma 9.4 there exists a* λ_2 *independent of* ε *such that*

$$\|\theta_\varepsilon\|_{L^{q_0}} \le \sqrt{\lambda_2}. \tag{9.131}$$

If moreover $1 < r < \frac{3}{2}$, *then*

$$\int_\Omega |\nabla\theta_\varepsilon|^r \, dx \le cL_\sigma^{r/(1-\sigma)}\big(1 + \|\theta_D\|^2_{W^{1,2}}\big), \quad \sigma = \frac{3-2r}{3-r}, \tag{9.132}$$

where L_σ *is the one in* (9.135) *below.*

Proof By Lemma 9.1 (which is valid for the second formula of (9.126)), we have

$$\|\theta_\varepsilon^-\|^2_{W^{1,2}} \le c\big(\|f_\varepsilon\|_{\mathbf{L}^t}\|v_\varepsilon\|_{\mathbf{V}} + \|g_R\|_{L^{4/3}(\Gamma_R)} + \|g\|_{L^{6/5}(\Omega)}\big). \tag{9.133}$$

Using (9.133) and arguing as in (9.61)–(9.89) yield

$$\begin{aligned}
&\int_\Omega \frac{|\nabla\theta_\varepsilon|^2}{(1+(\theta_\varepsilon - d_0)^+)^{1+\sigma}}\, dx \\
&\quad\le c\bigg[\Big(1+\frac{1}{\sigma}\Big)\Big(1+\|v_\varepsilon\|^2_{\mathbf{V}} + \|f_\varepsilon\|_{\mathbf{L}^t}\|v_\varepsilon\|_{\mathbf{V}} + \|g_R\|_{L^{4/3}(\Gamma_R)} + \|g\|_{L^{6/5}(\Omega)}\Big)^2 \\
&\qquad + \|\theta_D\|^2_{W^{1,2}} + \|v_\varepsilon\|^2_{\mathbf{V}} + \Big(\|f_\varepsilon\|_{\mathbf{L}^t}\|v_\varepsilon\|_{\mathbf{V}} + \|g_R\|_{L^{4/3}(\Gamma_R)} + \|g\|_{L^{6/5}(\Omega)}\Big)^2\|v_\varepsilon\|^2_{\mathbf{V}}\bigg].
\end{aligned} \tag{9.134}$$

Combining (9.134) with (9.130), we have

$$\begin{aligned}
&\int_\Omega \frac{|\nabla\theta_\varepsilon|^2}{(1+(\theta_\varepsilon - d_0)^+)^{1+\sigma}}\, dx \\
&\quad\le c\bigg[\Big(1+\frac{1}{\sigma}\Big)\Big(1+(1+\|f\|_{\mathbf{L}^t}+\|f_1\|_{\mathbf{V}^*})^2 + \|f\|_{\mathbf{L}^t}(1+\|f\|_{\mathbf{L}^t}+\|f_1\|_{\mathbf{V}^*}) \\
&\qquad + \|g_R\|_{L^{4/3}(\Gamma_R)} + \|g\|_{L^{6/5}(\Omega)}\Big)^2 + \|\theta_D\|^2_{W^{1,2}} + (1+\|f\|_{\mathbf{L}^t}+\|f_1\|_{\mathbf{V}^*})^2 \\
&\qquad + \Big(\|f\|_{\mathbf{L}^t}(1+\|f\|_{\mathbf{L}^t}+\|f_1\|_{\mathbf{V}^*}) + \|g_R\|_{L^{4/3}(\Gamma_R)} + \|g\|_{L^{6/5}(\Omega)}\Big)^2(1+\|f\|_{\mathbf{L}^t}+\|f_1\|_{\mathbf{V}^*})^2\bigg] \\
&\quad\equiv L_\sigma.
\end{aligned} \tag{9.135}$$

Using (9.135), in the same way as in (9.94) we have

$$\|\theta_\varepsilon\|_{L^{q_0}} \le c\left(L_{\sigma_0}^{1/(1-\sigma_0)}(1+\|\theta_D\|_{W^{1,2}}) + \|\theta_D\|_{W^{1,2}}\right) \equiv \overline{L}_{\sigma_0}, \tag{9.136}$$

where L_{σ_0} is the one with σ_0 instead of σ in L_σ of (9.135). Putting

$$\sqrt{\lambda_2} = \overline{L}_{\sigma_0}, \tag{9.137}$$

we get (9.131).

Now, for $1 < r < \frac{3}{2}$ putting $\sigma = \frac{3-2r}{3-r}$ and repeating the arguments of (9.91)–(9.93), we have (9.132). □

Fixing $\lambda = \lambda_2$, under the condition

$$\max\{|\alpha_0|, |\alpha_1|\}\sqrt{\lambda_2} \leq 1, \tag{9.138}$$

by (9.136) and (9.137) we have

$$\frac{\lambda_2}{\max\{\lambda_2, \|\theta_\varepsilon\|^2_{L^{q_0}}\}} = 1 \quad \forall\varepsilon > 0.$$

Therefore, by virtue of Lemmas 9.4 and 9.5 we have

Theorem 9.6 *If*

$$\max\{|\alpha_0|, |\alpha_1|\}\sqrt{\lambda_2} \leq 1,$$

then there exists a solution $(v_\varepsilon, \theta_\varepsilon) \in \mathbf{V} \times W^{1,2}(\Omega)$ *to the following problem*

$$\begin{cases} a_0(\theta_\varepsilon; v_\varepsilon, u) + a_2(v_\varepsilon, v_\varepsilon, u) + \big\langle \nabla\Phi_\varepsilon(v_\varepsilon), u\big\rangle - \big\langle (1-\alpha_0\theta_\varepsilon) f_\varepsilon, u\big\rangle = \langle f_1, u\rangle \\ \qquad\qquad\qquad\qquad\qquad\qquad\qquad\qquad\qquad\qquad \forall u \in \mathbf{V}, \\ b_0(\theta_\varepsilon; \theta_\varepsilon, \varphi) - \langle \gamma_\varepsilon(\theta_\varepsilon) v_\varepsilon, \nabla\varphi\rangle - \Big\langle \alpha_2\mu(\theta_\varepsilon) \dfrac{|\mathcal{E}(v_\varepsilon)|^2}{1+\varepsilon|\mathcal{E}(v_\varepsilon)|^2}, \varphi\Big\rangle - \langle \alpha_1\theta_\varepsilon f_\varepsilon \cdot v_\varepsilon, \varphi\rangle \\ \qquad\qquad\qquad\qquad\qquad\qquad\qquad\qquad = \langle f_2, \varphi\rangle \quad \forall\varphi \in W^{1,2}_{\Gamma_D}(\Omega), \end{cases}$$

$$\theta_\varepsilon - \theta_D \in W^{1,2}_{\Gamma_D}(\Omega),$$

and the solution satisfies:

$$\begin{aligned} &\|v_\varepsilon\|_{\mathbf{V}} \leq \frac{c}{2\mu_0}(1 + \|f\|_{\mathbf{L}^t} + \|f_1\|_{\mathbf{V}^*}), \\ &\|\theta_\varepsilon^-\|_{W^{1,2}_{\Gamma_D}(\Omega)} \leq c\Big(\|f\|_{\mathbf{L}^t}\|v_\varepsilon\|_{\mathbf{V}} + \|g_R\|_{L^{4/3}(\Gamma_R)} + \|g\|_{L^{6/5}(\Omega)}\Big), \\ &\int_\Omega |\nabla\theta|^r\, dx \leq cL_\sigma^{r/(1-\sigma)}\big(1 + \|\theta_D\|^2_{W^{1,2}}\big) \quad \forall r, 1 < r < \frac{3}{2}, \end{aligned}$$

where $\sigma = \frac{3-2r}{3-r}$.

Now repeating the arguments as in Sect. 9.3.3 with the solutions of Theorem 9.6, we complete the proof of Theorem 9.4. □

9.5 Bibliographical Remarks

The content of Chap. 9 is taken from [1].

Several papers concerning (9.1) have appeared.

In [2] the problem with $\alpha_0 = \alpha_1 = 0$ was studied under homogeneous Dirichlet boundary conditions for velocity and mixture of non-homogeneous Dirichlet condition and homogeneous Neumann condition for temperature.

In [3] when $|\alpha_0|$, $|\alpha_1|$ are small enough or $\alpha_0 \neq 0$, $\frac{\alpha_1}{\alpha_0} > 1$, the existence of a solution to the problems was shown under homogeneous Dirichlet boundary conditions for velocity and mixture of non-homogeneous Dirichlet and homogeneous Neumann conditions for temperature.

In [4] for problem on an open channel domain when $\alpha_0 = \alpha_1 = 0$, local solvability was studied under mixture of Dirichlet boundary condition of velocity and the free outflow condition for fluid and mixture of non-homogeneous Dirichlet and Neumann boundary conditions for temperature.

In [5] a modified problem on an open channel domain, where the buoyancy term $(1 - \alpha_0\theta)f$ is changed by $\rho(\theta)f$, $0 < \rho(\theta) < \rho_1(\text{const})$, $\alpha_1 = 0$ and the viscosity, specific heat and thermal conductivity are independent of temperature, was considered. For the problem under mixture of Dirichlet boundary condition of velocity and the free outflow condition for fluid and mixture of non-homogeneous Dirichlet and homogeneous Neumann boundary conditions for temperature it was proved that if the data of problem are small enough, then there exists a unique strong solution.

In [6] when the viscosity, specific heat and thermal conductivity are independent of temperature, the steady problem on 2-D bounded domain was considered. When the body force is small enough, the existence of a strong solution was proved under mixture of Dirichlet boundary condition of velocity, tangent stress and stress condition for fluid and mixture of non-homogeneous Dirichlet and homogeneous Neumann boundary conditions for temperature.

In several papers the non-Newtonian fluid were studied.

In [7] for non-Newtonian fluid with $\alpha_0 = \alpha_1 = 0$, under homogeneous Dirichlet boundary condition for velocity and mixture of homogeneous Dirichlet and Robin conditions for temperature existence of a solution was studied.

In [8] for the problems of a generalized Newtonian fluid with $\alpha_0 = \alpha_1 = 0$ regularity of weak solutions was proved under homogeneous Dirichlet boundary conditions of velocity and temperature. In [9] for the problems of a non-Newtonian fluid with $\alpha_0 = \alpha_1 = 0$ existence of a solutions was proved under homogeneous Dirichlet boundary conditions for velocity and temperature on a portion of boundary and a generalized Navier slip and Robin conditions for velocity and temperature on another portion of boundary.

In [10] for the problem of a non-Newtonian fluid with heat sources allowed in L^1 and even as measures, under homogeneous Dirichlet boundary conditions for velocity and Robin condition for temperature the existence of a distributional solution was shown for sufficiently small data.

But all results above exclude Newtonian fluid owing to conditions for nonlinear terms for strain.

References

1. T. Kim, D. Cao, Mixed boundary value problems of the system for steady flow of heat-conducting incompressible viscous fluids with dissipative heating. Meth. Appl. Anal. **27**, 87–124 (2020)
2. J. Naumann, Existence of weak solutions to the equations of stationary motion of heat-conducting Incompressible viscous fluids. Progress in Nonlinear Differential Equations and Their Applications, vol. 66 (Birkhäuser, 2005), pp. 373–390
3. J. Naumann, M. Pokorny, J. Wolf, On the existence of weak solutions to the equations of steady flow of heat-conducting fluids with dissipative heating. Nonlinear Anal. RWA **13**, 1600–1620 (2012)
4. M. Beneš, P. Kučera, On the Navier-Stokes flows for heat-conducting fluids with mixed boundary conditions. J. Math. Anal. Appl. **389**, 769–780 (2012)
5. M. Beneš, A note on regularity and uniqueness of natural convection with effects of viscous dissipation in 3D open channels. Z. Angew. Math. Phys. **65**, 961–975 (2014)
6. M. Beneš, A note on the regularity of thermally coupled viscous flows with critical growth in nonsmooth domains. Math. Meth. Appl. Sci. **36**, 1290–1300 (2013)
7. L. Consiglieri, Stationary weak solutions for a class of non-Newtonian fluids with energy transfer. Int. J. Non-Linear Mech. **32**, 961–972 (1997)
8. L. Consiglieri, T. Shilkin, Regularity of stationary weak solutions in the theory of generalized Newtonian fluid with energy transfer. J. Math. Sci. **115**, 2771–2788 (2003)
9. L. Consiglieri, A $(p-q)$ coupled system in elliptic nonlinear problems with nonstandard boundary conditions. J. Math. Anal. Appl. **340**, 183–196 (2008)
10. T. Roubíček, Steady-state buoyancy-driven viscous flow with measure data. Mathematica Bohemica **126**, 493–504 (2001)

Chapter 10
The Non-steady Equations for Heat-Conducting Fluids

In this chapter we are concerned with a non-steady system for motion of incompressible Newtonian heat-conducting fluids with mixed boundary conditions. The boundary condition for fluid is the case of total pressure and the boundary conditions for temperature may include Dirichlet, Neumann and Robin conditions together. On the basis of results of Sect. 3.1, we get a variational formulation for the problem. The variational formulation consists of a time-dependent variational inequality for velocity due to the boundary conditions of friction type and a variational equation for temperature.

Then we prove the existence of a solution to the problem. It is proved that if the buoyancy effect and energy dissipation effect due to expansion are small enough in accordance with data of problem, then there exists a solution with "defect measure".

10.1 Problem and Variational Formulation

10.1.1 Problem and Assumption

We are concerned with the following non-steady problem

$$\begin{cases} \dfrac{\partial v}{\partial t} - 2\nabla\cdot\big(\mu(\theta)\mathcal{E}(v)\big) + (v\cdot\nabla)v + \nabla p = (1-\alpha_0\theta)f, \\ \operatorname{div} v = 0, \\ \dfrac{\partial \theta}{\partial t} - \nabla\cdot(\kappa(\theta)\nabla\theta) + v\cdot\nabla\theta - \alpha_2\mu(\theta)\mathcal{E}(v):\mathcal{E}(v) = \alpha_1\theta f\cdot v + g, \\ v(0) = v_0,\ \ \theta(0) = \theta_0. \end{cases} \tag{10.1}$$

T. Kim and D. Cao, *Equations of Motion for Incompressible Viscous Fluids*, Advances in Mathematical Fluid Mechanics,
https://doi.org/10.1007/978-3-030-78659-5_10

Due to the dissipation of energy $\mu(\theta)\mathcal{E}(v):\mathcal{E}(v)$, it is more difficult to study the system (10.1) than the Boussinesq system.

Let Ω be as in Chap. 8. (See Remark 6.1.)

For temperature we are concerned with the boundary conditions

$$\begin{aligned}&(1)\quad \theta|_{\Gamma_D}=0,\\&(2)\quad \Big(\kappa(\theta)\frac{\partial\theta}{\partial n}+\beta(x)\theta\Big)\Big|_{\Gamma_R}=g_R(t,x),\\&\qquad \beta(x),\ g_R(t,x)-\text{given functions on }\Gamma_R,\ (0,T)\times\Gamma_R.\end{aligned}\tag{10.2}$$

The boundary condition for fluid is as in (7.4).

Let $\mathbf{V}$, $K(\Omega)$ be the same as in (5.5) and

$$\begin{aligned}&H:\text{completion in }\mathbf{L}^2(\Omega)\text{ of }\mathbf{V},\\&H_K:\text{closure in }\mathbf{L}^2(\Omega)\text{ of }K(\Omega),\\&K(Q)=\{u\in L^2(0,T;\mathbf{V}):u'\in L^2(0,T;\mathbf{V}^*);\,u_n|_{\Gamma_{10}}\ge 0,\ u_n|_{\Gamma_{11}}\le 0\},\\&(Q)=\{u\in\mathbf{C}^2(\bar Q):\operatorname{div}u=0,\,u|_{\Sigma_1}=0,\,u_\tau|_{(\Sigma_2\cup\Sigma_4\cup\Sigma_7)}=0,\,u\cdot n|_{(\Sigma_3\cup\Sigma_5)}=0\},\\&W^{1,p}_{\Gamma_D}(\Omega)=\{y\in W^{1,p}(\Omega):y|_{\Gamma_D}=0\},\\&C^1_{\Gamma_D}(\Omega)=\{y\in C^1(\bar\Omega):y|_{\Gamma_D}=0\}.\end{aligned}$$

We use inner products in (7.8) for $\mathbf{V}$ and $W^{1,2}_{\Gamma_D}(\Omega)$.

We use the following assumption.

Assumption 10.1 Assume the followings.

(1) $\Gamma_1\neq\emptyset$, $\Gamma_D\neq\emptyset$ and

$$\Gamma_R\subset\left(\cup_{i=1,3,5,8}\Gamma_i\right).\tag{10.3}$$

(2) (3) of Assumption 5.1 holds.

(3) For the functions of (10.1) $f\in L^\infty(Q)$, $g\in L^1\big(0,T;(W^{1,2}_{\Gamma_D})^*\big)$ and

$$\begin{aligned}&\mu\in C(\mathbb{R}),\ 0<\mu_0\le\mu(\xi)\le\mu_1<\infty\quad\forall\xi\in\mathbb{R},\\&\kappa\in C(\mathbb{R}),\ 0<\kappa_0\le\kappa(\xi)\le\kappa_1<\infty\quad\forall\xi\in\mathbb{R},\\&v_0\in H_K,\ \theta_0\in L^1(\Omega).\end{aligned}\tag{10.4}$$

(4) For the functions of (10.2), (7.4)

$$
\begin{aligned}
&g_R \in L^1(0, T; L^{4/3}(\Gamma_R)),\\
&\beta_1 \geq \beta(x) \geq 0, \;\; \beta_1 - \text{a constant}, \;\; \beta(x) - \text{measurable},\\
&\phi_i \in L^2(0, T; H^{-\frac{1}{2}}(\Gamma_i)), i = 2, 4, 7, \;\; \phi_i \in L^2(0, T; \mathbf{H}^{-\frac{1}{2}}(\Gamma_i)), i = 3, 5, 6,\\
&\alpha_{ij} \in L^\infty(\Gamma_5).
\end{aligned} \tag{10.5}
$$

10.1.2 Variational Formulation for Problem

In this section we first give variational formulations for the problem above.

Taking $(v \cdot \nabla)v = \mathrm{rot}\, v \times v + \frac{1}{2}\mathrm{grad}|v|^2$ into account, by (7.9)–(7.11), (8.8), we can see that smooth solutions (v, p, θ) of problem (10.1), (10.2), (7.4) satisfy the following.

$$
\begin{cases}
\left(\dfrac{\partial v}{\partial t}, u\right) + 2(\mu(\theta)\mathcal{E}(v), \mathcal{E}(u)) + \langle \mathrm{rot} v \times v, u\rangle + 2(\mu(\theta)k(x)v, u)_{\Gamma_2}\\
\qquad + 2(\mu(\theta)S\tilde{v}, \tilde{u})_{\Gamma_3} + 2(\alpha(x)v, u)_{\Gamma_5} + (\mu(\theta)k(x)v, u)_{\Gamma_7}\\
\qquad - 2(\mu(\theta)\varepsilon_{n\tau}(v), u)_{\Gamma_8} + (p + \dfrac{1}{2}|v|^2 - 2\mu(\theta)\varepsilon_{nn}(v), u_n)_{\Gamma_9\cup\Gamma_{10}\cup\Gamma_{11}}\\
\quad = \langle (1-\alpha_0\theta)f, u\rangle + \displaystyle\sum_{i=2,4,7} \langle \phi_i, u_n\rangle_{\Gamma_i} + \sum_{i=3,5,6} \langle \phi_i, u\rangle_{\Gamma_i} \quad \forall u \in \mathbf{V},\\
\left(\dfrac{\partial \theta}{\partial t}, \varphi\right) + (\kappa(\theta)\nabla\theta, \nabla\varphi) - (\theta v, \nabla\varphi) - (\alpha_2\mu(\theta)|\mathcal{E}(v)|^2, \varphi) + (\beta\theta, \varphi)_{\Gamma_R}\\
\qquad - (\alpha_1\theta f \cdot v, \varphi) = \langle g_R, \varphi\rangle_{\Gamma_R} + \langle g, \varphi\rangle \quad \forall \varphi \in W^{1,\infty}_{\Gamma_D}(\Omega),\\
|\sigma^t_\tau(\theta, v)| \leq g_\tau, \;\; \sigma^t_\tau(\theta, v)\cdot v_\tau + g_\tau|v_\tau| = 0 \;\; \text{on } \Gamma_8,\\
|\sigma^t_n(\theta, v, p)| \leq g_n, \;\; \sigma^t_n(\theta, v, p)v_n + g_n|v_n| = 0 \;\; \text{on } \Gamma_9,\\
\sigma^t_n(\theta, v, p) + g_{+n} \geq 0, \;\; (\sigma^t_n(\theta, v, p) + g_{+n})v_n = 0 \;\; \text{on } \Gamma_{10},\\
\sigma^t_n(\theta, v, p) - g_{-n} \leq 0, \;\; (\sigma^t_n(\theta, v, p) - g_{-n})v_n = 0 \;\; \text{on } \Gamma_{11}.
\end{cases} \tag{10.6}
$$

Define $a_0(\theta; \cdot, \cdot)$, $a_1(\cdot, \cdot, \cdot)$ and $f_1 \in \mathbf{V}^*$ by

$$
\begin{aligned}
&a_0(\theta; w, u) = 2(\mu(\theta)\mathcal{E}(w), \mathcal{E}(u)) + 2(\mu(\theta)k(x)w, u)_{\Gamma_2} + 2(\mu(\theta)S\tilde{w}, \tilde{u})_{\Gamma_3}\\
&\qquad + 2(\alpha(x)w, u)_{\Gamma_5} + (\mu(\theta)k(x)w, u)_{\Gamma_7} \quad \forall w, u \in \mathbf{V}, \theta \in W^{1,2}(\Omega),\\
&a_1(v, u, w) = \langle \mathrm{rot}\, v \times u, w\rangle \quad \forall v, u, w \in \mathbf{V},\\
&\langle f_1, u\rangle = \sum_{i=2,4,7} \langle \phi_i, u_n\rangle_{\Gamma_i} + \sum_{i=3,5,6} \langle \phi_i, u\rangle_{\Gamma_i} \quad \forall u \in \mathbf{V}.
\end{aligned} \tag{10.7}
$$

Define $b_0(\tilde{\theta}; \cdot, \cdot)$ and $f_2 \in (W^{1,2}_{\Gamma_D}(\Omega))^*$ by

$$
\begin{aligned}
&b_0(\tilde{\theta};\theta,\varphi) = (\kappa(\tilde{\theta})\nabla\theta,\nabla\varphi) + (\beta(x)\theta,\varphi)_{\Gamma_R} \quad \forall\theta,\tilde{\theta}\in W^{1,2}(\Omega),\ \varphi\in W^{1,2}_{\Gamma_D}(\Omega),\\
&\langle f_2,\varphi\rangle = \langle g_R,\varphi\rangle_{\Gamma_R} + \langle g,\varphi\rangle \quad \forall\varphi\in W^{1,2}_{\Gamma_D}(\Omega).
\end{aligned}
\tag{10.8}
$$

By (10.4) and (10.5),

$$
f_1\in L^2(0,T;\mathbf{V}^*),\quad f_2\in L^1(0,T;(W^{1,2}_{\Gamma_D})^*). \tag{10.9}
$$

Then, taking into account

$$
\sigma^t_\tau(\theta,v) = 2\mu(\theta)\varepsilon_{n\tau}(v),\quad \sigma^t_n(\theta,v,p) = -(p+\frac{1}{2}|v|^2) + 2\mu(\theta)\varepsilon_{nn}(v)
$$

and (10.6), we introduce the following variational formulation for problem (10.1), (10.2), (7.4).

Problem VE. Find $v\in K(Q)$, $\theta\in L^\infty(0,T;L^1(\Omega))\cap L^r(0,T;W^{1,r}_{\Gamma_D}(\Omega))$ for all $r\in[1,5/4)$ and $(\sigma^t_\tau,\sigma^t_n,\sigma^t_{+n},\sigma^t_{-n})\in \mathbf{L}^2_\tau(\Gamma_8)\times L^2(\Gamma_9)\times H^{-\frac{1}{2}}(\Gamma_{10})\times H^{-\frac{1}{2}}(\Gamma_{11})$ for a.e. $t\in(0,T)$ such that $v(0)=v_0$, $\theta(0)=\theta_0$ and

$$
\begin{cases}
\left\langle \dfrac{\partial v}{\partial t},u\right\rangle + a_0(\theta;v,u) + a_1(v,v,u) - (\sigma^t_\tau,u_\tau)_{\Gamma_8} - (\sigma^t_n,u_n)_{\Gamma_9}\\
\qquad - \langle\sigma^t_{+n},u_n\rangle_{\Gamma_{10}} - \langle\sigma^t_{-n},u_n\rangle_{\Gamma_{11}} - \langle f-\alpha_0\theta f,u\rangle = \langle f_1,u\rangle \quad \forall u\in\mathbf{V},\\
\left\langle \dfrac{\partial\theta}{\partial t},\varphi\right\rangle + b_0(\theta;\theta,\varphi) - \langle\theta v,\nabla\varphi\rangle - \langle\alpha_2\mu(\theta)|\mathcal{E}(v)|^2,\varphi\rangle - \langle\alpha_1\theta f\cdot v,\varphi\rangle\\
\qquad\qquad\qquad = \langle f_2,\varphi\rangle \quad \forall\varphi\in W^{1,\infty}_{\Gamma_D}(\Omega),\\
|\sigma^t_\tau|\le g_\tau,\ \sigma^t_\tau\cdot v_\tau + g_\tau|v_\tau| = 0 \quad \text{on } \Gamma_8,\\
|\sigma^t_n|\le g_n,\ \sigma^t_n v_n + g_n|v_n| = 0 \quad \text{on } \Gamma_9,\\
\sigma^t_{+n} + g_{+n}\ge 0,\ \langle\sigma^t_{+n}+g_{+n},v_n\rangle_{\Gamma_{10}} = 0 \quad \text{on } \Gamma_{10},\\
\sigma^t_{-n} - g_{-n}\le 0,\ \langle\sigma^t_{-n}-g_{-n},v_n\rangle_{\Gamma_{11}} = 0 \quad \text{on } \Gamma_{11},
\end{cases}
\tag{10.10}
$$

where $\mathbf{L}^2_\tau(\Gamma_8)$ is the subspace of $\mathbf{L}^2(\Gamma_8)$ consisting of functions such that $(u,n)_{\mathbf{L}^2(\Gamma_8)}=0$.

Remark 10.1 As in Remark 8.3, for solutions smooth enough Problem VE is equivalent to the problem (10.1), (10.2), (7.4).

We will find another variational formulation consisting of a variational inequality and a variational equation, which is equivalent to Problem VE if the solution is smooth enough (see Remark 10.2).

For fixed θ, let us consider the problem

$$\begin{cases} \left\langle \dfrac{\partial v}{\partial t}, u \right\rangle + a_0(\theta; v, u) + a_1(v, v, u) - (\sigma_\tau, u_\tau)_{\Gamma_8} - (\sigma_n, u_n)_{\Gamma_9} \\ \qquad - \langle \sigma_{+n}, u_n \rangle_{\Gamma_{10}} - \langle \sigma_{-n}, u_n \rangle_{\Gamma_{11}} - \langle (f - \alpha_0 \theta f, u \rangle = \langle f_1, u \rangle, \\ |\sigma_\tau^t| \le g_\tau, \ \sigma_\tau^t \cdot v_\tau + g_\tau |v_\tau| = 0 \quad \text{on } \Gamma_8, \\ |\sigma_n^t| \le g_n, \ \sigma_n^t v_n + g_n |v_n| = 0 \quad \text{on } \Gamma_9, \\ \sigma_{+n}^t + g_{+n} \ge 0, \ \langle \sigma_{+n}^t + g_{+n}, v_n \rangle_{\Gamma_{10}} = 0 \quad \text{on } \Gamma_{10}, \\ \sigma_{-n}^t - g_{-n} \le 0, \ \langle \sigma_{-n}^t - g_{-n}, v_n \rangle_{\Gamma_{11}} = 0 \quad \text{on } \Gamma_{11}. \end{cases} \tag{10.11}$$

Subtracting the first formula of (10.11) with $u = v$ from the first one of (10.11), we get

$$\begin{aligned} &\left\langle \frac{\partial v}{\partial t}, u - v \right\rangle + a_0(\theta; v, u - v) + a_1(v, v, u - v) - (\sigma_\tau^t, u_\tau - v_\tau)_{\Gamma_8} - (\sigma_n^t, u_n - v_n)_{\Gamma_9} \\ &- \langle \sigma_{+n}^t, u_n - v_n \rangle_{\Gamma_{10}} - \langle \sigma_{-n}^t, u_n - v_n \rangle_{\Gamma_{11}} - \langle f - \alpha_0 \theta f, u - v \rangle = \langle f_1, u - v \rangle \quad \forall u \in \mathbf{V}. \end{aligned} \tag{10.12}$$

Let $\Phi : \mathbf{V} \to \overline{\mathbb{R}}$ be the functional defined by (5.22). Then, the functional Φ is proper, convex, lower weak semi-continuous and nonnegative.

Define a functional $\Psi(u)$ by

$$\Psi(u) = \begin{cases} \displaystyle\int_0^T \Phi(u(t))\, dt & if\ \Phi(u(t)) \in L^1(0, T), \\ +\infty & otherwise. \end{cases} \tag{10.13}$$

In the same way as in Problem I of Chap. 5, we get from (10.7) and (10.12)

$$\begin{aligned} \left\langle \frac{\partial v}{\partial t}, u - v \right\rangle + a_0(\theta; v, u - v) + a_1(v, v, u - v) + \Phi(u) - \Phi(v) \\ \ge \langle (1 - \alpha_0 \theta) f, u - v \rangle + \langle f_1, u - v \rangle. \end{aligned} \tag{10.14}$$

If $u - v \in (Q)$, then

$$\begin{aligned} \int_0^T \left\langle \frac{\partial v}{\partial t}, u - v \right\rangle dt &= \int_0^T \left\langle \frac{\partial (v - u)}{\partial t}, v - u \right\rangle dt + \int_0^T \left\langle \frac{\partial u}{\partial t}, u - v \right\rangle dt \\ &= \int_0^T \left\langle \frac{\partial u}{\partial t}, u - v \right\rangle dt + \frac{1}{2}(\|v_0 - u(0)\|^2 - \|v(T) - u(T)\|^2). \end{aligned} \tag{10.15}$$

Define operators $A(\theta) : \mathbf{V} \to \mathbf{V}^*$ and $B : \mathbf{V} \times \mathbf{V} \to \mathbf{V}^*$, respectively, by

$$\begin{aligned} &\langle A(\theta) v, u \rangle = a_0(\theta; v, u) \quad \forall w, u \in \mathbf{V}, \\ &\langle B(v, u), w \rangle = a_1(v, u, w) \quad \forall v, u, w \in \mathbf{V}. \end{aligned} \tag{10.16}$$

If $v \in L^2(0,T;\mathbf{V}) \cap L^\infty(0,T;\mathbf{L}^2)$, then $v \in L^4(0,T;\mathbf{L}^3)$ (see (1.21)) and $B(v,v) \in L^{\frac{4}{3}}(0,T;\mathbf{V}^*)$. Thus, when $u \in L^4(0,T;\mathbf{V})$, the integral $\int_0^T \langle B(v,v), u-v\rangle\, dt$ is meaningful since $\langle B(v,v), v\rangle = 0$.

Combining (10.14), (10.15), and neglecting $\frac{1}{2}(\|w(T)-u(T)\|^2)$, we come to the following variational inequality corresponding to problem (10.11) (see (6.11), or (1.7) of [1], (2.1) of [2]).

Find $v \in L^2(0,T;\mathbf{V}) \cap L^\infty(0,T;H)$ such that

$$\begin{aligned}&\int_0^T \langle u' + A(\theta)v(t) + B(v(t),v(t)) - (1-\alpha_0\theta)f - f_1, u(t)-v(t)\rangle\, dt + \Psi(u) - \Psi(v)\\ &\qquad \geq -\frac{1}{2}\|v_0 - u(0)\|^2 \quad \forall u \in L^4(0,T;\mathbf{V}) \text{ with } u' \in L^2(0,T;\mathbf{V}^*).\end{aligned} \tag{10.17}$$

If $(v,\theta,\sigma_\tau,\sigma_n,\sigma_{+n},\sigma_{-n})$ is a solution of Problem VE, then (v,θ) satisfies the problem above.

Therefore, we have the following variational formulation of the problem which consists of a variational inequality for velocity and a variational equation for temperature.

Problem VI. Find $(v,\theta) \in \big(L^\infty(0,T;H) \cap L^2(0,T;\mathbf{V})\big) \times \big(L^\infty(0,T;L^1(\Omega)) \cap L^r(0,T;W^{1,r}_{\Gamma_D})\big)$ $\forall r \in [1,5/4)$ such that

$$\begin{cases}\displaystyle\int_0^T \langle u' + A(\theta)v(t) + B(v(t),v(t)) - (1-\alpha_0\theta)f - f_1, u(t)-v(t)\rangle\, dt + \Psi(u) - \Psi(v)\\ \qquad \geq -\dfrac{1}{2}\|v_0 - u(0)\|^2 \quad \forall u \in L^4(0,T;\mathbf{V}) \text{ with } u' \in L^2(0,T;\mathbf{V}^*),\\ \displaystyle\int_0^T \Big[\Big\langle -\theta, \frac{\partial\varphi}{\partial t}\Big\rangle + b_0(\theta;\theta,\varphi) - \langle\theta v, \nabla\varphi\rangle - \langle\alpha_2\mu(\theta)|\mathcal{E}(u)|^2, \varphi\rangle - \langle\alpha_1\theta f\cdot v, \varphi\rangle\\ \qquad - \langle f_2,\varphi\rangle\Big]\, dt = \langle\theta_0(x), \varphi(x,0)\rangle \quad \forall\varphi \in C^1(0,T;C^1_{\Gamma_D}(\Omega)) \text{ with } \varphi(\cdot,T) = 0.\end{cases} \tag{10.18}$$

Remark 10.2 If the solutions to Problem VI is smooth such that $v \in L^2(0,T;\mathbf{V})$, $v' \in L^2(0,T;\mathbf{V}^*)$, then v satisfies (10.14), which is equivalent to

$$\begin{aligned}&\langle v'(t) + A(\theta)v(t) + B(v(t),v(t)) - (1-\alpha_0\theta)f - f_1, u - v(t)\rangle + \Phi(u) - \Phi(v(t)) \geq 0\\ &\qquad\qquad \text{for a.e. } t \in [0,T], \quad \forall u \in K(\Omega)\end{aligned} \tag{10.19}$$

(Remark, p. 114 of [1]). In (10.14) putting $\langle F_1, u-v\rangle = \big\langle -\frac{\partial v}{\partial t} - (1-\alpha_0\theta)f - f_1, u-v\big\rangle$, by Theorem 5.4 we can get the existence of $(\sigma_\tau,\sigma_n,\sigma_{+n},\sigma_{-n}) \in \mathbf{L}^2_\tau(\Gamma_8) \times L^2(\Gamma_9) \times H^{-1/2}(\Gamma_{10}) \times H^{-1/2}(\Gamma_{11})$ for a.e. $t \in (0,T)$ such that $(v,\theta,\sigma_\tau,\sigma_n,\sigma_{+n},\sigma_{-n})$ satisfies (10.10).

It is desirable to get a solution satisfying (10.18), but we can only prove the existence of (v, θ) with a "defect measure" in the second equation of (10.18)

The main result of this chapter is the following

Theorem 10.1 *Let Assumption 10.1 be satisfied. If* $|\alpha_0| + |\alpha_1|$ *is small enough in accordance with the date of the problem, then there exists* $(v, \theta) \in \big(L^\infty(0, T; H) \cap L^2(0, T; \mathbf{V})\big) \times \big(L^\infty(0, T; L^1(\Omega)) \cap L^r(0, T; W^{1,r}_{\Gamma_D})\big)$ $\forall r \in [1, 5/4)$ *and a Radon measure* μ *such that*

$$\begin{cases} \displaystyle\int_0^T \langle u' + A(\theta)v(t) + B(v(t), v(t)) - (1 - \alpha_0\theta) f - f_1, u(t) - v(t)\rangle\, dt + \Psi(u) - \Psi(w) \\ \qquad\qquad \displaystyle \geq -\frac{1}{2}\|v_0 - u(0)\|^2 \quad \forall u \in L^4(0, T; \mathbf{V}) \text{ with } u' \in L^2(0, T; \mathbf{V}^*), \\ \displaystyle\int_0^T \Big[\Big\langle -\theta, \frac{\partial\varphi}{\partial t}\Big\rangle + b_0(\theta; \theta, \varphi) - \langle\theta v, \nabla\varphi\rangle - \langle\alpha_2\mu(\theta)|\mathcal{E}(u)|^2, \varphi\rangle - \langle\alpha_1\theta f \cdot v, \varphi\rangle \\ \qquad\qquad \displaystyle - \langle f_2, \varphi\rangle\Big]\, dt = \langle\theta_0(x), \varphi(x, 0)\rangle + \int_0^T \varphi\, d\mu \\ \qquad\qquad \forall\varphi \in C^1(0, T; C^1_{\Gamma_D}(\Omega)) \text{ with } \varphi(\cdot, T) = 0. \end{cases} \tag{10.20}$$

Moreover, (v, θ) *satisfies the estimate*

$$\begin{aligned} &\sup_{t\in[0,T]} \|v(t)\|^2 + \|v\|^2_{L^2(0,T;\mathbf{V})} + ess\ \sup_{t\in[0,T]} \|\theta(t)\|_{L^1(\Omega)} + \int_Q |\nabla\theta|^r\, dx dt + \int_Q \frac{|\nabla\theta|^2}{(1 + |\theta|)^{1+\delta}}\, dx dt \\ &\quad \leq C_1\Big(\|v_0\|, \|f\|_\infty, \sum_{i=2,4,7} \|\phi_i\|_{L_2(0,T;H^{-\frac{1}{2}}(\Gamma_i))'}, \sum_{i=3,5,6} \|\phi_i\|_{L_2(0,T;H^{-\frac{1}{2}}(\Gamma_i))'}, \\ &\qquad \delta, |\alpha|, \|\theta_0\|_{L^1(\Omega)}, \|g_R\|_{L^1(0,T;L^{4/3}(\Gamma_R))}, \|g\|_{L^1(0,T;(W^{1,2}_{\Gamma_D})^*)}\Big) \\ &\qquad \forall 1 \leq r < \frac{5}{4}, \quad \forall 0 < \delta < \frac{5 - 4r}{3}. \end{aligned} \tag{10.21}$$

10.2 Existence of a Solution

10.2.1 Existence of a Solution to an Approximate Problem

We first consider a problem approximating (10.18).

For every $0 < \varepsilon < 1$, let Φ_ε be the Moreau-Yosida approximation of Φ and $\nabla\Phi_\varepsilon$ be Fréchet derivative of Φ_ε.

Let $v_0 \in \mathbf{W}^{1,6}(\Omega) \cap K(\Omega)$, $\theta_0 \in W^{1,6}_{\Gamma_D}(\Omega)$ and $0 < \varepsilon \le 1$. Also assume that $g_R \in L^2(0,T; L^{4/3}(\Gamma_R))$, $g \in L^2(0,T;(W^{1,2}_{\Gamma_D})^*)$, and so $f_2 \in L^2(0,T;(W^{1,2}_{\Gamma_D})^*)$.

Problem VEA. Find $v \in L^6(0,T;\mathbf{V}) \cap C([0,T];H)$, $\theta \in L^2(0,T;W^{1,2}_{\Gamma_D}) \cap C([0,T];L^2(\Omega))$ such that

$$\begin{cases} \Big\langle \dfrac{\partial v}{\partial t}, u\Big\rangle + 2\Big\langle [\mu(\theta)+\varepsilon\|v\|^4_{\mathbf{V}}]\mathcal{E}(v), \mathcal{E}(u)\Big\rangle + \langle \mathrm{rot} v \times v, u\rangle + 2(\mu(\theta)k(x)v,u)_{\Gamma_2} \\ \qquad + 2(\mu(\theta)S\tilde{v},\tilde{u})_{\Gamma_3} + 2(\alpha(x)v,u)_{\Gamma_5} + (\mu(\theta)k(x)v,u)_{\Gamma_7} + \langle \nabla\Phi_\varepsilon(v(t)), u\rangle \\ = \Big\langle (1-\dfrac{\alpha_0\theta}{1+\varepsilon\theta^2}) f, u\Big\rangle + \langle f_1, u\rangle \qquad \forall u \in L^6(0,T;\mathbf{V}), \\ \Big\langle \dfrac{\partial \theta}{\partial t}, \varphi\Big\rangle + (\kappa(\theta)\nabla\theta, \nabla\varphi) + (\beta(x)\theta,\varphi)_{\Gamma_R} - \langle v\theta, \nabla\varphi\rangle - \Big\langle \alpha_2\mu(\theta)\dfrac{|\mathcal{E}(v)|^2}{1+\varepsilon|\mathcal{E}(v)|^2}, \varphi\Big\rangle \\ \qquad = \Big\langle \dfrac{\alpha_1\theta}{1+\varepsilon\theta^2} f\cdot v, \varphi\Big\rangle + \langle f_2, \varphi\rangle \qquad \forall \varphi \in L^2(0,T;W^{1,2}_{\Gamma_D}(\Omega)), \\ v(0) = v_0, \quad \theta(0) = \theta_0. \end{cases} \tag{10.22}$$

Let

$$\mathcal{V} := L^6(0,T;\mathbf{V}) \times L^2(0,T;W^{1,2}_{\Gamma_D}),$$
$$\|(\cdot,\cdot)\|_{\mathcal{V}} = \Big(\|\cdot\|^2_{L^6(0,T;\mathbf{V})} + \|\cdot\|^2_{L^2(0,T;W^{1,2}_{\Gamma_D})}\Big)^{1/2}.$$

Then

$$\mathcal{V}^* = L^{6/5}(0,T;\mathbf{V}^*) \times L^2\big(0,T;(W^{1,2}_{\Gamma_D})^*\big),$$
$$\|(\cdot,\cdot)\|_{\mathcal{V}^*} = \Big(\|\cdot\|^2_{L^{6/5}(0,T;\mathbf{V}^*)} + \|\cdot\|^2_{L^2\left(0,T;(W^{1,2}_{\Gamma_D})^*\right)}\Big)^{1/2}.$$

Theorem 10.2 *There exists a solution $(v_\varepsilon, \theta_\varepsilon) \in \mathcal{V}$ to Problem VEA.*

Proof In (10.22) let us make changes of the unknown functions by $v = e^{k_1 t}\hat{w}, \theta = \hat{\theta}$, where k_1 is a constant to be determined later (in (10.37)). Owing to the changes, differently from Ch. 9 we need not assume that $\Gamma_{2j}, \Gamma_{3j}, \Gamma_{7j}$ are convex and the matrix α is positive. Then we have the following problem equivalent to Problem VEA.

Problem VEA'. Find $\hat{w} \in L^6(0,T;\mathbf{V}) \cap C([0,T];H)$, $\hat{\theta} \in L^2(0,T;W^{1,2}_{\Gamma_D}) \cap C([0,T];L^2(\Omega))$ such that

$$
\begin{cases}
\Big\langle \dfrac{\partial \hat{w}}{\partial t}, u\Big\rangle + 2\Big\langle [\mu(\hat{\theta}) + e^{4k_1 t}\varepsilon\|\hat{w}\|_{\mathbf{V}}^4]\mathcal{E}(\hat{w}), \mathcal{E}(u)\Big\rangle + e^{k_1 t}\langle \mathrm{rot}\hat{w}\times\hat{w}, u\rangle \\
\qquad + k_1(\hat{w}, u) + 2(\mu(\hat{\theta})k(x)\hat{w}, u)_{\Gamma_2} + 2(\mu(\hat{\theta})S\tilde{\hat{w}}, \tilde{u})_{\Gamma_3} \\
\qquad + 2(\alpha(x)\hat{w}, u)_{\Gamma_5} + (\mu(\hat{\theta})k(x)\hat{w}, u)_{\Gamma_7} + e^{-k_1 t}\langle \nabla\Phi_\varepsilon(e^{k_1 t}\hat{w}(t)), u\rangle \\
\qquad = e^{-k_1 t}\Big\langle (1 - \dfrac{\alpha_0\hat{\theta}}{1+\varepsilon\hat{\theta}^2} f), u\Big\rangle + e^{-k_1 t}\langle f_1, u\rangle \\
\qquad\qquad \forall u \in L^6(0,T;\mathbf{V}), \\
\Big\langle \dfrac{\partial \hat{\theta}}{\partial t}, \varphi\Big\rangle + (\kappa(\hat{\theta})\nabla\hat{\theta}, \nabla\varphi) + (\beta(x)\hat{\theta}, \varphi)_{\Gamma_R} - e^{k_1 t}\langle \hat{w}\hat{\theta}, \nabla\varphi\rangle \\
\qquad - \Big\langle \alpha_2\mu(\hat{\theta})\dfrac{|\mathcal{E}(\hat{w})|^2}{e^{-2k_1 t} + \varepsilon|\mathcal{E}(\hat{w})|^2}, \varphi\Big\rangle \\
\qquad = e^{2k_1 t}\Big\langle \dfrac{\alpha_1\hat{\theta}}{1+\varepsilon\hat{\theta}^2} f\cdot\hat{w}, \varphi\Big\rangle + \langle f_2, \varphi\rangle \\
\qquad\qquad \forall\varphi \in L^2(0,T;W^{1,2}_{\Gamma_D}(\Omega)), \\
\hat{w}(0) = v_0, \quad \hat{\theta}(0) = \theta_0.
\end{cases}
\tag{10.23}
$$

Define an operator $\mathscr{L} : D(\mathscr{L}) \to \mathscr{V}^*$ by

$$
\begin{aligned}
&D(\mathscr{L}) := \big\{(v,\theta) \in \mathscr{V}, (v',\theta') \in \mathscr{V}^*, v(0) = 0, \theta(0) = 0\big\},\\
&\mathscr{L}(v,\theta) = (v',\theta'),\\
&\|(\cdot,\cdot)\|_{D(\mathscr{L})} := (\|\cdot,\cdot\|_{\mathscr{V}} + \|\cdot,\cdot\|_{\mathscr{V}^*}).
\end{aligned}
$$

Then, $\mathscr{L}$ is a linear, maximal monotone operator (see Example 1.1).

Putting

$$
\hat{w} = w + v_0, \quad \hat{\theta} = \theta + \theta_0,
$$

we will consider an evolution problem with zero initial conditions.

Define an operator $\mathscr{A}_\varepsilon : D(\mathscr{L}) \to \mathscr{V}^*$ by

$$
\begin{aligned}
\big\langle \mathscr{A}_\varepsilon(w,\theta), (u,\phi)\big\rangle = \int_0^T \Big[& 2\Big\langle [\mu(\hat{\theta}) + e^{4k_1 t}\varepsilon\|\hat{w}\|_{\mathbf{V}}^4]\mathcal{E}(\hat{w}), \mathcal{E}(u)\Big\rangle \\
& + e^{k_1 t}\langle \mathrm{rot}\hat{w}\times\hat{w}, u\rangle + k_1(\hat{w}, u) + 2(\mu(\hat{\theta})k(x)\hat{w}, u)_{\Gamma_2} \\
& + 2(\mu(\hat{\theta})S\tilde{\hat{w}}, \tilde{u})_{\Gamma_3} + 2(\alpha(x)\hat{w}, u)_{\Gamma_5} + (\mu(\hat{\theta})k(x)\hat{w}, u)_{\Gamma_7} \\
& + e^{-k_1 t}\langle \nabla\Phi_\varepsilon(e^{k_1 t}\hat{w}(t)), u\rangle + e^{-k_1 t}\Big\langle \frac{\alpha_0\hat{\theta}}{1+\varepsilon\hat{\theta}^2} f, u\Big\rangle \\
& + (\kappa(\hat{\theta})\nabla\hat{\theta}, \nabla\phi) + (\beta(x)\hat{\theta}, \phi)_{\Gamma_R} - e^{k_1 t}\langle \hat{w}\hat{\theta}, \nabla\phi\rangle \\
& - \Big\langle \alpha_2\mu(\hat{\theta})\frac{|\mathcal{E}(\hat{w})|^2}{e^{-2k_1 t} + \varepsilon|\mathcal{E}(\hat{w})|^2}, \phi\Big\rangle - e^{2k_1 t}\Big\langle \frac{\alpha_1\hat{\theta}}{1+\varepsilon\hat{\theta}^2} f\cdot\hat{w}, \phi\Big\rangle\Big]\, dt \\
& \qquad \forall (w,\theta) \in D(\mathscr{L}),\ (u,\phi) \in \mathscr{V}.
\end{aligned}
\tag{10.24}
$$

Estimate (10.47) to be given later shows that this operator is well-defined. Then, the existence of a solution to Problem VEA' is equivalent to one of a solution to

$$\mathscr{L}(w,\theta)+\mathscr{A}_\varepsilon(w,\theta)=\mathscr{F},\quad \mathscr{F}=\begin{pmatrix} e^{-k_1t}(f+f_1)\\ f_2\end{pmatrix}.$$

Applying Theorem 1.47, we will prove existence of a solution to the equation above. To this end, we need to check that $\mathscr{A}_\varepsilon$ satisfies assumptions of Theorem 1.47.

(i) Let us prove property (iii) of Theorem 1.47 for $\mathscr{A}_\varepsilon(w,\theta)$:

$$\frac{1}{\|(w,\theta)\|_{\mathscr{V}}}\big\langle\mathscr{A}_\varepsilon(w,\theta),(w,\theta)\big\rangle\to\infty\quad\text{as } \|(w,\theta)\|_{\mathscr{V}}\to\infty. \tag{10.25}$$

We have that

$$\begin{aligned}\big\langle\mathscr{A}_\varepsilon(w,\theta),(w,\theta)\big\rangle=\int_0^T\Big[&2\big\langle[\mu(\hat\theta)+e^{4k_1t}\varepsilon\|\hat w\|^4_{\mathbf{V}}]\mathcal{E}(\hat w),\mathcal{E}(w)\big\rangle\\ &+e^{k_1t}\langle\mathrm{rot}\hat w\times\hat w,w\rangle+k_1(\hat w,w)+2(\mu(\hat\theta)k(x)\hat w,w)_{\Gamma_2}\\ &+2(\mu(\hat\theta)S\tilde{\hat w},\tilde w)_{\Gamma_3}+2(\alpha(x)\hat w,w)_{\Gamma_5}+(\mu(\hat\theta)k(x)\hat w,w)_{\Gamma_7}\\ &+e^{-k_1t}\langle\nabla\Phi_\varepsilon(e^{k_1t}\hat w(t)),w\rangle+e^{-k_1t}\Big\langle\frac{\alpha_0\hat\theta}{1+\varepsilon\hat\theta^2}f,w\Big\rangle\\ &+(\kappa(\hat\theta)\nabla\hat\theta,\nabla\theta)+(\beta(x)\hat\theta,\theta)_{\Gamma_R}-e^{k_1t}\langle\hat w\hat\theta,\nabla\theta\rangle\\ &-\Big\langle\alpha_2\mu(\hat\theta)\frac{|\mathcal{E}(\hat w)|^2}{e^{-2k_1t}+\varepsilon|\mathcal{E}(\hat w)|^2},\theta\Big\rangle-e^{2k_1t}\Big\langle\frac{\alpha_1\hat\theta}{1+\varepsilon\hat\theta^2}f\cdot\hat w,\theta\Big\rangle\Big]\,dt.\end{aligned} \tag{10.26}$$

Let us now estimate each term on the right hand side above. By Hölder's inequality we have

$$\begin{aligned}&\int_0^T2\big\langle[\mu(\hat\theta)+e^{4k_1t}\varepsilon\|\hat w\|^4_{\mathbf{V}}]\mathcal{E}(\hat w),\mathcal{E}(w)\big\rangle\,dt\\ &\ge 2\mu_0\|w\|^2_{L^2(0,T;\mathbf{V})}+\int_0^T2\big(\mu(\hat\theta)\mathcal{E}(\hat v_0),\mathcal{E}(w)\big)\,dt\\ &\quad+\varepsilon\int_0^T2\big\langle e^{4k_1t}\|\hat w\|^4_{\mathbf{V}}\mathcal{E}(\hat w),\mathcal{E}(\hat w)\big\rangle\,dt-\varepsilon\int_0^T2\big\langle e^{4k_1t}\|\hat w\|^4_{\mathbf{V}}\mathcal{E}(\hat w),\mathcal{E}(v_0)\big\rangle\,dt\\ &\ge\mu_0\|w\|^2_{L^2(0,T;\mathbf{V})}+2\varepsilon\|w+v_0\|^6_{L^6(0,T;\mathbf{V})}-2c\varepsilon\|w+v_0\|^5_{L^6(0,T;\mathbf{V})}\|v_0\|_{L^6(0,T;\mathbf{V})}-kT\|v_0\|^2\\ &\ge\mu_0\|w\|^2_{L^2(0,T;\mathbf{V})}+\varepsilon\|w\|^6_{L^6(0,T;\mathbf{V})}-K_1,\end{aligned} \tag{10.27}$$

where K_1 depends on $\|v_0\|_{\mathbf{V}},\varepsilon,T$. By the fact that $\Gamma_{2j},\Gamma_{3j},\Gamma_{7j}$ are in $C^{2,1}$ and Assumption 10.1, there exists a constant M such that

$$\|S(x)\|_\infty,\ \|k(x)\|_\infty,\ \|\alpha\|_{L^\infty(\Gamma_5)}\le M. \tag{10.28}$$

Thus,

$$\Big|\int_0^T \Big(2\big(\mu(\hat\theta)k(x)w, w\big)_{\Gamma_2} + 2\big(\mu(\hat\theta)S\tilde w, \tilde w\big)_{\Gamma_3} + 2\big(\alpha(x)w, w\big)_{\Gamma_5} + \big(\mu(\hat\theta)k(x)w, w\big)_{\Gamma_7}\Big)\,dt\Big| \\ \le \frac{1}{8}\mu_0\|w\|^2_{L^2(0,T;\mathbf{V})} + k_{11}\|w\|^2_{L^2(0,T;H)} \tag{10.29}$$

(see Theorem 1.27) and

$$\Big|\int_0^T \Big(2(\mu(\hat\theta)k(x)v_0, w)_{\Gamma_2} + 2(\mu(\hat\theta)S\tilde v_0, \tilde w)_{\Gamma_3} + 2(\alpha(x)v_0, w)_{\Gamma_5} + (\mu(\hat\theta)k(x)v_0, w)_{\Gamma_7}\Big)\,dt\Big| \\ \le \frac{1}{8}\mu_0\|w\|^2_{L^2(0,T;\mathbf{V})} + K_2, \tag{10.30}$$

where K_2 depends on $\|v_0\|_{\mathbf{V}}$, μ_0, μ_1, T. Since $\langle \operatorname{rot} w \times v, v\rangle = 0$,

$$\Big|\int_0^T e^{k_1t}\langle \operatorname{rot}\hat w \times \hat w, w\rangle\,dt\Big| = \Big|\int_0^T e^{k_1t}\big[\langle \operatorname{rot} w \times v_0, w\rangle + \langle \operatorname{rot} v_0 \times v_0, w\rangle\big]\,dt\Big| \\ \le \frac{1}{8}\mu_0\|w\|^2_{L^2(0,T;\mathbf{V})} + K_3 + k_{12}\|w\|^2_{L^2(0,T;H)}, \tag{10.31}$$

where K_3 depends on $\|v_0\|_{\mathbf{W}^{1,6}(\Omega)}$, μ_0, μ_1, T.

Let us estimate

$$\int_0^T e^{-k_1t}\langle \nabla\Phi_\varepsilon(e^{k_1t}\hat w(t)), w\rangle\,dt = \int_0^T e^{-k_1t}\langle \nabla\Phi_\varepsilon(e^{k_1t}\hat w(t)), e^{k_1t}\hat w - e^{k_1t}v_0\rangle\,dt.$$

Since Φ_ε is convex, continuous and Fréchet differentiable, by (1.37) we have

$$\langle \nabla\Phi_\varepsilon(e^{k_1t}\hat w(t)), e^{k_1t}v_0 - e^{k_1t}\hat w\rangle \le \Phi_\varepsilon(e^{k_1t}v_0) - \Phi_\varepsilon(e^{k_1t}\hat w(t)) \le \Phi(e^{k_1t}v_0).$$

Thus,

$$\int_0^T e^{-k_1t}\langle \nabla\Phi_\varepsilon(e^{k_1t}\hat w(t)), w\rangle\,dt \ge -\int_0^T e^{-k_1t}\Phi(e^{k_1t}v_0)\,dt = -K_4, \qquad K_4 \ge 0. \tag{10.32}$$

Also, we have

$$\begin{aligned}
&\int_0^T \big[(\kappa(\theta)\nabla\theta, \nabla\theta)\big]\,dt \ge \kappa_0\|\theta\|^2_{L^2(0,T;W^{1,2}_{\Gamma_D})},\\
&\Big|\int_0^T \big[(\kappa(\theta)\nabla\theta_0, \nabla\theta)\big]\,dt\Big| \le \frac{\kappa_0}{6}\|\theta\|^2_{L^2(0,T;W^{1,2}_{\Gamma_D})} + K_5,\\
&\int_0^T (\beta(x)\theta, \theta)_{\Gamma_R}\,dt \ge 0,\\
&\Big|\int_0^T (\beta(x)\theta_0, \theta)_{\Gamma_R}\,dt\Big| \le \frac{\kappa_0}{6}\|\theta\|^2_{L^2(0,T;W^{1,2}_{\Gamma_D})} + K_6,
\end{aligned} \tag{10.33}$$

where K_5, K_6 depend on $\|\theta_0\|_{W^{1,6}(\Omega)}$ and T. Since $\langle w\theta, \nabla\theta\rangle = 0$ (see (8.8)),

$$\begin{aligned}\Big|\int_0^T -e^{k_1 t}\langle \hat{w}\hat{\theta}, \nabla\theta\rangle\, dt\Big| &= \Big|\int_0^T -e^{k_1 t}\langle \hat{w}\theta_0, \nabla\theta\rangle\, dt\Big| \\ &= \Big|\int_0^T -e^{k_1 t}\big[\langle w\theta_0, \nabla\theta\rangle + \langle v_0\theta_0, \nabla\theta\rangle\big]\, dt\Big| \\ &\le \frac{\kappa_0}{6}\|\theta\|^2_{L^2(0,T;W^{1,2}_{\Gamma_D})} + K_7 + k_{13}\|w\|^2_{L^2(0,T;H)}.\end{aligned} \tag{10.34}$$

It is easy to get

$$\int_0^T \Big\langle \alpha_2\mu(\hat{\theta})\frac{|\mathcal{E}(\hat{w})|^2}{e^{-2k_1 t} + \varepsilon|\mathcal{E}(\hat{w})|^2}, \theta\Big\rangle dt \le \frac{c}{\varepsilon}\|\theta\|_{L^2(0,T;L^2(\Omega))}. \tag{10.35}$$

Since

$$\frac{\xi}{1+\varepsilon\xi^2} \le \frac{1}{2\sqrt{\varepsilon}} \quad \forall \xi \in [0,\infty),$$

we have

$$\begin{aligned}&\Big|\int_0^T e^{-k_1 t}\Big\langle \frac{\alpha_0\hat{\theta}}{1+\varepsilon\hat{\theta}^2} f, w\Big\rangle dt\Big| \le \frac{c}{\sqrt{\varepsilon}}\|w\|_{L^2(0,T;H)}, \\ &\int_0^T e^{2k_1 t}\Big\langle \frac{\alpha_1\hat{\theta}}{1+\varepsilon\hat{\theta}^2} f\cdot\hat{w}, \theta\Big\rangle dt\Big| \le \frac{K_8}{\varepsilon} + \frac{k_{14}}{\varepsilon}\|w\|^2_{L^2(0,T;H)} + \frac{\kappa_0}{6}\|\theta\|^2_{L^2(0,T;W^{1,2}_{\Gamma_D})}.\end{aligned} \tag{10.36}$$

Taking

$$k_1 = k_{11} + k_{12} + k_{13} + \frac{k_{14}}{\varepsilon}, \tag{10.37}$$

by (10.27)–(10.36) we have from (10.26)

$$\begin{aligned}&\big\langle \mathscr{A}_\varepsilon(w,\theta), (w,\theta)\big\rangle \\ &\quad\ge \min\Big\{\frac{\mu_0}{2}e^{-k_1 T}, \varepsilon, \frac{\kappa_0}{2}\Big\}\Big(\|w\|^2_{L^2(0,T;\mathbf{V})} + \|w\|^6_{L^6(0,T;\mathbf{V})} + \|\theta\|^2_{L^2(0,T;W^{1,2}_{\Gamma_D})}\Big) \\ &\qquad - \sum_{i=1}^7 K_i - \frac{K_8}{\varepsilon} - \frac{c}{\sqrt{\varepsilon}}\|w\|_{L^2(0,T;H)} - \frac{c}{\varepsilon}\|\theta\|_{L^2(0,T;L^2(\Omega))} \quad \forall (w,\theta)\in D(\mathscr{L}),\end{aligned} \tag{10.38}$$

which implies (10.25).

(ii) Let us prove property (ii) of Theorem 1.47 for $\mathscr{A}_\varepsilon(w,\theta)$.

Taking into account (10.28) and applying an inequality $|a+b|^p \le 2^p(|a|^p + |b|^p)$, $p \in (1,\infty)$, we have

$$\frac{1}{\|u\|_{L^6(0,T;\mathbf{V})}}\Big|\int_0^T \Big[2\Big([\mu(\hat\theta)+e^{4k_1t}\varepsilon\|\hat w\|_{\mathbf{V}}^4]\mathcal{E}(\hat w),\mathcal{E}(u)\Big)+k_1(\hat w,u)+2(\mu(\hat\theta)k(x)\hat w,u)_{\Gamma_2}$$
$$+2(\mu(\hat\theta)S\tilde{\hat w},\tilde u)_{\Gamma_3}+2(\alpha(x)\hat w,u)_{\Gamma_5}+(\mu(\hat\theta)k(x)\hat w,u)_{\Gamma_7}\Big]dt\Big| \tag{10.39}$$
$$\le c\big(\|\hat w\|_{L^2(0,T;\mathbf{V})}+\|\hat w\|^6_{L^6(0,T;\mathbf{V})}\big)\le c\big(K+\|w\|_{L^2(0,T;\mathbf{V})}+\|w\|^6_{L^6(0,T;\mathbf{V})}\big),$$

where K depends on v_0.

By (1.37) and $\Phi(0_{\mathbf{V}})=0$, we know that $\Phi_\varepsilon(0_{\mathbf{V}})=0$. Also, since Φ_ε is convex, continuous and Fréchet differentiable, we have $\nabla\Phi_\varepsilon(0_{\mathbf{V}})=0_{\mathbf{V}^*}$. Then, since the operator $\nabla\Phi_\varepsilon$ is Lipschitz continuous with the constant ε^{-1} (see Remark 1.16), we have

$$\frac{1}{\|u\|_{L^6(0,T;\mathbf{V})}}\Big|\int_0^T e^{-k_1t}\langle\nabla\Phi_\varepsilon(e^{k_1t}\hat w(t)),u\rangle\,dt\Big|$$
$$=\frac{1}{\|u\|_{L^6(0,T;\mathbf{V})}}\Big|\int_0^T e^{-k_1t}\langle\nabla\Phi_\varepsilon(e^{k_1t}\hat w(t))-\nabla\Phi_\varepsilon(0_{\mathbf{V}}),u\rangle\,dt\Big| \tag{10.40}$$
$$\le\frac{1}{\|u\|_{L^6(0,T;\mathbf{V})}}\int_0^T\frac{1}{\varepsilon}\|\hat w(t)\|_{\mathbf{V}}\|u\|_{\mathbf{V}}\,dt$$
$$\le\frac{c}{\varepsilon}\|\hat w\|_{L^2(0,T;\mathbf{V})}\le\frac{c}{\varepsilon}\Big(K+\|w\|_{L^2(0,T;\mathbf{V})}\Big).$$

Similarly, we have

$$\frac{1}{\|\phi\|_{L^2(0,T;W^{1,2}_{\Gamma_D})}}\Big|\int_0^T\Big[(\kappa(\hat\theta)\nabla\hat\theta,\nabla\phi)+(\beta(x)\hat\theta,\phi)_{\Gamma_R}-\Big\langle\alpha_2\mu(\hat\theta)\frac{|\mathcal{E}(\hat w)|^2}{e^{-2k_1t}+\varepsilon|\mathcal{E}(\hat w)|^2},\phi\Big\rangle\Big]dt\Big|$$
$$\le c\Big(\|\hat\theta\|_{L^2(0,T;W^{1,2}_{\Gamma_D})}+\frac{1}{\varepsilon}\Big)$$
$$\le c\Big(K+\|\theta\|_{L^2(0,T;W^{1,2}_{\Gamma_D})}+\frac{1}{\varepsilon}\Big), \tag{10.41}$$

where K depends on θ_0.

On the other hand, taking into account $w(0)=0$, $\theta(0)=0$ and applying (1.27), for $(w,\theta)\in D(\mathscr{L})$ we have

$$(w,\theta)\in C([0,T];H)\times C([0,T];L^2(\Omega)),$$
$$\|w\|_{C([0,T];H)}\le c\|w'\|^{1/2}_{L^{6/5}(0,T;\mathbf{V}^*)}\|w\|^{1/2}_{L^6(0,T;\mathbf{V})}, \tag{10.42}$$
$$\|\theta\|_{C([0,T];H)}\le c\|\theta'\|^{1/2}_{L^2\left(0,T;(W^{1,2}_{\Gamma_D})^*\right)}\|\theta\|^{1/2}_{L^2(0,T;W^{1,2}_{\Gamma_D})}.$$

Also, by Theorem 1.12

$$\|w\|_{\mathbf{L}^3(\Omega)}\le c\|w\|^{1/2}_{\mathbf{L}^2(\Omega)}\|\nabla w\|^{1/2}_{\mathbf{L}^2(\Omega)}\qquad\forall w\in\mathbf{V},$$
$$\|\theta\|_{L^4(\Omega)}\le c\|\theta\|^{1/4}_{L^2(\Omega)}\|\nabla\theta\|^{3/4}_{\mathbf{L}^2(\Omega)}\qquad\forall\theta\in W^{1,p}_{\Gamma_D}(\Omega). \tag{10.43}$$

Applying (10.42) and (10.43), by Hölder's inequality with exponents 4, 4/3 and Young's inequality we have

$$\begin{aligned}
\Big|\int_0^T \Big[e^{k_1 t}\langle \mathrm{rot}\hat{w}\times\hat{w}, u\rangle\Big]dt\Big| &\le c\int_0^T \|\mathrm{rot}\hat{w}\|_{\mathbf{L}^2}\|\hat{w}\|_{\mathbf{L}^3}\|u\|_{\mathbf{L}^6}\,dt\\
&\le c\|\hat{w}\|^{1/2}_{C([0,T];H)}\int_0^T \|\nabla\hat{w}\|^{3/2}_{\mathbf{L}^2}\|u\|_{\mathbf{V}}\,dt\\
&\le c\|w'\|^{1/4}_{L^{6/5}(0,T;\mathbf{V}^*)}\|\hat{w}\|^{1/4}_{L^6(0,T;\mathbf{V})}\|\hat{w}\|^{3/2}_{L^6(0,T;\mathbf{V})}\|u\|_{L^{4/3}(0,T;\mathbf{V})}\\
&\le c\|w'\|^{1/4}_{L^{6/5}(0,T;\mathbf{V}^*)}\|\hat{w}\|^{7/4}_{L^6(0,T;\mathbf{V})}\|u\|_{L^6(0,T;\mathbf{V})}\\
&\le \Big(\frac{1}{4}\|w'\|_{L^{6/5}(0,T;\mathbf{V}^*)}+K\|\hat{w}\|^{7/3}_{L^6(0,T;\mathbf{V})}\Big)\|u\|_{L^6(0,T;\mathbf{V})}.
\end{aligned}\tag{10.44}$$

Applying Hölder's inequality with exponents 8, 8/3, 2 and Young's inequality with exponents 4, 4/3, in the same way we have

$$\begin{aligned}
&\Big|\int_0^T e^{k_1 t}\langle\hat{w}\hat{\theta},\nabla\phi\rangle\,dt\Big| \le c\int_0^T \|\hat{w}\|_{\mathbf{L}^4}\|\hat{\theta}\|_{\mathbf{L}^4}\|\nabla\phi\|_{\mathbf{L}^2}\,dt\\
&\le c\|\hat{w}\|^{1/4}_{C([0,T];H)}\|\hat{\theta}\|^{1/4}_{C([0,T];L^2)}\int_0^T \|\nabla\hat{w}\|^{3/4}_{\mathbf{L}^2}\|\nabla\hat{\theta}\|^{3/4}_{\mathbf{L}^2}\|\phi\|_{W^{1,2}_{\Gamma_D}}\,dt\\
&\le c\|\hat{w}'\|^{1/8}_{L^{6/5}(0,T;\mathbf{V}^*)}\|\hat{w}\|^{1/8}_{L^6(0,T;\mathbf{V})}\|\hat{\theta}'\|^{1/8}_{L^2(0,T;(W^{1,2}_{\Gamma_D})^*)}\|\hat{\theta}\|^{1/8}_{L^2(0,T;W^{1,2}_{\Gamma_D})}\|\hat{w}\|^{3/4}_{L^6(0,T;\mathbf{V})}\\
&\qquad\times\|\hat{\theta}\|^{3/4}_{L^2(0,T;W^{1,2}_{\Gamma_D})}\|\phi\|_{L^2(0,T;W^{1,2}_{\Gamma_D})}\\
&\le c\|\hat{w}'\|^{1/8}_{L^{6/5}(0,T;\mathbf{V}^*)}\|\hat{w}\|^{7/8}_{L^6(0,T;\mathbf{V})}\|\hat{\theta}'\|^{1/8}_{L^2(0,T;(W^{1,2}_{\Gamma_D})^*)}\|\hat{\theta}\|^{7/8}_{L^2(0,T;W^{1,2}_{\Gamma_D})}\|\phi\|_{L^2(0,T;W^{1,2}_{\Gamma_D})}\\
&\le \Big(\delta\|w'\|^{1/2}_{L^{6/5}(0,T;\mathbf{V}^*)}+K_\delta\|\hat{w}\|^{7/6}_{L^6(0,T;\mathbf{V})}\Big)\Big(\delta\|\theta'\|^{1/2}_{L^2(0,T;(W^{1,2}_{\Gamma_D})^*)}+K_\delta\|\hat{\theta}\|^{7/6}_{L^2(0,T;W^{1,2}_{\Gamma_D})}\Big)\\
&\qquad\times\|\phi\|_{L^2(0,T;W^{1,2}_{\Gamma_D})}\\
&\le \frac{1}{2}\Big[\Big(\delta\|w'\|^{1/2}_{L^{6/5}(0,T;\mathbf{V}^*)}+K_\delta\|\hat{w}\|^{7/6}_{L^6(0,T;\mathbf{V})}\Big)^2\\
&\quad+\Big(\delta\|\theta'\|^{1/2}_{L^2(0,T;(W^{1,2}_{\Gamma_D})^*)}+K_\delta\|\hat{\theta}\|^{7/6}_{L^2(0,T;W^{1,2}_{\Gamma_D})}\Big)^2\Big]\cdot\|\phi\|_{L^2(0,T;W^{1,2}_{\Gamma_D})}\\
&\le\Big[\delta^2\|w'\|_{L^{6/5}(0,T;\mathbf{V}^*)}+K^2_\delta\|\hat{w}\|^{7/3}_{L^6(0,T;\mathbf{V})}+\delta^2\|\theta'\|_{L^2(0,T;(W^{1,2}_{\Gamma_D})^*)}+K^2_\delta\|\hat{\theta}\|^{7/3}_{L^2(0,T;W^{1,2}_{\Gamma_D})}\Big]\\
&\qquad\times\|\phi\|_{L^2(0,T;W^{1,2}_{\Gamma_D})}.
\end{aligned}\tag{10.45}$$

Also,

$$
\begin{aligned}
&\left| \int_0^T e^{-k_1 t} \left\langle \frac{\alpha_0 \hat{\theta}}{1+\varepsilon \hat{\theta}^2} f, u \right\rangle dt \right| \le \frac{c}{\sqrt{\varepsilon}} \|u\|_{L^2(0,T;H)}, \\
&\int_0^T e^{2k_1 t} \left\langle \frac{\alpha_1 \hat{\theta}}{1+\varepsilon \hat{\theta}^2} f \cdot \hat{w}, \phi \right\rangle dt \Big| \le \frac{c}{\sqrt{\varepsilon}} (K + \|w\|_{L^2(0,T;H)}) \|\phi\|_{L^2(0,T;W^{1,2}_{\Gamma_D})}.
\end{aligned}
\tag{10.46}
$$

Owing to (10.39)–(10.41), (10.44), (10.35), (10.46) and (10.45) with $\delta = \frac{1}{2}$, we have from (10.24)

$$
\begin{aligned}
\|\mathscr{A}_\varepsilon(w,\theta)\|_{\mathscr{V}^*} &= \sup_{\|(u,\phi)\|_{\mathscr{V}}=1} \langle \mathscr{A}_\varepsilon(w,\theta), (u,\phi) \rangle \\
&\le c\Big(K + \frac{K}{\varepsilon} + \frac{1}{\varepsilon}\|w\|_{L^2(0,T;\mathbf{V})} + \|w\|_{L^2(0,T;\mathbf{V})} + \|w\|^6_{L^6(0,T;\mathbf{V})} + \|\theta\|_{L^2(0,T;W^{1,p}_{\Gamma_D})} \\
&\quad + K\|\hat{w}\|^{7/3}_{L^6(0,T;\mathbf{V})} + K\|\hat{w}\|^{7/3}_{L^6(0,T;\mathbf{V})} + \frac{1}{\varepsilon}\|\theta\|_{L^2(0,T;L^2(\Omega))} + K\|\hat{\theta}\|^{7/3}_{L^2(0,T;W^{1,2}_{\Gamma_D})} \Big) \\
&\quad + \frac{1}{2}\big(\|w'\|_{L^{6/5}(0,T;\mathbf{V}^*)} + \|\theta'\|_{L^2(0,T;(W^{1,2}_{\Gamma_D})^*)} \big) \\
&\le c\Big(K + \frac{K}{\varepsilon} + \frac{1}{\varepsilon}\|w\|_{L^2(0,T;\mathbf{V})} + \frac{1}{\varepsilon}\|\theta\|_{L^2(0,T;L^2(\Omega))} + \|w\|^6_{L^6(0,T;\mathbf{V})} + \|\theta\|_{L^2(0,T;W^{1,p}_{\Gamma_D})} \\
&\quad + \|w\|^{7/3}_{L^6(0,T;\mathbf{V})} + \|\theta\|^{7/3}_{L^2(0,T;W^{1,2}_{\Gamma_D})} \Big) + \frac{1}{2}\|\mathscr{L}(w,\theta)\|_{\mathscr{V}^*},
\end{aligned}
\tag{10.47}
$$

which shows the property (ii) for $\mathscr{A}_\varepsilon(w,\theta)$.

(iii) Let us prove property (i) of Theorem 1.47 for $\mathscr{A}_\varepsilon(w,\theta)$.

Let $\{(w_k,\theta_k)\} \subset D(\mathscr{L})$ be a sequence such that

$$
\begin{aligned}
&(w_k,\theta_k) \rightharpoonup (w,\theta) \in D(\mathscr{L}) \quad \text{in } \mathscr{V}, \\
&\mathscr{L}(w_k,\theta_k) \rightharpoonup \mathscr{L}(w,\theta) \quad \text{in } \mathscr{V}^*, \\
&\limsup_{k\to\infty} \langle \mathscr{A}_\varepsilon(w_k,\theta_k), (w_k,\theta_k) - (w,\theta) \rangle \le 0.
\end{aligned}
\tag{10.48}
$$

By taking a subsequence if necessary, we may assume that

$$
\begin{aligned}
&w_k \to w \quad \text{in } L^6(0,T;\mathbf{L}^s(\Omega)) \ (1 \le s < 6) \text{ and a.e. in } Q \\
&\theta_k \to \theta \quad \text{in } L^2(0,T;L^s(\Omega)) \ (1 \le s < 6) \text{ and a.e. in } Q
\end{aligned}
\tag{10.49}
$$

when $k \to \infty$. Then, by (10.24) we have

$$
\begin{aligned}
&c\min\{\mu_0,\kappa_0\}\int_Q \left(\|w_k-w\|_{\mathbf{V}}^2+\|\theta_k-\theta\|_{W^{1,2}_{\Gamma_D}}\right)dt\\
&=c\min\{\mu_0,\kappa_0\}\int_Q \left(\|\hat{w}_k-\hat{w}\|_{\mathbf{V}}^2+\|\hat{\theta}_k-\hat{\theta}\|_{W^{1,2}_{\Gamma_D}}\right)dt\\
&\le \int_0^T 2\Big([\mu(\hat{\theta}_k)+e^{4k_1t}\varepsilon\|\hat{w}_k\|_{\mathbf{V}}^4]\mathcal{E}(\hat{w}_k),\mathcal{E}(w_k-w)\Big)\,dt\\
&\quad-\int_0^T 2\Big(\mu(\hat{\theta}_k)\mathcal{E}(\hat{w}),\mathcal{E}(w_k-w)\Big)\,dt-\int_0^T 2e^{4k_1t}\varepsilon\|\hat{w}_k\|_{\mathbf{V}}^4\langle\mathcal{E}(\hat{w}_k),\mathcal{E}(w_k-w)\rangle\,dt\\
&\quad+\int_0^T (\kappa(\hat{\theta}_k)\nabla\hat{\theta}_k,\nabla\theta_k-\nabla\theta)\,dt-\int_0^T (\kappa(\hat{\theta}_k)\nabla\hat{\theta},\nabla\theta_k-\nabla\theta)\,dt\\
&\le \langle\mathscr{A}_\varepsilon(\hat{w}_k,\hat{\theta}_k),(w_k-w,\theta_k-\theta)\rangle-\int_0^T 2\Big(\mu(\hat{\theta}_k)\mathcal{E}(\hat{w}),\mathcal{E}(w_k-w)\Big)\,dt\\
&\quad+\int_0^T 2e^{4k_1t}\varepsilon\Big\langle\|\hat{w}_k\|_{\mathbf{V}}^4\mathcal{E}(\hat{w}),\mathcal{E}(w-w_k)\Big\rangle\,dt-\int_0^T e^{k_1t}\langle\mathrm{rot}\hat{w}_k\times\hat{w}_k,w_k-w\rangle\,dt\\
&\quad-\int_0^T e^{-k_1t}\langle\nabla\Phi_\varepsilon(e^{k_1t}\hat{w}_k(t)),w_k-w\rangle\,dt-\int_0^T e^{-k_1t}\Big\langle\frac{\alpha_0\hat{\theta}_k}{1+\varepsilon\hat{\theta}_k^2}f,w_k-w\Big\rangle\,dt\\
&\quad-\int_0^T (\kappa(\hat{\theta}_k)\nabla\hat{\theta},\nabla\theta_k-\nabla\theta)\,dt-\int_0^T\Big[(k_1(\hat{w}_k,w_k-w)+2(\mu(\hat{\theta}_k)k(x)\hat{w}_k,w_k-w)_{\Gamma_2}\\
&\qquad+2(\mu(\hat{\theta}_k)S\tilde{\hat{w}}_k,\tilde{w}_k-\tilde{w})_{\Gamma_3}+2(\alpha(x)\hat{w}_k,w_k-w)_{\Gamma_5}\\
&\qquad+(\mu(\hat{\theta}_k)k(x)\hat{w}_k,w_k-w)_{\Gamma_7}+(\beta(x)\hat{\theta}_k,\theta_k-\theta)_{\Gamma_R}-e^{k_1t}\langle\hat{w}_k\hat{\theta}_k,\nabla(\theta_k-\theta)\rangle\\
&\qquad-\Big\langle\alpha_2\mu(\hat{\theta}_k)\frac{|\mathcal{E}(\hat{w}_k)|^2}{e^{-2k_1t}+\varepsilon|\mathcal{E}(\hat{w}_k)|^2},\theta_k-\theta\Big\rangle-e^{2k_1t}\Big\langle\frac{\alpha_1\hat{\theta}_k}{1+\varepsilon\hat{\theta}_k^2}f\cdot\hat{w}_k,\theta_k-\theta\Big\rangle\Big]\,dt,
\end{aligned}
\tag{10.50}
$$

where k_1 is the one in (10.37) and c is a positive constant.

Now let us estimate all terms except the first one on the right hand side of (10.50).

$$
\begin{aligned}
&\limsup_{k\to\infty}\int_0^T 2\Big(e^{4k_1t}\varepsilon\|\hat{w}_k\|_{\mathbf{V}}^4\mathcal{E}(\hat{w}_k),\mathcal{E}(w-w_k)\Big)\,dt\\
&\qquad=\limsup_{k\to\infty}\int_0^T 2\Big(e^{4k_1t}\varepsilon\|\hat{w}_k\|_{\mathbf{V}}^4\mathcal{E}(\hat{w}_k),\mathcal{E}(\hat{w}-\hat{w}_k)\Big)\,dt\\
&\qquad\le\int_0^T 2e^{4k_1t}\varepsilon\limsup_{k\to\infty}\|\hat{w}_k\|_{\mathbf{V}}^4\cdot\limsup_{k\to\infty}\Big(\mathcal{E}(\hat{w}_k),\mathcal{E}(\hat{w}-\hat{w}_k)\Big)\,dt \qquad (10.51)\\
&\qquad\le\int_0^T 2e^{4k_1t}\varepsilon\limsup_{k\to\infty}\|\hat{w}_k\|_{\mathbf{V}}^4\cdot\Big(\|\mathcal{E}(\hat{w})\|^2-\liminf_{k\to\infty}\|\mathcal{E}(\hat{w}_k)\|^2\Big)\,dt\\
&\qquad\le\int_0^T 2e^{4k_1t}\varepsilon\limsup_{k\to\infty}\|\hat{w}_k\|_{\mathbf{V}}^4\cdot\Big(\|\mathcal{E}(\hat{w})\|^2-\|\mathcal{E}(\hat{w})\|^2\Big)\,dt=0.
\end{aligned}
$$

Taking into account Corollary 1.1, by (10.48) we know

$$\int_0^T 2\Big(\mu(\hat{\theta}_k)\mathcal{E}(\hat{w}), \mathcal{E}(w_k - w)\Big)\,dt \to 0,$$
$$\int_0^T \big(\kappa(\hat{\theta}_k)\nabla\hat{\theta}, \nabla\theta_k - \nabla\theta\big)\,dt \to 0. \tag{10.52}$$

Also, by (10.48) and (10.49) we have that

$$\int_0^T e^{k_1 t}\langle \mathrm{rot}\hat{w}_k \times \hat{w}_k, w_k - w\rangle\,dt \le c\|\hat{w}_k\|_{\mathbf{L}^4(Q)}\|\nabla\hat{w}_k\|_{\mathbf{L}^2(Q)}\|w_k - w\|_{\mathbf{L}^4(Q)} \to 0 \tag{10.53}$$

as $k \to \infty$. Since $\nabla\Phi_\varepsilon$ is monotone, we have

$$\begin{aligned}
&-\int_0^T e^{-k_1 t}\langle \nabla\Phi_\varepsilon(e^{k_1 t}\hat{w}_k(t)), w_k - w\rangle\,dt\\
&\quad= -\int_0^T e^{-2k_1 t}\langle \nabla\Phi_\varepsilon(e^{k_1 t}\hat{w}_k(t)), e^{k_1 t}\hat{w}_k - e^{k_1 t}\hat{w}\rangle\,dt\\
&\quad= -\int_0^T e^{-2k_1 t}\langle \nabla\Phi_\varepsilon(e^{k_1 t}\hat{w}_k(t)) - \nabla\Phi_\varepsilon(e^{k_1 t}\hat{w}), e^{k_1 t}\hat{w}_k - e^{k_1 t}\hat{w}\rangle\,dt\\
&\qquad -\int_0^T e^{-2k_1 t}\langle \nabla\Phi_\varepsilon(e^{k_1 t}\hat{w}), e^{k_1 t}\hat{w}_k - e^{k_1 t}\hat{w}\rangle\,dt\\
&\quad\le -\int_0^T e^{-2k_1 t}\langle \nabla\Phi_\varepsilon(e^{k_1 t}\hat{w}), e^{k_1 t}w_k - e^{k_1 t}w\rangle\,dt \to 0 \quad \text{as } k \to \infty.
\end{aligned} \tag{10.54}$$

By (10.49) we have that

$$\int_0^T e^{-k_1 t}\Big\langle \frac{\alpha_0\hat{\theta}_k}{1+\varepsilon\hat{\theta}_k^2} f, w_k - w\Big\rangle\,dt \to 0,$$
$$\int_0^T e^{2k_1 t}\Big\langle \frac{\alpha_1\hat{\theta}_k}{1+\varepsilon\hat{\theta}_k^2} f\cdot\hat{w}_k, \theta_k - \theta\Big\rangle\,dt \to 0 \tag{10.55}$$

as $k \to \infty$. By taking a sequence if necessary, we get from (10.48)

$$\begin{aligned}
w_k &\to w \quad \text{in } L^6(0,T;\mathbf{W}^{\alpha,2}(\Omega))\ (2/3 < \alpha < 1),\\
\theta_k &\to \theta \quad \text{in } L^2(0,T;W^{\alpha,2}(\Omega))\ (2/3 < \alpha < 1).
\end{aligned}$$

Then, by the trace theorem and Remark 3.4 we have

$$
\begin{aligned}
&-\int_0^T \Big[(k_1(\hat{w}_k, w_k - w) + 2(\mu(\hat{\theta}_k)k(x)\hat{w}_k, w_k - w)_{\Gamma_2} + 2(\mu(\hat{\theta}_k)S\tilde{\hat{w}}_k, \tilde{w}_k - \tilde{w})_{\Gamma_3} \\
&\quad + 2(\alpha(x)\hat{w}_k, w_k - w)_{\Gamma_5} + (\mu(\hat{\theta}_k)k(x)\hat{w}_k, w_k - w)_{\Gamma_7} + (\beta(x)\hat{\theta}_k, \theta_k - \theta)_{\Gamma_R} \\
&\quad - \Big\langle \alpha_2\mu(\hat{\theta}_k)\frac{|\mathcal{E}(\hat{w}_k)|^2}{e^{-2k_1 t} + \varepsilon|\mathcal{E}(\hat{w}_k)|^2}, \theta_k - \theta\Big\rangle\Big]\, dt \to 0 \quad \text{as } k \to \infty.
\end{aligned}
\tag{10.56}
$$

Now let us prove that by taking subsequences there holds

$$
\begin{aligned}
&\int_0^T e^{k_1 t}\langle \hat{w}_k\hat{\theta}_k, \nabla(\theta_k - \theta)\rangle\, dt = \int_0^T e^{k_1 t}\langle(\hat{w}_k - \hat{w})\hat{\theta}_k, \nabla(\theta_k - \theta)\rangle\, dt \\
&\quad + \int_0^T e^{k_1 t}\langle \hat{w}(\hat{\theta}_k - \hat{\theta}), \nabla(\theta_k - \theta)\rangle\, dt + \int_0^T e^{k_1 t}\langle \hat{w}\hat{\theta}, \nabla(\theta_k - \theta)\rangle\, dt \to 0 \\
&\qquad\qquad \text{as } k \to \infty.
\end{aligned}
\tag{10.57}
$$

Since $\{\hat{w}_k\}$ is bounded in $L^6(0, T; \mathbf{V})$, taking into account (10.48) and applying Theorem 1.39, we have that the set is relatively compact in the space $L^5(0, T\ \mathbf{H}^{9/10}(\Omega)) \subset \mathbf{L}^5(Q)$, where imbedding $\mathbf{H}^{9/10}(\Omega) \hookrightarrow \mathbf{L}^5(\Omega)$ was used.
Since $\{\theta_k\}$ is bounded in $L^2(0, T; W^{1,2}(\Omega)) \cap L^\infty(0, T; L^2(\Omega))$, by Theorem 1.33 and (1.22) it is bounded in

$$
L^{10/3}(0, T; W^{3/5,2}(\Omega)) \subset L^{10/3}(Q)
$$

as well and thus by Theorem 1.39 the set $\{\theta_k\}$ is also relatively compact in $L^3(Q)$. Consequently, there exist the subsequences such that

$$
\begin{aligned}
&\hat{w}_k \to \hat{w} \text{ in } \mathbf{L}^5(Q), \\
&\{\theta_k\} \text{ is bounded in } L^{10/3}(Q), \\
&\theta_k \to \theta \text{ in } L^3(Q).
\end{aligned}
$$

Applying Hölder's inequalities with exponents 5, 10/3, 2 for the first term and with exponents 6, 3, 2 for the second term of the right hand side of (10.57), we get convergence to zero of the two terms. The convergence to zero of the last term of the right hand side of (10.57) is obvious.

Taking into account (10.51)–(10.57) and Assumption (10.48), we have from (10.50)

$$
\begin{aligned}
c\min\{\mu_0, \kappa_0\} \lim_{k\to\infty} \sup \int_Q \big(\|w_k - w\|_{\mathbf{V}}^2 + \|\theta_k - \theta\|_{W^{1,2}_{\Gamma_D}}\big)\, dt \\
\leq \lim_{k\to\infty} \sup \big\langle \mathscr{A}_\varepsilon(\hat{w}_k, \hat{\theta}_k), (w_k - w, \theta_k - \theta)\big\rangle \leq 0,
\end{aligned}
$$

which implies that

$$\begin{aligned}
&w_k \to w \quad \text{in } L^2(0,T;\mathbf{V}),\\
&\theta_k \to \theta \quad \text{in } L^2(0,T;W^{1,2}_{\Gamma_D}),\\
&\nabla w_k \to \nabla w \quad \text{a.e. in } Q,\\
&\nabla \theta_k \to \nabla \theta \quad \text{a.e. in } Q.
\end{aligned} \tag{10.58}$$

When $(u,\phi) \in \mathscr{V}$, by definition of $\mathscr{A}$

$$\begin{aligned}
&\Big\langle \mathscr{A}_\varepsilon(w_k,\theta_k),(w_k-u,\theta_k-\phi)\Big\rangle\\
&= \int_0^T \Big[2\Big\langle [\mu(\hat{\theta}_k) + e^{4k_1 t}\varepsilon \|\hat{w}\|^4_{\mathbf{V}}\mathcal{E}(\hat{w}_k), \mathcal{E}(w_k-u)\Big\rangle + e^{k_1 t}\langle \mathrm{rot}\hat{w}_k \times \hat{w}_k, w_k-u\rangle\\
&\quad + k_1(\hat{w}_k, w_k-u) + 2(\mu(\hat{\theta}_k)k(x)\hat{w}_k, w_k-u)_{\Gamma_2} + 2(\mu(\hat{\theta}_k)S\tilde{\hat{w}}_k, \tilde{w}_k-\tilde{u})_{\Gamma_3}\\
&\quad + 2(\alpha(x)\hat{w}_k, w_k-u)_{\Gamma_5} + (\mu(\hat{\theta}_k)k(x)\hat{w}_k, w_k-u)_{\Gamma_7}\\
&\quad + e^{-k_1 t}\langle \nabla\Phi_\varepsilon(e^{k_1 t}\hat{w}_k(t)), w_k-u\rangle - e^{-k_1 t}\Big\langle \frac{\alpha_0\hat{\theta}_k}{1+\varepsilon\hat{\theta}_k^2} f, w_k-u\Big\rangle\\
&\quad + (\kappa(\hat{\theta}_k)\nabla\hat{\theta}_k, \nabla(\theta_k-\phi)) + (\beta(x)\hat{\theta}_k, \theta_k-\phi)_{\Gamma_R} - e^{k_1 t}\langle \hat{w}_k\hat{\theta}_k, \nabla(\theta_k-\phi)\rangle\\
&\quad - \Big\langle \alpha_2\mu(\hat{\theta}_k)\frac{|\mathcal{E}(\hat{w}_k)|^2}{e^{-2k_1 t}+\varepsilon|\mathcal{E}(\hat{w}_k)|^2}\theta_k - \phi\Big\rangle - e^{2k_1 t}\Big\langle \frac{\alpha_1\hat{\theta}_k}{1+\varepsilon\hat{\theta}_k^2} f\cdot\hat{w}_k, \theta_k-\phi\Big\rangle\Big]\,dt.
\end{aligned} \tag{10.59}$$

By Corollary 1.1 and (10.58), we have

$$\lim_{k\to\infty}\int_0^T 2\big(\mu(\hat{\theta}_k)\mathcal{E}(\hat{w}_k), \mathcal{E}(w_k-u)\big)\,dt = \int_0^T 2\big(\mu(\hat{\theta})\mathcal{E}(\hat{w}), \mathcal{E}(w-u)\big)\,dt. \tag{10.60}$$

The sequence $\{e^{4k_1 t}\|\hat{w}_k\|^4_{\mathbf{V}}\varepsilon_{ij}(\hat{w}_k)\}$ is bounded in $L^{6/5}(0,T;L^2(\Omega))$ and $\varepsilon_{ij}(u) \in L^6(0,T;L^2(\Omega))$, and by taking a subsequence and denoting with the same subindex, we see that

$$\lim_{k\to\infty}\int_0^T 2e^{4k_1 t}\Big\langle \|\hat{w}_k\|^4_{\mathbf{V}}\mathcal{E}(\hat{w}_k), \mathcal{E}(\hat{u})\Big\rangle\,dt \to \int_0^T 2e^{4k_1 t}\Big\langle \|\hat{w}\|^4_{\mathbf{V}}\mathcal{E}(\hat{w}), \mathcal{E}(\hat{u})\Big\rangle\,dt. \tag{10.61}$$

Since $\liminf_{k\to\infty}\|\hat{w}_k\|^4_{\mathbf{V}}\big(\mathcal{E}(\hat{w}_k),\mathcal{E}(\hat{w}_k)\big) \geq \|\hat{w}\|^4_{\mathbf{V}}\big(\mathcal{E}(\hat{w}),\mathcal{E}(\hat{w})\big)$ (see Exercises 2, (c), p. 173 of [3]), by (10.60) and (10.61) we have

$$\begin{aligned}
&\liminf_{k\to\infty}\int_0^T 2\Big\langle [\mu(\hat{\theta}_k)+e^{4k_1 t}\varepsilon\|\hat{w}_k\|^4_{\mathbf{V}}\mathcal{E}(\hat{w}_k), \mathcal{E}(w_k-u)\Big\rangle\,dt\\
&\qquad\geq \int_0^T 2\Big\langle [\mu(\hat{\theta}) + e^{4k_1 t}\varepsilon\|\hat{w}\|^4_{\mathbf{V}}]\mathcal{E}(\hat{w}), \mathcal{E}(w-u)\Big\rangle\,dt.
\end{aligned} \tag{10.62}$$

Since $e^{k_1 t}\langle \mathrm{rot}\hat{w}_k \times \hat{w}_k, w_k-u\rangle = e^{k_1 t}\langle \mathrm{rot}\hat{w}_k \times \hat{w}_k, -v_0-u\rangle$, by (10.58) we can see that

$$\int_0^T e^{k_1 t}\langle \mathrm{rot}\hat{w}_k \times \hat{w}_k, w_k - u\rangle\, dt \to \int_0^T e^{k_1 t}\langle \mathrm{rot}\hat{w} \times \hat{w}, w - u\rangle\, dt. \tag{10.63}$$

Also, by (10.58) we can see that

$$-\int_0^T e^{-k_1 t}\langle \nabla\Phi_\varepsilon(e^{k_1 t}\hat{w}_k(t)), w_k - u\rangle\, dt \to -\int_0^T e^{-k_1 t}\langle \nabla\Phi_\varepsilon(e^{k_1 t}\hat{w}(t)), w - u\rangle\, dt. \tag{10.64}$$

Since

$$\frac{\alpha_0\hat{\theta}_k}{1+\varepsilon\hat{\theta}_k^2} \to \frac{\alpha_0\hat{\theta}}{1+\varepsilon\hat{\theta}^2} \text{ a.e. in } Q \quad\text{and}\quad \frac{\alpha_0\hat{\theta}_k}{1+\varepsilon\hat{\theta}_k^2} \text{ is bounded in } L^2(0,T;L^2),$$

we have

$$\frac{\alpha_0\hat{\theta}_k}{1+\varepsilon\hat{\theta}_k^2} \rightharpoonup \frac{\alpha_0\hat{\theta}}{1+\varepsilon\hat{\theta}^2} \text{ in } L^2(0,T;L^2) \tag{10.65}$$

(see Lemma 1.2). Owing to (10.58) and (10.65), we have

$$\begin{aligned}
&\int_0^T e^{-k_1 t}\Big\langle \frac{\alpha_0\hat{\theta}_k}{1+\varepsilon\hat{\theta}_k^2} f, w_k - u\Big\rangle dt \\
&\quad= \int_0^T e^{-k_1 t}\Big\langle \frac{\alpha_0\hat{\theta}_k}{1+\varepsilon\hat{\theta}_k^2} f, w_k - w\Big\rangle dt + \int_0^T e^{-k_1 t}\Big\langle \frac{\alpha_0\hat{\theta}_k}{1+\varepsilon\hat{\theta}_k^2} f, w - u\Big\rangle dt \\
&\quad\to \int_0^T e^{-k_1 t}\Big\langle \frac{\alpha_0\hat{\theta}}{1+\varepsilon\hat{\theta}^2} f, w - u\Big\rangle dt.
\end{aligned} \tag{10.66}$$

Using (10.58) and (10.65), we have

$$\begin{aligned}
&\int_0^T e^{2k_1 t}\Big\langle \frac{\alpha_1\hat{\theta}_k}{1+\varepsilon\hat{\theta}_k^2} f\cdot\hat{w}_k, \theta_k - \phi\Big\rangle dt = \int_0^T e^{2k_1 t}\Big\langle \frac{\alpha_1\hat{\theta}_k}{1+\varepsilon\hat{\theta}_k^2} f\cdot\hat{w}_k, \theta_k - \theta\Big\rangle dt \\
&\quad+ \int_0^T e^{2k_1 t}\Big\langle \frac{\alpha_1\hat{\theta}_k}{1+\varepsilon\hat{\theta}_k^2} f\cdot\hat{w}_k - w, \theta\Big\rangle dt + \int_0^T e^{2k_1 t}\Big\langle \frac{\alpha_1\hat{\theta}_k}{1+\varepsilon\hat{\theta}_k^2} f\cdot\hat{w}, \theta - \phi\Big\rangle dt \\
&\quad\to \int_0^T e^{2k_1 t}\Big\langle \frac{\alpha_1\hat{\theta}}{1+\varepsilon\hat{\theta}^2} f\cdot\hat{w}, \theta - \phi\Big\rangle dt.
\end{aligned} \tag{10.67}$$

By Corollary 1.1 and (10.58)

$$(\kappa(\hat{\theta}_k)\nabla\hat{\theta}_k, \nabla(\theta_k - \phi)) \to (\kappa(\hat{\theta})\nabla\hat{\theta}, \nabla(\theta - \phi)). \tag{10.68}$$

It is easy to prove convergence of other terms in the right hand side of (10.59).

Thus, by (10.60)–(10.64) and (10.66)–(10.68), we have the existence of a subsequence $\{(v_k, \theta_k)\}$ such that

$$\liminf_{k\to\infty} \big\langle \mathscr{A}_\varepsilon(w_k, \theta_k), (w_k - u, \theta_k - \phi)\big\rangle \geq \big\langle \mathscr{A}_\varepsilon(w, \theta), (w - u, \theta - \phi)\big\rangle, \tag{10.69}$$

by which property (i) of Theorem 1.47 for $\mathscr{A}_\varepsilon(w, \theta)$ is proved.

Therefore, by Theorem 1.47 there exists a solution to Problem VEA', which gives us the conclusion of Theorem 10.2. □

10.2.2 Estimates of Solutions to the Approximate Problem

Let $0 < \varepsilon < 1$. For $v_0 \in H_K$, $\theta_0 \in L^1(\Omega)$ and $f_2 \in L^1(0, T; (W^{1,2}_{\Gamma_D})^*)$, there exists $v_{0\varepsilon} \in \mathbf{W}^{1,6}(\Omega) \cap K(\Omega)$, $\theta_{0\varepsilon} \in W^{1,6}_{\Gamma_D}(\Omega)$ and $f_{2\varepsilon} \in L^2\big(0, T; (W^{1,2}_{\Gamma_D})^*\big)$ such that

$$\|v_0 - v_{0\varepsilon}\|_H \leq \varepsilon, \quad \|\theta_0 - \theta_{0\varepsilon}\|_{L^1(\Omega)} \leq \varepsilon, \quad \|f_2 - f_{2\varepsilon}\|_{L^1(0,T;(W^{1,2}_{\Gamma_D})^*)} \leq \varepsilon. \tag{10.70}$$

Let $(v_\varepsilon(t), \theta_\varepsilon(t))$ be a solution to Problem VEA with the initial function $(v_{0\varepsilon}, \theta_{0\varepsilon})$ and $f_{2\varepsilon} \in L^2(0, T; (W^{1,2}_{\Gamma_D})^*)$ instead of f_2.

Define $A_\varepsilon(\theta) : \mathbf{V} \to \mathbf{V}^*$ by

$$\langle A_\varepsilon(\theta)v, w\rangle = \langle A(\theta)v, w\rangle + 2\varepsilon\|v\|^4_{\mathbf{V}}\big\langle \mathcal{E}(v), \mathcal{E}(w)\big\rangle \quad \forall v, w \in \mathbf{V},$$

where $A(\theta)$ is the one in (10.16). Then, we get from (10.22)

$$\begin{aligned} \Big\langle \frac{\partial v_\varepsilon(t)}{\partial t}, u(t)\Big\rangle &+ \langle A_\varepsilon(\theta_\varepsilon)v_\varepsilon(t) + B(v_\varepsilon(t), v_\varepsilon(t)), u(t)\rangle + \langle \nabla\Phi_\varepsilon(v_\varepsilon(t)), u(t)\rangle \\ &= -\Big\langle \frac{\alpha_0\theta}{1+\varepsilon\theta^2} f, u(t)\Big\rangle + \langle f + f_1, u(t)\rangle \quad \forall u \in L^6(0, T; \mathbf{V}). \end{aligned} \tag{10.71}$$

By (10.28) there exists c_* such that

$$\begin{aligned} &\big|\nu(k(x)z, z)_{\Gamma_2} + 2\nu(S\tilde{z}, \tilde{z})_{\Gamma_3} + (\alpha(x)z, z)_{\Gamma_5} + \nu(k(x)z, z)_{\Gamma_7}\big| \\ &\qquad \leq \mu_0\|z\|^2_{\mathbf{V}} + c_*\|z\|^2 \quad \forall z \in \mathbf{V} \end{aligned} \tag{10.72}$$

(see Theorem 1.27).

Remark 10.3 Note that c_* depends on the shape of Γ_i, $i = 2, 3, 7$, and the norm of matrix α on Γ_5, and so for the fixed domain Ω it depends only on the norm of α.

By (10.5), (10.7), (10.15) and (10.72), we have

$$\begin{aligned} &\langle A_\varepsilon(\theta_\varepsilon)z, z\rangle \geq \mu_0\|z\|^2_{\mathbf{V}} + 2\varepsilon\|z\|^6_{\mathbf{V}} - c_*\|z\|^2 \quad \forall z \in \mathbf{V}, \\ &|\langle A_\varepsilon(\theta_\varepsilon)z, w\rangle| \leq c_2\big(\|z\|_{\mathbf{V}} + \varepsilon\|z\|^5_{\mathbf{V}}\big)\|w\|_{\mathbf{V}} \quad \forall z, w \in \mathbf{V}. \end{aligned} \tag{10.73}$$

Putting $u = v_\varepsilon$ in (10.71) and taking into account (10.73), we have from (10.71)

$$
\begin{aligned}
\frac{1}{2}\|v_\varepsilon(t)\|^2 &+ \int_0^t \big(\mu_0\|v_\varepsilon\|_{\mathbf{V}}^2 + 2\varepsilon\|v_\varepsilon\|_{\mathbf{V}}^6\big)\,ds + \int_0^t \langle\nabla\Phi_\varepsilon(v_\varepsilon(s)), v_\varepsilon(s)\rangle\,ds \\
&\le \frac{1}{2}\|v_{0\varepsilon}\|^2 + c_*\int_0^t \|v_\varepsilon\|^2\,ds + c\int_0^t (\|f\|_\infty^2 + \|f_1\|_{\mathbf{V}^*}^2)\,ds \qquad (10.74)\\
&\quad + |\alpha_0|\|f\|_\infty \int_0^t\int_\Omega |\theta_\varepsilon||v_\varepsilon|\,dxds.
\end{aligned}
$$

Since Φ_ε is convex, continuous and Fréchet differentiable, we have

$$
\Phi_\varepsilon(u) - \Phi_\varepsilon(v_\varepsilon(t)) \ge \langle\nabla\Phi_\varepsilon(v_\varepsilon(t)), u - v_\varepsilon(t)\rangle \quad \forall u\in\mathbf{V}, \qquad (10.75)
$$

and so by $\Phi_\varepsilon(0_{\mathbf{V}}) = 0$

$$
0 \le \Phi_\varepsilon(v_\varepsilon(t)) \le \langle\nabla\Phi_\varepsilon(v_\varepsilon(t)), v_\varepsilon(t)\rangle. \qquad (10.76)
$$

From (10.74) and (10.76) we have

$$
\begin{aligned}
\|v_\varepsilon(t)\|^2 &+ 2\int_0^t \big(\mu_0\|v_\varepsilon\|_{\mathbf{V}}^2 + 2\varepsilon\|v_\varepsilon\|_{\mathbf{V}}^6\big)\,ds + 2\int_0^t \Phi_\varepsilon(v_\varepsilon(t))\,ds \\
&\le \|v_{0\varepsilon}\|^2 + 2c\|f\|_\infty^2 T + 2c\|f_1\|_{L^2(0,T;\mathbf{V}^*)}^2 + 2c_*\int_0^t \|v_\varepsilon(s)\|^2\,ds \\
&\quad + |\alpha_0|\|f\|_\infty \int_0^t\int_\Omega |\theta_\varepsilon||v_\varepsilon|\,dxds \quad \forall t\in[0,T].
\end{aligned}
\qquad (10.77)
$$

For fixed $0 < \delta < 1$, define

$$
\begin{aligned}
\Phi(\xi) &:= \Big(1 - \frac{1}{(1+|\xi|)^\delta}\Big)\operatorname{sign}\xi, \quad \xi\in\mathbb{R},\\
\Psi(\xi) &:= \begin{cases} \int_0^\xi \Phi(\tau)\,d\tau & \text{for } \xi \ge 0,\\ -\int_\xi^0 \Phi(\tau)\,d\tau & \text{for } \xi < 0.\end{cases}
\end{aligned}
$$

Then

$$
\begin{aligned}
&\Phi'(\xi) = \frac{\delta}{(1+|\xi|)^{1+\delta}} \quad \forall\xi\in\mathbb{R},\\
&|\Phi(\xi)| \le 1, \quad 0 \le \Phi'(\xi) \le \delta \quad \forall\xi\in\mathbb{R},\\
&\Psi'(\xi) = \Phi(\xi),\\
&\Psi(\xi) = |\xi| + \frac{1}{1-\delta}\big(1 - (1+|\xi|)^{1-\delta}\big) \quad \forall\xi\in\mathbb{R},\\
&\frac{|\xi|}{2} - \frac{2^{(1-\delta)/\delta}}{1-\delta} \le \Psi(\xi) \le |\xi| \quad \forall\xi\in\mathbb{R}.
\end{aligned}
\qquad (10.78)
$$

From (10.22) we get

$$\begin{aligned}\Big\langle\frac{\partial\theta_\varepsilon}{\partial t},\varphi\Big\rangle&+(\kappa(\theta_\varepsilon)\nabla\theta_\varepsilon,\nabla\varphi)+(\beta(x)\theta_\varepsilon,\varphi)_{\Gamma_R}-\langle v_\varepsilon\theta_\varepsilon,\nabla\varphi\rangle-\Big\langle\alpha_2\mu(\theta_\varepsilon)\frac{|\mathcal{E}(v_\varepsilon)|^2}{1+\varepsilon|\mathcal{E}(v_\varepsilon)|^2},\varphi\Big\rangle\\&=\Big\langle\frac{\alpha_1\theta_\varepsilon}{1+\varepsilon\theta_\varepsilon^2}f\cdot v_\varepsilon,\varphi\Big\rangle+\langle f_{2\varepsilon},\varphi\rangle\quad\forall\varphi\in L^2(0,T;W^{1,2}_{\Gamma_D}(\Omega)).\end{aligned}\tag{10.79}$$

Since $\theta_\varepsilon\in L^2(0,T;W^{1,2}_{\Gamma_D}(\Omega))\cap C([0,T];L^2(\Omega))$, we have

$$\begin{aligned}&\Phi(\theta_\varepsilon(t))\in W^{1,2}_{\Gamma_D}(\Omega),\quad\Psi(\theta_\varepsilon(t))\in W^{1,2}_{\Gamma_D}(\Omega)\quad\text{for a.e. } t\in[0,T],\\&\langle v_\varepsilon\theta_\varepsilon,\nabla\Phi(\theta_\varepsilon(t))\rangle=-\langle(v_\varepsilon\cdot\nabla\theta_\varepsilon),\Phi(\theta_\varepsilon(t))\rangle\\&\qquad=-\langle v_\varepsilon,\nabla\Psi(\theta_\varepsilon(t))\rangle\\&\qquad=\langle\operatorname{div}v_\varepsilon,\Psi(\theta_\varepsilon(t))\rangle=0,\end{aligned}\tag{10.80}$$

where $v\cdot n|_{\Gamma_R}=0$ (see (10.3)) and $\Psi(\theta_\varepsilon)=0$ on Γ_D were used.
Also

$$\begin{aligned}&\int_0^t\Big\langle\frac{\partial\theta_\varepsilon}{\partial t},\Phi(\theta_\varepsilon)\Big\rangle ds=\int_\Omega\Psi(\theta_\varepsilon(t))\,dx-\int_\Omega\Psi(\theta_{0\varepsilon})\,dx\quad\forall t\in[0,T],\\&(\kappa(\theta_\varepsilon)\nabla\theta_\varepsilon,\nabla\Phi(\theta_\varepsilon))=\int_\Omega\kappa(\theta_\varepsilon)|\nabla\theta_\varepsilon|^2\Phi'(\theta_\varepsilon)\,dx,\\&(\beta(x)\theta_\varepsilon,\Phi(\theta_\varepsilon))_{\Gamma_R}\ge0.\end{aligned}\tag{10.81}$$

Putting $\varphi(t)=\Phi(\theta_\varepsilon(t))$ in (10.79) and using (10.80) and (10.81), we have

$$\begin{aligned}\int_\Omega\Psi(\theta_\varepsilon(t))\,dx&+\int_0^t\int_\Omega\kappa(\theta_\varepsilon)|\nabla\theta_\varepsilon|^2\Phi'(\theta_\varepsilon)\,dx+\int_0^t(\beta(x)\theta_\varepsilon,\Phi(\theta_\varepsilon))_{\Gamma_R}\,ds\\&=\int_\Omega\Psi(\theta_{0\varepsilon})\,dx+\int_0^t\Big\langle\alpha_2\mu(\theta_\varepsilon)\frac{|\mathcal{E}(v_\varepsilon)|^2}{1+\varepsilon|\mathcal{E}(v_\varepsilon)|^2},\Phi(\theta_\varepsilon)\Big\rangle ds\\&\quad+\int_0^t\langle\frac{\alpha_1\theta_\varepsilon}{1+\varepsilon\theta_\varepsilon^2}f\cdot v_\varepsilon,\Phi(\theta_\varepsilon)\rangle\,ds+\int_0^t\langle f_{2\varepsilon},\Phi(\theta_\varepsilon)\rangle\,ds.\end{aligned}\tag{10.82}$$

By virtue of (10.78) and (10.81), we have from (10.82)

$$\begin{aligned}\frac12\|\theta_\varepsilon(t)\|_{L^1(\Omega)}&-\frac{2^{(1-\delta)/\delta}}{1-\delta}\text{meas}\,\Omega+\delta\int_0^t\int_\Omega\kappa_0\frac{|\nabla\theta_\varepsilon|^2}{(1+|\theta_\varepsilon|)^{1+\delta}}\,dxds\\&\le\|\theta_{0\varepsilon}\|_{L^1(\Omega)}+\alpha_2\mu_1\int_0^t\|v_\varepsilon\|^2_{\mathbf{V}}\,ds\\&\quad+|\alpha_1|\|f\|_\infty\int_0^t\int_\Omega|\theta_\varepsilon||v_\varepsilon|\,dxds+c\int_0^t\|f_{2\varepsilon}\|_{(W^{1,2}_{\Gamma_D})^*}\,ds.\end{aligned}\tag{10.83}$$

Multiplying (10.83) by $\frac{\mu_0}{\alpha_2\mu_1}$ and adding the resulting formula to (10.77), we obtain

$$\begin{aligned}&\|v_\varepsilon(t)\|^2+\|\theta_\varepsilon(t)\|_{L^1(\Omega)}+\int_0^t\left(\mu_0\|v_\varepsilon\|_{\mathbf{V}}^2+2\varepsilon\|v_\varepsilon\|_{\mathbf{V}}^6\right)ds\\&\qquad+\int_0^t\Phi_\varepsilon(v_\varepsilon(t))\,ds+\delta\int_0^t\int_\Omega\frac{|\nabla\theta_\varepsilon|^2}{(1+|\theta_\varepsilon|)^{1+\delta}}\,dxdt\\&\le c\Lambda_\delta+c_*\int_0^t\|v_\varepsilon(s)\|^2\,ds+c_1(|\alpha_0|+|\alpha_1|)\|f\|_\infty\int_0^t\int_\Omega|\theta_\varepsilon||v_\varepsilon|\,dxds),\\&\qquad\qquad 0<\delta<1,\end{aligned}\tag{10.84}$$

where

$$\Lambda_\delta=c_\delta+\|v_0\|^2+\|\theta_0\|_{L^1(\Omega)}+\|f\|_\infty^2T+\|f_1\|^2_{L^2(0,T;\mathbf{V}^*)}+\|f_2\|_{L^1(0,T;(W^{1,2}_{\Gamma_D})^*)}.\tag{10.85}$$

By Hölder's inequality and Young's inequality,

$$\begin{aligned}&c_1(|\alpha_0|+|\alpha_1|)\|f\|_\infty\int_0^t\int_\Omega|\theta_\varepsilon||v_\varepsilon|\,dxds\\&\qquad\le c_1(|\alpha_0|+|\alpha_1|)\|f\|_\infty\Big(\int_0^t\|\theta_\varepsilon\|^2_{L^{6/5}}\,ds\Big)^{1/2}\Big(\int_0^t\|v_\varepsilon\|^2_{\mathbf{V}}\,ds\Big)^{1/2}\\&\qquad\le c_1(\mu_0)(|\alpha_0|+|\alpha_1|)\|f\|_\infty\int_0^t\|\theta_\varepsilon\|^2_{L^{6/5}}\,ds+\frac{\mu_0}{2}\int_0^t\|v_\varepsilon\|^2_{\mathbf{V}}\,ds.\end{aligned}\tag{10.86}$$

By Theorem 1.12 with $\theta=\frac{13}{18}$, we have

$$\|\theta_\varepsilon\|_{L^{6/5}}\le\|\theta_\varepsilon\|_{L^1}^{13/18}\|\theta_\varepsilon\|_{L^{5/2}}^{5/18}.\tag{10.87}$$

Taking into account (10.86) and (10.87), we deduce from (10.84)

$$\begin{aligned}&\|v_\varepsilon(t)\|^2+\|\theta_\varepsilon(t)\|_{L^1(\Omega)}+\int_0^t\left(\frac{\mu_0}{2}\|v_\varepsilon\|_{\mathbf{V}}^2+2\varepsilon\|v_\varepsilon\|_{\mathbf{V}}^6\right)ds\\&\quad+\int_0^t\Phi_\varepsilon(v_\varepsilon(t))\,ds+\delta\int_0^t\int_\Omega\frac{|\nabla\theta_\varepsilon|^2}{(1+|\theta_\varepsilon|)^{1+\delta}}\,dxdt\\&\le c\Lambda_\delta+c_*\int_0^t\|v_\varepsilon(s)\|^2\,ds+c_1(\mu_0)(|\alpha_0|+|\alpha_1|)\|f\|_\infty\int_0^t\|\theta_\varepsilon(s)\|^{13/9}_{L^1}\|\theta_\varepsilon(s)\|^{5/9}_{L^{5/2}}\,ds.\end{aligned}\tag{10.88}$$

For $0<\delta<1$ define

$$\eta_\varepsilon:=\frac{|\theta_\varepsilon|}{(1+|\theta_\varepsilon|)^{(1+\delta)/2}}\quad\text{a.e. in }Q.$$

Then,

$$(1+|\theta_\varepsilon|)^{(1-\delta)/2}\le 1+\eta_\varepsilon,\quad|\nabla\eta_\varepsilon|\le\frac{|\nabla\theta_\varepsilon|}{(1+|\theta_\varepsilon|)^{(1+\delta)/2}}\quad\text{a.e. in }Q.\tag{10.89}$$

Taking $\delta = \frac{1}{6}$, we have from the first inequality of (10.89)

$$\|\theta_\varepsilon(s)\|_{L^{5/2}}^{5/6} \le c_6(1 + \|\eta_\varepsilon\|_{L^6}^2).$$

Thus, we have from (10.88)

$$\begin{aligned}
&\|v_\varepsilon(t)\|^2 + \|\theta_\varepsilon(t)\|_{L^1(\Omega)} + \int_0^t \big(\frac{\mu_0}{2}\|v_\varepsilon\|_{\mathbf{V}}^2 + 2\varepsilon\|v_\varepsilon\|_{\mathbf{V}}^6\big)\,ds \\
&\quad + \int_0^t \Phi_\varepsilon(v_\varepsilon(t))\,ds + \frac{1}{6}\int_0^t\int_\Omega \frac{|\nabla\theta_\varepsilon|^2}{(1+|\theta_\varepsilon|)^{7/6}}\,dxdt \\
&\le c\Lambda_{1/6} + c_*\int_0^t \|v_\varepsilon(s)\|^2\,ds \\
&\quad + c_1(\mu_0)(|\alpha_0| + |\alpha_1|)\|f\|_\infty \int_0^t \|\theta_\varepsilon(s)\|_{L^1}^{13/9}(1 + \|\eta_\varepsilon\|_{L^6}^2)^{2/3}\,ds.
\end{aligned} \tag{10.90}$$

By Gronwall's inequality, we have from (10.90)

$$\|v_\varepsilon(t)\|^2 \le \Big(c\Lambda_{1/6} + c_1(|\alpha_0| + |\alpha_1|)\|f\|_\infty \int_0^t \|\theta_\varepsilon(s)\|_{L^1}^{13/9}(1 + \|\eta_\varepsilon\|_{L^6}^2)^{2/3}\,ds\Big)e^{c_* t}. \tag{10.91}$$

It follows from (10.90) and (10.91) that

$$\begin{aligned}
&\|v_\varepsilon(t)\|^2 + \|\theta_\varepsilon(t)\|_{L^1(\Omega)} + \int_0^t \big(\frac{\mu_0}{2}\|v_\varepsilon\|_{\mathbf{V}}^2 + 2\varepsilon\|v_\varepsilon\|_{\mathbf{V}}^6\big)\,ds \\
&\quad + \int_0^t \Phi_\varepsilon(v_\varepsilon(t))\,ds + \frac{1}{6}\int_0^t\int_\Omega \frac{|\nabla\theta_\varepsilon|^2}{(1+|\theta_\varepsilon|)^{7/6}}\,dxdt \\
&\le c\Lambda_{1/6}(1 + c_* e^{c_* T}T) \\
&\quad + (1 + c_* e^{c_* T}T)c_1(|\alpha_0| + |\alpha_1|)\|f\|_\infty \int_0^t \|\theta_\varepsilon(s)\|_{L^1}^{13/9}(1 + \|\eta_\varepsilon(s)\|_{L^6}^2)^{2/3}\,ds.
\end{aligned} \tag{10.92}$$

Put

$$\phi(t) := \|\theta_\varepsilon\|_{C([0,t];L^1)} + \|\eta_\varepsilon\|_{L^2(0,t;L^6)}^2.$$

Taking into account (10.89) and using Hölder's inequality with exponents $3/2$, 3, we have from (10.92)

$$\begin{aligned}
\phi(t) &\le \|\theta_\varepsilon\|_{C([0,t];L^1)} + c\int_0^t\int_\Omega \frac{|\nabla\theta_\varepsilon|^2}{(1+|\theta_\varepsilon|)^{7/6}}\,dxds \\
&\le c\Lambda_{1/6}(1 + c_* e^{c_* T}T) \\
&\quad + (1 + c_* e^{c_* T}T)c_1(|\alpha_0| + |\alpha_1|)\|f\|_\infty T^{1/3}\|\theta_\varepsilon(s)\|_{C[0,t];L^1)}^{13/9}\big(t + \|\eta_\varepsilon\|_{L^2(0,t;L^6)}^2\big)^{2/3},
\end{aligned}$$

and so

$$t+\phi(t)\le K_0+(1+c_*e^{c_*T}T)c_5(|\alpha_0|+|\alpha_1|)\|f\|_\infty T^{1/3}\big(t+\phi(t)\big)^{19/9}, \tag{10.93}$$

where K_0 is $c\Lambda_{1/6}(1+c_*e^{c_*T}T)+T$. By (10.85), without loss of generality, we may assume that $\phi(0)\le K_0$. If

$$(1+c_*e^{c_*T}T)c(|\alpha_0|+|\alpha_1|)\|f\|_\infty T^{1/3}\le \frac{1}{3}(2K_0)^{(1-19/9)}, \tag{10.94}$$

then by Lemma 1 of Appendix of [4] we have from (10.93)

$$\phi(t)\le 2c\Lambda_{1/6}(1+c_*e^{c_*T}T)+T \quad \forall t\in[0,T]. \tag{10.95}$$

By virtue of (10.92) and (10.95), using (10.94) we have

$$\|v_\varepsilon(t)\|^2+\|\theta_\varepsilon(t)\|_{L^1(\Omega)}+\int_0^t\big(\frac{\mu_0}{2}\|v_\varepsilon\|^2_{\mathbf{V}}+2\varepsilon\|v_\varepsilon\|^6_{\mathbf{V}}\big)\,ds+\int_0^t\int_\Omega\frac{|\nabla\theta_\varepsilon|^2}{(1+|\theta_\varepsilon|)^{1+\delta}}\,dxdt \le C(\Lambda,c_*) \tag{10.96}$$

and

$$\int_0^T \Phi_\varepsilon(v_\varepsilon(t))\,dt\le C(\Lambda,c_*), \tag{10.97}$$

where $C(\Lambda,c_*)$ is independent of ε, Φ. Since Φ is nonnegative, by (1.36) and (10.97) we have

$$\int_0^T \|v_\varepsilon(t)-J_\varepsilon(v_\varepsilon(t))\|^2_{\mathbf{V}}\,dt\le c_7\varepsilon \tag{10.98}$$

with c_7 independent of ε.

Taking $u(t)=\big(v_\varepsilon(t)-v_\varepsilon(s)\big)$, $s\in(0,T)$, in (10.71), we then have

$$\begin{aligned}&\frac{1}{2}\frac{d\|v_\varepsilon(t)-v_\varepsilon(s)\|^2}{dt}\\&\quad+\Big\langle A_\varepsilon(\theta_\varepsilon)v_\varepsilon(t)+B(v_\varepsilon(t),v_\varepsilon(t))+\frac{\alpha_0\theta_\varepsilon}{1+\varepsilon\theta_\varepsilon^2}f-f-f_1,v_\varepsilon(t)-v_\varepsilon(s)\Big\rangle\\&=\langle\nabla\Phi_\varepsilon(v_\varepsilon(t)),v_\varepsilon(s)-v_\varepsilon(t)\rangle\le\Phi_\varepsilon(v_\varepsilon(s))-\Phi_\varepsilon(v_\varepsilon(t))\le\Phi_\varepsilon(v_\varepsilon(s)).\end{aligned} \tag{10.99}$$

Then by virtue of (10.73), we have from (10.99)

$$\begin{aligned}\frac{1}{2}\frac{d\|v_\varepsilon(t)-v_\varepsilon(s)\|^2}{dt}\le&\,\Phi_\varepsilon(v_\varepsilon(s))+\langle A_\varepsilon v_\varepsilon(t),v_\varepsilon(s)\rangle+\langle B(v_\varepsilon(t),v_\varepsilon(t)),v_\varepsilon(s)\rangle\\&+\Big\langle-\frac{\alpha_0\theta_\varepsilon}{1+\varepsilon\theta_\varepsilon^2}f+f+f_1,v_\varepsilon(t)-v_\varepsilon(s)\Big\rangle+c_*\|v_\varepsilon(t)\|^2,\end{aligned} \tag{10.100}$$

where $\langle B(v_\varepsilon(t),v_\varepsilon(t)),v_\varepsilon(t)\rangle=0$ was used.

Let us integrate each term of (10.100) respectively first with respect to t from s to $s+h$ and then with respect to s from 0 to T, where $v_\varepsilon(t)=0$ when $t\in(T,T+h)$.

$$\int_0^T\int_s^{s+h}\frac{d\|v_\varepsilon(t)-v_\varepsilon(s)\|^2}{dt}\,dtds=\int_0^T\|v_\varepsilon(s+h)-v_\varepsilon(s)\|^2\,ds. \tag{10.101}$$

By (10.97)

$$\int_0^T\int_s^{s+h}\Phi_\varepsilon(v_\varepsilon(s))\,dtds\le h\int_0^T\Phi_\varepsilon(v_\varepsilon(s))\,ds\le c_6h. \tag{10.102}$$

By (10.96) and (10.73), we have

$$\begin{aligned}&\int_0^T\int_s^{s+h}\langle A_\varepsilon(\theta)v_\varepsilon(t),v_\varepsilon(s)\rangle\,dtds\\&\le c\int_0^T\|v_\varepsilon(s)\|_{\mathbf{V}}\Big(\int_s^{s+h}(\|v_\varepsilon(t)\|_{\mathbf{V}}+\varepsilon\|v_\varepsilon(t)\|_{\mathbf{V}}^5)dt\Big)\,ds\\&\le c\int_0^T\|v_\varepsilon(s)\|_{\mathbf{V}}(\sqrt{h}\|v_\varepsilon\|_{L^2(0,T;\mathbf{V})})\,ds+c\int_0^T\|v_\varepsilon(s)\|_{\mathbf{V}}\Big(\varepsilon h^{1/6}\|v_\varepsilon\|^5_{L^6(0,T;\mathbf{V})}\Big)\,ds\\&\le c_8h^{1/6},\end{aligned} \tag{10.103}$$

where c_8 is independent of ε. Since $\|w\|_{\mathbf{L}^3}\le K\|w\|^{\frac12}_{\mathbf{L}^2}\|w\|^{\frac12}_{\mathbf{L}^6}$,

$$\begin{aligned}|\langle B(v,w),z\rangle|=|\langle\operatorname{rot}v\times w,z\rangle|&\le K\|\operatorname{rot}v\|_{\mathbf{L}^2}\|w\|_{\mathbf{L}^3}\|z\|_{\mathbf{L}^6}\\&\le K\|v\|_{\mathbf{V}}\|w\|^{\frac12}\|w\|^{\frac12}_{\mathbf{V}}\|z\|_{\mathbf{V}},\end{aligned} \tag{10.104}$$

and so by (10.96) we have

$$\begin{aligned}&\int_0^T\int_s^{s+h}\langle B(v_\varepsilon(t),v_\varepsilon(t))v_\varepsilon(s)\rangle\,dtds\\&\le K\int_0^T\int_s^{s+h}\|v_\varepsilon(t)\|^{\frac32}_{\mathbf{V}}\|v_\varepsilon(t)\|^{\frac12}\|v_\varepsilon(s)\|_{\mathbf{V}}\,dtds\\&\le K\int_0^T\|v_\varepsilon(s)\|_{\mathbf{V}}\Big(\int_s^{s+h}\|v_\varepsilon(t)\|^2_{\mathbf{V}}\,dt\Big)^{\frac34}\Big(\int_s^{s+h}\|v_\varepsilon(t)\|^2\,dt\Big)^{\frac14}\,ds\\&\le c_9h^{\frac14},\end{aligned} \tag{10.105}$$

where $\|v_\varepsilon(t)\|^2\le C(\Lambda,c_*)$ (see (10.96)) was used. Also, by (10.96) we have

$$
\begin{aligned}
&\int_0^T\int_s^{s+h}\langle (f+f_1)(t), v_\varepsilon(t)\rangle\,dtds \le \int_0^T |\langle (f+f_1)(t), v_\varepsilon(t)\rangle| \Big(\int_{t-h}^t ds\Big)dt \le Kh,\\
&\int_0^T\int_s^{s+h}\langle (f+f_1)(t), -v_\varepsilon(s)\rangle\,dtds \le K\int_0^T \|v_\varepsilon(s)\|_{\mathbf{V}}\int_s^{s+h}\|(f+f_1)(t)\|_{\mathbf{V}^*}\,dtds \\
&\qquad\qquad \le c_9\sqrt{h}.
\end{aligned}
\tag{10.106}
$$

In the same way we get

$$
\int_0^T\int_s^{s+h}\Big\langle \frac{\alpha_0\theta_\varepsilon}{1+\varepsilon\theta_\varepsilon^2}f,\, v_\varepsilon(t)-v_\varepsilon(s)\Big\rangle\,dtds \le Kh + c_9\sqrt{h}, \tag{10.107}
$$

$$
\int_0^T\int_s^{s+h} c_*\|v_\varepsilon(t)\|^2\,dtds \le Kh. \tag{10.108}
$$

By virtue of (10.100)–(10.108), uniformly with respect to ε

$$
\int_0^T \|v_\varepsilon(s+h)-v_\varepsilon(s)\|^2\,ds \le O(h^{\frac{1}{6}}), \tag{10.109}
$$

which implies that the set $\{v_\varepsilon\}$ is relatively compact in $L^2(0,T;\mathbf{W}^{\frac{9}{10},2}(\Omega))$ (see Theorem 1.38).

Therefore, by (10.96), there exists v and a subsequence $\{v_{\varepsilon_k}\}$ such that as $\varepsilon_k\to 0$,

$$
\begin{aligned}
v_{\varepsilon_k} &\overset{*}{\rightharpoonup} v \ \text{ in } L^\infty(0,T;H),\\
v_{\varepsilon_k} &\to v \ \text{ in } L^2(0,T;\mathbf{W}^{\frac{9}{10},2}(\Omega)),\\
v_{\varepsilon_k} &\rightharpoonup v \ \text{ in } L^2(0,T;\mathbf{V}).
\end{aligned}
\tag{10.110}
$$

We will show that θ_ε is relative compact in $L^r(0,T;W^{\alpha,r})$ $\forall r\,(1<r<5/4)$, $\forall\alpha\,(0<\alpha<1)$. To this end, we need to get an estimate on $\nabla\theta_\varepsilon$. Let $1\le r<3/2$. By (10.96), inequalities $|a+b|^p\le 2^p(|a|^p+|b|^p)$, $|a|+|b|\le(|a|^{\frac{1}{p}}+|b|^{\frac{1}{p}})^p$, $p\in(1,\infty)$, and Hölder's inequality with exponents $\frac{2}{r}$, $\frac{2}{2-r}$, we have

$$
\begin{aligned}
\int_Q|\nabla\theta_\varepsilon|^r\,dxdt &\le \Big(\int_Q\frac{|\nabla\theta_\varepsilon|^2}{(1+|\theta_\varepsilon|)^{1+\delta}}\,dxdt\Big)^{r/2}\Big(\int_Q(1+|\theta_\varepsilon|)^{r(1+\delta)/(2-r)}\,dxdt\Big)^{(2-r)/2}\\
&\le 2^{r(1+\delta)/2}C(\Lambda,c_*)^{r/2}\Big[(\text{mes } Q)^{(2-r)/2}+\Big(\int_Q|\theta_\varepsilon|^{r(1+\delta)/(2-r)}\,dxdt\Big)^{(2-r)/2}\Big],
\end{aligned}
\tag{10.111}
$$

where $C(\Lambda,c_*)$ is the one in (10.96).

To use the property $W^{1,r}(\Omega)\subset L^q(\Omega)$, let us take q such that $\frac{1}{r}-\frac{1}{3}=\frac{1}{q}$, i.e. $q=\frac{3r}{3-r}$. Let us take $\delta_0>0$ such that

$$
1<r(1+\delta_0)/(2-r)<q,
$$

which holds if $0 < \delta_0 < (3-2r)/(3-r)$. Set $s = r(1+\delta_0)/(2-r)$ and take λ such that

$$\frac{1}{s} = \frac{\lambda}{1} + \frac{1-\lambda}{q}.$$

Then, $\lambda = \frac{q-s}{s(q-1)}$, $0 < \lambda < 1$ and by Theorem 1.12

$$\|\eta\|_{L^s} \le c\|\eta\|_{L^1}^{\lambda}\|\eta\|_{L^q}^{1-\lambda} \le c\|\eta\|_{L^1}^{\lambda}\|\nabla\eta\|_{\mathbf{L}^r}^{1-\lambda} \quad \forall \eta \in W^{1,r}_{\Gamma_D}. \tag{10.112}$$

Therefore, by (10.96) and (10.112) we have

$$\int_Q |\theta_\varepsilon|^{r(1+\delta_0)/(2-r)}\, dxdt \le cC(\Lambda, c_*)^{\lambda s}\int_0^T \left(\int_\Omega |\nabla\theta_\varepsilon|^r\, dx\right)^{(1-\lambda)s/r} dt.$$

We will fix r and δ_0 so that

$$1 \le r < \frac{5}{4}, \quad 0 < \delta_0 < \frac{5-4r}{3},$$

which ensure that $0 < \delta_0 < (3-2r)/(3-r)$.

Taking into account $\lambda = \frac{q-s}{s(q-1)}$, $q = \frac{3r}{3-r}$ and $s = \frac{r(1+\delta_0)}{2-r}$, we have

$$(1-\lambda)s/r = \frac{1}{r}\frac{q(s-1)}{q-1} = \frac{3(2r-2+r\delta_0)}{(4r-3)(2-r)} < \frac{3\left(2r-2+r\frac{5-4r}{3}\right)}{(4r-3)(2-r)} = 1.$$

Then, by Hölder's inequality we have

$$\int_Q |\theta_\varepsilon|^{r(1+\delta_0)/(2-r)}\, dxdt \le cC(\Lambda, c_*)^{\lambda s}\left(\int_0^T\int_\Omega |\nabla\theta_\varepsilon|^r\, dx\, dt\right)^{(1-\lambda)s/r},$$

$$\left(\int_Q |\theta_\varepsilon|^{r(1+\delta_0)/(2-r)}\, dxdt\right)^{(2-r)/2} \le cC(\Lambda, c_*)^{\lambda s(2-r)/2}\left(\int_Q |\nabla\theta_\varepsilon|^r\, dxdt\right)^{(1-\lambda)s(2-r)/2r}, \tag{10.113}$$

where $\frac{(1-\lambda)s(2-r)}{2r} = \frac{1}{2}(1-\lambda)(1+\delta_0) < 1$. Then, it follows from Young's inequality and (10.113) that

$$2^{r(1+\delta)/2}C(\Lambda, c_*)^{r/2}\left(\int_Q |\theta_\varepsilon|^{r(1+\delta_0)/(2-r)}\, dxdt\right)^{(2-r)/2} \le \frac{1}{2}\int_Q |\nabla\theta_\varepsilon|^r\, dxdt + R(C(\Lambda, c_*)), \tag{10.114}$$

where $R(\cdot)$ is a nonnegative continuous function on $[0, \infty)$. Substituting (10.114) into (10.111) with $\delta = \delta_0$, we have

$$\int_Q |\nabla\theta_\varepsilon|^r\, dxdt \le cC(\Lambda, c_*)^{r/2} + 2R(C(\Lambda, c_*)), \quad 1 < r < \frac{5}{4}, \quad 0 < \delta_0 < \frac{5-4r}{3}. \tag{10.115}$$

Taking $r = 6/5$, we have from (10.115)

$$\|\theta_\varepsilon\|^{\frac{6}{5}}_{L^{6/5}(0,T;L^2)} \le cC(\Lambda, c_*)^{r/2} + 2R(C(\Lambda, c_*)). \tag{10.116}$$

From (10.79) we will get an estimate on θ'_ε. Let $1 \le r < \frac{5}{4}$ and $r' := r/(r-1) > 5$. By Hölder's inequality,

$$\begin{aligned} &|(\kappa(\theta_\varepsilon)\nabla\theta_\varepsilon, \nabla\varphi)| \le \kappa_1 \|\nabla\theta_\varepsilon\|_{\mathbf{L}^r}\|\nabla\varphi\|_{\mathbf{L}^{r'}}, \\ &|(\beta(x)\theta_\varepsilon, \varphi)_{\Gamma_R}| \le \beta_1\|\theta_\varepsilon\|_{L^1(\Gamma_R)}\|\varphi\|_{L^\infty(\Gamma_R)} \le c\|\theta_\varepsilon\|_{W^{1,r}_{\Gamma_D}(\Omega)}\|\varphi\|_{W^{1,r'}_{\Gamma_D}(\Omega)}. \end{aligned} \tag{10.117}$$

In the last inequality a trace theorem (see Theorem 1.25) was used. As in [5] let us estimate $\langle v_\varepsilon\theta_\varepsilon, \nabla\varphi\rangle$. Let r satisfy $\frac{21}{20} < r < \frac{5}{4}$. Then there exists $\frac{4}{3} < \sigma < q$ such that

$$\frac{q(\sigma-1)}{\sigma(q-1)} = \frac{5r}{8},$$

where $q = \frac{3r}{3-r}$. Define

$$\rho := \frac{4\sigma}{3\sigma-4}, \quad \lambda := \frac{q-\sigma}{\sigma(q-1)}.$$

Then,

$$\frac{1}{4} + \frac{1}{\sigma} + \frac{1}{\rho} = 1, \quad \frac{1}{\sigma} = \frac{\lambda}{1} + \frac{1-\lambda}{q}, \quad 1-\lambda = \frac{5r}{8}.$$

Taking into account

$$\begin{aligned} &\|\theta_\varepsilon\|_{L^\sigma} \le c\|\theta_\varepsilon\|^\lambda_{L^1}\|\theta_\varepsilon\|^{1-\lambda}_{L^q} \le c\|\theta_\varepsilon\|^\lambda_{L^1}\|\nabla\theta_\varepsilon\|^{1-\lambda}_{\mathbf{L}^r}, \\ &\|v_\varepsilon\|_{\mathbf{L}^4(\Omega)} \le c\|v_\varepsilon\|^{1/4}_{\mathbf{L}^2(\Omega)}\|\nabla v_\varepsilon\|^{3/4}_{\mathbf{L}^2(\Omega)}, \end{aligned}$$

we have

$$\begin{aligned} |\langle v_\varepsilon\theta_\varepsilon, \nabla\varphi\rangle| &\le c\|v_\varepsilon\|_{\mathbf{L}^4}\|\theta_\varepsilon\|_{\mathbf{L}^\sigma}\|\nabla\varphi\|_{\mathbf{L}^\rho} \\ &\le c\|v_\varepsilon\|^{1/4}_{L^\infty(0,T;H)}\|\theta_\varepsilon\|^\lambda_{L^\infty(0,T;L^1)}\|\nabla v_\varepsilon\|^{3/4}_{\mathbf{L}^2(\Omega)}\|\nabla\theta_\varepsilon\|^{1-\lambda}_{\mathbf{L}^r}\|\varphi\|_{W^{1,\rho}_{\Gamma_D}}, \end{aligned} \tag{10.118}$$

which is the desired one. Also,

$$\left|\left\langle \alpha_2\mu(\theta_\varepsilon)\frac{|\mathcal{E}(v_\varepsilon)|^2}{1+\varepsilon|\mathcal{E}(v_\varepsilon)|^2}, \varphi\right\rangle\right| \le c\mu_1\|v_\varepsilon\|^2_{\mathbf{V}}\max_{x\in\Omega}|\varphi| \le c\|v_\varepsilon\|^2_{\mathbf{V}}\|\varphi\|_{W^{1,r'}_{\Gamma_D}} \quad \forall\varphi \in W^{1,r'}_{\Gamma_D}(\Omega), \tag{10.119}$$

$$|\langle \frac{\alpha_1 \theta_\varepsilon}{1+\varepsilon\theta_\varepsilon^2} f \cdot v_\varepsilon, \varphi\rangle| \le c\|\theta_\varepsilon\| \|f\|_\infty \|v_\varepsilon\| \max_{x\in\Omega} |\varphi| \le c\|\theta_\varepsilon\| \|f\|_\infty \|v_\varepsilon\| \|\varphi\|_{W^{1,r'}_{\Gamma_D}}$$
$$\forall \varphi \in W^{1,r'}_{\Gamma_D}(\Omega). \tag{10.120}$$

Let $\tau := \max\{r', \rho\}$. Then, by (10.117)–(10.120), we have from (10.79)

$$\Big|\Big\langle \frac{\partial \theta_\varepsilon}{\partial t}, \varphi\Big\rangle\Big| \le c\big(\|\nabla\theta_\varepsilon\|_{\mathbf{L}^r} + \|v_\varepsilon\|^{1/4}_{L^\infty(0,T;H)} \|\theta_\varepsilon\|^{\lambda}_{L^\infty(0,T;L^1)} \|\nabla v_\varepsilon\|^{3/4}_{\mathbf{L}^2(\Omega)} \|\nabla\theta_\varepsilon\|^{1-\lambda}_{\mathbf{L}^r} + \|v_\varepsilon\|^2_{\mathbf{V}}$$
$$+ \|\theta_\varepsilon\| \|f\|_\infty \|v_\varepsilon\| + \|f_{2\varepsilon}\|_{(W^{1,2}_{\Gamma_D})^*}\big) \|\varphi\|_{W^{1,\tau}_{\Gamma_D}} \quad \forall \varphi \in W^{1,\tau}_{\Gamma_D}(\Omega).$$

Hence, taking into account (10.115) and (10.116), we see that

$$\theta'_\varepsilon \in L^1\big(0,T; (W^{1,\tau}_{\Gamma_D})^*\big)$$

and

$$\|\theta'_\varepsilon\|_{L^1\left(0,T;(W^{1,\tau}_{\Gamma_D})^*\right)}$$
$$\le c\Big(\|\theta_\varepsilon\|_{L^1\left(0,T;W^{1,r}_{\Gamma_D}\right)} + \|v_\varepsilon\|^{1/4}_{L^\infty(0,T;H)} \|\theta_\varepsilon\|^{\lambda}_{L^\infty(0,T;L^1)} \|v_\varepsilon\|^{3/4}_{L^2(0,T;\mathbf{V})} \|\theta_\varepsilon\|^{\frac{5r}{8}}_{L^r(0,T;W^{1,r}_{\Gamma_D})}$$
$$+ \|v_\varepsilon\|^2_{L^2(0,T;\mathbf{V})} + \|\theta_\varepsilon\|_{L^1(0,T;L^2)} \|f\|_\infty \|v_\varepsilon\|_{L^\infty(0,T;H)} + \|f_{2\varepsilon}\|_{L^1(0,T;(W^{1,2}_{\Gamma_D})^*)}\Big), \tag{10.121}$$

where $\frac{21}{20} < r < \frac{5}{4}$. Therefore, $\|\theta'_\varepsilon\|_{L^1\left(0,T;(W^{1,\tau}_{\Gamma_D})^*\right)}$ is bounded.

Thus, by (10.115) $\{\theta_\varepsilon\}$ is relatively compact in $L^r(0,T;W^{\alpha,r})$ $\forall r\,(1<r<5/4)$, $\forall \alpha\,(0<\alpha<1)$ (see Theorem 1.39). Therefore, there exists θ and a subsequence $\{\theta_{\varepsilon_k}\}$ such that as $\varepsilon_k \to 0$

$$\begin{aligned}
&\theta_{\varepsilon_k} \overset{*}{\rightharpoonup} \theta \quad \text{in } L^\infty(0,T;L^1(\Omega)),\\
&\theta_{\varepsilon_k} \rightharpoonup \theta \quad \text{in } L^r(0,T;W^{1,r}_D(\Omega))\ \forall r\,(1<r<5/4),\\
&\theta_{\varepsilon_k} \to \theta \quad \text{in } L^r(0,T;W^{\alpha,r}(\Omega))\ \forall r\,(1<r<5/4),\ \forall\alpha\,(\frac{1}{r}<\alpha<1)\\
&\qquad \text{and a.e. in } Q.
\end{aligned} \tag{10.122}$$

10.2.3 Passing to the Limit

By passing to limit of solutions in Theorem 10.2 we will prove Theorem 10.1. Owing to (10.110) and (10.122), we can extract subsequences, which are denoted as before, such that as $\varepsilon_k \to 0$

$$\begin{aligned}
&v_{\varepsilon_k} \rightharpoonup v \quad \text{in } L^2(0,T;\mathbf{V}),\\
&v_{\varepsilon_k} \overset{*}{\rightharpoonup} v \quad \text{in } L^\infty(0,T;H),\\
&v_{\varepsilon_k} \to v \quad \text{in } L^2(0,T;\mathbf{W}^{\frac{9}{10},2}(\Omega)),\\
&\theta_{\varepsilon_k} \rightharpoonup \theta \quad \text{in } L^r(0,T;W_D^{1,r}(\Omega)) \quad \forall r\,(1<r<5/4),\\
&\theta_{\varepsilon_k} \to \theta \quad \text{in } L^r(0,T;W^{\alpha,r}(\Omega)) \quad \forall r\,(1<r<5/4),\ \forall\alpha\,\Big(\frac{1}{r}<\alpha<1\Big)\\
&\qquad \text{and a.e. in } Q.
\end{aligned} \tag{10.123}$$

Subtracting (10.71) with $u = v_{\varepsilon_k}$ from (10.71), we have

$$\begin{aligned}
&\int_0^T \Big\langle \frac{\partial v_{\varepsilon_k}}{\partial t} + A_{\varepsilon_k}(\theta_{\varepsilon_k})v_{\varepsilon_k}(t) + B(v_{\varepsilon_k}(t), v_{\varepsilon_k}(t)) + \nabla\Phi_{\varepsilon_k}(v_{\varepsilon_k}), u(t) - v_{\varepsilon_k}(t)\Big\rangle\, dt\\
&\qquad = \int_0^T \Big\langle -\frac{\alpha_0\theta_{\varepsilon_k}}{1+\varepsilon_k\theta_{\varepsilon_k}^2} f + f + f_1, u(t) - v_{\varepsilon_k}(t)\Big\rangle\, dt \qquad \forall u \in L^6(0,T;\mathbf{V}).
\end{aligned} \tag{10.124}$$

Since

$$\begin{aligned}
&\int_0^T \langle v_{\varepsilon_k}'(t), u(t) - v_{\varepsilon_k}(t)\rangle\, dt\\
&\quad = \int_0^T \langle u'(t), u(t) - v_{\varepsilon_k}(t)\rangle\, dt - \frac{1}{2}\|v_{\varepsilon_k}(T) - u(T)\|^2 + \frac{1}{2}\|v_{\varepsilon_k}(0) - u(0)\|^2\\
&\qquad \forall u \in L^6(0,T;\mathbf{V}) \text{ with } u' \in L^2(0,T;\mathbf{V}^*),
\end{aligned}$$

we have from (10.124)

$$\begin{aligned}
&\int_0^T \Big\langle u'(t) + A_{\varepsilon_k}(\theta_{\varepsilon_k})v_{\varepsilon_k}(t) + B(v_{\varepsilon_k}(t), v_{\varepsilon_k}(t)) + \nabla\Phi_{\varepsilon_k}(v_{\varepsilon_k}), u(t) - v_{\varepsilon_k}(t)\Big\rangle\, dt\\
&\quad - \int_0^T \Big\langle -\frac{\alpha_0\theta_{\varepsilon_k}}{1+\varepsilon\theta_{\varepsilon_k}^2} f + f + f_1, u(t) - v_{\varepsilon_k}(t)\Big\rangle\, dt \ge -\frac{1}{2}\|v_{\varepsilon_k}(0) - u(0)\|^2\\
&\qquad \forall u \in L^6(0,T;\mathbf{V}) \text{ with } u' \in L^2(0,T;\mathbf{V}^*).
\end{aligned} \tag{10.125}$$

Taking into account (10.75) and $\big(\varepsilon_k \||v_{\varepsilon_k}\|_{\mathbf{V}}^4 \mathcal{E}(v_{\varepsilon_k}), \mathcal{E}(v_{\varepsilon_k})\big) \ge 0$, we have from (10.125)

$$\begin{aligned}
&\int_0^T \langle u'(t) + A(\theta_{\varepsilon_k})v_{\varepsilon_k}(t) + B(v_{\varepsilon_k}(t), v_{\varepsilon_k}(t), u(t) - v_{\varepsilon_k}(t)\rangle\, dt\\
&\quad + \int_0^T (\varepsilon_k \||v_{\varepsilon_k}\|_{\mathbf{V}}^4 \mathcal{E}(v_{\varepsilon_k}), \mathcal{E}(u))\, dt - \int_0^T \Big\langle -\frac{\alpha_0\theta_{\varepsilon_k}}{1+\varepsilon\theta_{\varepsilon_k}^2} f + f + f_1, u(t) - v_{\varepsilon_k}(t)\Big\rangle\, dt\\
&\quad + \int_0^T (\Phi_{\varepsilon_k}(u(t)) - \Phi_{\varepsilon_k}(v_{\varepsilon_k}(t)))\, dt \ge -\frac{1}{2}\|v_{\varepsilon_k}(0) - u(0)\|^2.
\end{aligned} \tag{10.126}$$

Since $\Phi(u) \geq \Phi_{\varepsilon_k}(u)$ and $\Phi(J_{\varepsilon_k} v_{\varepsilon_k}(t)) \leq \Phi_{\varepsilon_k}(v_{\varepsilon_k}(t))$ (see (1.37)), we have from (10.126)

$$\int_0^T \langle u'(t) + A(\theta_{\varepsilon_k}) v_{\varepsilon_k}(t) + B(v_{\varepsilon_k}(t), v_{\varepsilon_k}(t)), u(t) - v_{\varepsilon_k}(t)\rangle \, dt$$
$$+ \int_0^T \big(\varepsilon_k \| |v_{\varepsilon_k}\|_{\mathbf{V}}^4 \mathcal{E}(v_{\varepsilon_k}), \mathcal{E}(u)\big) \, dt - \int_0^T \Big\langle - \frac{\alpha_0 \theta_{\varepsilon_k}}{1 + \varepsilon \theta_{\varepsilon_k}^2} f + f + f_1, u(t) - v_{\varepsilon_k}(t)\Big\rangle dt$$
$$+ \Psi(u) - \int_0^T \Phi(J_{\varepsilon_k} v_{\varepsilon_k}(t))) \, dt \geq -\frac{1}{2} \|v_{\varepsilon_k}(0) - u(0)\|^2. \tag{10.127}$$

Taking into account (10.123), by Corollary 1.1 we get as $\varepsilon_k \to 0$

$$\int_0^T \langle A(\theta_{\varepsilon_k}) v_{\varepsilon_k}(t), u(t)\rangle \, dt \equiv \int_0^T a_0(\theta_{\varepsilon_k}; v_{\varepsilon_k}, u) \, dt \to \int_0^T \langle A(\theta) v(t), u(t)\rangle \, dt. \tag{10.128}$$

Owing to (10.123), as $\varepsilon_k \to 0$

$$v_{\varepsilon_k} \to v \quad \text{in } L^2(0, T; \mathbf{L}^2(\partial\Omega)).$$

Thus, by Lemma 1.3 we have

$$2(\mu(\theta_{\varepsilon_k}) k(x) v_{\varepsilon_k}, v_{\varepsilon_k})_{\Gamma_2} + 2(\mu(\theta_{\varepsilon_k}) S \tilde{v}_{\varepsilon_k}, \tilde{v}_{\varepsilon_k})_{\Gamma_3} + 2(\alpha(x) v_{\varepsilon_k}, v_{\varepsilon_k})_{\Gamma_5} + (\mu(\theta_{\varepsilon_k}) k(x) v_{\varepsilon_k}, v_{\varepsilon_k})_{\Gamma_7}$$
$$\to 2(\mu(\theta) k(x) v, v)_{\Gamma_2} + 2(\mu(\theta) S \tilde{v}, \tilde{v})_{\Gamma_3} + 2(\alpha(x) v, v)_{\Gamma_5} + (\mu(\theta) k(x) v, v)_{\Gamma_7}.$$

Therefore, taking into account

$$\liminf_{\varepsilon_k \to 0} 2(\mu(\theta_{\varepsilon_k}) \mathcal{E}(v_{\varepsilon_k}), \mathcal{E}(v_{\varepsilon_k})) \geq 2(\mu(\theta) \mathcal{E}(v), \mathcal{E}(v)),$$

we have

$$\liminf_{\varepsilon_k \to 0} \langle A(\theta_{\varepsilon_k}) v_{\varepsilon_k}(t), v_{\varepsilon_k}(t)\rangle \, dt \geq \langle A(\theta) v(t), v(t)\rangle \, dt. \tag{10.129}$$

By (10.128) and (10.129), we have

$$\liminf_{\varepsilon_k \to 0} \int_0^T \langle A(\theta_{\varepsilon_k}) v_{\varepsilon_k}(t), u(t) - v_{\varepsilon_k}(t)\rangle \, dt \leq \int_0^T \langle A(\theta) v(t), u(t) - v(t)\rangle \, dt. \tag{10.130}$$

By (10.96) we have

$$\begin{aligned}\Big|\int_0^T \varepsilon_k\|v_{\varepsilon_k}\|_{\mathbf{V}}^4\big(\mathcal{E}(v_{\varepsilon_k}),\mathcal{E}(u)\big)\,dt\Big| &\le c{\varepsilon_k}^{1/6}\int_0^T {\varepsilon_k}^{5/6}\|v_{\varepsilon_k}\|_{\mathbf{V}}^5\|u(t)\|_{\mathbf{V}}\,dt\\ &\le c{\varepsilon_k}^{1/6}\Big(\int_0^T \big({\varepsilon_k}^{5/6}\|v_{\varepsilon_k}\|_{\mathbf{V}}^5\big)^{6/5}dt\Big)^{5/6}\Big(\int_0^T \|u(t)\|_{\mathbf{V}}^6\,dt\Big)^{1/6}\\ &\le c{\varepsilon_k}^{1/6}C(\Lambda,c^*)^{5/6}\|u\|_{L^6(0,T;\mathbf{V})}\to 0 \quad\text{as } \varepsilon_k\to 0.\end{aligned} \tag{10.131}$$

By (10.98), we have that $J_{\varepsilon_k}v_{\varepsilon_k}\rightharpoonup v$ in $L^2(0,T;\mathbf{V})$ as $\varepsilon_k\to 0$. Since the functional $\Phi:\mathbf{V}\to\overline{\mathbb{R}}$ is weakly lower semi-continuous, we have

$$\liminf_{\varepsilon_k\to 0}\int_0^T \Phi(J_{\varepsilon_k}v_{\varepsilon_k}(t))\,dt \ge \int_0^T \Phi(v(t))\,dt \equiv \Psi(v). \tag{10.132}$$

It is easy to prove that

$$\int_0^T \langle B(v_{\varepsilon_k}(t),v_{\varepsilon_k}(t)),u(t)\rangle\,dt \to \int_0^T \langle B(v(t),v(t)),u(t)\rangle\,dt \tag{10.133}$$

as $\varepsilon_k\to 0$ (see (6.46)–(6.49)).

Let us prove that

$$\int_0^T \Big\langle\frac{\alpha_0\theta_{\varepsilon_k}}{1+\varepsilon_k\theta_{\varepsilon_k}^2}f,u(t)-v_{\varepsilon_k}(t)\Big\rangle dt \to \int_0^T \langle\alpha_0\theta f,u(t)-v(t)\rangle\,dt\to 0 \quad\text{as } \varepsilon_k\to 0. \tag{10.134}$$

By (10.123) and Lemma 1.5, we have

$$\begin{aligned}&\int_0^T \Big\langle\Big(\frac{\alpha_0\theta_{\varepsilon_k}}{1+\varepsilon_k\theta_{\varepsilon_k}^2}-\alpha_0\theta\Big)f,u(t)\Big\rangle dt\\ &\quad=\int_0^T \Big\langle\frac{\alpha_0(\theta_{\varepsilon_k}-\theta)}{1+\varepsilon_k\theta_{\varepsilon_k}^2}f,u(t)\Big\rangle dt+\int_0^T \Big\langle\Big(\frac{1}{1+\varepsilon_k\theta_{\varepsilon_k}^2}-1\Big)\alpha_0\theta f,u(t)\Big\rangle dt\to 0\end{aligned} \tag{10.135}$$

and

$$\begin{aligned}\int_0^T \Big(\Big\langle\frac{\alpha_0\theta_{\varepsilon_k}}{1+\varepsilon_k\theta_{\varepsilon_k}^2}f,v_{\varepsilon_k}(t)\Big\rangle-\big\langle\alpha_0\theta f,v(t)\big\rangle\Big)dt&=\int_0^T \Big\langle\frac{\alpha_0(\theta_{\varepsilon_k}-\theta)}{1+\varepsilon_k\theta_{\varepsilon_k}^2}f,v_{\varepsilon_k}(t)\Big\rangle dt\\ +\int_0^T \Big\langle\frac{\alpha_0\theta}{1+\varepsilon_k\theta_{\varepsilon_k}^2}f,v_{\varepsilon_k}(t)-v(t)\Big\rangle dt&+\int_0^T \Big\langle\Big(\frac{1}{1+\varepsilon_k\theta_{\varepsilon_k}^2}-1\Big)\alpha_0\theta f,v(t)\Big\rangle dt\to 0.\end{aligned} \tag{10.136}$$

By (10.135) and (10.136), we get (10.134).

Therefore, by (10.130)–(10.133) and (10.134), we have from (10.127)

$$\int_0^T \langle u'(t) + A(\theta)v(t) + B(v(t), v(t)) - (1-\alpha_0\theta)f - f_1, u(t) - v(t)\rangle\, dt \\ + \Psi(u) - \Psi(v) \ge -\frac{1}{2}\|v_0 - u(0)\|^2. \tag{10.137}$$

Since $\|B(v(t), v(t))\|_{\mathbf{V}^*} \le K\|v(t)\|_{\mathbf{V}}^{\frac{3}{2}}\|v(t)\|^{\frac{1}{2}}$ (see (10.104)) and $v \in L^\infty(0,T;L^2)$, we have $B(v,v) \in L^{\frac{4}{3}}(0,T;\mathbf{V}^*)$. Therefore, by density of the set $\{L^6(0,T;\mathbf{V}) : u' \in L^2(0,T;\mathbf{V}^*)\}$ in $\{L^4(0,T;\mathbf{V}) : u' \in L^2(0,T;\mathbf{V}^*)\}$, (10.137) is valid for all $u \in \{L^4(0,T;\mathbf{V}) : u' \in L^2(0,T;\mathbf{V}^*)\}$.

Thus, the first formula of (10.20) is valid.

For any $\varphi \in C^1([0,T]; C^1_{\Gamma_D}(\Omega))$ with $\varphi(T) = 0$, we have from (10.79)

$$\begin{aligned} &\langle\theta_{\varepsilon_k}(t), \varphi(t)\rangle - \int_0^t \langle\theta_{\varepsilon_k}, \frac{\partial\varphi}{\partial t}\rangle ds + \int_0^t (\kappa(\theta_{\varepsilon_k})\nabla\theta_{\varepsilon_k}, \nabla\varphi)\, ds + \int_0^t (\beta(x)\theta_{\varepsilon_k}, \varphi)_{\Gamma_R}\, ds \\ &\quad - \int_0^t \langle v_{\varepsilon_k}\theta_{\varepsilon_k}, \nabla\varphi\rangle\, ds - \int_0^t \Big\langle \alpha_2\mu(\theta_{\varepsilon_k})\frac{|\mathcal{E}(v_{\varepsilon_k})|^2}{1+\varepsilon|\mathcal{E}(v_{\varepsilon_k})|^2}, \varphi\Big\rangle ds \\ &= \langle\theta_{\varepsilon_k}(0), \varphi(0)\rangle + \int_0^t \Big\langle \frac{\alpha_1\theta_{\varepsilon_k}}{1+\varepsilon\theta^2_{\varepsilon_k}} f \cdot v_{\varepsilon_k}, \varphi\Big\rangle ds + \int_0^t \langle f_{2\varepsilon_k}, \varphi\rangle ds \quad \forall t \in [0,T]. \end{aligned} \tag{10.138}$$

By Corollary 1.1 we have

$$\int_0^t (\kappa(\theta_{\varepsilon_k})\nabla\theta_{\varepsilon_k}, \nabla\varphi)\, ds \to \int_0^t (\kappa(\theta)\nabla\theta, \nabla\varphi)\, ds \quad \text{as } \varepsilon_k \to 0. \tag{10.139}$$

Since $v_{\varepsilon_k} \to v$ in $L^2(0,T;\mathbf{W}^{\frac{9}{10},2}(\Omega))$ and $\mathbf{W}^{\frac{9}{10},2}(\Omega) \subset \mathbf{L}^5(\Omega)$ (see Theorem 1.20), we have

$$v_{\varepsilon_k} \to v \quad \text{in } L^2(0,T;\mathbf{L}^5(\Omega)). \tag{10.140}$$

Since $\{\theta_{\varepsilon_k}\} \subset \big(L^{6/5}(0,T;W^{1,6/5}(\Omega)) \cap L^\infty(0,T;L^1(\Omega))\big)$ and $W^{1,6/5}(\Omega) \subset L^2(\Omega)$, by complex interpolation with exponent $\frac{7}{12}$ we have

$$\{\theta_{\varepsilon_k}\} \subset L^2(0,T;L^{\frac{24}{19}}(\Omega)) \subset L^2(0,T;L^{\frac{5}{4}}(\Omega)) \tag{10.141}$$

(see Theorem 1.33). Taking into account (10.140) and (10.141), we have

$$\int_0^t \langle v_{\varepsilon_k}\theta_{\varepsilon_k}, \nabla\varphi\rangle\, ds \to \int_0^t \langle v\theta, \nabla\varphi\rangle\, ds \quad \text{as } \varepsilon_k \to 0. \tag{10.142}$$

Since

$$\int_0^t \Big\langle \frac{\mu(\theta_{\varepsilon_k})|\mathcal{E}(v_{\varepsilon_k})|^2}{1+\varepsilon_k|\mathcal{E}(v_{\varepsilon_k})|^2}, \varphi\Big\rangle ds = \int_0^t \Big\langle \mu(\theta_{\varepsilon_k})\frac{|\mathcal{E}(v_{\varepsilon_k})|\sqrt{\varphi}}{\sqrt{1+\varepsilon_k|\mathcal{E}(v_{\varepsilon_k})|^2}}, \frac{|\mathcal{E}(v_{\varepsilon_k})|\sqrt{\varphi}}{\sqrt{1+\varepsilon_k|\mathcal{E}(v_{\varepsilon_k})|^2}}\Big\rangle ds,$$

by Lemma 1.4 and Corollary 1.2, we have

$$\liminf_{\varepsilon_k\to 0}\int_\Omega\Big\langle\frac{\mu(\theta_{\varepsilon_k})|\mathcal{E}(v_{\varepsilon_k})|^2}{1+\varepsilon_k|\mathcal{E}(v_{\varepsilon_k})|^2},\varphi\Big\rangle ds\geq\int_0^t\Big\langle\mu(\theta)|\mathcal{E}(v)|^2,\varphi\Big\rangle ds. \tag{10.143}$$

In the same way as in (10.136) we have

$$\int_0^t\Big\langle\frac{\alpha_1\theta_{\varepsilon_k}}{1+\varepsilon_k\theta_{\varepsilon_k}^2}f\cdot v_{\varepsilon_k},\varphi\Big\rangle ds\to\int_0^t\langle\alpha_1\theta f\cdot v,\varphi\rangle\,ds. \tag{10.144}$$

Therefore taking into account (10.139), (10.123) and (10.142)–(10.144), we have from (10.138)

$$\begin{aligned}&\big\langle\theta(t),\varphi(t)\big\rangle-\int_0^t\langle\theta,\frac{\partial\varphi}{\partial t}\rangle\,ds+\int_0^t(\kappa(\theta)\nabla\theta,\nabla\varphi)\,ds\\&\quad+\int_0^t(\beta(x)\theta,\varphi)_{\Gamma_R}\,ds-\int_0^t\langle v\theta,\nabla\varphi\rangle\,ds\\&\geq\int_0^t\alpha_2\big\langle\mu(\theta)|\mathcal{E}(v)|^2,\varphi\big\rangle ds+\big\langle\theta(0),\varphi(0)\big\rangle+\int_0^t\langle\alpha_1\theta f\cdot v,\varphi\rangle\,ds+\int_0^t\big\langle f_2,\varphi\big\rangle ds\\&\quad\text{for a.e. } t\in[0,T]\quad\forall\varphi\in C^1([0,T];C^1_{\Gamma_D}(\Omega))\text{ with }\varphi(T)=0,\ \varphi\geq 0.\end{aligned} \tag{10.145}$$

Now, we will prove that θ satisfies the second equation of (10.20) with "defect measure".

By (10.96), we have that $\alpha_2\mu(\theta_{\varepsilon_k})|\mathcal{E}(v_{\varepsilon_k})|^2\in L^1(Q)$, and these give continuous linear functionals on $C(\overline{Q})$. Since $\overline{Q}$ is compact, every $\alpha_2\mu(\theta_{\varepsilon_k})|\mathcal{E}(v_{\varepsilon_k})|^2$ is identified with a Radon measure (see Theorem 7.10.4, p.111 of [6]). Since $\{\alpha_2\mu(\theta_{\varepsilon_k})|\mathcal{E}(v_{\varepsilon_k})|^2\}$ is bounded in $L^1(Q)$ and $C(\overline{Q})$ is separable, there exists a $*-$weak convergent subsequence and a positive Radon measure $\mu_0\in\mathscr{M}(\bar{Q})$ such that as $\varepsilon_k\to 0$

$$\int_0^T\alpha_2\big\langle\mu(\theta_{\varepsilon_k})|\mathcal{E}(v_{\varepsilon_k})|^2,\varphi\big\rangle dt\to\int_{\bar{Q}}\varphi\,d\mu_0\quad\forall\varphi\in C(\overline{Q})$$

and

$$\begin{aligned}&-\int_0^T\langle\theta,\frac{\partial\varphi}{\partial t}\rangle\,dt+\int_0^T(\kappa(\theta)\nabla\theta,\nabla\varphi)\,dt+\int_0^T(\beta(x)\theta,\varphi)_{\Gamma_R}\,dt-\int_0^T\langle v\theta,\nabla\varphi\rangle\,dt\\&=\int_{\bar{Q}}\varphi\,d\mu_0+\big\langle\theta(0),\varphi(0)\big\rangle+\int_0^t\big\langle f_2,\varphi\big\rangle ds\quad\forall\varphi\in C^1([0,T];C^1_{\Gamma_D}(\Omega)),\ \varphi(\cdot,T)=0.\end{aligned} \tag{10.146}$$

From (10.145) and (10.146), we have

$$\int_{\bar{Q}}\varphi\,d\mu_0\geq\int_0^T\alpha_2\big\langle\mu(\theta)|\mathcal{E}(v)|^2,\varphi\big\rangle dt\quad\forall\varphi\in C^1([0,T];C^1_{\Gamma_D}(\Omega)),\ \varphi\geq 0,\ \varphi(\cdot,T)=0.$$

Define

$$\mu=\mu_0-\alpha_2\mu(\theta)|\mathcal{E}(v)|^2\lambda_4|_{\mathscr{B}(\overline{Q})},$$

where λ_4 is the Lebesgue measure in $\mathbb{R}^4$ and $\mathscr{B}(\overline{Q})$ is the $\sigma-$algebra of Borel subsets of $\overline{Q}$. Then, the second equation of (10.20) holds with a "defect measure" μ.

To get estimate (10.21), we use (10.96) and (10.115).

Let us prove

$$\int_Q \frac{|\nabla\theta|^2}{(1+|\theta|)^{1+\delta}}\,dxdt \le \liminf_{\varepsilon\to 0}\int_Q \frac{|\nabla\theta_{\varepsilon_k}|^2}{(1+|\theta_{\varepsilon_k}|)^{1+\delta}}\,dxdt \quad 0<\delta<1. \tag{10.147}$$

Put

$$z_{\varepsilon_k,i} := \frac{\partial\theta_{\varepsilon_k}/\partial x_i}{(1+|\theta_{\varepsilon_k}|)^{(1+\delta)/2}} \quad \text{a.e. in } Q \ (i=1,2,3).$$

By (10.96) $\{z_{\varepsilon_k}\}$ is bounded in $L^2(Q)$, and by choosing a subsequence if necessary, we can assume as $k\to+\infty$ that

$$z_{\varepsilon_k,i} \rightharpoonup z_i \quad \text{in } L^2(Q). \tag{10.148}$$

To get (10.147), it is enough to prove that

$$z_i = \frac{\partial\theta/\partial x_i}{(1+|\theta|)^{(1+\delta)/2}}. \tag{10.149}$$

By (10.122) as $k\to+\infty$

$$\partial\theta_{\varepsilon_k}/\partial x_i \rightharpoonup \partial\theta/\partial x_i \quad \text{in } L^r(Q),\ 1<r<\frac{5}{4},$$
$$\frac{1}{(1+|\theta_{\varepsilon_k}|)^{(1+\delta)/2}} \to \frac{1}{(1+|\theta|)^{(1+\delta)/2}} \quad \text{in } L^{r'}(Q).$$

Then,

$$\frac{\partial\theta_{\varepsilon_k}/\partial x_i}{(1+|\theta_{\varepsilon_k}|)^{1+\delta}} \to \frac{\partial\theta/\partial x_i}{(1+|\theta|)^{1+\delta}} \quad \text{in } L^1(Q),$$

from which we get

$$\frac{\partial\theta_{\varepsilon_k}/\partial x_i}{(1+|\theta_{\varepsilon_k}|)^{1+\delta}} \to \frac{\partial\theta/\partial x_i}{(1+|\theta|)^{1+\delta}} \quad \text{a.e. in } Q \tag{10.150}$$

for a subsequence if necessary. By (10.148) and (10.150), we get (10.149) (see Lemma 1.1).

By virtue of (10.96), (10.115), (10.123) and (10.147), we get (10.21). The main result is thus proved.

Remark 10.4 The existence of a solution to the non-steady problem (10.1) with mixed boundary conditions including the static pressure is open yet.

10.3 Bibliographical Remarks

The content of Chap. 10 is taken from [7].

The system (10.1) was studied in several papers. For the equation with $\alpha_0 = \alpha_1 = 0$, the existence of solution to the problem was studied in [5] under homogeneous Dirichlet boundary conditions for velocity and temperature, while in [8] existence of a solution under homogeneous Dirichlet boundary conditions for velocity and homogeneous Neumann condition for temperature was studied. In [4] when $|\alpha_0|, |\alpha_1|$ are small enough in accordance with data of the problem, the existence of a solution was discussed under homogeneous Dirichlet boundary conditions for velocity and mixture of homogeneous Dirichlet condition and homogeneous Neumann condition for temperature. All papers mentioned above concern with homogeneous Dirichlet boundary condition for velocity. The definitions of solutions in [4, 5, 8] for temperature θ include the "defect measures" due to bad smoothness of the term for dissipation of energy.

In [9] for the problem with $\alpha_0 = \alpha_1 = 0$, the existence of solution to the problem was studied under Navier slip boundary conditions for velocity and homogeneous Dirichlet condition for temperature. But, the definition of solution for temperature is inequality instead of equality (see (11) in [9]), and so in essence the solution satisfies an equality with a "defect measure" as [4, 5, 8, 10].

Dealing with the problem $\alpha_0 = \alpha_1 = 0$ and $g = 0$, in [11, 12] instead of the equation for temperature of (10.1) the equation of total energy

$$\frac{\partial}{\partial t}\left(\theta + \frac{|v|^2}{2}\right) + \operatorname{div}\left((\theta + \frac{|v|^2}{2} + p)v\right) - \operatorname{div}\left(\kappa(\theta)\nabla\theta\right) = \operatorname{div}\left(\mu(\theta)\mathcal{E}(v)v\right) + f\cdot v \quad (10.151)$$

is used. Under Navier slip boundary conditions for velocity and homogeneous Neumann condition for temperature in [11] and under periodic boundary conditions for velocity and temperature in [12], the existence of solutions to the problems is proved, but unlike [4, 5, 8, 10] the equation (10.151) is satisfied in the sense of distribution without the defect measures. It is not only reason of success in [11] to use the equation (10.151) to avoid the term for dissipation of energy in temperature equation of (10.1). Using smoothness of eigenvectors of a steady linear problem, they obtained the existence and estimates of more smooth solutions to approximate equations, which gives possibility to avoid the term for dissipation of energy.

In [13] for the 2-D problem with $\alpha_0 = \alpha_1 = 0$ existence of a strong solution was proved under homogeneous Dirichlet boundary conditions for velocity and temperature when the ratio between the derivatives of the viscosity and the thermal conductivity functions and their lower bounds is small enough.

[14–17] were concerned with (10.1) with $\alpha_1 = 0$. In [14] the problem under nonhomogeneous Dirichlet boundary conditions for velocity and temperature in time dependent domain was studied and existence of a local-in-time solution or existence of solution on the given interval for small data was established. In [15] existence of a strong solution and periodic solution for 2-D problem was studied under the boundary conditions and domain as above. In [16] under homogeneous Dirichlet boundary conditions for velocity and temperature existence of a strong solution and periodic solution were studied when data of the problem are small enough. [17] deals with the initial boundary value problem on a time-dependent domain with homogeneous Dirichlet boundary condition for velocity and temperature. The existence and uniqueness of local weak solutions, and as well as the existence of a global weak solution for small initial data are proved.

The problems with mixture of Dirichlet boundary condition of velocity and outflow condition for fluid were studied in [18–20]. In [18] when the viscosity, specific heat and thermal conductivity are independent of temperature, for non-steady problem on an open 2-D channel domain the local-in-time existence of a solution was proved under mixed boundary conditions above for fluid and mixture of non-homogeneous Dirichlet and homogeneous Neumann boundary conditions for temperature. In [20] on exterior-like domains, which is surrounded by a solid surface and a disjoint artificial boundary outside of the solid, a modified problem, where the buoyancy term $(1-\alpha_0\theta)f$ is changed by $\rho(\theta)$, $(0<\rho(\theta)<\rho_1)$, $\alpha_1=0$ and specific heat is independent of temperature, was studied. For the problem the local-in-time existence of a solution was proved under mixed boundary conditions above for fluid and mixture of homogeneous Dirichlet and Neumann boundary conditions for temperature. In [19] another modified problem on 3-D pipes, where the buoyancy term and $\alpha_1\theta f\cdot v$ are replaced, respectively, by $\rho(\theta)f$ and $\alpha_1\rho(\theta)f\cdot v$, $(0<\rho_0<\rho(\theta)<\rho_1)$, and viscosity, specific heat and thermal conductivity are independent of temperature, was considered. When the data of problem are small enough, the existence, regularity and uniqueness of a solution were proved under mixed boundary conditions above for fluid and mixture of Robin and homogeneous Neumann boundary conditions for temperature. In the papers with mixed boundary conditions for fluid above the assumptions for shape of domain is essential because in such cases the solutions to the corresponding steady Stokes problem belong to more smooth space than $W^{1,2}(\Omega)$ which is used for estimations of solutions.

For a system for non-homogeneous heat-conducting viscous incompressible fluid we refer to [21, 22].

There are several papers concerning with the systems for heat conducting non-Newtonian fluid corresponding to (10.1). In [23] when $\alpha_0=\alpha_1=0$, under homogeneous Dirichlet boundary condition for velocity and mixture of homogeneous Dirichlet and Robin conditions for temperature existence of a solution was obtained. In [24] when $\alpha_0\geq 0, \alpha_1\geq 0$ are small enough in accordance with data of the problem, under homogeneous Dirichlet boundary condition for velocity and homogeneous Neumann boundary condition for temperature the existence of a solution was proved. In [25] when $\alpha_0=\alpha_1=0$, under mixture of Navier slip condition and homogeneous Dirichlet boundary condition for velocity and mixture of a generalized Robin conditions and the homogeneous Dirichlet boundary conditions for energy, the existence of a solution to the problem was established. However, due to the conditions for nonlinear terms of strain the results of [23–25] exclude Newtonian fluid.

References

1. J. Naumann, On evolution inequalities of Navier-Stokes type in three dimensions. Ann. Math. Pure Appl. **124**(4), 107–125 (1980)
2. D.S. Konovalova, Subdifferential boundary value problems for the nonstationary Navier-Stokes and Stokes equations. Differ. Equ. **36**, 875–885 (2000)
3. H. Amann, J. Escher, *Analysis I* (Birkhäuser, 2005)
4. J. Naumann, J. Wolf, Existence of weak solutions to the equations of natural convection with dissipative heating, *Advances in Mathematical Fluid Mechanics* (Springer, 2010), pp. 367–384
5. J. Naumann, On the existence of weak solutions to the equations of non-stationary motion of heat-conducting incompressible viscous fluids. Math. Methods Appl. Sci. **29**, 1883–1906 (2006)
6. V.I. Bogachev, *Measure Theory I, II* (Springer, 2007)
7. T. Kim, D. Cao, A non-steady system with friction boundary conditions for flow of heat-conducting incompressible viscous fluids. J. Math. Anal. Appl. **484**, 123676 (2020)
8. J. Naumann, An existence theorem for weak solutions to the equations of non-stationary motion of heat-conducting incompressible viscous fluids. J. Nonlinear Convex Anal. **7**(3), 483–497 (2006)
9. L. Consiglieri, Existence of proper weak solutions to the Navier-Stokes-Fourier system. Funkc. Ekvacioj **55**, 347–365 (2012)
10. J. Naumann, M. Pokorny, J. Wolf, On the existence of weak solutions to the equations of steady flow of heat-conducting fluids with dissipative heating. Nonlinear Anal.: RWA **13**, 1600–1620 (2012)
11. M. Bulíček, E. Feireisl, J. Málek, A Navier-Stokes-Fourier system for incompressible fluids with temperature dependent material coefficients. Nonlinear Anal.: RWA **10**, 992–1015 (2009)
12. E. Feireisl, J. Málek, On the Navier-Stokes equations with temperature -dependent transport coefficients. Differ. Equ. Nonlinear Mech. **(2006)** 14 pp. (electronic) Art. ID 90616
13. L. Consiglieri, Regularity for the Navier-Stokes-Fourier system. Differ. Equ. Appl. **1**, 583–604 (2009)
14. H. Inoue, On heat convection equations with dissipative terms in time dependent domains. Nonlinear Anal. **30**, 4441–4448 (1997)
15. H. Inoue, Strong solutions of two dimensional heat convection equations with dissipating terms. Tokyo J. Math. **22**, 445–472 (1999)
16. H. Inoue, Global and periodic solutions of heat convection equations with internal friction terms. Nonlinear Anal. **47**, 2881–2886 (2001)
17. G. Łukaszewicz, P. Krzyżanowski, On the heat convection equations with dissipation term in regions with moving boundaries. Math. Methods Appl. Sci. **20**(4), 347–368 (1997)
18. M. Beneš, Strong solutions to non-stationary channel flows of heat-conducting viscous incompressible fluids with dissipative heating. Acta Appl. Math. **116**, 237–254 (2011)
19. M. Beneš, I. Pazaninb, On existence, regularity and uniqueness of thermally coupled incompressible flows in a system of three dimensional pipes. Nonlinear Anal. **149**, 56–80 (2017)
20. M. Beneš, J. Tichy, On coupled Navier-Stokes and energy equations in exterior-like domains. Comput. Math. Appl. **70**, 2867–2882 (2015)
21. Y. Liu, Global strong solutions for the incompressible nematic liquid crystal flows with density-dependent viscosity coefficient. J. Math. Anal. Appl. **462**, 1387–1397 (2018)
22. W. Wang, H. Yu, P. Zhang, Global strong solutions for 3D viscous incompressible heat conducting Navier-Stokes flows with the general external force. Math. Methods Appl. Sci. **41**, 4589–4601 (2018)
23. L. Consiglieri, Weak solutions for a class of non-Newtonian fluids with energy transfer. J. Math. Fluid Mech. **2**, 267–293 (2000)
24. J. Nečas, T. Roubíček, Buoyancy-driven viscous flow with L^1-data. Nonlinear Anal. **46**, 737–755 (2001)
25. L. Consiglieri, Friction boundary conditions on thermal incompressible viscous flows. Ann. Math. Pure Appl. **187**, 647–665 (2008)

Index

T. Kim and D. Cao, *Equations of Motion for Incompressible Viscous Fluids*, Advances in Mathematical Fluid Mechanics,
https://doi.org/10.1007/978-3-030-78659-5

The manufacturer's authorised representative in the EU is Springer Nature Customer Service Centre GmbH, Europaplatz 3, 69115 Heidelberg, Germany. If you have any concerns regarding our products, please contact ProductSafety@springernature.com

Printed and bound by CPI Group (UK) Ltd, Croydon, CR0 4YY

30/07/2026

02181999-0004